Classroom Manual for

Medium/Heavy Duty Truck Steering and Suspension

Online Services

Delmar Online

To access a wide variety of Delmar products and services on the World Wide Web, point your browser to:

http://www.Delmar.com

or email: info@delmar.com

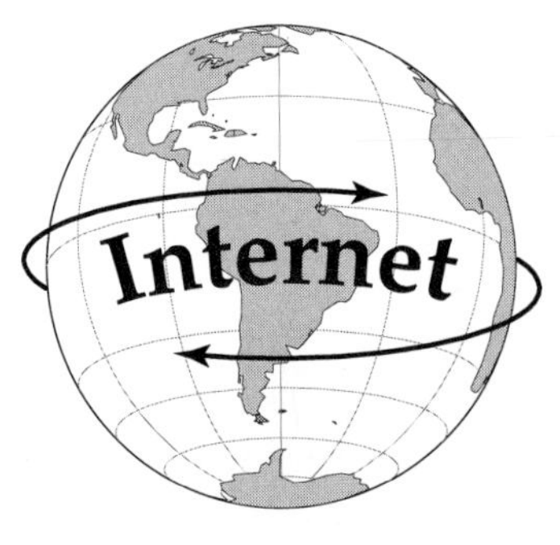

I(T)P®

Classroom Manual for

Medium/Heavy Duty Truck Steering and Suspension

Don Knowles
ASE Medium/Heavy Duty Truck Master Technician
Moose Jaw, Saskatchewan
CANADA

Jack Erjavec
Series Advisor
Columbus State Community College
Columbus, Ohio

Delmar Publishers

an International Thomson Publishing company 

Albany • Bonn • Boston • Cincinnati • Detroit • London • Madrid
Melbourne • Mexico City • New York • Pacific Grove • Paris • San Francisco
Singapore • Tokyo • Toronto • Washington

NOTICE TO THE READER

Publisher does not warrant or guarantee any of the products described herein or perform any independent analysis in connection with any of the product information contained herein. Publisher does not assume, and expressly disclaims, any obligation to obtain and include information other than that provided to it by the manufacturer.

The reader is expressly warned to consider and adopt all safety precautions that might be indicated by the activities herein and to avoid all potential hazards. By following the instructions contained herein, the reader willingly assumes all risks in connection with such instructions.

The publisher makes no representation or warranties of any kind, including but not limited to, the warranties of fitness for particular purpose or merchantability, nor are any such representations implied with respect to the material set forth herein, and the publisher takes no responsibility with respect to such material. The publisher shall not be liable for any special, consequential, or exemplary damages resulting, in whole or part, from the readers' use of, or reliance upon, this material.

Illustration courtesy of: Bruce Kaiser
Cover Design: Michele Canfield/Cheri Plasse

Delmar Staff
Publisher: Alar Elken
Acquisitions Editor: Vernon R. Anthony
Developmental Editor: Catherine A. Wein
Production Editor: Mary Ellen Black
Project Editor: Megeen Mulholland
Production Coordinator: Karen Smith
Art and Design Coordinator: Michele Canfield/Cheri Plasse
Editorial Assistant: Betsy Hough

Printed in the United States of America

For more information, contact:
Delmar Publishers
3 Columbia Circle, Box 15015
Albany, New York 12212-5015

International Thomson Publishing Europe
Berkshire House
168-173 High Holborn
London, WC1V 7AA
United Kingdom

ITP, Nelson Australia
102 Dodds Street
South Melbourne
Victoria 3205, Australia

Nelson Canada
1120 Birchmont Road
Scarborough, Ontario
M1K 5G4, Canada

ITP Spain/Paraninfo
Calle Magallanes, 25
28015-Madrid, España

International Thomson Publishing France
Tour Maine-Montparnasse
33 Avenue du Maine
75755 Paris Cedex 15, France

International Thomson Editores
Seneca 53
Colonia Polanco
11560 Mexico D. F. Mexico

International Thomson Publishing GmbH
Königswinterer Straße 418
53227 Bonn
Germany

International Thomson Publishing Asia
60 Albert Street
#05-10 Albert Complex
Singapore 189969

International Thomson Publishing—Japan
Hirakawa-cho Kyowa Building, 3F
2-2-1 Hirakawa-cho
Chiyoda-ku, Tokyo 102
Japan

2 3 4 5 6 7 8 9 10 XXX 11 10 09 08 07 06

Library of Congress Cataloging-in-Publication Data
Knowles, Don.
Classroom manual and shop manual for medium/heavy duty truck steering and suspension / Don Knowles, Jack Erjavec.
p. cm. — (Today's technician)
Includes bibliographical references and index.
ISBN 0-8273-7284-1 (alk. paper)
1. Trucks—Steering-gear. I. Erjavec, Jack. II. Title.
III. Series.
TL259.K56 1998
629.2'47—dc21 98-23983
CIP

CONTENTS

PREFACE

Unlike yesterday's mechanic, the technician of today and for the future must know the underlying theory of all systems and be able to service and maintain those systems. Today's technician must also know how these individual systems interact with each other. Standards and expectations have been set for today's technician, and these must be met in order to keep the world's medium and heavy duty trucks running efficiently and safely.

The *Today's Technician* series, by Delmar Publishers, features textbooks that cover all mechanical and electrical systems of medium and heavy duty trucks. Principal titles correspond with the eight major areas of ASE (National Institute for Automotive Service Excellence) certification.

Each title is divided into two manuals: a Classroom Manual and a Shop Manual. Dividing the material into two manuals provides the reader with the information needed to begin a successful career as a medium and heavy duty truck technician without interrupting the learning process by mixing cognitive and performance-based learning objectives.

Each Classroom Manual contains the principles of operation for each system and subsystem. It also discusses the design variations used by different manufacturers. The Classroom Manual is organized to build upon basic facts and theories. The primary objective of this manual is to allow the reader to gain an understanding of how each system and subsystem operates. This understanding is necessary to diagnose the complex truck systems.

The understanding acquired by using the Classroom Manual is required for competence in the skill areas covered in the Shop Manual. All of the high priority skills, as identified by ASE, are explained in the Shop Manual. The Shop Manual also includes step-by-step instructions for diagnostic and repair procedures. Photo Sequences are used to illustrate many of the common service procedures. Other common procedures are listed and are accompanied with fine-line drawings and photographs that allow the reader to visualize and conceptualize the finest details of the procedure. The Shop Manual also contains the reasons for performing the procedures, as well as when that particular service is appropriate.

The two manuals are designed to be used together and are arranged in corresponding chapters. Not only are the chapters in the manuals linked together, the contents of the chapters are also linked. Both manuals contain clear and thoughtfully selected illustrations. Many of the illustrations are original drawings or photos prepared for inclusion in this series. This means that the art is a vital part of each manual.

The page layout is designed to include information that would otherwise break up the flow of information presented to the reader. The main body of the text includes all of the "need-to-know" information and illustrations. In the side margins are many of the special features of the series. Items such as definitions of new terms, common trade jargon, tools lists, and cross-references are placed in the margin, out of the normal flow of information so as not to interrupt the thought process of the reader.

Jack Erjavec, Series Advisor

Classroom Manual

To stress the importance of safe work habits, the Classroom Manual dedicates one full chapter to safety. Included in this chapter are common safety practices, safety equipment, and safe handling of hazardous materials and wastes. This includes information on MSDS sheets and OSHA regulations. Other features of this manual include:

Cognitive Objectives

These objectives define the contents of the chapter and define what the student should have learned upon completion of the chapter.

Each topic is divided into small units to promote easier understanding and learning.

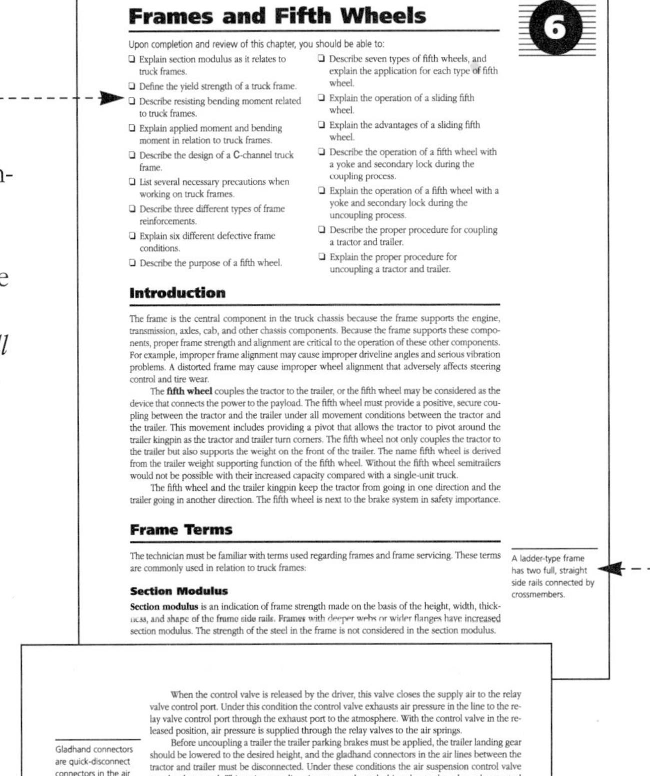

CHAPTER 6

Frames and Fifth Wheels

Upon completion and review of this chapter, you should be able to:

- ❑ Explain section modulus as it relates to truck frames.
- ❑ Define the yield strength of a truck frame.
- ❑ Describe resisting bending moment related to truck frames.
- ❑ Explain applied moment and bending moment in relation to truck frames.
- ❑ Describe the design of a C-channel truck frame.
- ❑ List several necessary precautions when working on truck frames.
- ❑ Describe three different types of frame reinforcements.
- ❑ Explain six different defective frame conditions.
- ❑ Describe the purpose of a fifth wheel.
- ❑ Describe seven types of fifth wheels, and explain the application for each type of fifth wheel.
- ❑ Explain the operation of a sliding fifth wheel.
- ❑ Explain the advantages of a sliding fifth wheel.
- ❑ Describe the operation of a fifth wheel with a yoke and secondary lock during the coupling process.
- ❑ Explain the operation of a fifth wheel with a yoke and secondary lock during the uncoupling process.
- ❑ Describe the proper procedure for coupling a tractor and trailer.
- ❑ Explain the proper procedure for uncoupling a tractor and trailer.

Introduction

The frame is the central component in the truck chassis because the frame supports the engine, transmission, axles, cab, and other chassis components. Because the frame supports these components, proper frame strength and alignment are critical to the operation of these other components. For example, improper frame alignment may cause improper driveline angles and serious vibration problems. A distorted frame may cause improper wheel alignment that adversely affects steering control and tire wear.

The **fifth wheel** couples the tractor to the trailer, or the fifth wheel may be considered as the device that connects the power to the payload. The fifth wheel must provide a positive, secure coupling between the tractor and the trailer under all movement conditions between the tractor and the trailer. This movement includes providing a pivot that allows the tractor to pivot around the trailer kingpin as the tractor and trailer turn corners. The fifth wheel not only couples the tractor to the trailer but also supports the weight on the front of the trailer. The name fifth wheel is derived from the trailer weight supporting function of the fifth wheel. Without the fifth wheel semitrailers would not be possible with their increased capacity compared with a single-unit truck.

The fifth wheel and the trailer kingpin keep the tractor from going in one direction and the trailer going in another direction. The fifth wheel is next to the brake system in safety importance.

Frame Terms

The technician must be familiar with terms used regarding frames and frame servicing. These terms are commonly used in relation to truck frames:

A ladder-type frame has two full, straight side rails connected by crossmembers.

Section Modulus

Section modulus is an indication of frame strength made on the basis of the height, width, thickness, and shape of the frame side rails. Frames with deeper webs or wider flanges have increased section modulus. The strength of the steel in the frame is not considered in the section modulus.

Marginal Notes

New terms are pulled out and defined. Common trade jargon also appears in the margin and gives some of the common terms used for components. This allows the reader to speak and understand the language of the trade, especially when conversing with an experienced technician.

When the control valve is released by the driver, this valve closes the supply air to the relay valve control port. Under this condition the control valve exhausts air pressure in the line to the relay valve control port through the exhaust port to the atmosphere. With the control valve in the released position, air pressure is supplied through the relay valves to the air springs.

Gladhand connectors are quick-disconnect connectors in the air lines between the tractor and trailer.

Before uncoupling a trailer the trailer parking brakes must be applied, the trailer landing gear should be lowered to the desired height, and the gladhand connectors in the air lines between the tractor and trailer must be disconnected. Under these conditions the air suspension control valve may be depressed. This action supplies air pressure through this valve to the relay valve control ports. When the relay valves receive this pressure signal, these valves exhaust air pressure from the air springs to the atmosphere. The air spring deflation prevents a sudden rise in the truck frame height when the trailer is uncoupled. This sudden rise in truck frame height could damage the shock absorbers.

CAUTION: Be sure no one is under or near the tractor frame when deflating the air springs. The frame may drop suddenly, resulting in personal injury.

WARNING: Never operate the tractor until the air suspension control valve is released and the air suspension has returned to normal ride height. Operating the tractor with the air springs deflated or partially deflated causes rough riding, erratic vehicle handling, and air suspension component damage.

See Chapter 8 in the Shop Manual for tractor rear axle air suspension system maintenance, diagnosis, and service.

When the control valve handle is released, the air passage is closed through this valve to the relay valve control ports. This action allows air pressure to flow through the relay valves to the air springs and return the rear suspension to the normal ride height. When coupling a trailer to the tractor, leave the control valve in the released or off position.

In some air suspension systems a **dump valve** is combined with each height control valve (Figure 8-15). The dump valve operation is basically the same as the relay valve operation. The dump valve may be operated by air pressure from a manually operated valve. Some dump valves are operated by an electrical switch.

Volvo Optimized Air Suspension (VOAS) System

The Volvo optimized air suspension (VOAS) system is a modified version of the previous air suspension on rear axles of Volvo GM heavy duty trucks (Figure 8-16). A number of modifications in the VOAS system provide improved ride, reduced unsprung weight, increased durability, and better alignment ability. The VOAS system is available on all Volvo GM on-highway tractors. On 6 × 4

Figure 8-15 Height control valve with integral dump valve. (Courtesy of Neway Anchorlock International, Inc.)

224

Cautions and Warnings

Throughout the text, cautions are given to alert the reader to potentially hazardous materials or unsafe conditions. Warnings are also given to advise the student of things that can go wrong if instructions are not followed or if a nonacceptable part or tool is used.

References to the Shop Manual

Reference to the appropriate topic in the Shop Manual is given whenever necessary. Although the chapters of the two manuals are synchronized, material covered in other chapters of the Shop Manual may be fundamental to the topic discussed in the Classroom Manual.

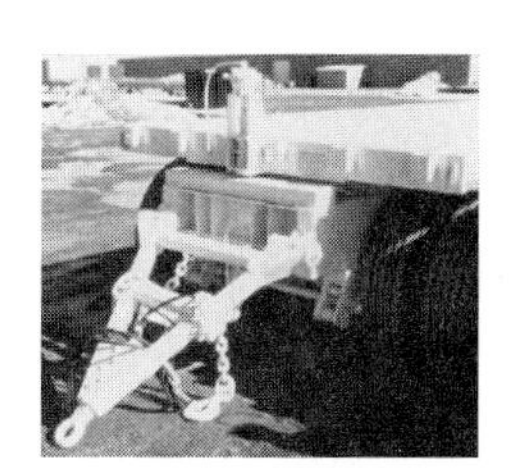

Figure 6-43 Converter dolly with a draw bar. (Courtesy of Dorsey Trailers Inc.)

5. Raise the safety latch, and pull the operating handle fully forward. When this handle is hard to pull, back the tractor slightly to relieve the kingpin to jaw force.
6. Release the tractor parking brakes and drive the tractor forward away from the trailer.

Pintle Hooks and Draw Bars

See Chapter 6 in the Shop Manual for draw bar and pintle hook diagnosis and service.

When one trailer is connected behind another trailer, a dolly with a **draw bar** may be used to couple the rearmost trailer to the forward trailer in an A-train configuration. A heavy steel ring on the front of the draw bar is connected to a **pintle hook** that is bolted to the forward trailer. The pintle hook is securely bolted at the centerline of the forward trailer. The steel ring is securely bolted to the draw bar (Figure 6-43). The pintle hook has a spring-loaded latch that prevents the steel ring on the draw bar from disengaging from the pintle hook.

A BIT OF HISTORY

Before 1910 Sternberg and later Sterling trucks had wood frame side members reinforced by a large steel angle section. After 1910 these trucks had channel steel frame side members reinforced with a wood insert. These wood inserts were made from white oak, and the total thickness of the frame side members was 2 in (5.08 cm) (Figure 6-44). All Sterling trucks with a payload capacity more than 1500 lb (680.4 kg) had frame side members with wood inserts. The advantages claimed for this type of frame design were:

1. Improved shock-absorbing qualities
2. Increased rigidity and strength
3. Improved sound-deadening qualities
4. Lower maintenance costs
5. Improved driver comfort

166

A Bit of History

This feature gives the student a sense of the evolution of truck systems. This feature not only contains nice-to-know information, but also should spark some interest in the subject matter.

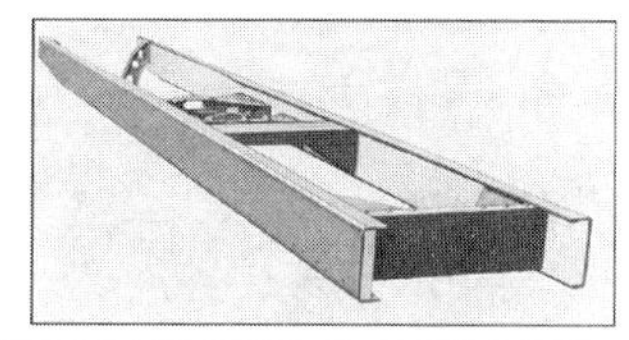

Figure 6-44 Sterling truck frame with wood reinforced side members. Reprinted with permission from SAE book SP-941 © 1993 Society of Automotive Engineers, Inc.

Summary

- ❑ Section modulus is an indication of frame strength based on the height, width, thickness, and shape of the frame side rails.
- ❑ Yield strength is a measure of the steel strength used in the frame.
- ❑ RBM is the most accurate indication of frame strength. RBM is calculated by multiplying the yield strength and the section modulus.
- ❑ Frame area is the total cross section of the frame rail in square inches.
- ❑ The C-channel is the most common truck frame design, but I-beam and box frame designs are used in some applications.
- ❑ Crossmembers are mounted between the frame rails.
- ❑ Aluminum alloy frames are used in some trucks. These frames are lighter, but they must be thicker to provide adequate strength.
- ❑ Frame reinforcements are installed on the frame in the area of greatest load concentration to increase the RBM.
- ❑ Frame sag occurs when the frame or one side rail is bent downward from the original position
- ❑ Buckle is a frame condition that refers to the frame or one side rail that is bent upward from the original position.
- ❑ A diamond frame condition is present when one frame rail is moved rearward or forward in relation to the opposite frame rail.
- ❑ Frame twist occurs when the end of one frame rail is bent upward or downward the opposite frame rail.
- ❑ Sidesway occurs when one or both frame rails are bent inward or outward.
- ❑ Tracking is the alignment of the truck axles with each other.
- ❑ A fifth wheel is used to couple the tractor to the trailer kingpin.
- ❑ Types of fifth wheels include, semioscillating, fully-oscillating, rigid, nontilt conv lizing, compensating, and elevating.

Terms to Know

Applied moment
Bending moment
Box frames
Buckle
C-channel
Compensating fifth wheel
Converter dolly
Crossmembers
Diamond frame
Draw bar
Elevating fifth wheel
Fifth wheel
Fifth wheel jaws
Fishplate
Frame reinforcements
Frame sag
Frame twist
Fully-oscillating fifth

Summaries

Each chapter concludes with summary statements that contain the important topics of the chapter. These are designed to help the reader review the contents.

Terms to Know

A list of new terms appears next to the Summary. Definitions for these terms can be found in the Glossary at the end of the manual.

Section modulus
Semioscillating fifth wheel
Sidesway
Sliding fifth wheel
Stabilized fifth wheel
Stationary fifth wheel
Tracking
Trailer bolster plate
Yield strength

- ❑ A fifth wheel may be stationary or sliding.
- ❑ During the tractor-to-trailer coupling process the trailer height should be adjusted so the trailer bolster plate makes initial contact with the fifth wheel at a point 8 in (20.32 cm) to the rear of the center on the fifth wheel mounting bracket.
- ❑ After the tractor is coupled to the trailer, the driver should get under the tractor and trailer and use a flashlight to visually inspect the position of the kingpin in the fifth wheel jaws.

Review Questions

Short Answer Essays

1. Explain the term section modulus in relation to truck frames.
2. Explain yield strength as it relates to truck frames.
3. Explain the most accurate way of calculating frame strength.
4. Describe bending moment in relation to truck frames.
5. Describe the parts of a C-channel.
6. Explain three requirements for a truck frame.
7. Describe three precautions that must be observed when working on truck frames.
8. Explain the purpose of a fifth wheel.
9. Explain the proper procedure for coupling a tractor and trailer
10. Describe a high hitch condition when coupling a tractor and trailer, and explain how this condition occurs.

Fill-in-The-Blanks

1. A semioscillating fifth wheel oscillates around and axis that is perpendicular to the tractor ________________.
2. A fully oscillating fifth wheel provides front-to-rear oscillation and ________________ oscillation.
3. A nontilt convertible fifth wheel can be converted from a rigid fifth wheel to a ________________ fifth wheel.
4. On a stabilizing fifth wheel the top part of the fifth wheel rotates with the ________________.
5. An elevating fifth wheel may be operated by ________________ pressure or ________________ pressure.
6. The locking mechanism on a sliding fifth wheel may be operated mechanically or by ________________ pressure.
7. The locking mechanism on a sliding fifth wheel should not be released when the tractor is ________________.
8. Safety factor on a truck frame is the amount of ________________ that can be safely absorbed by the truck frame members.
9. The web on a C-channel frame is the area between the ________________.
10. A fishplate is a heavy steel frame reinforcement bolted to the ________________ of the truck frame.

168

Review Questions

Short answer essay, fill-in-the-blank, and multiple-choice type questions follow each chapter. These questions are designed to accurately assess the student's competence in the stated objectives at the beginning of the chapter.

Shop Manual

To stress the importance of safe work habits, the Shop Manual also dedicates one full chapter to safety. Other important features of this manual include:

Performance Objectives

These objectives define the contents of the chapter and define what the student should have learned upon completion of the chapter. These objectives also correspond with the list of required tasks for ASE certification. *Each ASE task is addressed.*

Although this textbook is not designed to simply prepare someone for the certification exams, it is organized around the ASE task list. These tasks are defined generically when the procedure is commonly followed and specifically when the procedure is unique for specific vehicle models. Imported and domestic model trucks are included in the procedures.

CHAPTER 8

Air Suspension System Diagnosis and Service

Upon completion and review of this chapter, you should be able to:

- ❑ Perform preoperational inspections on tractor air suspension systems.
- ❑ Perform pretrip and regular inspections on tractor air suspension systems.
- ❑ Measure and adjust tractor air suspension ride height.
- ❑ Perform preoperational inspections on trailer air suspension systems.
- ❑ Perform pretrip and regular inspections on trailer air suspension systems.
- ❑ Measure and adjust trailer air suspension ride height.
- ❑ Perform preventive maintenance on height control valves.
- ❑ Diagnose height control valves.
- ❑ Diagnose and service pressure protection valves.
- ❑ Perform hand control valve adjustments.
- ❑ Remove and replace air springs.
- ❑ Remove and replace main support (rigid) beams.
- ❑ Remove and replace main support (rigid) beam bushings.
- ❑ Remove and replace frame brackets.
- ❑ Align rear axles
- ❑ Perform lift axle adjustment and service.
- ❑ Diagnose air suspension system problems.

Basic Tools
Basic technician's tool set
Service manuals
Tape measure
Floor jack
Safety stands
Drill bits

Tractor Air Suspension System Preoperational Inspection

CAUTION: When completing any tractor air suspension inspection, always apply the tractor parking brakes and block the tractor wheels. If this action is not taken, the tractor may roll ahead or backward causing personal injury or tractor damage.

Before placing a tractor into service a preoperational air suspension system inspection must be completed. The preoperational inspection includes these items identified by the numbers in Figure 8-1.

1. Be sure the air pressure in the tractor air brake system is greater than 70 psi (482.65 kPa). With the tractor engine shut off, place a soap and water solution on all air line connections in the air suspension system to check for leaks. If bubbles appear at any air line connection, the air leak must be repaired.
2. Measure the minimum clearance around each air spring. This clearance should be 1¾ in (44.45 mm).
3. Inspect the shock absorbers for proper installation and any indication of fluid leaks.

SERVICE TIP: Torque specifications on many air suspension systems are based on clean lubricated threads. Always consult the suspension or truck manufacturer's recommended torque conditions and specified torques.

4. Be sure the shock absorber mounting bolts are tightened to the specified torque.
5. Check the torque on the air spring mounting bolts.
6. Be sure the pivot bolt that retains the **rigid beam** in the front bracket is tightened to the specified torque.
7. Be sure the spacer washers are properly installed between the rigid beam and the front bracket.

A rigid beam may be called a main support beam.

301

Tools Lists

Each chapter begins with a list of the Basic Tools needed to perform the tasks included in the chapter. Whenever a Special Tool is required to complete a task, it is listed in the margin next to the procedure.

Marginal Notes

Page numbers for cross-referencing appear in the margin. Some of the common terms used for components, and other bits of information, also appear in the margin. This provides an understanding of the language of the trade and helps when conversing with an experienced technician.

Photo Sequences

Many procedures are illustrated in detailed Photo Sequences. These detailed photographs show the students what to expect when they perform particular procedures. They also can provide a student a familiarity with a system or type of equipment, which the school may not have.

Photo Sequence 2
Checking Steering Free-Play

A medium duty truck with a short-and-long arm (coil spring) front suspension is required for this sequence. It would be more convenient to position the truck on a lift with the tires supported on the lift or drive the truck on a wheel alignment ramp for this sequence.

P2-1 With the front wheels straight ahead, apply a light turning force to the steering wheel in both directions until turning resistance is felt. Position a tape measure above the steering wheel to measure the steering free-play.

P2-2 Determine the specified steering wheel free-play in the truck manufacturer's service manual. When the steering wheel free-play is more than specified, continue the steering diagnosis to locate the exact cause of the steering looseness.

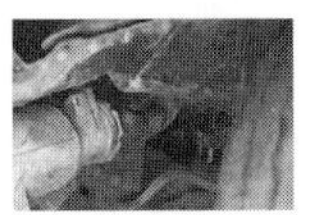

P2-3 While an assistant turns the steering wheel in both directions with a light turning force, inspect the U joint(s) or flexible coupling in the steering shaft for looseness.

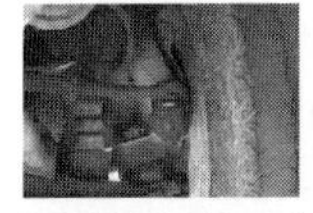

P2-4 While an assistant turns the steering wheel in both directions with a light turning force, inspect the outer tie rod ends for looseness.

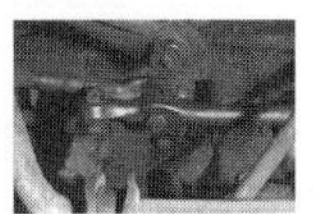

P2-5 While an assistant turns the steering wheel in both directions with a light turning force, inspect the inner ends of the tie rods for looseness.

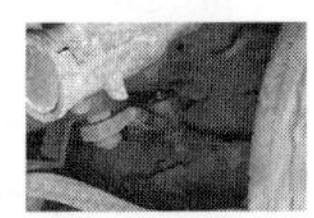

P2-6 While the steering wheel is turned in both directions with a light turning force, inspect the ball stud between the pitman arm and center link for looseness.

128

Service Tips

Whenever a short-cut or special procedure is appropriate, it is described in the text. These tips are generally those things commonly done by experienced technicians.

Cautions and Warnings

Throughout the text, cautions are given to alert the reader to potentially hazardous materials or unsafe conditions. Warnings are also given to advise the student of things that can go wrong if instructions are not followed or if a nonacceptable part or tool is used.

2. Install the bolt through the front spring bracket and the spring bushing. Tighten this bolt to the specified torque.
3. Install the rear shackle spacer washers and plates. Tighten the pinch bolts in the shackle plates to the specified torque.
4. Use a feeler gauge to measure the clearance between the spacer washers and the shackle plates. If this clearance is not within the truck manufacturer's specifications, remove the shackle plates and install shims with the proper thickness to provide the specified clearance.
5. Install the upper plate over the center of the spring and position the caster shim and shock absorber plate under the center of the spring.
6. Raise the axle with a floor jack until the axle contacts the shock absorber plate. If the shock absorber plate is not positioned between the spring and the axle, the axle contacts the **caster shim.** Be sure the head of the center bolt extends into the opening in the axle to position the axle properly on the spring.

SERVICE TIP: The thick part of the caster shim faces toward the rear of the axle.

7. Install the U-bolt nuts and washers and tighten these nuts enough to hold the spring in the proper position on the axle.
8. Lift the front axle with a floor jack and remove the safety stands. Lower the front tires onto the shop floor and remove the floor jack.

WARNING: If the spring U-bolts are not tightened properly, the axle may shift on the spring. This action may shear off the center bolt. This condition causes improper axle position on the spring, resulting in serious steering problems such as pulling to one side and pulling while braking.

CAUTION: The torque on all the spring U-bolts should be checked at regular intervals as part of the preventive maintenance program.

9. Tighten the U-bolt nuts to the specified torque. Initially tighten these U-bolt nuts to one-third of the specified torque in a specific tightening sequence. Continue tightening these nuts gradually in the same sequence until the specified torque is reached.

When it is necessary to remove the axle without removing the springs, remove the tires, wheels, and hubs. Remove the brake components as explained previously in this chapter. Support the axle with two floor jacks and remove the U-bolts, nuts, shock absorber plate, upper place, and caster shim. Lower the two floor jacks at the same time, until the jacks are fully lowered. Pull the floor jacks forward to remove the axle from under the truck.

Shock Absorber Visual Inspection

Shock absorbers should be inspected for loose mounting bolts and worn mounting bushings. If these components are loose, rattling noise is evident, and replacement of the bushings and bolts is necessary. In some shock absorbers the bushing is permanently mounted in the shock, and the complete unit must be replaced if the bushing is worn. When the mounting bushings are worn, the

choice type question that includes the word *except* in the question heading. In this type of question the technician selects one incorrect response out of four responses.

ASE regulations demand that each technician must have 2 years of working experience in the truck service industry before taking a certification test or tests. However, relevant formal training may be substituted for one of the years of working experience. Contact ASE for details regarding this substitution. The contents of the medium/heavy duty truck suspension and steering test are listed in Table 2-1.

Shops that employ ASE-certified technicians display an official **ASE blue seal of excellence.** This blue seal increases the customer's awareness of the shop's commitment to quality service and the competency of certified technicians.

ASE School Bus Technician Certification

ASE also provides a school bus certification program. Three areas of certification in the school bus certification program are as follows:

1. Brakes
2. Suspension and steering
3. Electric/electronic systems

ASE Automotive Technician Certification

ASE also provides a certification program for automotive technicians in these areas:

1. Engine repair (test A1)
2. Automatic transmission/transaxle (test A2)
3. Manual drivetrain and axles (test A3)
4. Suspension and steering (test A4)
5. Brakes (test A5)
6. Electrical/electronic systems (test A6)
7. Heating and air conditioning (test A7)
8. Engine performance (test A8)

ASE certification is available for autobody technicians. ASE certification is also provided in specialty areas such as advanced engine performance, alternate fuels, and engine machinist.

See Chapter 2 in the Classroom Manual for basic theories.

CUSTOMER CARE: Some truck service centers have a policy of performing some minor service as an indication of their appreciation to the customer. This service may include cleaning all the windows or vacuuming the floors before the truck is returned to the customer.

References to the Classroom Manual

Reference to the appropriate topic in the Classroom Manual is given whenever necessary. Although the chapters of the two manuals are synchronized, material covered in other chapters of the Classroom Manual may be fundamental to the topic discussed in the Shop Manual.

Customer Care

This feature highlights those little things a technician can do or say to enhance customer relations.

Job Sheets

Located at the end of each chapter, the Job Sheets provide a format for students to perform procedures covered in the chapter. A reference to the ASE Task addressed by the procedure is referenced on the Job Sheet.

JOB SHEET 4

4

Name ____________________ Date ____________________

Demonstrate Proper Use of a Micrometer and a Dial Indicator

On completion of this job sheet, you should be able to use a dial indicator and a micrometer properly.

NATEF/ASE Correlation

This job sheet is related to the NATEF and ASE medium/heavy duty truck brakes task list content area:

A. Air Brakes Diagnosis and Repair, 2. Mechanical/Foundation, Task: 11. Inspect and measure brake drums or rotors; determine needed repairs.
B. Hydraulic Brakes Diagnosis and Repair, 2. Mechanical System, Task: 2. Inspect, reface, or replace brake drums and rotors.

Tools and Materials

Truck with front disk brakes
Floor jack and safety stands
Micrometer
Dial indicator

Procedure

1. Raise the front axle with a floor jack and lower the axle so it is resting securely on safety stands. Remove the floor jack.

CAUTION: If the front axle is not resting securely on the safety stands or the safety stand legs are not resting squarely on the shop floor, the truck may fall off these stands, resulting in personal injury.

2. With the front wheel removed and the vehicle chassis supported on safety stands, adjust the front wheel bearings on the side where the rotor is being measured.
3. Mount a dial indicator so the stem contacts the rotor surface approximately 1 in (2.54 cm) from the outer edge of the rotor.
4. Turn the rotor for one revolution and observe the runout on the dial indicator. Rotor runout ____________________
5. Compare the runout indicated on the dial indicator to the runout specifications in the service manual.
   Specified rotor runout ____________________
6. Use a micrometer to measure the rotor thickness approximately 1 in (2.54 cm) from the outer edge of the rotor.
   Rotor thickness ____________________
7. Compare the rotor thickness to the minimum discard thickness stamped on the rotor.
   Discard thickness ____________________

55

Case Studies

Case Studies concentrate on the ability to properly diagnose the systems. Each chapter ends with a case study in which a vehicle has a problem, and the logic used by a technician to solve the problem is explained.

CASE STUDY

A customer brought his tractor into the shop with a complaint of steering pull to the left and rear wheel hop. The customer indicated the wheel hop occurred on the right dual wheels on the forward axle in the tandem rear axle. The tractor had an equalizing beam rear suspension. The technician road tested the tractor and verified the customer complaint. During a prealignment inspection the technician did not find any problems.

The technician placed the tractor on the wheel aligner and connected the wheel sensors on the computer aligner to the wheel rims. When the alignment readings were taken on the front and rear axles, the front suspension angles were within specifications except the right front wheel caster, which had 2 degrees more positive caster than specified. The rear suspension angles were read on both rear axles. The alignment angles on the rearmost axle were within specifications. On the forward rear axle the computer wheel aligner indicated the thrust angle was not within specifications, the right wheel had excessive toe-out, and the left wheel indicated excessive toe-in.

Terms To Know

Bump steer	Computer wheel aligners	Torque steer
Camber	Kingpin inclination	Trammel gauge
Camber and caster gauge	Laser aligners	Wheel hop
Caster	Memory steer	Wheel sensor
Commercial vehicle inspectors	Toe	

Terms to Know

Terms in this list can be found in the Glossary at the end of the manual.

ASE Style Review Questions

1. While discussing wheel hop on a drive axle:
Technician A says wheel hop may be caused by loose torque rod bushings.
Technician B says wheel hop may be caused by improper rear springs for the vehicle load.
Who is correct?
A. A only C. Both A and B
B. B only D. Neither A nor B
2. Normal steering wheel returnability after a 180-degree turn should allow the steering wheel to return to within:
A. 0 degrees to 10 degrees of the center position.
B. 10 degrees to 20 degrees of the center position.
C. 20 degrees to 30 degrees of the center position.
D. 30 degrees to 60 degrees of the center position.
3. The steering on a tractor pulls to the left while driving straight ahead and there is no indication of excessive tire tread wear:
Technician A says the left front wheel may have more positive caster compared with the right front wheel.
Technician B says the left front wheel may have excessive positive camber.
Who is correct?
A. A only C. Both A and B
B. B only D. Neither A nor B
4. The right front wheel on an I-beam front suspension has excessive negative camber. The cause of this problem could be:
A. Improper shim thickness between the underside ...

ASE Style Review Questions

Each chapter contains ASE style review questions that reflect the performance objectives listed at the beginning of the chapter. These questions can be used to review the chapter as well as to prepare for the ASE certification exam.

Diagnostic Chart

Chapters include detailed diagnostic charts linked with the appropriate ASE task. These charts list common problems and most probable causes. They also list a page reference in the Classroom Manual for better understanding of the system's operation and a page reference in the Shop Manual for details on the procedure necessary for correcting the problem.

Appendix A

ASE Practice Examination

1. A truck exhibits excessive steering wheel free-play. *Technician A* says that either a worn steering shaft universal joint or an out-of-adjustment steering gear could cause this free-play. *Technician B* says that the cause could be either a bent idler arm, or no lubricant in the steering gear. Who is correct?
A. A only C. Both A and B
B. B only D. Neither A nor B
2. A driver complains that his heavy-duty truck requires excessive steering effort and that the steering wheel return is too fast. Which of these is the most likely cause?
A. Incorrect turning angle
B. Too much negative camber
C. Too much positive caster
D. Underinflated tires
3. Which of these will result in excessive play in the mounted steering column assembly on a truck equipped with a tilt steering column?
A. Faulty anti-lash spring in the centering spheres
B. Upper bearing race missing
C. Upper bearing not seating in the race
D. Loose support lock shoe pin
4. A truck is experiencing high steering effort. *Technician A* says a sheared shift tube could be the cause. *Technician B* says the column assembly is misaligned. Who is correct?
A. A only C. Both A and B
B. B only D. Neither A nor B
5. Which of these components should a technician check if a wheel and tire do not return to the straight-ahead position during a front axle and linkage-binding test?
A. King pin bearings
B. Front suspension springs
C. Rear suspension springs
D. Shock absorbers
6. The rear wheel bearings on a truck are going to be adjusted. *Technician A* says the rear wheel bearings should never be preloaded on rear axles. *Technician B* says the rear wheel lock washers may be tang type or dowel type. Who is correct?
A. A only C. Both A and B
B. B only D. Neither A nor B
7. *Technician A* says mismatched tire sizes can cause high axle lubricant temperature. *Technician B* says overinflated tires cause excessive tire wear. Who is correct?
A. A only C. Both A and B
B. B only D. Neither A nor B
8. Which of the following conditions can cause steering wheel shimmy?
A. Too high a toe-in setting
B. Incorrect dynamic wheel balance
C. An excessive load in the truck
D. Too high a toe-out setting
9. All of the following statements about wheel and tire removal are true except:
A. The stud directly across from the number one stud is tightened second.
B. With the stud nuts slightly loose on demountable rims, strike the wheel clamps with a hammer to loosen these clamps.
C. Remove the wheel nuts before loosening the wheel clamps.
D. Different makes of rim components should not be interchanged.
10. Two technicians are discussing front and rear wheel bearing diagnosis. Technician A says that growling noise produced by a bad front wheel bearing is most noticeable while turning a corner. Technician B says the growling noise produced by a bad rear wheel bearing is most noticeable at high speeds. Who is correct?
A. A only C. Both A and B
B. B only D. Neither A nor B
11. *Technician A* says a metal burr behind one of the bearing races could cause misalignment wear on a front wheel bearing. *Technician B* says fine abrasives could cause brinelling on a front wheel bearing. Who is correct?
A. A only C. Both A and B
B. B only D. Neither A nor B

377

ASE Practice Examination

A 50 question ASE practice exam, located in the appendix, is included to test students on the content of the complete Shop Manual.

Reviewers

I would like to extend a special thank you to those who saw things I overlooked and for their contributions:

Alan B. Clark
Lane Community College
Eugene, OR

David Ferri
Nashville Auto-Diesel College
Nashville, TN

Michael Henich
Linn Benton Community College
Albany, OR

Winston A. Ingraham
University College of Cape Breton
Sydney, Nova Scotia
CANADA

John Kershaw
GM Training Center
Moorestown, NJ

George Liidemann
Centennial College of Applied Arts and Technology
Scarborough, Ontario
CANADA

Rory Perrodin
Barton County Community College
Great Bend, KS

Joe Sechrest
Forshyth Technical Community College
Winston Salem, NC

Randy Turnage
Wilson Technical Community College
Wilson, NC

W. Scott Welch
Pennsylvania College of Technology
Williamsport, PA

Contributing Companies

I would also like to thank these companies who provided technical information and art for this edition:

Accuride Corporation
Air-O-Matic Power Steering Div., Sycon Corp.
Allied Signal Automotive Truck Brake Systems
American Automobile Manufacturer's Association
American Steel Foundries
Ammco Tools, Inc.
ASE (National Institute for Automotive Service Excellence)
Bear Automotive Service Equipment Company
Bee Line Company
Bostrom Seating, Inc.
Bridgestone/Firestone, Inc.
Chrysler Corporation
Commercial Vehicle Safety Alliance
Cooper Moog Automotive
CR Services
CRC Industries, Inc.
Dorsey Trailers, Inc.
DuPont Automotive Products
Eaton Corporation
Ford Motor Company
Freightliner Corporation
General Fire Extinguisher Corporation
Heavy Duty Trucking
Hendrickson International
Hennessy Industries, Inc.
Holland Hitch Company
Hunter Engineering Company
John Deere & Company
Kleer-Flo Company
Mac Tools, Inc.
Mack Trucks, Inc.
The Maintenance Council, American Trucking Association
MPSI
Navistar International Transportation Corporation
Neway Anchorlok International, Inc.
OTC, a division of SPX Corporation
P&H Handling, a Harnishfeger Industries Co.
Power Packer
Reyco Industries, Inc.
Rockwell International
SAE International
Sealed Power Corporation
The Sherwin Williams Company
Siebe North
SKF USA Inc.
Snap-on Tools Corporation
Sun Electric Corporation
The Timken Company
Volvo/GM Heavy Truck Corporation
Weaver Division, Walter Kidde & Co. Inc.
Western Emergency Equipment

Safety Practices

Upon completion and review of this chapter, you should be able to:

- ❑ Recognize shop hazards and take the necessary steps to avoid personal injury or property damage.
- ❑ Explain the purposes of the Occupational Safety and Health Act.
- ❑ Identify the necessary steps for personal safety in the truck shop.
- ❑ Describe the reasons for prohibiting drug and alcohol use in the truck shop.
- ❑ Explain the steps required to provide electrical safety in the truck shop.
- ❑ Define the steps required to provide safe handling and storage of gasoline and diesel fuel.
- ❑ Describe the necessary housekeeping safety steps.
- ❑ Explain the essential general truck shop safety practices.
- ❑ Define the steps required to provide fire safety in the truck shop.
- ❑ Describe typical fire extinguisher operating procedure.
- ❑ Explain four different types of fires, and the type of fire extinguisher required for each type of fire.
- ❑ Describe three other pieces of truck shop safety equipment other than fire extinguishers.
- ❑ Follow proper safety precautions while handling hazardous waste materials.
- ❑ Dispose of hazardous waste materials in accordance with state and federal regulations.

Introduction

Safety is extremely important in the truck shop. The knowledge and practice of safety precautions prevent serious personal injury and expensive property damage. Medium/heavy duty truck students and technicians must be familiar with shop hazards and all types of safety, including personal safety, gasoline and diesel fuel handling, housekeeping, general shop safety, fire, and hazardous material. The first step in providing a safe shop is learning about all types of safety precautions. The second and most important step in this process, however, is applying our knowledge of safety precautions while working in the shop. We must actually develop safe working habits in the shop from our understanding of various safety precautions. When shop employees have a careless attitude toward safety, accidents are more likely to occur. All shop personnel must develop a serious attitude toward safety in the shop. This results in shop personnel who will learn and adopt all shop safety rules.

Shop personnel must be familiar with their rights regarding hazardous waste disposal. These rights are explained in the right-to-know laws. Secondly, shop personnel must be familiar with hazardous materials in the truck shop, and the proper disposal methods for these materials according to state and federal regulations.

A BIT OF HISTORY

In 1975, 102,508 diesel powered trucks were sold in the United States, and 84,878 of these trucks had a gross vehicle weight rating (GVWR) of more than 33,000 pounds. In 1994, total sales of diesel powered trucks in the United States were 439,730, and 189,692 of these trucks had a GVWR of more than 33,000 pounds.

Occupational Safety and Health Act

OSHA regulates working conditions in the United States.

The **Occupational Safety and Health Act (OSHA)** was passed by the United States government in 1970. The purposes of this legislation are these:

1. To assist and encourage the citizens of United States in their efforts to ensure safe and healthful working conditions by providing research, information, education, and training in the field of occupational safety and health.
2. To ensure safe and healthful working conditions for working men and women by authorizing enforcement of the standards developed under the Act.

Because approximately 25% of workers are exposed to health and safety hazards on the job, OSHA is necessary to monitor, control, and educate workers regarding health and safety in the work place. Employers and employees should be familiar with work place hazardous materials information systems (WHMIS).

Shop Hazards

Shop hazards must be recognized and avoided to prevent personal injury.

Service technicians and students encounter many hazards in a truck shop. Basic shop safety rules and procedures must be followed to avoid personal injury. Some of the safety concerns in a truck shop are:

1. Flammable liquids such as gasoline and paint must be handled and stored properly.
2. Flammable materials such as oily rags must be stored properly to avoid a fire hazard.
3. Batteries contain a corrosive sulfuric acid solution and produce explosive hydrogen gas while charging.
4. Loose sewer and drain covers may cause foot or toe injuries.
5. Caustic liquids, such as those in hot cleaning tanks, are harmful to the skin and eyes.
6. High-pressure air in the shop's compressed air system can be dangerous if it penetrates the skin and enters the bloodstream.
7. Frayed cords on electric equipment and lights may result in severe electrical shock.
8. Hazardous waste material, such as batteries and the caustic cleaning solution from a hot or cold cleaning tank, must be handled properly to avoid harmful effects.
9. Carbon monoxide and particulates from truck exhaust are harmful to human beings.
10. Loose clothing or long hair may become entangled in rotating parts on equipment or vehicles, resulting in serious injury.
11. Dust and vapors generated during some repair jobs are harmful. Asbestos dust generated during brake lining service and clutch service is a contributor to lung cancer.
12. High noise levels from shop equipment such as an air chisel may be harmful to the ears.
13. Oil, grease, water, or parts cleaning solutions on shop floors may cause someone to slip and fall, resulting in serious injury.

Safety in the Truck Shop

Personal injury, vehicle damage, and property damage must be avoided by following safety rules regarding personal protection, substance abuse, electrical safety, gasoline safety, housekeeping safety, fire safety, and general shop safety.

Each person in a truck shop must follow certain basic shop safety rules to remove the danger from shop hazards. When all personnel in the shop follow these basic shop safety rules, personal injury, vehicle damage, and property damage may be prevented.

Personal Safety

Personal safety is the responsibility of each technician in the shop. Always follow these safety practices:

1. Always use the correct tool for the job. If the wrong tool is used it may slip and cause hand injury.
2. Follow the truck manufacturer's recommended service procedures.
3. Always wear eye protection such as safety glasses, or a face shield (Figure 1-1).
4. Wear protective gloves when cleaning parts in hot or cold tanks, and when handling hot parts such as exhaust manifolds.
5. Do not smoke when working on a truck. A spark from a cigarette or lighter may ignite flammable materials in the work area.
6. When working on a running engine, keep hands and tools away from rotating parts.
7. Do not wear loose clothing, and keep long hair tied behind your head. Loose clothing or long hair is easily entangled in rotating parts.
8. Wear safety shoes or boots.
9. Do not wear watches, jewelry, or rings when working on a truck. Severe burns occur when jewelry makes contact between an electric terminal and ground.
10. Always place a shop exhaust hose on the truck exhaust pipe if the engine is running, and be sure the exhaust fan is running. Carbon monoxide in the vehicle exhaust can be harmful or fatal.
11. Be sure that the shop has adequate ventilation.
12. Make sure the work area has adequate lighting.
13. Use trouble lights with steel or plastic cages around the bulb. If an unprotected bulb breaks, it may ignite flammable materials in the area.
14. When servicing a truck, always apply the parking brake.
15. Avoid working on a truck parked on an incline.
16. Never work under a truck unless the vehicle chassis is supported securely on safety stands.
17. When one end of a truck is raised, place wheel chocks on both sides of the wheels remaining on the floor.
18. Be sure that you know the location of shop first aid kits and eye wash fountains.
19. Familiarize yourself with the location of all shop fire extinguishers.
20. Do not use any type of open flame heater to heat the work area.
21. Collect oil, fuel, brake fluid, and other liquids in the proper safety containers.

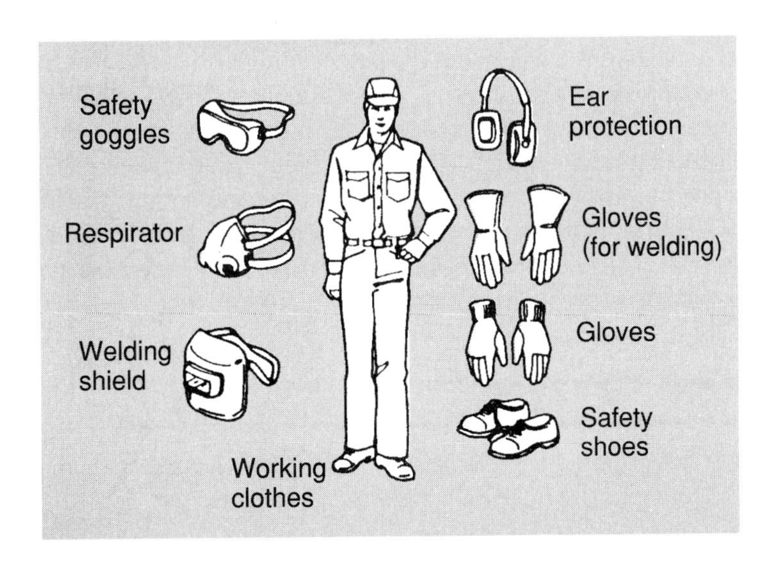

Figure 1-1 Shop safety equipment including safety goggles, respirator, welding shield, proper work clothes, ear protection, welding gloves, work gloves, and safety shoes. (Courtesy of Oldsmobile Division, General Motors Corporation.)

22. Use only approved cleaning fluids and equipment. Do not use gasoline to clean parts.
23. Obey all state and federal safety, fire, and hazardous material regulations.
24. Always operate equipment according to the equipment manufacturer's recommended procedure.
25. Do not operate equipment unless you are familiar with the correct operating procedure.
26. Do not leave running equipment unattended.
27. Do not use electrical equipment, including trouble lights, with frayed cords.
28. Be sure the safety shields are in place on rotating equipment.
29. Before operating electrical equipment, be sure the power cord has a ground connection.
30. When working in an area where extreme noise levels are encountered, wear ear plugs or covers.
31. Always wear boots or shoes that provide adequate foot protection. Heavy-duty work boots or shoes with steel toe caps are best for working in the truck shop. Footwear must protect against heavy falling objects, flying sparks, and corrosive liquids. Soles on footwear must protect against punctures by sharp objects. Running shoes and street shoes are not recommended in the shop.
32. Wear a respirator to protect your lungs when working in dusty conditions.
33. When working on, around, or under a truck or tractor, disconnect the batteries and tag the driver's door, steering wheel, or start button so that no one starts the vehicle. The tag should read, "Do not start."
34. If a truck or tractor has a brake failure or the parking brake chambers have been caged for towing the vehicle, the unit should be tagged on the door, steering wheel, and start button, "Do not start or move—No Brakes."

Smoking, Alcohol, and Drugs in the Shop

TRADE JARGON: The improper or excessive use of alcoholic beverages, drugs, or both may be referred to as substance abuse.

Do not smoke when working in the shop. If the shop has designated smoking areas, smoke only in these areas. Do not smoke in customers' trucks. A non-smoker may not appreciate cigarette odor in his or her truck. A spark from a cigarette or lighter may ignite flammable materials in the work place. The use of drugs or alcohol must be avoided while working in the shop. Even a small amount of drugs or alcohol affects reaction time. In an emergency situation, slow reaction time may cause personal injury. If a heavy object falls off the work bench, and your reaction time is slowed by drugs, or alcohol, you may not get your foot out of the way in time, resulting in foot injury. When a fire starts in the work place and you are a few seconds slower getting a fire extinguisher into operation because of alcohol or drug use, it could make the difference between extinguishing a fire, and having expensive fire damage or injury.

Lifting and Carrying

Truck service jobs often require heavy lifting. You should know your maximum weight-lifting ability, and should not attempt to lift more than this weight. If a heavy part exceeds your weight-lifting ability, have a co-worker help with the lifting job. Follow these steps when lifting or carrying an object:

1. If the object is to be carried be sure your path is free from loose parts or tools.
2. Position your feet close to the object and position your back reasonably straight for proper balance (Figure 1-2).

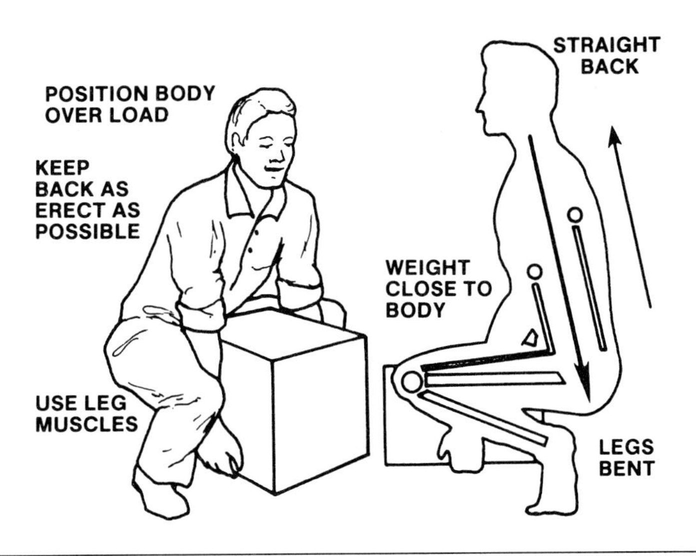

Figure 1-2 Proper lifting procedure.

3. Your back and elbows should be kept as straight as possible. Continue to bend your knees until your hands reach the best lifting location on the object to be lifted.
4. Be certain the container is in good condition. If a container falls apart during the lifting operation, parts may drop out of the container, resulting in foot injury or part damage.
5. Maintain a firm grip on the object, and do not attempt to change your grip while lifting is in progress.
6. Straighten your legs to lift the object, and keep the object close to your body. Use leg muscles rather than back muscles.
7. If you have to change direction of travel, turn your whole body instead of twisting it.
8. Do not bend forward to place an object on a work bench or table. Position the object on the front surface of the work bench and slide it back. Do not pinch your fingers under the object while setting it on the front of the bench.
9. If the object must be placed on the floor or a low surface, bend your legs to lower the object. Do not bend your back forward, because this movement strains back muscles.
10. When a heavy object must be placed on the floor, locate suitable blocks under the object to prevent jamming your fingers under the object.

Hand Tool Safety

Many shop accidents are caused by improper use, and care, of hand tools. These hand tool safety steps must be followed:

1. Maintain tools in good condition and keep them clean. Worn tools may slip and result in hand injury. If a hammer with a loose head is used, the head may fly off and cause personal injury or vehicle damage. Your hand may slip off a greasy tool, and this action may cause part of your body to hit the vehicle. For example, your head may hit the vehicle hood.
2. Use of the wrong tool for the job may cause damage to the tool, fastener, or your hand, if the tool slips. If you use a screwdriver for a chisel or pry bar, the blade may shatter, causing serious personal injury.
3. Use sharp pointed tools with caution. Always check your pockets before sitting on the vehicle seat. A screwdriver, punch, or chisel in the back pocket may put an expensive tear in the upholstery. Do not lean over fenders with sharp tools in your pockets.

4. Tool tips that are intended to be sharp should be kept in a sharp condition. Sharp tools, such as chisels, will do the job faster with less effort.

Truck and Air Suspension Safety Precautions

In an air suspension system the air pressure is supplied from one of the air reservoirs in the air brake system to the suspension system. Therefore, some of the safety precautions related to air suspension systems also apply to air brake systems. When performing any service on a truck, or a truck with an air suspension system, always observe these safety precautions:

1. Block the truck wheels before working on the vehicle.
2. Do not raise or lower a tilt cab with the engine running (Figure 1-3). This action may cause components to contact rotating belts and pulleys.
3. Prior to removing air suspension system components, raise the truck frame with a floor jack to remove the load from the air springs. Lower the frame so it is supported on safety stands. Open the air brake system reservoir draincocks and relieve the air pressure to zero before disconnecting the air suspension system components.
4. Shut off the engine when working on the truck. Only run the engine when instructed to do so in a service procedure recommended by the truck manufacturer. If the engine must be running, be very careful not to contact rotating or hot components.
5. Follow the truck manufacturer's recommended service procedures in the service manual.
6. Do not disconnect any air line or hose under air pressure. The hose or line may whip around, causing personal injury.
7. Never allow pressures to exceed the maximum specified pressures in the air suspension system, and always wear safety glasses.

Figure 1-3 Do not raise or lower a tilt cab with the engine running. (Courtesy of General Motors Corporation.)

8. Do not attempt to remove or service a component until your are familiar with the required service procedures.
9. Be sure all lines and components are equivalent to the original lines or components.
10. Components with striped or damaged threads must be replaced.
11. When opening reservoir draincocks never place your hand or any part of your body near the draincock.
12. Do not look into jets or passages when blowing them out with compressed air.

Electrical Equipment Safety

1. Frayed cords on electrical equipment such as shop lights, drills, grinders, wheel aligners, wheel balancers, overhead hoists and trolleys, and cleaning equipment must be replaced or repaired immediately.
2. All electric cords from lights and electric equipment must have a ground connection. The ground connector is the round terminal in a three-prong electrical plug. Do not use a two-prong adapter to plug in a three-prong electrical cord. Three-prong electrical outlets should be mandatory in all shops.
3. Do not leave electrical equipment running and unattended.

Gasoline and Diesel Fuel Safety

Gasoline is a very explosive liquid. One exploding gallon of gasoline has a force equal to 14 sticks of dynamite. It is the expanding vapors from gasoline that are extremely dangerous. These vapors are present even in cold temperatures. Vapors formed in gasoline tanks on many trucks are controlled, but vapors from a gasoline storage can may escape from the can, resulting in a hazardous situation. Therefore, gasoline storage containers must be placed in a well-ventilated space. Although diesel fuel is not as volatile as gasoline, the same basic rules apply to diesel fuel and gasoline storage.

Approved gasoline storage cans have a flash-arresting screen at the outlet (Figure 1-4). These screens prevent external ignition sources from igniting the gasoline within the can while the gasoline is being poured. Follow these safety precautions regarding gasoline containers:

1. Always use approved gasoline containers that are painted red for proper identification.

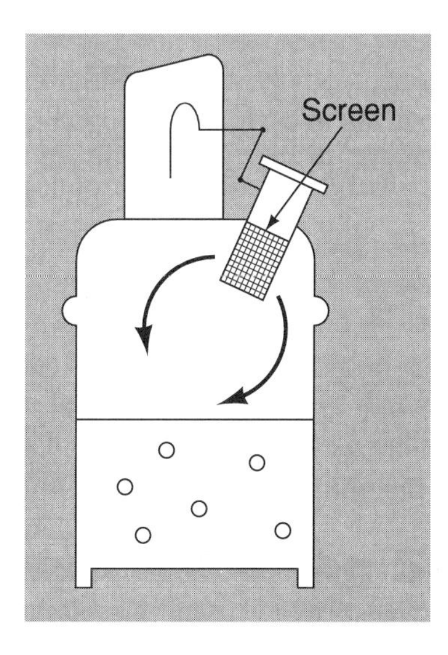

Figure 1-4 Approved gasoline container.

2. Do not completely fill gasoline containers. Always leave the level of gasoline at least one inch from the top of the container. This allows expansion of the gasoline at higher temperatures. If gasoline containers are completely full, the gasoline will expand when the temperature increases. This expansion forces gasoline from the can and creates a dangerous spill.
3. If gasoline or diesel fuel containers must be stored, place them in a designated storage locker or facility. Do not store gasoline containers in your home.
4. When a gasoline container must be transported, be sure it is secured against upsets.
5. Do not store a partially filled gasoline container for long periods of time, because it may give off vapors and produce a potential danger.
6. Never leave gasoline containers open except while filling or pouring gasoline from the container.
7. Do not prime an engine with gasoline while cranking the engine.
8. Never use gasoline as a cleaning agent.
9. Always connect a ground strap to containers when filling or transferring fuel or other flammable products from one container to another. This prevents static electricity that could result in explosion and fire.

Housekeeping Safety

1. Keep shop floors clean. Always clean shop floors immediately after a spill.
2. Store paint and other flammable liquids in a closed steel cabinet (Figure 1-5).
3. Oily rags must be stored in approved, covered, garbage containers (Figure 1-6). A slow generation of heat occurs from oxidation of oil on these rags. Heat may continue to be generated until the ignition temperature is reached. The oil and the rags then begin to burn, causing a fire. This action is called spontaneous combustion. However, if the oily rags are in an airtight, approved container, the fire cannot get enough oxygen to cause burning.
4. Keep the shop neat and clean. Always pick up tools and parts, and do not leave creepers lying on the floor.
5. Keep the workbenches clean. Do not leave heavy objects, such as used parts, on the bench after you are finished with them.

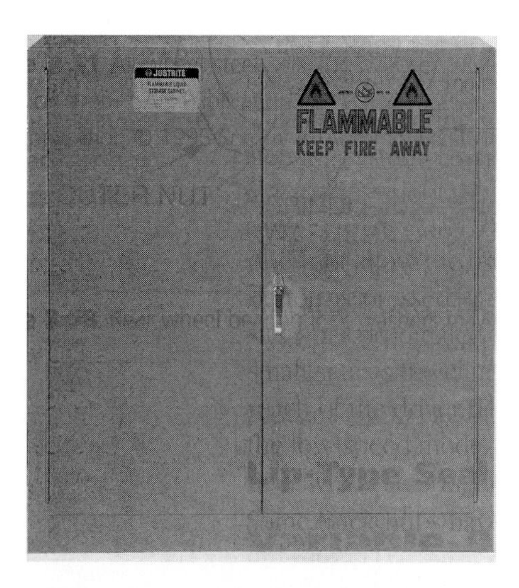

Figure 1-5 Paints and combustible material containers must be kept in an approved safety cabinet. (Courtesy of Sherwin-Williams Co.)

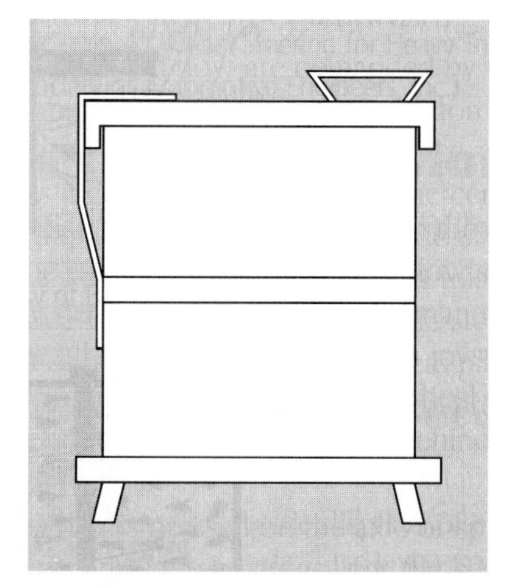

Figure 1-6 Oily rags must be stored in approved airtight containers.

General Truck Shop Safety

Safety in the truck shop is extremely important. When shop safety rules are observed, injuries are reduced and shop productivity is increased because employees are not absent from work because of injuries. Observing shop safety rules involves teamwork. Everyone in the shop must obey shop safety rules to provide a safe working area. If one employee does not obey safety rules, another employee may be injured. For example, if an employee leaves a hydraulic jack handle sticking out from under a vehicle, another employee may trip over this handle and injure his or her back. These general shop safety rules must be observed in the truck shop:

1. All sewer covers must fit properly and be kept securely in place.
2. Always wear a face shield, protective gloves, and protective clothing when necessary. Gloves should be worn when working with solvents and caustic solutions, handling hot metal, or grinding metal. Various types of protective gloves are available. Shop coats and coveralls are the most common types of protective clothing.
3. Never direct high pressure air from an air gun against human flesh. If this is done, air may penetrate the skin and enter the bloodstream, causing serious health problems or even death. Always keep air hoses in good condition. If an end blows off an air hose, the hose may whip around and result in personal injury. Use only Occupational Safety and Health Act (OSHA) approved air gun nozzles.
4. Handle all hazardous waste materials according to state and federal regulations. (These regulations are explained later in this chapter.)
5. Always place a shop exhaust hose on the exhaust pipe of a truck if the engine is running in the shop, and be sure the shop exhaust fan is turned on.
6. Keep hands, long hair, and tools away from rotating parts on running engines such as fan blades and belts. Remember that an electric-drive fan may start turning at any time.
7. When servicing brakes or clutches from manual transmissions, always clean asbestos dust from these components with an approved asbestos dust vacuum cleaner or an approved parts washer.
8. Always use the correct tool for the job. For example, never strike a hardened steel component, such as a piston pin, with a steel hammer. This type of component may shatter, and fragments can penetrate eyes or skin.
9. Follow the truck manufacturer's recommended service procedures.
10. Be sure the shop has adequate ventilation.
11. Make sure the work area has adequate lighting.
12. Use trouble lights with steel or plastic cages around the bulb. If an unprotected bulb breaks, it may ignite flammable materials in the area.
13. When servicing a truck always apply the parking brake.
14. Avoid working on a truck parked on an incline.
15. *Never work under a truck unless the truck chassis is supported securely on safety stands.*
16. When one end of a truck is raised, place wheel chocks on both sides of the wheels remaining on the floor.
17. Be sure that you know the location of shop first-aid kits, eye wash fountains, and fire extinguishers.
18. Collect oil, fuel, brake fluid, and other liquids in the proper safety containers.
19. Use only approved cleaning fluids and equipment. Do not use gasoline to clean parts.
20. Obey all state and federal safety, fire, and hazardous material regulations.
21. Always operate equipment according to the equipment manufacturer's recommended procedure.

22. Do not operate equipment unless you are familiar with the correct operating procedure.
23. Do not leave running equipment unattended.
24. Be sure the safety shields are in place on rotating equipment.
25. All shop equipment must have regularly scheduled maintenance and adjustment.
26. Some shops have safety lines around equipment. Always work within these lines when operating equipment.
27. Be sure that shop heating equipment is well ventilated.
28. Do not run in the shop or engage in horseplay.
29. Post emergency phone numbers near the phone. These numbers should include a doctor, ambulance, fire department, hospital, and police.
30. Do not place hydraulic jack handles where someone can trip over them.
31. Keep aisles clear of debris.

Fire Safety

Observing fire safety rules may prevent a fire in the truck shop. Following the proper fire safety rules and procedures may also make the difference between extinguishing a fire with minimum damage and having a fire get out of control causing very extensive damage. Follow these fire safety rules and procedures in the truck shop:

1. Familiarize yourself with the location and operation of all shop fire extinguishers.
2. If a fire extinguisher is used, report this to management so the extinguisher can be recharged.
3. Do not use any type of open flame heater to heat the work area.
4. Do not turn on the ignition switch or crank the engine with a fuel line disconnected.
5. Store all combustible materials, such as gasoline, paint, and oily rags, in approved safety containers.
6. Clean up gasoline, diesel fuel, oil, or grease spills immediately.
7. Always wear clean shop clothes. Do not wear oil-soaked clothes.
8. Do not allow sparks and flames near batteries.
9. Welding tanks must be securely fastened in an upright position.
10. Do not block doors, stairways, or exits.
11. Do not smoke when working on vehicles.
12. Do not smoke or create sparks near flammable materials or liquids.
13. Store combustible shop supplies such as paint in a closed steel cabinet.
14. Keep gasoline and diesel fuel in approved safety containers.
15. If a fuel tank is removed from a vehicle, do not drag the tank on the shop floor.
16. Know the approved fire escape route from your classroom or shop to the outside of the building.
17. If a fire occurs, do not open doors or windows. This action creates extra draft, which makes the fire worse.
18. Do not put water on a gasoline fire, because the water will make the fire worse.
19. Call the fire department as soon as a fire begins, and then attempt to extinguish the fire.
20. If possible, stand 6 to 10 feet from the fire, and aim the fire extinguisher nozzle at the base of the fire with a sweeping action.
21. If a fire produces a lot of smoke in the room, remain close to the floor to obtain oxygen and avoid breathing smoke.
22. If the fire is too hot, or the smoke makes breathing difficult, get out of the building.
23. Do not re-enter a burning building.

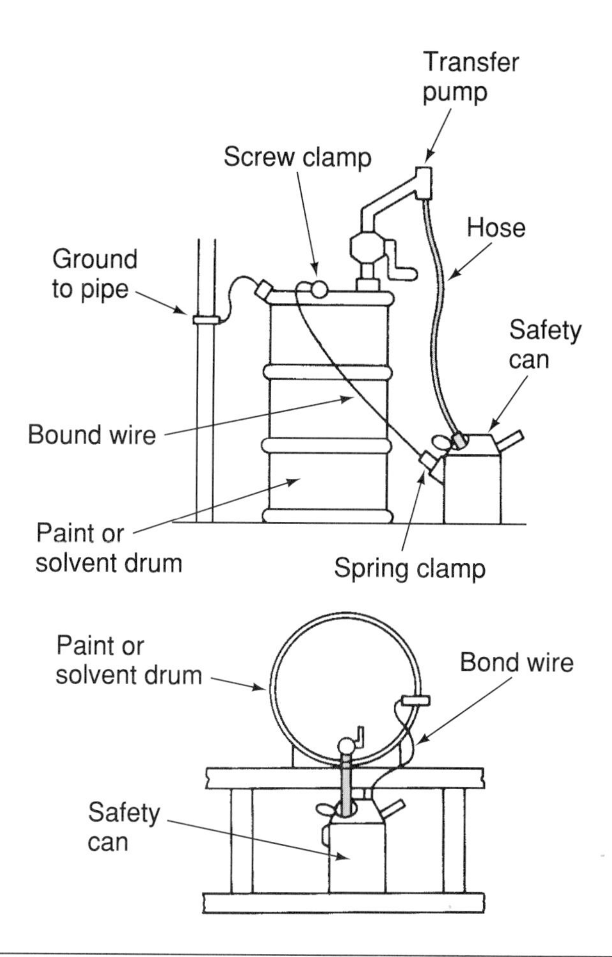

Figure 1-7 Safe procedures for flammable liquid transfer. (Courtesy of Du Pont Co.)

24. Keep solvent containers covered except when pouring from one container to another. When flammable liquids are transferred from bulk storage, the bulk container should be grounded to a permanent shop fixture such as a metal pipe. During this transfer process, the bulk container should be grounded to the portable container (Figure 1-7). These ground wires prevent the buildup of a static electric charge which could result in a spark and disastrous explosion. Always discard or clean empty solvent containers, because fumes in these containers are a fire hazard.

25. Familiarize yourself with different types of fires and fire extinguishers, and know the type of extinguisher to use on each fire.

Truck Shop Safety Equipment

Fire Extinguishers

Fire extinguishers are one of the most important pieces of safety equipment. All shop personnel must know the location of the fire extinguishers in the shop. If you have to waste time looking for an extinguisher after a fire starts, the fire could get out of control before you get the extinguisher into operation. Fire extinguishers should be located where they are easily accessible at all times. Everyone working in the shop must know how to operate the fire extinguishers. Several different types of fire extinguishers exist, but their operation usually involves these steps:

Shop safety equipment must be easily accessible and in good working condition.

1. Get as close as possible to the fire without jeopardizing your safety.
2. Grasp the extinguisher firmly and aim the extinguisher at the fire.
3. Pull a pin from the extinguisher handle.
4. Squeeze the handle to dispense the contents of the extinguisher.

5. Direct the fire extinguisher nozzle at the base of the fire, and dispense the contents of the extinguisher with a sweeping action back and forth across the fire. Most extinguishers discharge their contents in 8 to 25 seconds.
6. Always be sure the fire is extinguished.
7. Always keep an escape route open behind you so a quick exit is possible if the fire gets out of control.

A decal on each fire extinguisher identifies the type of chemical in the extinguisher and provides operating information (Figure 1-8). Shop personnel should be familiar with the following types of fires and fire extinguishers:

1. Class A fires are those involving ordinary combustible materials such as paper, wood, clothing, and textiles. **Multipurpose dry chemical extinguishers** are used on these fires.
2. Class B fires involve the burning of flammable liquids such as gasoline, diesel fuel, oil, paint, solvents, and greases. These fires may be extinguished with multipurpose dry chemical extinguishers. **Fire extinguishers containing halogen, or halon,** may be used to extinguish class B fires. The chemicals in this type of extinguisher attach to the hydrogen, hydroxide, and oxygen molecules to stop the combustion process almost instantly. However, the resultant gases from the use of halogen-type extinguishers are very toxic and harmful to the operator of the extinguisher. Halon fire extinguishers are now illegal because of chlorofluorocarbon (CFC) regulations.
3. Class C fires involve the burning of electrical equipment, such as wires, motors, and switches. These fires may be extinguished with multipurpose dry chemical extinguishers.
4. Class D fires involve the combustion of metal chips, turnings, and shavings. Dry chemical extinguishers are the only type of extinguisher recommended for these fires.

Additional information regarding types of extinguishers for various types of fires is provided in Table 1-1.

Causes of Eye Injuries

Eye injuries may occur in various ways in a truck shop. Some of the common eye accidents are these:

1. Thermal burns from excessive heat
2. Irradiation burns from excessive light, such as from an arc welder
3. Chemical burns from strong liquids such as battery electrolyte
4. Foreign material in the eye
5. Penetration of the eye by a sharp object
6. A blow from a blunt object

Wearing safety glasses and observing shop safety rules will prevent most eye accidents.

Eyewash Fountains

If a chemical gets in your eyes it must be washed out immediately to prevent a chemical burn. An eyewash fountain is the most effective way to wash the eyes. An eyewash fountain is similar to a drinking water fountain, but the eyewash fountain has water jets placed throughout the fountain top. Every shop should be equipped with some eyewash facility (Figure 1-9). Be sure you know the location of the eyewash fountain in the shop.

Safety Glasses and Face Shields

The mandatory use of eye protection with safety glasses or a face shield is one of the most important safety rules in a truck shop. Face shields protect the face, and safety glasses protect the eyes. When grinding, safety glasses must be worn; a face shield can also be worn. Many shop insurance policies require the use of eye protection in the shop. Some medium/heavy truck technicians have been blinded in one or both eyes because they did not bother to wear safety glasses. All safety

Figure 1-8 Types and sizes of fire extinguishers. (Courtesy of General Fire Extinguisher Corporation.)

FIRES	TYPE	USE	OPERATION
A CLASS A FIRES ORDINARY COMBUSTIBLE MATERIALS SUCH AS WOOD, PAPER, TEXTILES, AND SO FORTH. REQUIRES...COOLING-QUENCHING	**FOAM** SOLUTION OF ALUMINUM SULPHATE AND BICARBONATE OF SODA	OK FOR **AB** NOT FOR **C**	FOAM: DIRECT STREAM INTO THE BURNING LIQUID. ALLOW FOAM TO FALL LIGHTLY ON FIRE
B CLASS B FIRES FLAMMABLE LIQUIDS, GREASES, GASOLINE, OILS, PAINTS, AND SO FORTH. REQUIRES...BLANKETING OR SMOTHERING	**CARBON DIOXIDE** CARBON DIOXIDE GAS UNDER PRESSURE	NOT FOR **A** OK FOR **BC**	CARBON DIOXIDE: DIRECT DISCHARGE AS CLOSE TO FIRE AS POSSIBLE. FIRST AT EDGE OF FLAMES AND GRADUALLY FORWARD AND UPWARD
C CLASS C FIRES ELECTRICAL EQUIPMENT, MOTORS, SWITCHES, AND SO FORTH. REQUIRES...A NONCON-DUCTING AGENT	**DRY CHEMICAL**	MULTI-PURPOSE TYPE: OK FOR **ABC** ORDINARY BC TYPE: NOT FOR **A**; OK FOR **BC**	DRY CHEMICAL: DIRECT STREAM AT BASE OF FLAMES, USE RAPID LEFT-TO-RIGHT MOTION TOWARD FLAMES
	SODA-ACID BICARBONATE OF SODA SOLU-TION AND SULPHURIC ACID	OK FOR **A** NOT FOR **BC**	SODA-ACID: DIRECT STREAM AT BASE OF FLAME

Table 1-1 Fire extinguisher selection. (Courtesy of General Fire Extinguisher Corporation.)

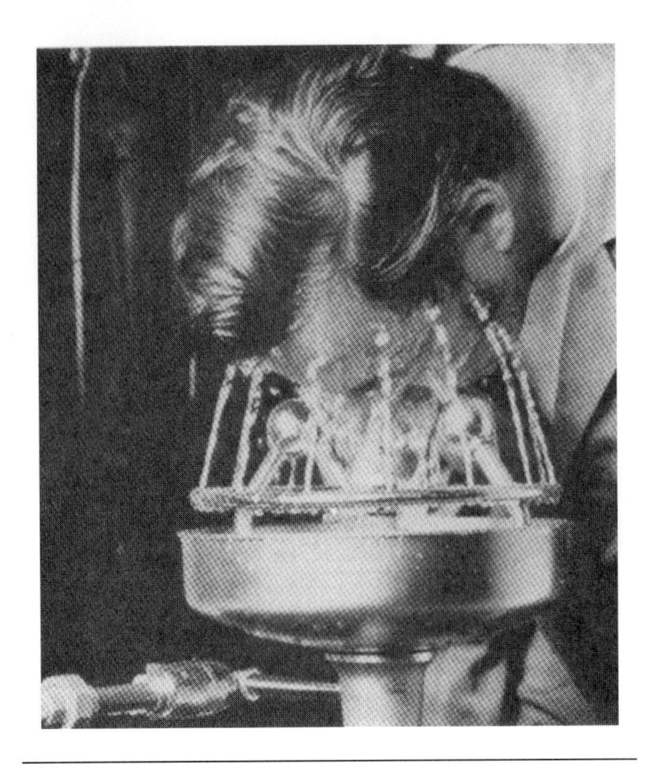

Figure 1-9 Eyewash fountain. (Courtesy of Western Emergency Equipment.)

Figure 1-10 Safety glasses with side protection must be worn in the truck shop. (Courtesy of Siebe North, Inc.)

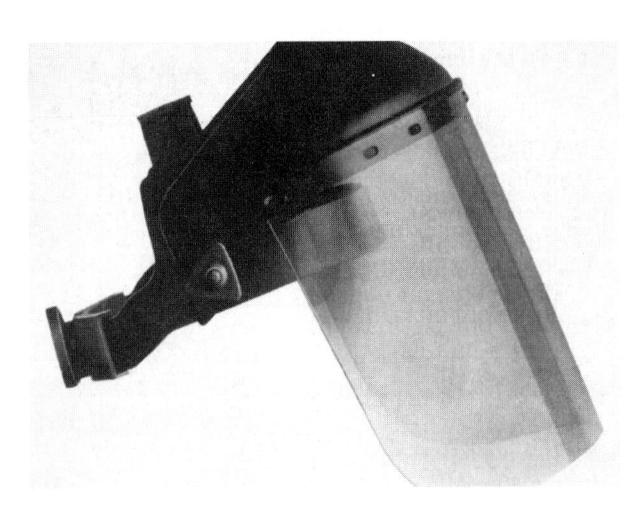

Figure 1-11 Face shield. (Courtesy of Siebe North, Inc.)

glasses must be equipped with safety glass, and they should provide some type of side protection (Figure 1-10). When selecting a pair of safety glasses, find one that feels comfortable on your face. If they are uncomfortable, you may tend to take them off, leaving the eyes unprotected. A face shield should be worn when handling hazardous chemicals or when using an electric grinder or buffer (Figure 1-11).

See chapter 1 in the Shop Manual for safety rules related to tool use and equipment operation in the shop.

First-Aid Kits

First-aid kits should be clearly identified and conveniently located (Figure 1-12). These kits contain such items as bandages and ointment required for minor cuts. All shop personnel must be familiar with the location of first-aid kits. At least one of the shop personnel should have basic first-aid training, and this person should be in charge of administering first aid and keeping first-aid kits filled.

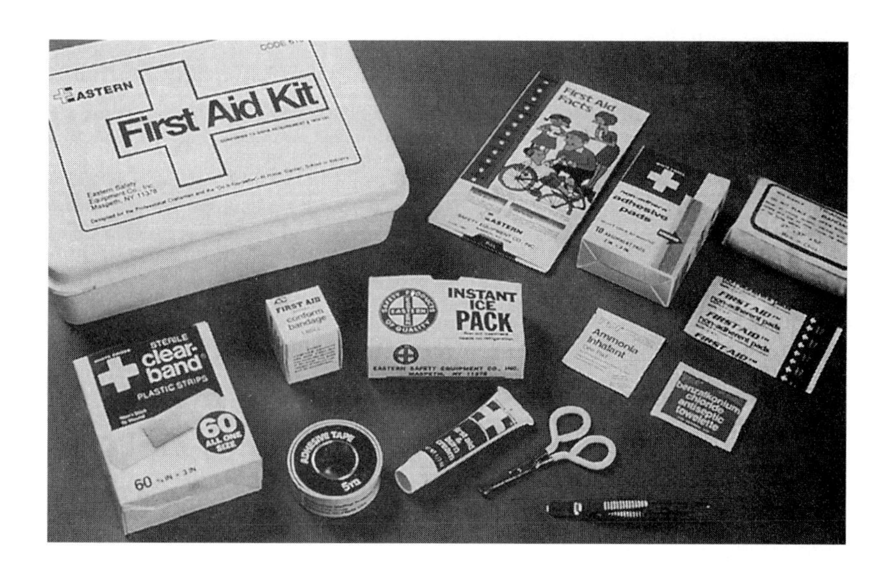

Figure 1-12 First-aid kit.

Hazardous Waste Disposal

CAUTION: When handling hazardous waste material, always wear proper protective clothing and equipment, as detailed in the right-to-know laws. This includes respirator equipment (Figure 1-13). All recommended procedures must be followed accurately. Personal injury may result from improper clothing, equipment, and procedures when handling hazardous materials.

Hazardous waste materials in truck shops are chemicals or components that the shop no longer needs, that pose a danger to the environment and people if they are disposed of in ordinary

Figure 1-13 Wear recommended safety clothing and equipment when hazardous materials are handled. (Courtesy of Du Pont Co.)

garbage cans or sewers. It should be noted, however, that no material is considered hazardous waste until the shop has finished using it and is ready to dispose of it. **The Environmental Protection Agency (EPA)** publishes a list of hazardous materials which is included in the Code of Federal Regulations. Waste is considered hazardous if it is included on the EPA list of hazardous materials, or if it has one or more of these characteristics:

1. **Reactive.** Any material that reacts violently with water or other chemicals is considered hazardous. If a material releases cyanide gas, hydrogen sulphide gas, or similar gases when exposed to low pH acid solutions, it is hazardous.
2. **Corrosive.** If a material burns the skin or dissolves metals and other materials it is considered hazardous.
3. **Toxic.** Materials are hazardous if they leach one or more of eight heavy metals in concentrations greater than 100 times the primary drinking water standard.
4. **Ignitable.** A liquid is hazardous if it has a flash point below 140° F (60° C), and a solid is hazardous if it ignites spontaneously.

The Resource Conservation and Recovery Act (RCRA) states that hazardous material users are responsible for hazardous materials from the time they become a waste until the proper waste disposal is completed.

WARNING: Hazardous waste disposal laws include serious penalties for anyone responsible for breaking these laws.

Federal and state laws control the disposal of hazardous waste materials. Every shop employee must be familiar with these laws. Hazardous waste disposal laws include the **Resource Conservation and Recovery Act (RCRA).** This law basically states that hazardous material users are responsible for hazardous materials from the time they become a waste until the proper waste disposal is completed. Many truck shops hire independent hazardous waste haulers to dispose of hazardous waste material (Figure 1-14). The shop owner or manager should have a written contract with the hazardous waste hauler. Rather than have hazardous waste material hauled to an approved hazardous waste disposal site, a shop may choose to recycle the material in the shop. In this case, the user must store hazardous waste material properly and safely, and be responsible for the transportation of this material until it arrives at an approved hazardous waste disposal site and is processed according to the law.

The RCRA controls these types of automotive waste:

1. Paint and body repair products waste
2. Solvents for parts and equipment cleaning
3. Batteries and battery acid
4. Mild acids used for metal cleaning and preparation
5. Waste oil and engine coolants or antifreeze
6. Air-conditioning refrigerants
7. Engine oil filters

Figure 1-14 Hazardous waste hauler. (Courtesy of Du Pont Co.)

Never under any circumstances use any of the following methods to dispose of hazardous waste material:

1. Pour hazardous wastes on weeds to kill them
2. Pour hazardous wastes on gravel streets to prevent dust
3. Throw hazardous wastes in a dumpster
4. Dispose of hazardous wastes anywhere but an approved disposal site
5. Pour hazardous wastes down sewers, toilets, sinks, or floor drains

The **right-to-know laws** state that employees have a right to know when the materials they use at work are hazardous. The right-to-know laws started with the **Hazard Communication Standard** published by OSHA in 1983. This document was originally intended for chemical companies and manufacturers that required employees to handle hazardous materials in their work situations. At present most states have established their own right-to-know laws. The federal courts have decided to apply these laws to all companies, including truck service shops. Under the right-to-know laws the employer has three responsibilities regarding the handling of hazardous materials by their employees. First, all employees must be trained about the types of hazardous materials they will encounter in the work place. The employees must be informed about their rights under legislation regarding the handling of hazardous materials. All hazardous materials must be properly labelled, and information about each hazardous material must be posted on **material safety data sheets (MSDS)** available from the manufacturer (Figure 1-15). In Canada, MSDS sheets are called **work place hazardous materials information systems (WHMIS).**

Material safety data sheets (MSDS) provide extensive information about hazardous materials.

The employer has a responsibility to place MSDS sheets where they are easily accessible by all employees. The MSDS sheets provide extensive information about the hazardous material such as:

1. Chemical name
2. Physical characteristics
3. Protective equipment required for handling

CRC MATERIAL SAFETY DATA SHEET SILOO

CRC Industries, Inc. • 885 Louis Drive • Warminster, PA 18974 • (215) 674-4300

PRODUCT NAME CLEAN-R-CARB (AEROSOL) #-MSDS05079

PRODUCT- 5079,5079T,5081,5081T

(Page 1 of 2)

1. INGREDIENTS	CAS #	ACGIH TLV	OSHA PEL	OTHER LIMITS	%
Acetone	67-64-1	750 ppm	750 ppm		2-5
Xylene	1330-20-7	100 ppm	100 ppm		68-75
2-Butoxy Ethanol	111-76-2	25 ppm	25 ppm	(skin)	3-5
Methanol	67-56-1	200 ppm	200 ppm		3-5
Detergent	-	NA	NA		0-1
Propane	74-98-6	NA	1000 ppm		10-20
Isobutane	75-28-5	NA	NA	1000ppm	10-20

2. PHYSICAL DATA : (without propellent)

Specific Gravity : 0.865

Vapor Pressure : ND

% Volatile : > 99

Boiling Point : 176 F initial

Evaporation Rate : Moderately fast

Freezing Point : ND

Vapor Density : ND

Appearance and Odor:

pH: NA

A clear colorless liquid, aromatic odor

Solubility : Partially soluble in water.

3. FIRE AND EXPLOSION DATA

Flashpoint : -40 F Method : TCC

Flammable Limits : propellent LEL:1.8 UEL:9.5

Extinguishing Media : CO2, dry chemical, foam

Unusual Hazards : Aerosol cans may explode when heated above 120 F.

4. REACTIVITY AND STABILITY

Stability : Stable

Hazardous decomposition products

: CO2, carbon monoxide (thermal)

Materials to avoid : Strong oxidizing agents and sources of ignition.

5. PROTECTION INFORMATION

Ventilation : Use mechanical means to insure vapor conc. is below TLV.

Respiratory : Use self-contained breathing apparatus above TLV.

Gloves : Solvent resistant Eye & Face : Safety glasses

Other Protective Equipment: Not normally required for aerosol product usage.

Figure 1-15 Material safety data sheets (MSDS) inform employees about hazardous materials. (Courtesy of Storm Vulcan Co.)

4. Explosion and fire hazards
5. Other incompatible materials
6. Health hazards such as signs and symptoms of exposure, medical conditions aggravated by exposure, and emergency and first-aid procedures
7. Safe handling precautions
8. Spill and leak procedures

Second, the employer has a responsibility to make sure that all hazardous materials are properly labelled. The label information must include health, fire, and reactivity hazards posed by the material, along with the protective equipment necessary to handle the material. The manufacturer must supply all warning and precautionary information about hazardous materials, and this information must be read and understood by employees before they handle the material.

Third, employers are responsible for maintaining permanent files regarding hazardous materials. These files must include information on hazardous materials in the shop, proof of employee training programs, and information about accidents such as spills or leaks of hazardous materials. The employers' files must also include proof that employees' requests for hazardous material information such as MSDS sheets have been met. A general right-to-know compliance procedure manual must be maintained by the employer.

Antifreeze and Solvent Handling

Used antifreeze and spent solvent are hazardous waste materials. These materials may be stored on site in durable, labelled containers, and then transported to approved waste disposal sites. An antifreeze recycler may be used to recycle antifreeze (Figure 1-16). After the antifreeze is recycled, a small amount of antifreeze additive is added to maintain the proper additive balance.

It is the responsibility of the repair shop to determine whether their spent solvent is hazardous waste. Waste solvents that are considered hazardous waste have a flash point below 140° F (60° C). Hot water or aqueous parts cleaners may be used to avoid disposing of spent solvent as hazardous waste. Solvent-type parts cleaners with filters are available to greatly extend solvent life and reduce spent solvent disposal costs (Figure 1-17). Solvent reclaimers that clean and restore the solvent so it lasts indefinitely are available (Figure 1-18).

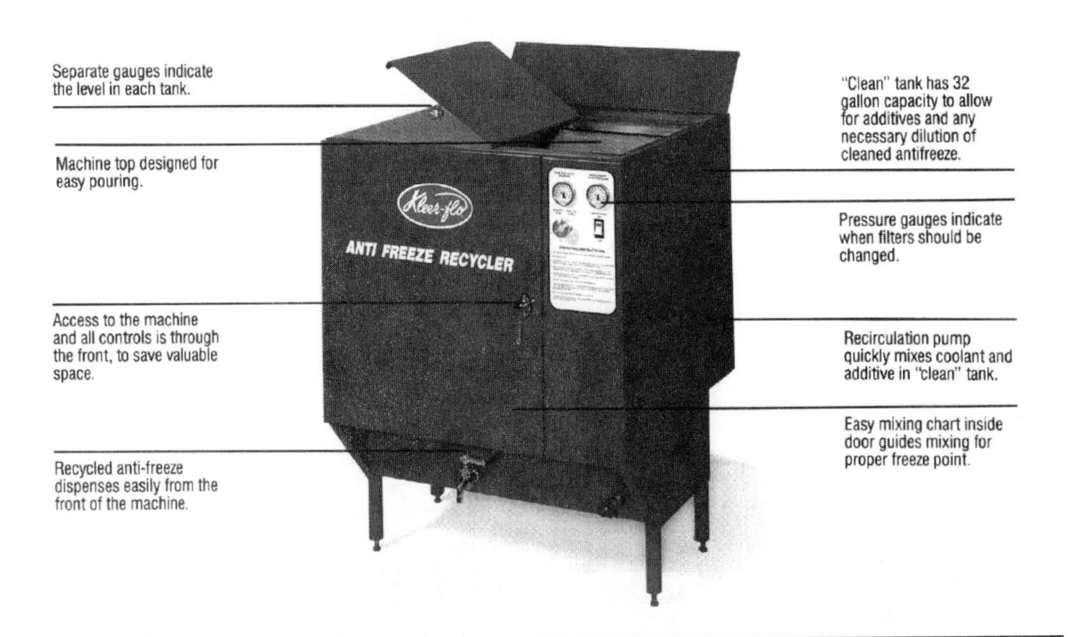

Figure 1-16 Antifreeze recycler. (Courtesy of Kleer-Flo Company.)

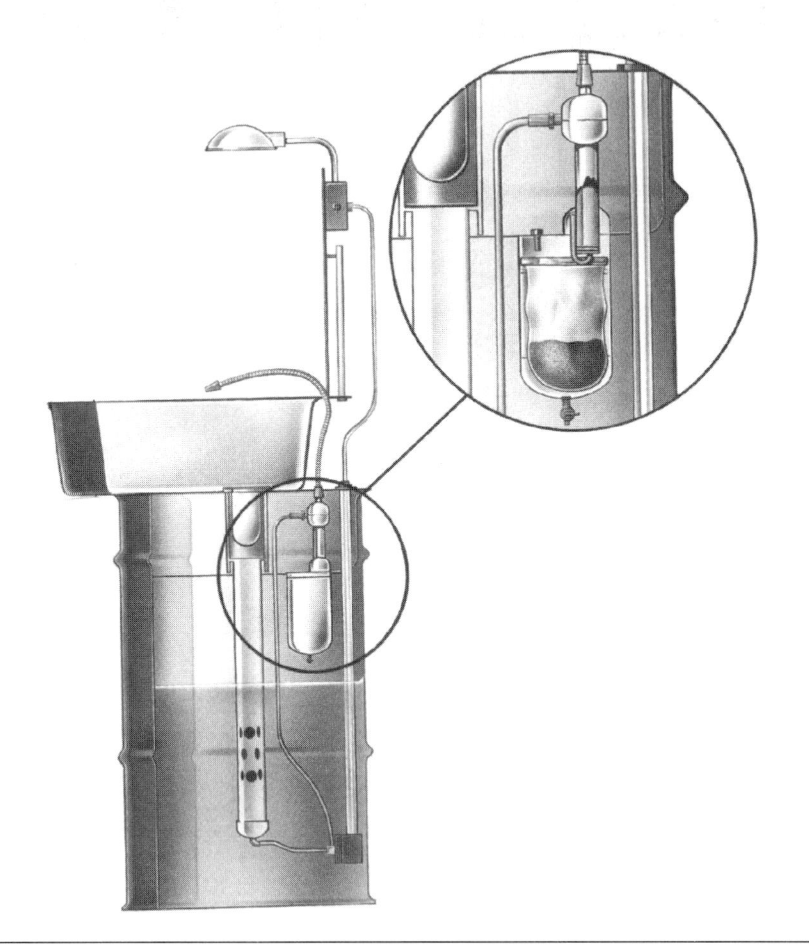

Figure 1-17 Solvent parts washer with filter. (Courtesy of Kleer-Flo Company.)

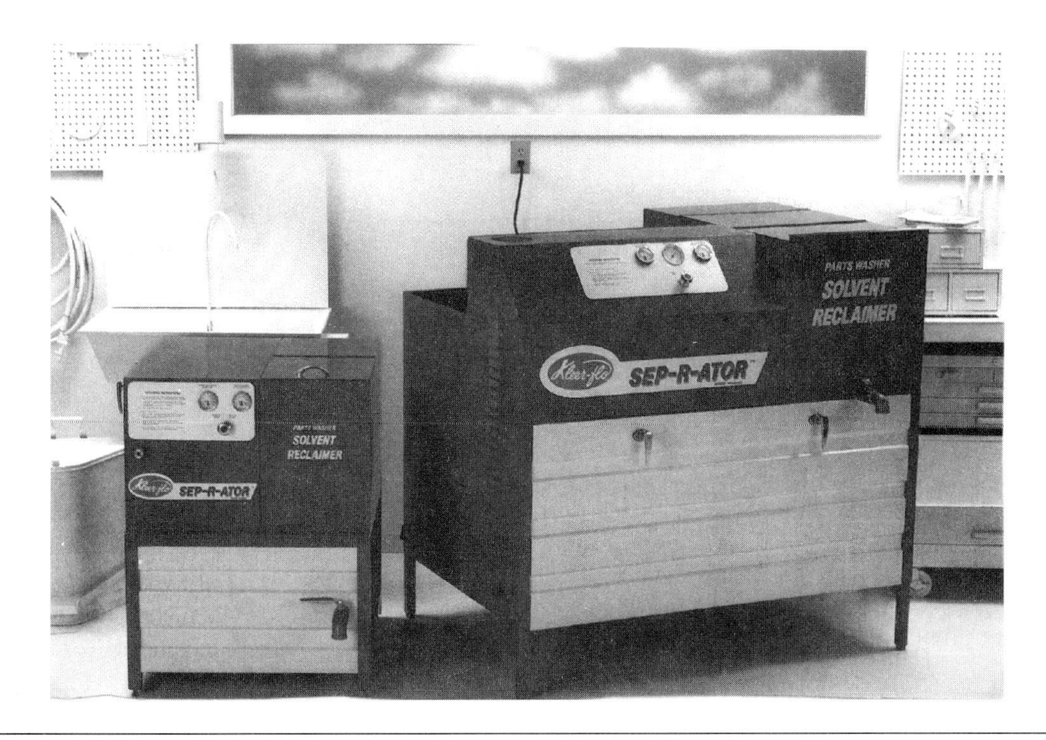

Figure 1-18 Solvent reclaimer. (Courtesy of Kleer-Flo Company.)

Terms to Know

Corrosive

Environmental Protection Agency (EPA)

Halogen, and halon, fire extinguishers

Hazard Communication Standard

Ignitable

Material safety data sheets (MSDS)

Multipurpose dry chemical fire extinguisher

Occupational Safety and Health Administration (OSHA)

Reactive

Resource Conservation and Recovery Act (RCRA)

Right-to-know laws

Toxic

Work place hazardous materials information systems (WHMIS)

Summary

- ❑ The United States Occupational Safety and Health Act of 1970 assured safe and healthful working conditions and authorized enforcement of safety standards.
- ❑ Many hazardous materials and conditions can exist in a truck shop, including flammable liquids and materials, corrosive acid solutions, loose sewer covers, caustic liquids, high-pressure air, frayed electric cords, hazardous waste materials, carbon monoxide, improper clothing, harmful vapors, high noise levels, and spills on shop floors.
- ❑ Work place hazardous material information systems (WHMIS) provide information regarding hazardous materials, labelling and handling.
- ❑ The danger regarding hazardous conditions and materials can be avoided by applying the necessary safety precautions. These precautions encompass all areas of safety, including personal safety, gasoline and diesel fuel safety, housekeeping safety, general shop safety, fire safety, and hazardous waste handling safety.
- ❑ The truck shop must supply the necessary shop safety equipment, and all shop personnel must be familiar with the location and operation of this equipment. Shop safety equipment includes gasoline and diesel fuel safety cans, steel storage cabinets, combustible material containers, fire extinguishers, eye-wash fountains, safety glasses and face shields, first-aid kits, and hazardous waste disposal containers.

Review Questions

Short Answer Essays

1. Explain the purposes of the Occupational Safety and Health Act.
2. Define twelve shop hazards, and explain why each hazard is dangerous.
3. Describe five steps that are necessary for personal protection in the truck shop.
4. Explain why smoking is dangerous in the shop.
5. Describe the danger in drug or alcohol use in the shop.
6. Explain three safety precautions related to electrical safety in the shop.
7. Define six essential safety precautions regarding gasoline handling.
8. Describe five steps required to provide housekeeping safety in the shop.
9. List twenty rules related to general shop safety.
10. Describe how to put out a fire with a fire extinguisher.

Fill-in-the-Blanks

1. The poisonous gas in vehicle exhaust is ____________ ____________.
2. Heavy-duty boots with ____________ toe caps are best for working in the shop.
3. One gallon of gasoline has a force equal to ____________ sticks of dynamite.
4. Breathing asbestos dust may cause ____________ ____________.
5. Class C fires involve the burning of ____________ equipment.
6. Irradiation eye burns may be caused by excessive light from an ____________ ____________.

7. Hazardous wastes in a truck shop include (1) ____________________
(2) ____________________ (3) ____________________ (4) ____________________
(5) ____________________ (6) ____________________.

8. The right-to-know laws state that employees have a right to know when the ____________________ they handle at work are____________________.

9. Material safety data sheets (MSDS) supply specific information regarding ____________________ ____________________.

10. Hazardous materials must never be dumped in ____________________ ____________________ ____________________ or ____________________.

ASE Style Review Questions

1. While discussing shop hazards:
Technician A says high pressure air from an air gun may penetrate the skin.
Technician B says air in the bloodstream may be fatal.
Who is correct?
A. A only **C.** Both A and B
B. B only **D.** Neither A nor B

2. When lifting heavy objects:
A. Bend your back to grasp the object.
B. Place your feet as far as possible from the object.
C. Straighten your back to lift the object.
D. Be sure the container is in good condition prior to lifting.

3. While discussing fire fighting:
Technician A says water should be sprayed on a gasoline fire.
Technician B says if a fire occurs inside a building, the doors and windows should be opened.
Who is correct?
A. A only **C.** Both A and B
B. B only **D.** Neither A nor B

4. While discussing hazardous waste disposal:
Technician A says the right-to-know laws require employers to train employees regarding hazardous waste materials.
Technician B says the right-to-know laws do not require employers to keep permanent records regarding hazardous waste materials.
Who is correct?
A. A only **C.** Both A and B
B. B only **D.** Neither A nor B

5. While discussing hazardous materials:
Technician A says a solid that ignites spontaneously is considered a hazardous material.
Technician B says a liquid with a flash point below 140°F (60°C) is considered a hazardous material.
Who is correct?
A. A only **C.** Both A and B
B. B only **D.** Neither A nor B

6. While discussing hazardous waste disposal:
Technician A says certain types of hazardous waste may be poured down a floor drain.
Technician B says hazardous waste users are responsible for hazardous waste materials from the time they become waste until the proper waste disposal is completed.
Who is correct?
A. A only **C.** Both A and B
B. B only **D.** Neither A nor B

7. When handling hazardous materials in a truck shop:
A. Air conditioning refrigerants are not considered hazardous waste materials.
B. Hazardous waste materials must not be recycled in the shop.
C. MSDS sheets contain information about safe handling procedures for hazardous materials.
D. Used engine coolant must be hauled to a hazardous waste disposal site.

8. For electrical, gasoline, and diesel fuel safety:
A. A two-prong adapter may be used to plug in a three-prong electrical cord.
B. Electric cords from lights and electrical equipment do not require a ground connection.
C. When stored in the shop, gasoline or diesel fuel containers should be completely full.
D. Gasoline containers must be secured when they are transported.

9. All of these statements about housekeeping and general truck shop safety are false *except:*
A. An ankle injury may be caused by a loose sewer cover.
B. High pressure air may be used to clean your hands.
C. Clutch assemblies may be cleaned with compressed air.
D. A trouble light with a rough service bulb does not require a protective cage.

10. All these statements about safety when servicing air suspension systems are true *except:*

A. An air suspension line under pressure should not be disconnected.

B. An air suspension line may be replaced with a different size line.

C. Never allow air suspension system pressures to exceed maximum specified pressure.

D. Components with stripped or damaged threads should be replaced.

Basic Theories

Upon completion and review of this chapter, you should be able to:

- ❑ Explain Newton's Laws of Motion.
- ❑ Define work and force.
- ❑ List the most common types of energy and energy conversions.
- ❑ Define inertia and momentum.
- ❑ Define mass, weight, and volume.
- ❑ Define torque and power.
- ❑ Define friction and coefficient of friction.
- ❑ Define atmospheric pressure and vacuum.
- ❑ Explain the action of a tilted, rolling wheel.
- ❑ Explain why a bicycle front wheel is designed with the tire pivot point behind the point where the center line of the front forks meets the road.
- ❑ Explain the compressibility of gases and the noncompressibility of liquids.
- ❑ Describe how proper dynamic balance is obtained on wheels in motion.
- ❑ Explain Pascal's Law.
- ❑ Define the basic law of levers.

Introduction

An understanding of the basics is absolutely essential before a study of complex systems and components is attempted. Basic suspension and steering theories, such as energy, compressed air principles, principles of wheels in motion, and the balance of wheels in motion, must be understood before a study of suspension and steering systems. An understanding of principles involving liquids and gases, levers, heat, friction, work, force, power, torque, energy, and energy conversion is also essential before studying the complex suspension and steering systems of today. This chapter provides the necessary basic theories so the systems explained later in the book may be more easily understood.

A BIT OF HISTORY

Sales of medium duty and heavy duty trucks have fluctuated considerably in the United States during the 1980s and 1990s. Medium duty truck sales peaked at slightly more than 200,000 trucks per year in 1979. These truck sales hit a low of just more than 100,000 trucks per year in 1983 and 1991. Sales of medium duty trucks in 1994 were more than 150,000 trucks per year. Heavy duty truck sales peaked at just less than 200,000 trucks per year in 1979. These truck sales hit a low of approximately 80,000 trucks per year in 1982 (Figure 2-1). Heavy duty truck sales in 1996 were approximately 170,000 trucks per year. A definite connection seems to exist between the state of the economy and truck sales. During the years when medium duty and heavy duty truck sales were low, the economy was in recession.

Newton's Laws of Motion

First Law

Newton's first law of motion applies to inertia. Inertia may be defined as the property of matter by which it will remain at rest or in uniform motion in the same direction unless acted on by some external force. When a truck is parked on a level street, it remains stationary unless it is driven or pushed. If the gas pedal is depressed with the engine running and the transmission in drive, the engine delivers power to the drive wheels, and this force moves the truck.

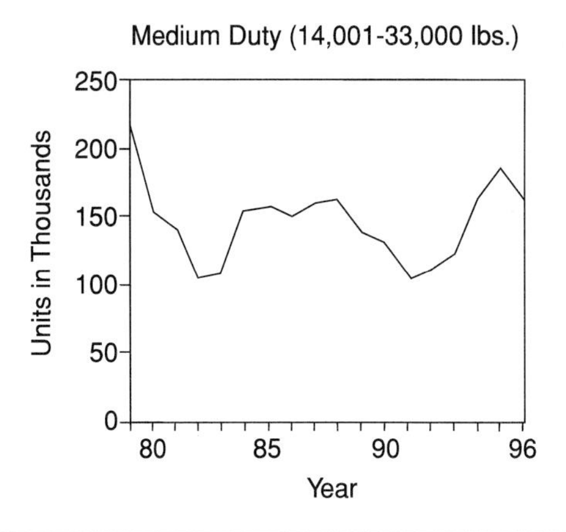

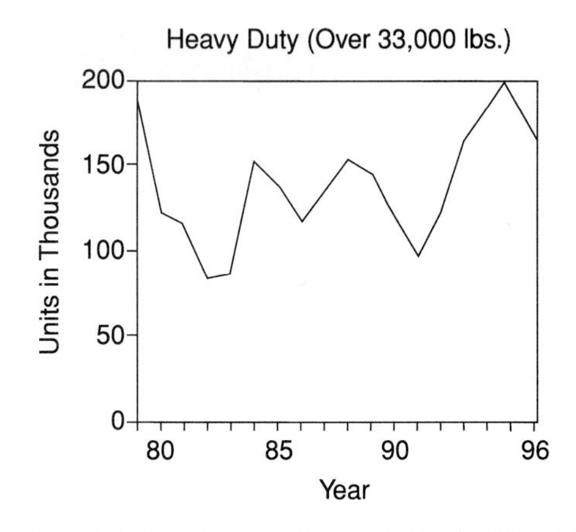

Figure 2-1 Medium duty and heavy duty truck sales in the United States. (Courtesy of American Automobile Mfrs. Association [AAMA].)

Second Law

A medium duty truck may be defined as a truck with a gross vehicle weight rating (GVWR) of 14,001 to 33,000 lb.

Newton's second law of motion applies to force. Force is equal to mass times acceleration. A body's acceleration is directly proportional to the force applied to it, and the body moves in a straight line away from the force. For example, if the engine power supplied to the drive wheels increases, the vehicle accelerates faster.

Third Law

A heavy duty truck may be defined as a truck with a GVWR in excess of 33,001 lb.

Newton's third law of motion applies to action and reaction. For every action there is an equal and opposite reaction. A practical application of this law occurs when the wheel on a vehicle strikes a hump in the road surface. This action drives the wheel and suspension upward with a certain force, and a specific amount of energy is stored in the spring. After this action occurs, the spring forces the wheel and suspension downward with a force equal to the initial upward force.

Work and Force

Work is defined as the result of a force allowed to act through a distance.

Force is equal to mass times acceleration. When a force moves a certain mass a specific distance, work is produced. When work is accomplished, the mass may be lifted or slid on a surface (Figure 2-2). Because force is measured in pounds and distance is measured in feet, the measurement for work is foot-pounds (ft-lb). In the metric system work is measured in Newton meters (Nm). If a force moves a 3,000-pound vehicle for 50 feet, 150,000 ft-lb of work is produced. Mechanical force acts on an object to start, stop, or change the direction of the object. It is possible to apply force to an object and not move the object. Under this condition no work is done. Work is only accomplished when an object is started, stopped, or redirected by mechanical force.

Energy, Heat, and Temperature

Energy may be defined as the ability to do work.

When **energy** is released to do work, it is called kinetic energy. This type of energy may also be referred to as energy in motion. Stored energy may be called potential energy. Energy is available in six forms:

1. Chemical energy is contained in the molecules of different atoms. In a truck, chemical energy is contained in the molecules of gasoline or diesel fuel and also in the molecules of electrolyte in the battery.
2. Electrical energy is required to move electrons through an electrical circuit. In a truck the battery is capable of producing electrical energy to start the engine, and the alter-

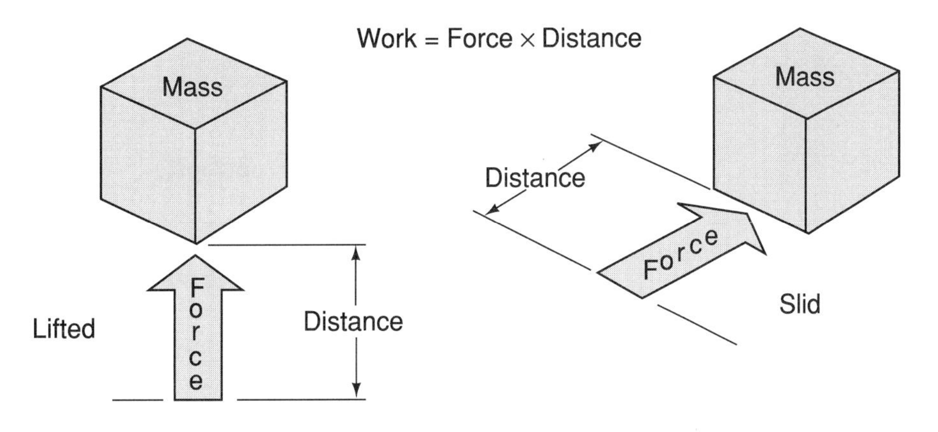

Figure 2-2 Work is accomplished when a mass is lifted or slid on a surface.

nator produces electrical energy to power the electrical accessories and recharge the battery.

3. Mechanical energy is defined as the ability to move objects. In a truck the battery supplies electrical energy to the starting motor, and this motor converts the electrical energy to mechanical energy to crank the engine.
4. Thermal energy may be defined as energy produced by heat. When gasoline burns, thermal energy is released.
5. Radiant energy is defined as light energy. In a truck radiant energy is produced by the lights.
6. Nuclear energy is defined as the energy within atoms when they are split apart or combined. Nuclear power plants generate electricity with this principle. This type of energy is not used in a truck.

Energy Conversion

Energy conversion occurs when one form of energy is changed to another form. Because energy is not always in the desired form, it must be converted to a form we can use. Some of the most common energy conversions in the trucking industry are these:

Chemical to Thermal Energy Conversion

Chemical energy in gasoline or diesel fuel is converted to thermal energy when the fuel burns in the engine cylinders.

Thermal to Mechanical Energy Conversion

Mechanical energy is required to rotate the drive wheels and move the vehicle. The piston and crankshaft in the engine and the drivetrain are designed to convert the thermal energy produced by the burning fuel into mechanical energy (Figure 2-3).

Electrical to Mechanical Energy Conversion

The windshield wiper motor converts electrical energy from the battery or alternator to mechanical energy to drive the windshield wipers.

Mechanical to Electrical Energy Conversion

The alternator is driven by mechanical energy from the engine, and the alternator converts this energy to electrical energy, which powers the electrical accessories on the vehicle.

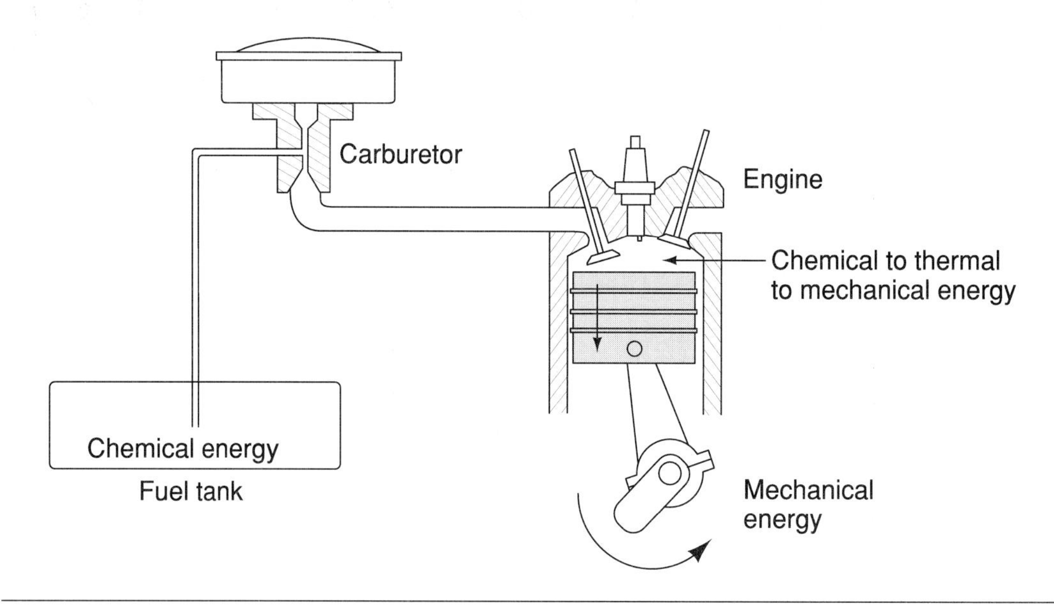

Figure 2-3 Thermal energy in the fuel is converted to mechanical energy in the engine cylinders. The piston, crankshaft, and drivetrain deliver this mechanical energy to the drive wheels.

Inertia

Inertia is defined as the tendency of an object at rest to remain at rest, or the tendency of an object in motion to stay in motion.

Inertia may be defined as the property of matter by which it will remain at rest or in uniform motion in the same direction unless acted on by some external force. Inertia is summarized in Newton's first law of motion. The inertia of an object at rest is called static inertia, whereas dynamic inertia refers to the inertia of an object in motion. Inertia exists in liquids, solids, and gases. When you push and move a parked vehicle, you overcome the static inertia of the vehicle. If you catch a ball in motion, you overcome the dynamic inertia of the ball.

Momentum

When a force overcomes static friction and causes an object to increase in speed, the object gains momentum.

Momentum is the product of an object's weight times its speed. An object loses momentum if another force overcomes the dynamic inertia of the moving object.

Mass, Weight, and Volume

Mass is a property of matter.

Weight is the measurement of the earth's gravitational pull on the object.

Volume is the length, width, and height of a space occupied by an object.

Mass may be defined as a quantity of matter cohering together in one body or quantity. **Weight** may also be defined as the force that is exerted on an object by gravity. The weight varies when the elevation of the object above sea level changes. The object's weight becomes less when the elevation above sea level increases.

A lawn mower is much easier to push than a 2,500-pound vehicle because the lawn mower has very little inertia compared with the vehicle. A spaceship might weigh 100 tons here on earth where it is affected by the earth's gravitational pull. In outer space beyond the earth's gravity and atmosphere, the spaceship is almost weightless. Here on earth mass and weight are measured in pounds and ounces in the English system. In the metric system mass and weight are measured in kilograms.

Volume may be defined as space occupied as in the volume of a container. Volume is a measurement of size, and it is related to mass and weight. For example, a pound of gold and a pound of feathers both have the same weight, but the pound of feathers occupies a much larger volume. In the English system volume is measured in cubic inches, cubic feet, cubic yards, or gallons. The measurement for volume in the metric system is cubic centimeters, or liters.

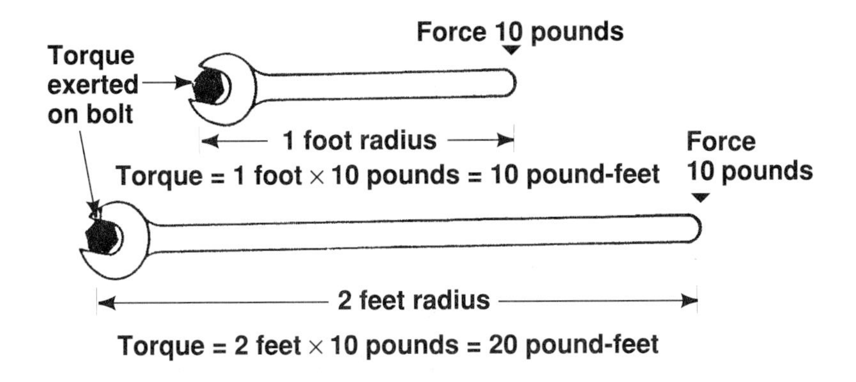

Figure 2-4 Force applied to a wrench produces torque. If the bolt turns, work is accomplished.

Torque

Torque is a force that produces a turning effort.

Torque is the turning effort caused by a force. Torque is a force that tries to turn or twist something around and is scientifically defined as the product of a force and the perpendicular distance between the line of action of the force and the axis of rotation. When you pull a wrench to tighten a bolt, you supply torque to the bolt. This torque, or twisting force, is calculated by multiplying the force and the radius. For example, if you supply 10 pounds (lb) (4.53 kilograms [kg]) of force on the end of a 2-ft (60.96-cm) wrench to tighten a bolt, the torque is 10 × 2 = 20 ft lb (27 Nm) (Figure 2-4). If the bolt turns during torque application, work is done. When a bolt does not rotate during torque application, no work is accomplished.

Power

Power is a measurement for the rate or speed at which work is done.

James Watt, a Scotsman, is credited with being the first person to calculate **power.** He measured the amount of work that a horse could do in a specific time. Watt calculated that a horse could move 330 lb (149.68 kg) for 100 ft (30.48 meters [m]) in 1 minute. If you multiply 330 lb (149.68 kg) by 100 ft (30.48 m), the answer is 33,000 ft lb (44,550 Nm) of work. Watt determined that one horse could do 33,000 ft lb (44,550 Nm) of work in 1 minute, thus 1 horsepower (hp) (0.746 kilowatts [kw]) is equal to 33,000 ft lb (44,550 Nm) per minute, or 550 ft lb (742.5 Nm) per second (Figure 2-5). Two hp (1.46 kw) could do this same amount of work in 30 seconds, or 4 hp (2.98 kw) would be capable of completing this work in 15 seconds. If you push a 3,000-pound (1,360 kg) car for 11 ft (3.35 m) in 15 seconds, you produce 4 horsepower. From this brief discussion about horsepower we can understand that as power increases, speed also increases, or the time to do work decreases.

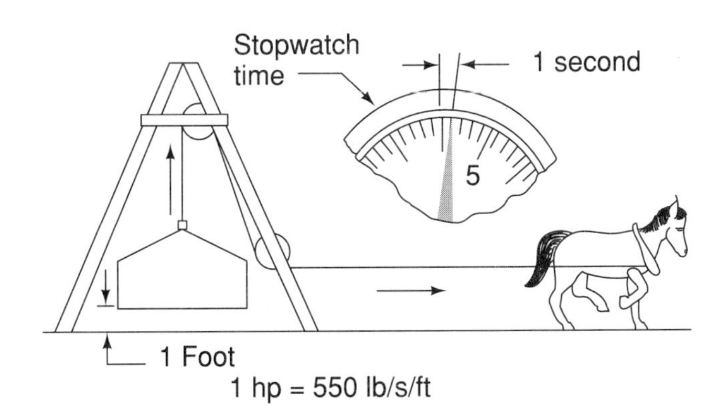

Figure 2-5 One horsepower is produced when 550 pounds are moved 1 foot in 1 second.

Figure 2-6 Truck with improved aerodynamic design and reduced Cd. (Courtesy of Volvo GM Heavy Trucks Corporation.)

Friction

Friction is the resistance to relative motion of two bodies in contact.

Friction may occur in solids, liquids, and gases. When a truck is driven down the road, friction occurs between the air and the truck's surface. This friction opposes the momentum of the moving truck. Because friction creates heat, some of the mechanical energy from the truck's momentum is changed to heat energy in the air and body components. The mechanical energy from the engine must overcome the vehicle inertia and the friction of the air striking the truck. Body design has a very dramatic effect on the amount of friction developed by the air striking the truck. The total resistance to motion caused by friction between a moving truck and the air is referred to as coefficient of drag (Cd). Many trucks now have an improved aerodynamic design and reduced Cd (Figure 2-6).

TRADE JARGON: Coefficient of drag (Cd) may also be called aerodynamic drag.

The study of Cd is very complicated and also very important. The amount of energy used to overcome air friction varies depending on the aerodynamic design of the truck. Therefore reducing a truck's Cd can be an effective method of improving fuel economy.

Coefficient of Friction

Coefficient of friction is the amount of force required to move a given weight when it is in contact with another surface.

Friction may be defined as the resistance to relative motion between two bodies in contact. The amount of friction developed between two bodies in contact is referred to as the coefficient of friction.

The coefficient of friction (COF) is expressed as the amount of force required to move one body when it is in contact with another body. If a 100-lb (45.36 kg) metal weight is resting on a metal surface and the pull required to move the weight is 60 lb (27.21 kg), the COF is 60% or .6 (Figure 2-7). When the pull required to move the 100-lb (45.36 kg) weight is 50 lb (22.68 kg), the COF is 50% or .5. If the pull required to move the 100-lb (45.36 kg) weight is 35 lb (15.87), the COF is 35% or .35.

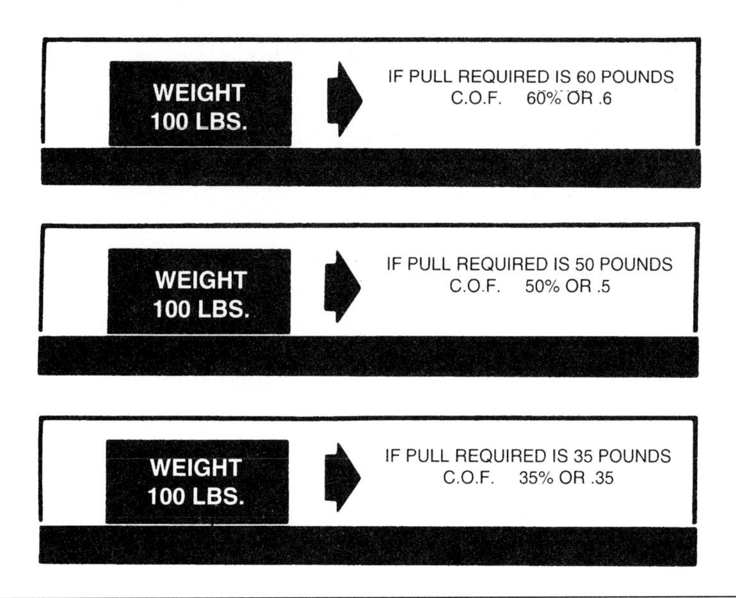

Figure 2-7 Coefficient of friction. (Courtesy of Allied Signal Truck Brake Systems Co.)

The COF varies depending on the condition of the contacting surfaces. For example, if the contacting surfaces on the 100-lb (45.36 kg) metal weight and the metal surface are dry, the COF is higher. If oil or grease is placed between the contacting surfaces on the 100-lb (45.36 kg) metal weight and the metal surface, the COF is greatly reduced.

In a truck steering system the rotational force exerted by the driver on the steering wheel must overcome the friction of the front tires on the road surface to turn the front wheels to the right or left. The driver's rotational force on the steering wheel may be assisted by hydraulic pressure from the power steering system.

Energy of Motion Changed to Heat Energy

When relative motion is present between two contacting bodies, heat is developed. The amount of heat developed depends on a number of factors, including the speed of relative motion and the weight of the body in motion. For example, the tires become hotter on a trailer with a 40,000-lb (18,144 kg) load compared with the tires on a trailer with a 20,000-lb (9,072 kg) load. When the tractor and trailer speed increases, the tire temperature also increases.

Basic Hydraulic Principles

Atmospheric Pressure

Atmosphere may be defined as the gaseous envelope surrounding any celestial body. The earth's atmosphere is the gaseous envelope surrounding the earth. We usually refer to the atmosphere as simply air. The atmosphere contains approximately 20% oxygen and 80% nitrogen. Because air is gaseous matter with mass and weight, it exerts pressure on the earth's surface. Imagine a 1 square inch column of air extending from the earth's surface to the outer edge of the atmosphere (Figure 2-8). Scientists have calculated that this column weighs 14.7 lb (101.35 kilopascals [kPa]) at sea level. Therefore we state that **atmospheric pressure** is 14.7 pounds per square inch (psi) (101.35 kPa) at sea level. On top of a 10,000-ft (3048 m) mountain the 1 inch square column of air extending to the outer edge of the atmosphere is shorter and therefore lighter, and thus atmospheric pressure decreases as altitude above sea level increases.

Pressure may be defined as a force exerted on a given surface area.

Atmospheric pressure may be defined as the total weight of the earth's atmosphere.

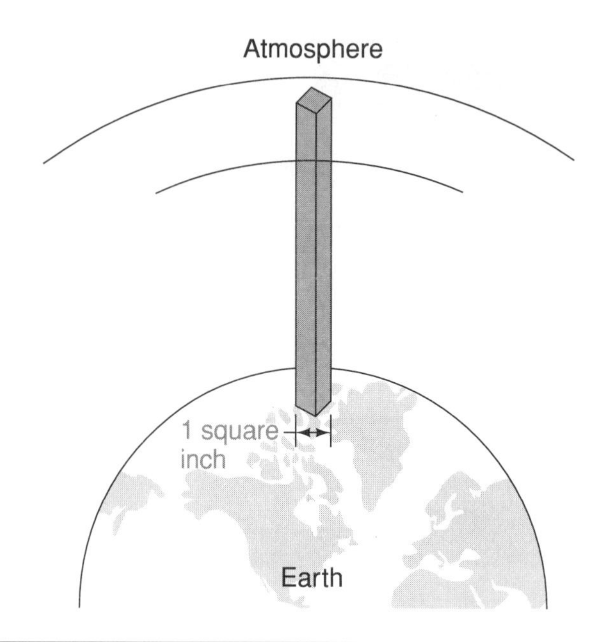

Figure 2-8 A column of air 1 inch square extending from the earth's surface at sea level to the outer edge of the atmosphere weighs 14.7 lb.

When dealing with a gas at a fixed or constant volume, pressure and temperature are directly related. If you increase one, you also increase the other.

An equal volume of hot air weighs less and exerts less pressure on the earth's surface than cold air.

Atmospheric pressure decreases as altitude increases.

Vacuum may be described as the absence of atmospheric pressure.

Liquids, solids, and gases tend to move from an area of high pressure to a low-pressure area.

Atmospheric Pressure and Temperature

When air becomes hotter, it expands, and this hotter air is lighter compared with an equal volume of cooler air. The hotter, lighter air exerts less pressure on the earth's surface compared with cooler air. If the temperature decreases, air contracts and becomes heavier. Therefore an equal volume of cooler air exerts more pressure on the earth's surface compared with hotter air.

Atmospheric Pressure and Height

When you climb above sea level, atmospheric pressure decreases. At 5,000 ft (1,524 m) above sea level, a 1 square inch column of air from the earth's surface to the outer edge of the atmosphere is 5,000 ft (1,524 m) shorter than the same column at sea level. Therefore the weight of this column of air is less at 5,000 ft (1,524 m) elevation than at sea level. As altitude continues to increase, atmospheric pressure continues to decrease. At an altitude of several hundred miles above sea level the earth's atmosphere ends, and only vacuum exists beyond that point.

Vacuum

When a space contains a **vacuum,** it contains a smaller quantity of air compared with the quantity of air the space is capable of containing with atmospheric pressure in the space. Vacuum could be measured in pounds per square inch, but inches of mercury (in Hg) are most commonly used for this measurement. Let us assume that a plastic "U" tube is partially filled with mercury, and atmospheric pressure is allowed to enter one end of the tube. If vacuum is supplied to the other end of the "U" tube, the mercury is forced downward by the atmospheric pressure. When this movement occurs, the mercury also moves upward on the side where the vacuum is supplied. If the mercury moves downward 10 in (25.4 centimeters [cm]) where the atmospheric pressure is supplied and upward 10 in (25.4 cm) where the vacuum is supplied, 20 in Hg (67.6 kPa) is supplied to the "U" tube. If a complete vacuum exists in a space, the pounds per square inch absolute (PSIA) is zero, and this is equal to 29.9 in Hg (101.06 kPa) (Figure 2-9).

Vacuum and atmospheric pressure are used in several truck systems. For example, atmospheric pressure is available outside the engine air intake. When a piston moves downward with an intake valve open, a vacuum is created in the cylinder above the piston. The air moves rapidly from the high pressure outside the air intake to the lower pressure in the cylinder (Figure 2-10).

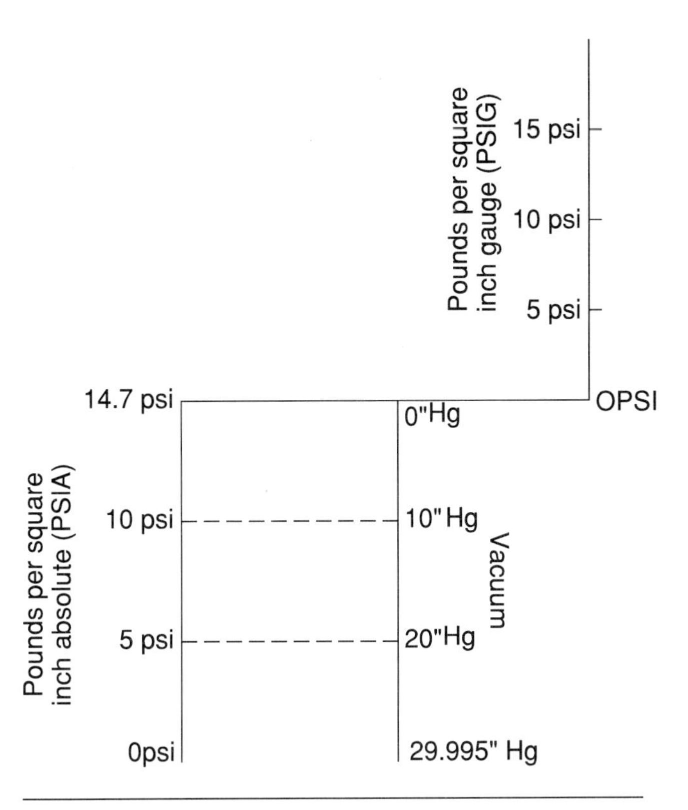

Figure 2-9 Atmospheric pressure and vacuum scales.

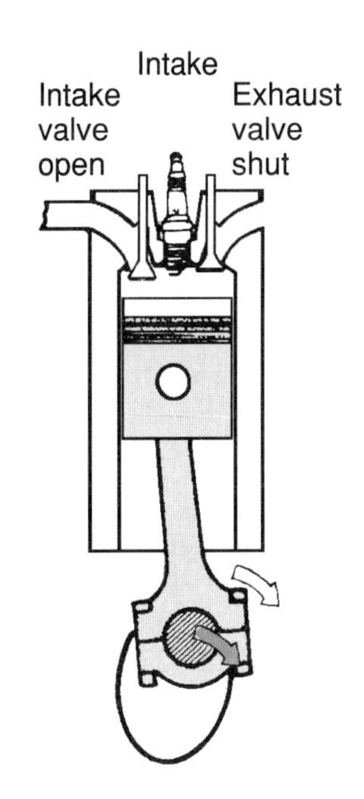

Figure 2-10 Air moves from a high pressure outside the air intake to a low pressure in the cylinder. (Courtesy of Sun Electric Corporation.)

Pumps use high and low pressure to move liquids or gases. For example, as a power steering pump rotates, it creates a high pressure at the pump outlet and a low pressure at the inlet. This pressure difference causes power steering fluid to flow through the power steering system.

When dealing with a constant mass of gas, pressure and volume are inversely proportional. If volume is decreased, pressure is increased.

Venturi

If a gas or a liquid is flowing through a pipe and the pipe diameter is narrow in one place, the flow speeds up in the narrow area. This narrow area in a pipe is called a venturi. The increased speed of liquid or gas flow in the venturi area causes a lower pressure in the narrow area. An increase in the speed of liquid or gas flow in a venturi causes a corresponding decrease in pressure. Some power steering pumps use a venturi principle to assist in the control of pump pressure (Figure 2-11).

Pressure and Compressibility

Because liquids and gases are both substances that flow, they may be classified as fluids. If a nail punctures a truck tire, the air escapes until the pressure in the tirc is equal to atmospheric pressure outside the tire. When the tire is repaired and inflated, air pressure is forced into the tire. If the tire is inflated to 32 psi (220.64 kPa), this pressure is applied to every square inch on the inner tire surface. Pressure is always supplied equally to the entire surface of a container. Because air is a gas, the molecules have plenty of space between them. When the tire is inflated, the pressure in the tire increases and the air molecules are squeezed closer together or compressed. Under this condition, the air molecules cannot move as freely, but extra molecules of air can still be forced into the tire. Therefore gases such as air are said to be **compressible.**

The air in the tire may be compared with a few balls on a billiard table without pockets. If a few more balls are placed on the table, the balls are closer together, but they can still move freely (Figure 2-12).

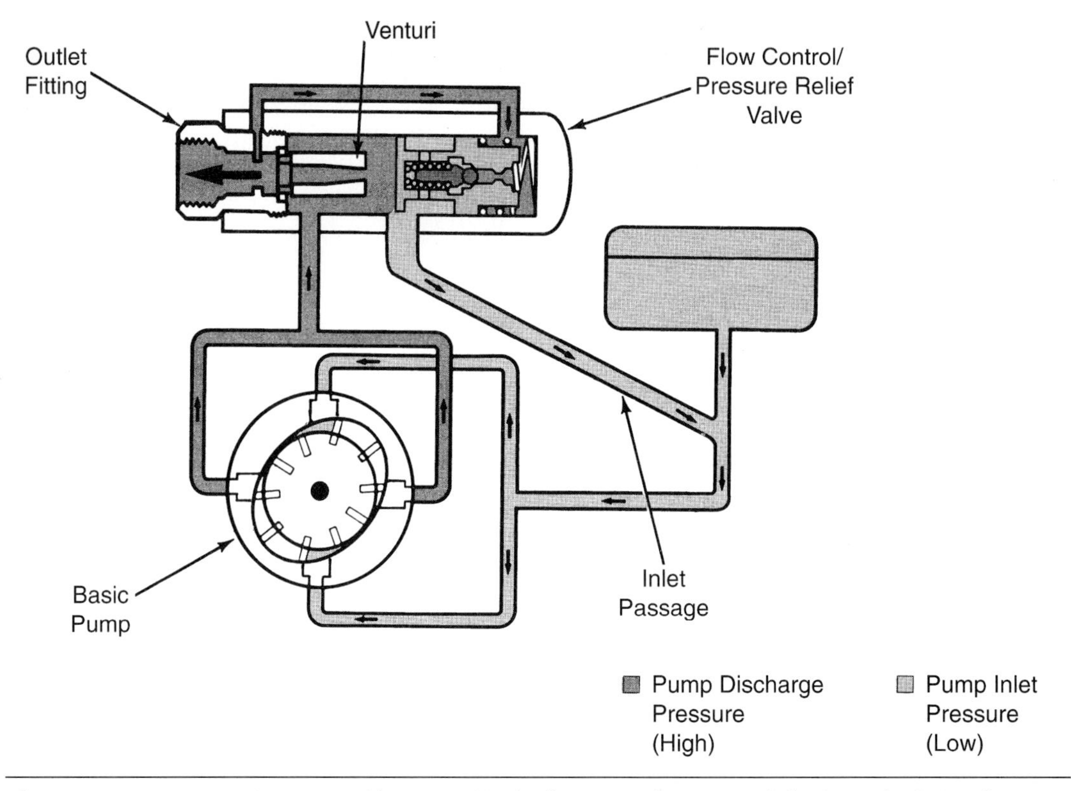

Figure 2-11 Power steering pump with a venturi in the flow control/pressure relief valve outlet fitting. (Courtesy of General Motors Corporation, Service Technology Group.)

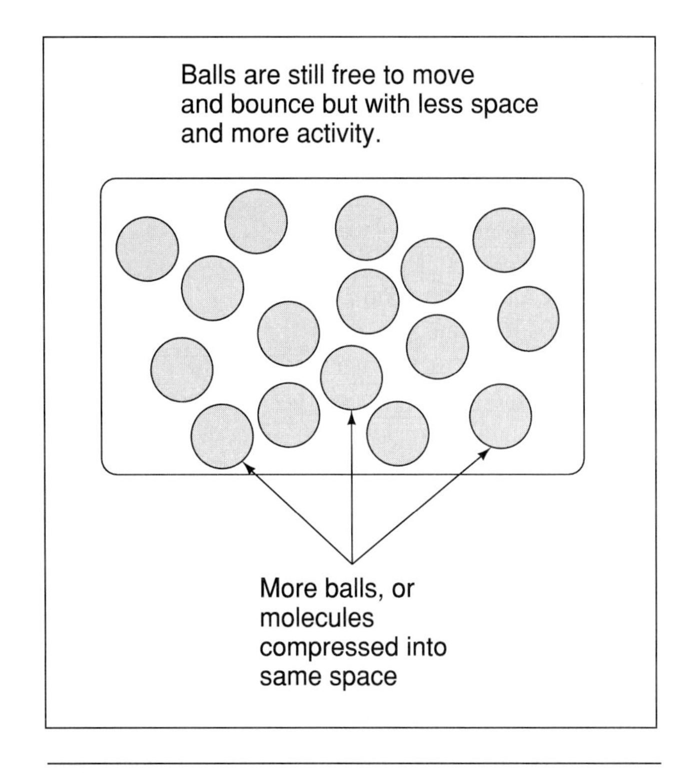

Figure 2-12 Gases can be compressed much like more balls may be placed on a billiard table containing a few balls. (Courtesy of Chrysler Corporation.)

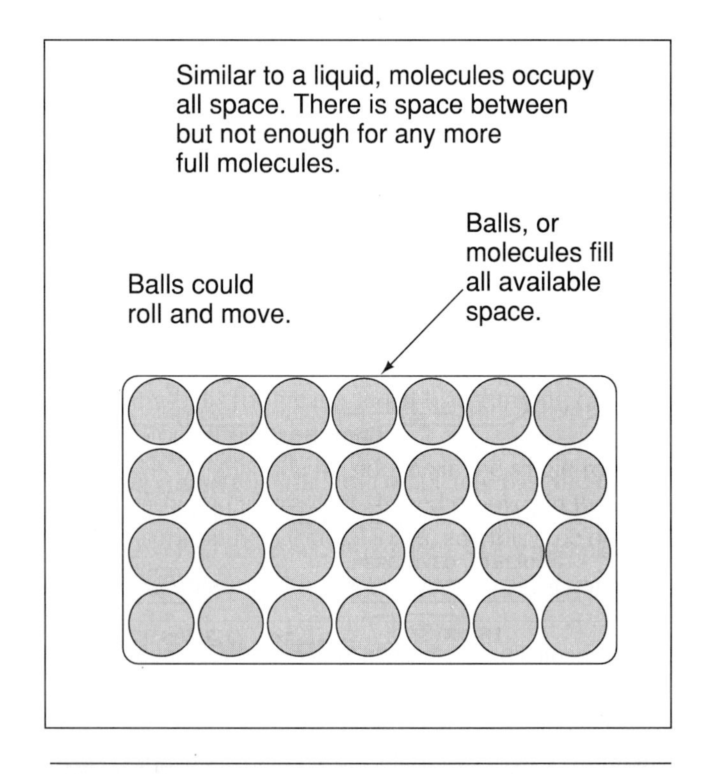

Figure 2-13 Liquids are noncompressible, just as more balls cannot be added to a billiard table with no pockets that is completely filled with balls. (Courtesy of Chrysler Corporation.)

If the vehicle is driven at high speed, friction between the road surface and the tires heats the tires and the air in the tires. The tire is flexing continually at the road surface, and this action produces heat. When air temperature increases, the pressure in the tire also increases. Conversely, a temperature decrease reduces pressure.

If 100 cubic feet (ft^3) (2.8 m^3) of air is forced into a large truck tire and the same amount of air is forced into a much smaller car tire, the pressure in the car tire is much greater.

Molecules in a liquid may be compared with a billiard table without pockets that is completely filled with balls. These balls can roll around, but no additional balls can be placed on the table because the balls cannot be compressed. Similarly, liquid molecules are **highly incompressible** (Figure 2-13).

Liquids are highly incompressible.

Liquid Flow

If a tube is filled with billiard balls and the outlet is open, more balls may be added to the inlet. When each ball is moved into the inlet, a ball is forced from the outlet. If the outlet is closed, no more balls can be forced in the inlet (Figure 2-14).

The billiard balls in the tube may be compared with molecules of power steering fluid in the line between the power steering pump and the power steering gear. Because highly incompressible fluid fills the line from this pump to the gear, the force developed by the power steering pump pressure is transmitted through the line to the steering gear (Figure 2-15).

This pressure is applied equally to every square inch in the power steering gear pistons. Pascal's law states, "Pressure on a confined fluid is transmitted equally in all directions and acts with equal force on equal areas." This statement means the pressure is the same throughout the entire power steering system.

If the diameter of each power steering gear piston is 3 in (7.62 cm), the area of each piston is $3 \times 3.142 = 9.42$ sq in (60.81 sq cm) (Figure 2-16). If the pressure supplied by the power steering pump is 500 psi (3,447.5 kPa), the force on each power steering gear piston is $500 \times 9.62 = 4{,}810$ psi (33,164.95 kPa). The combined force on the two power steering gear pistons is $4{,}810 \times 2 = 9{,}620$ psi (66,329 kPa). This pressure supplied to the power steering gear pistons helps to move these pistons, and this action helps the driver to turn the steering wheel. The front wheels are turned by the combined forces of the driver's rotational force on the steering wheel and the hydraulic force supplied by the power steering pump to the power steering gear pistons.

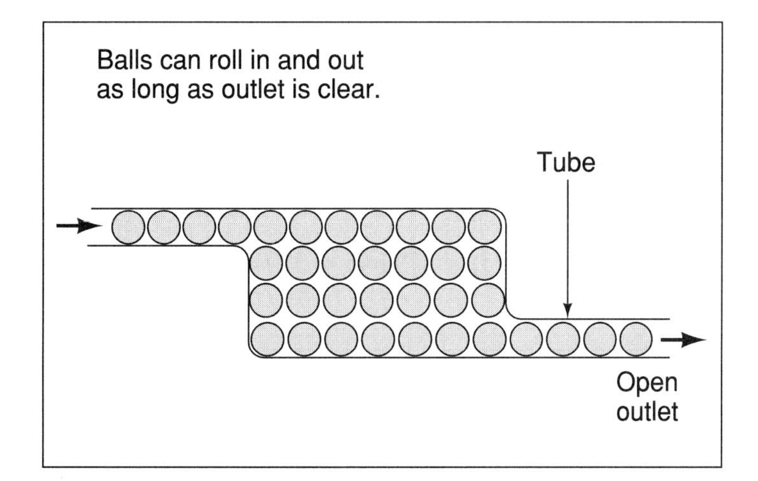

Figure 2-14 Billiard ball movement in a tube filled with balls compared with liquid flow. (Courtesy of Chrysler Corporation.)

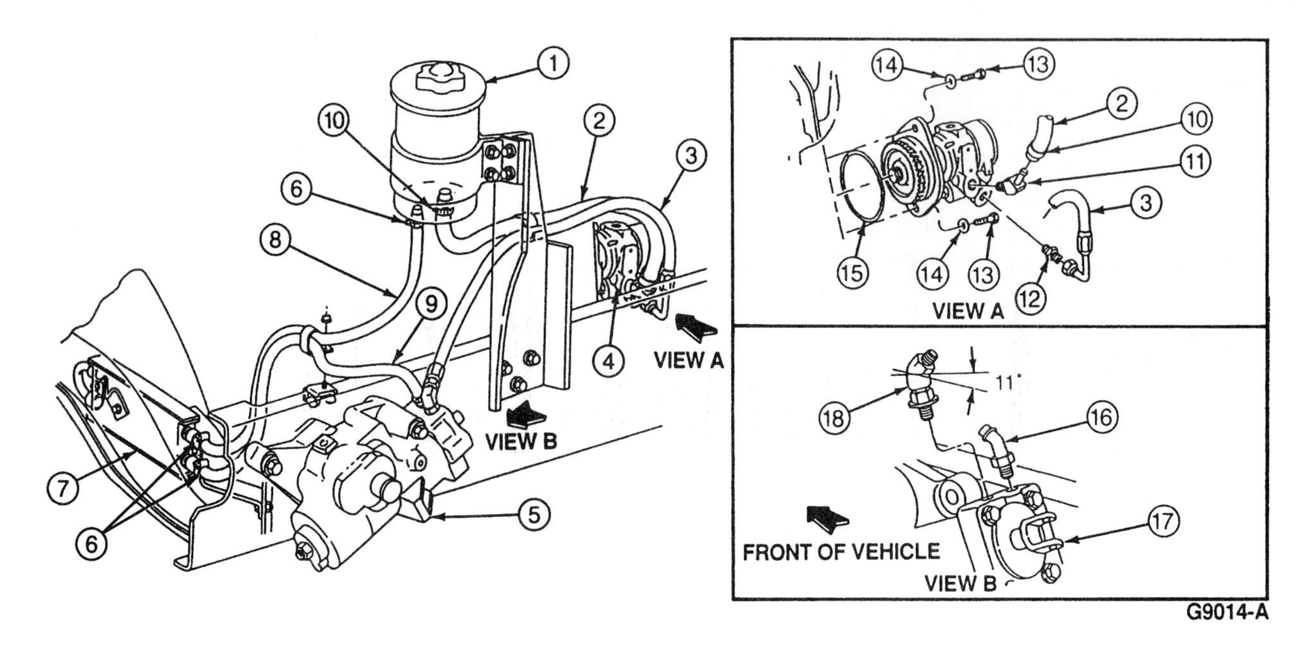

Item	Part Number	Description
1	3A697	Reservoir
2	381172-S160A	Hose, Reservoir to Pump
3	3A719	Hose, Pressure Hose — Pump to Steering Gear
4	3A674	Pump
5	3N503	Steering Gear
6	388651-S	Clamp
7	3D746	Oil Cooler — See Oil Cooler Installation Views
8	3881170-S250A	Hose, Oil Cooler to Reservoir

Item	Part Number	Description
9	3881170-S250A	Hose, Steering Gear to Cooler
10	388653-S	Clamp
11	3E599	Elbow
12	384074-S36	Adapter
13	58721-S2	Bolt
14	44881-S2	Washer
15	87093-S95	O-Ring
16	391068-S36	Nipple
17	3N503	Power Steering Gear
18	384285-S36	Elbow

Figure 2-15 Fluid pressure from the power steering pump is supplied through the high-pressure hose to the power steering gear. (Courtesy of Sterling Truck Corporation.)

Temperature

Temperature affects all liquids, solids, and gases. The volume of any matter increases as the temperature increases, and conversely the volume decreases in relation to a reduction in temperature. When the gases in an engine cylinder are burned, the sudden temperature increase causes rapid gas expansion which pushes the piston downward and causes engine rotation (Figure 2-17).

Principles Involving Tires and Wheels in Motion

A rolling wheel and tire assembly that is tilted always moves in the direction of the tilt.

If you roll a cone-shaped piece of metal on a smooth surface, the cone does not move straight ahead. Instead, the cone moves toward the direction of the tilt on the cone. When you are riding a bicycle and want to make a left turn, if you tilt the bicycle to the left it is much easier to complete the turn. The reason for this action is that a tilted, rolling wheel tends to move in the direction of the tilt. Similarly, if a tire and wheel assembly on a vehicle is tilted, the tire and wheel tend to move in the direction of the tilt (Figure 2-18). This principle is used in front wheel alignment.

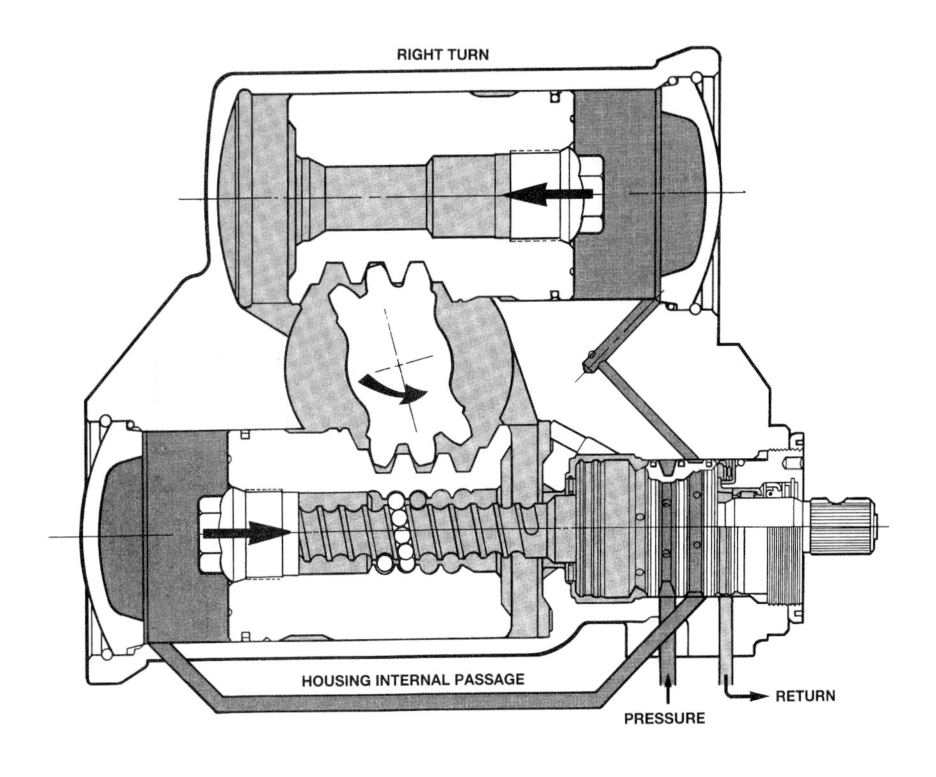

Figure 2-16 The pressure supplied to the power steering gear pistons helps to move these pistons, and this action helps the driver turn the steering wheel. (Courtesy of General Motors Corporation.)

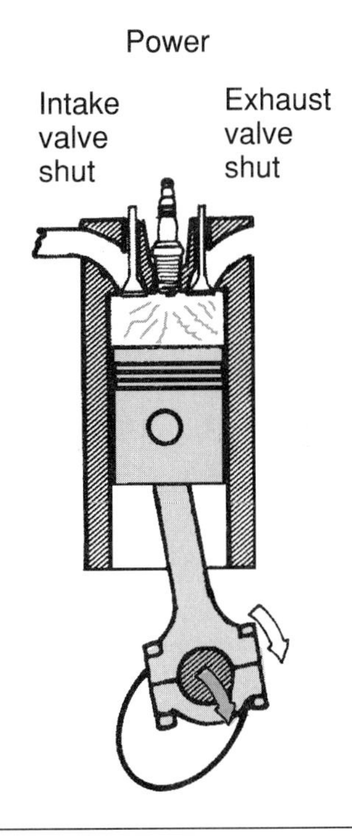

Figure 2-17 Hot, expanding gases push the piston downward and rotate the crankshaft. (Courtesy of Sun Electric Corporation.)

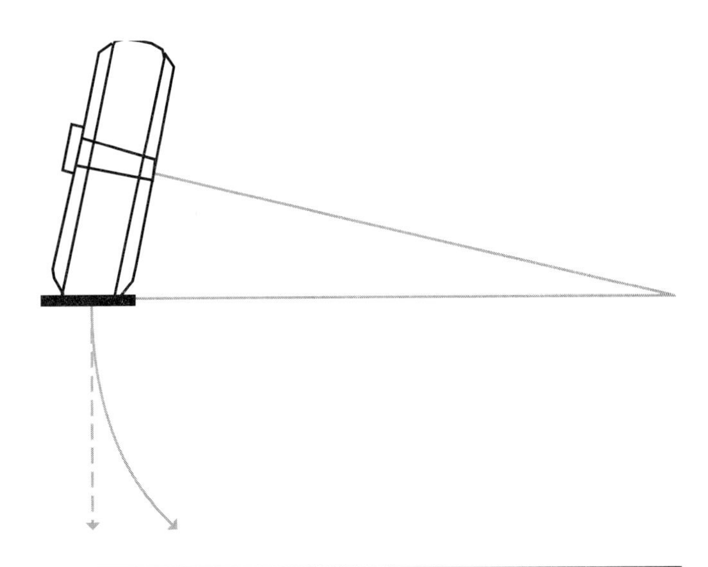

Figure 2-18 A tilted, rolling wheel tends to move in the direction of the tilt.

The casters on a piece of furniture are angled so the center of the caster wheel is some distance from the pivot center (Figure 2-19). When the furniture is moved, the casters turn on their pivots to bring the caster wheels into line with the pushing force on the furniture (Figure 2-20). This caster action causes the furniture to move easily in a straight line.

The weight of a bicycle rider is projected through the bicycle front forks to the road surface, and the tire pivots on the vertical center line of the wheel when the handlebars are turned. Notice the center line of the front forks is tilted rearward in relation to the vertical center line of the wheel (Figure 2-21). Because the tire pivot point is behind the front forks center line where the weight is projected against the road surface, the front wheel tends to return to the straight ahead position after a turn. The wheel also tends to remain in the straight ahead position as the bicycle is driven. This principle is applied in truck front wheel alignment.

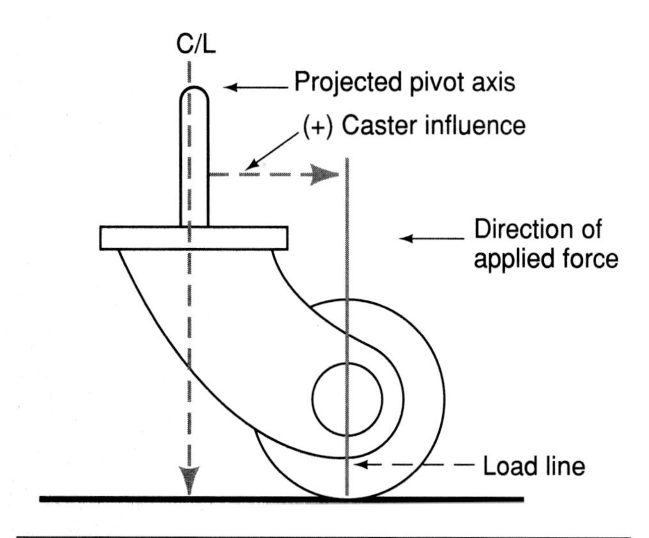

Figure 2-19 Distance between the wheel center and pivot center on a caster.

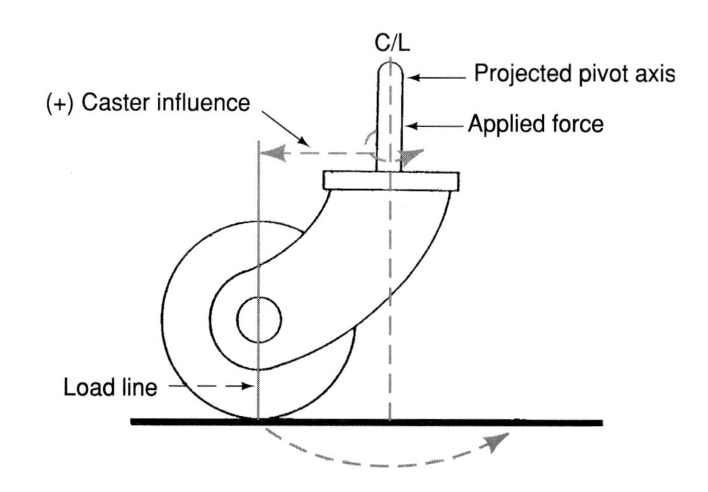

Figure 2-20 Caster aligned with pushing force provides straight ahead movement when furniture is pushed.

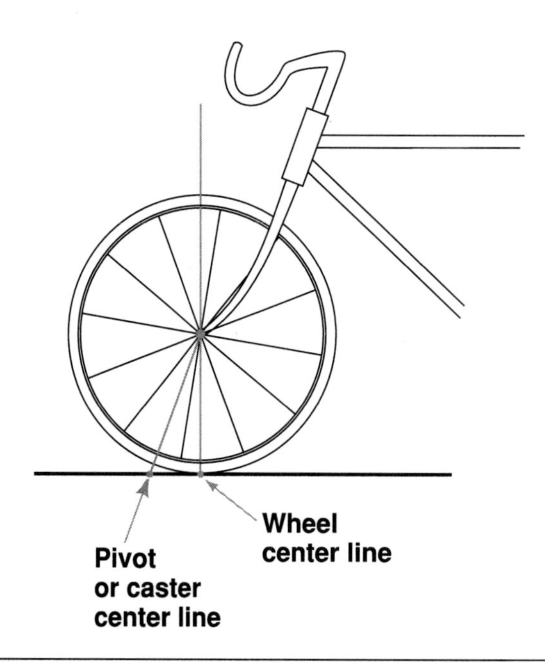

Figure 2-21 If a wheel center line and pivot point are behind the front fork center line pivot point, the wheel tends to return to the straight ahead position after a turn. The wheel also tends to remain in the straight ahead position as the bicycle is driven.

Principles Involving the Balance of Wheels in Motion

When the weight of a wheel and tire is distributed equally around the center of wheel rotation viewed from the side, the wheel and tire have proper **static balance.** Under this condition the wheel and tire assembly has no tendency to rotate by itself regardless of the wheel position. If the weight is not distributed equally around the center of wheel rotation, the wheel and tire are statically unbalanced (Figure 2-22). As the wheel and tire rotate, centrifugal force acts on this static imbalance and causes the wheel to "tramp" or "hop."

When a ball is rotated on the end of a string, the ball and string form an angle with the axis of rotation. If the rotational speed is increased, the ball and string form a 90-degree angle with the axis of rotation (Figure 2-23). Any weight will always tend to rotate at a 90-degree angle to the axis of rotation.

If two balls are positioned on a metal bar so their weight is equally distributed on the center line of the axis of rotation, the path of rotation remains at a 90-degree angle to the center line of the axis when the bar is rotated. Under this condition the metal bar and the balls are in **dynamic balance** (Figure 2-24).

If weights are placed on a metal bar so their weight is not equally distributed in relation to the center line of the rotational axis of rotation, the weights still tend to rotate at a 90-degree angle in relation to the rotational axis. This action forces the pivot out of its vertical axis (Figure 2-25). When the bar is rotated 180 degrees, the bar is forced out of its vertical axis in the opposite direction. If this condition is present, the bar has a wobbling action as it rotates. Under this condition the bar is said to have dynamic imbalance, but static balance is maintained.

Similarly, when the weight on a wheel and tire is not distributed equally on both sides of the tire center line viewed from the front, the wheel and tire are dynamically unbalanced. This condition produces a wobbling action as the wheel and tire rotate. The weight must be distributed equally in relation to the tire center line to provide proper dynamic balance (Figure 2-26). These principles are used in truck wheel balancing.

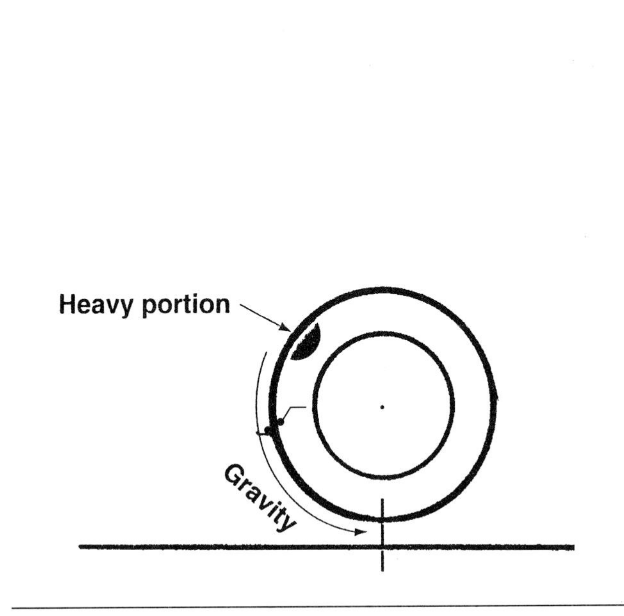

Figure 2-22 Static imbalance caused by unequal weight distribution around the center of wheel rotation.

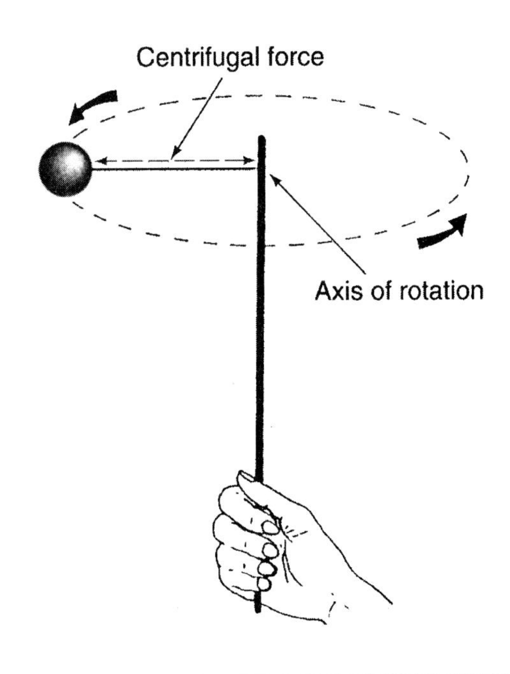

Figure 2-23 A weight tends to rotate at a 90-degree angle in relation to the axis of rotation.

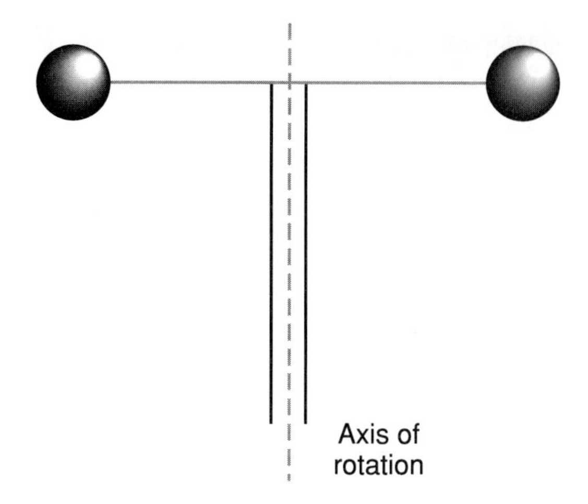

Figure 2-24 When weight on a metal bar is distributed equally on the center line of the axis of rotation, the bar and balls remain in dynamic balance during rotation.

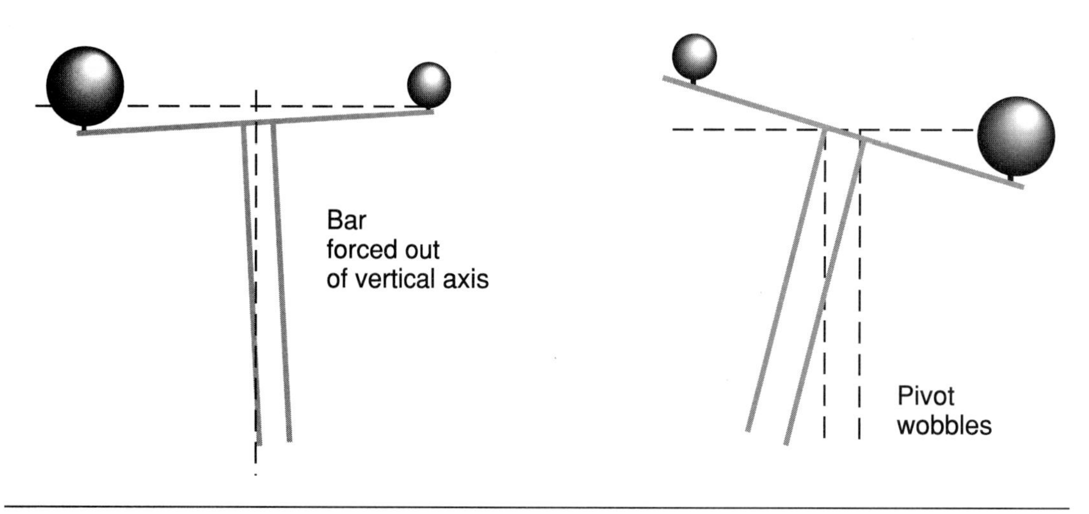

Figure 2-25 When the weight is not equally distributed in relation to the center line of the rotational axis, dynamic imbalance causes a wobbling action during rotation.

Compressed Air Fundamentals

Energy and Compressed Air

CAUTION: Always remember that compressed air contains a large amount of energy. Never direct compressed air against any part of your body. If compressed air penetrates the skin and enters the bloodstream, it may cause serious medical complications or death. Always observe all the truck manufacturer's recommended safety and service procedures in the service manual when servicing suspension and steering systems.

Air at atmospheric pressure may be called free air.

As mentioned previously, the normal atmospheric pressure around us is approximately 14.7 psi (101.28 kPa), depending on altitude, humidity, temperature and other factors. When discussing

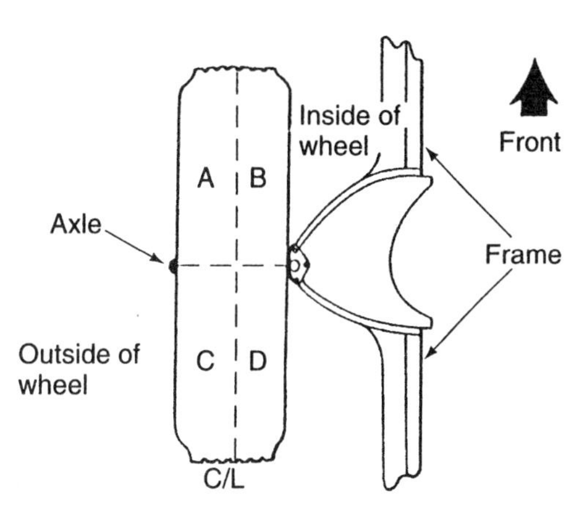

Figure 2-26 Weight distributed equally in relation to the tire and wheel center line.

compressed air, we ignore the 14.7 psi (101.28 kPa) atmospheric pressure and consider the atmosphere to contain **free air** under no pressure. For example, air pressure gauges read zero when connected only to atmospheric pressure. Compressed air may be defined as air that is forced into a smaller space than it would ordinarily occupy in its free or atmospheric state.

Compressed air may be defined as air that is forced into a smaller space than it would ordinarily occupy in its free or atmospheric state.

Air in its free or compressed state may be compared with a coil spring. When a coil spring is not compressed, it does not store any energy. Similarly, air in its atmospheric or free state does not store any energy (Figure 2-27). When a coil spring is compressed, it stores energy, and compressed air also stores a specific amount of energy (Figure 2-28). This energy in the compressed air can be used to do work.

Let us consider one reservoir containing air at atmospheric pressure and a second reservoir containing air at a pressure greater than atmospheric pressure. If a pipe is connected between these reservoirs, air flows from the reservoir with the higher pressure into the reservoir with the lower pressure until the pressure equalizes in the two reservoirs. The flow between the two reservoirs may be controlled by a shut-off valve (Figure 2-29).

If reservoir 1 has a volume of 8 ft^3 (0.22 m^3) and the compressor forces another 8 ft^3 into the reservoir, the reservoir pressure increases from zero to 14.7 psi (101.28 kPa). Each time a quantity

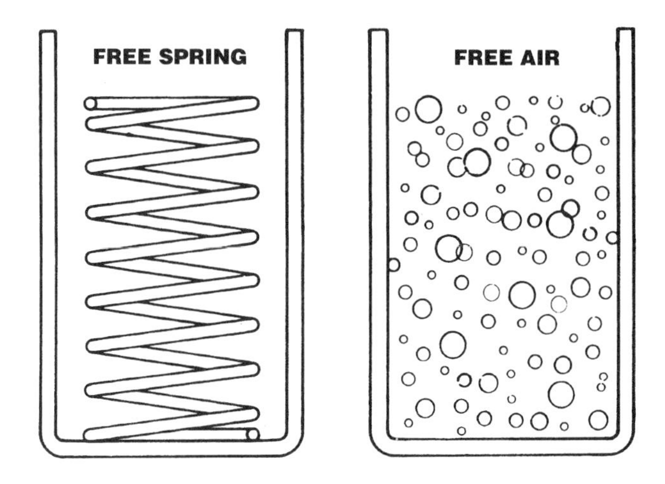

Figure 2-27 Air at atmospheric pressure or a coil spring that is not compressed do not store any energy. (Courtesy of Allied Signal Truck Brake Systems Co.)

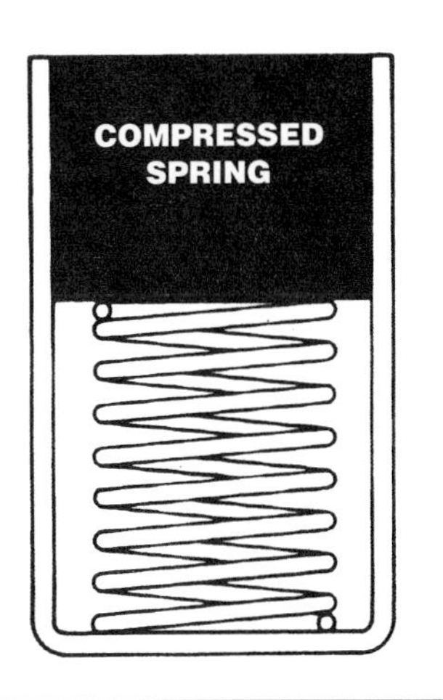

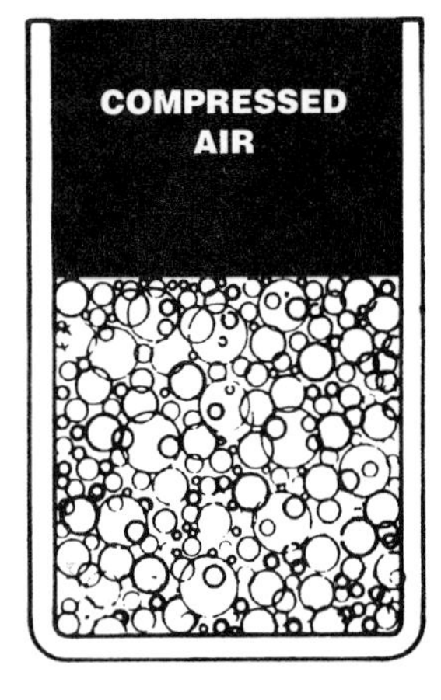

Figure 2-28 Compressed air or a compressed coil spring store energy. (Courtesy of Allied Signal Truck Brake Systems Co.)

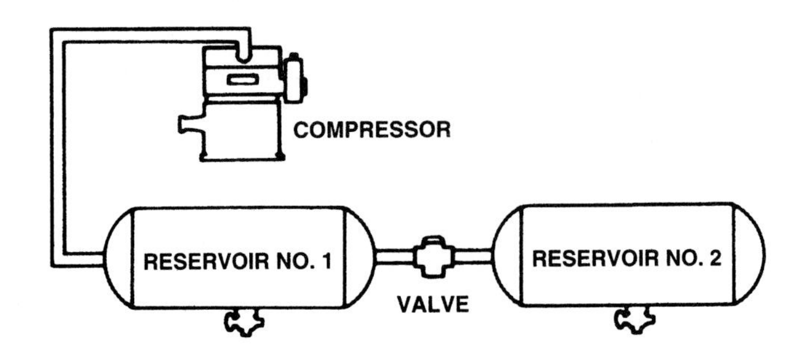

Figure 2-29 Compressor and reservoirs in an air brake system. (Courtesy of Allied Signal Truck Brake Systems Co.)

of free air equal to the volume of the reservoir is forced into the reservoir, the gauge pressure increases by 14.7 psi (101.28 kPa).

CAUTION: When removing air suspension components, always raise the vehicle frame with a floor jack and lower the frame onto safety stands so the vehicle weight is removed from the suspension system. Open the draincocks on all the air brake system reservoirs to release all the air pressure from the air suspension system. If air suspension system lines or fittings are loosened while under pressure, the line may whip around causing personal injury.

Air pressure is supplied from these reservoirs to the air brake system. Air pressure is also supplied from these same reservoirs to the air suspension system, air seat, cab air suspension system, and other air-operated systems on the tractor.

When compressed air is supplied on one side of a movable piston or flexible diaphragm in a sealed chamber, the air pressure causes the piston or flexible diaphragm to move until an equal force is supplied to the other side of the piston. If the piston has an area of 10 in^2 (64.5 cm^2) and the air pressure supplied to the piston is 10 psi (68.9 kPa), the piston force is 100 lb (45.3 kg) (Figure 2-30). The piston force may be increased by increasing the air pressure. The air pressure supplied to the piston or flexible diaphragm is equal in all directions. Pascal's law states, "that pressure on a confined fluid transmits force equally in all directions and at right angles to all surfaces."

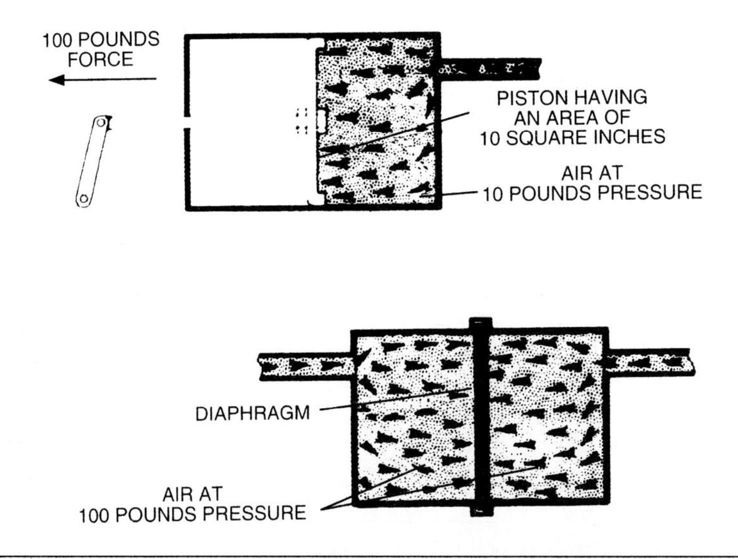

Figure 2-30 Air pressure supplied to a diaphragm or piston in a sealed chamber. (Courtesy of Allied Signal Truck Brake Systems Co.)

When air at equal pressure is supplied to each side of the piston or flexible diaphragm, the piston or diaphragm does not move. If air pressure is supplied to one side of the piston or flexible diaphragm and an equal mechanical force is supplied to the opposite side, again no piston or diaphragm movement occurs.

WARNING: Never supply more than the specified air pressure to air suspension system components. This action may damage these components or cause personal injury.

Levers

Because various levers are used in truck suspension systems, a basic knowledge of lever operation is necessary to understand air brake operation. A lever may be defined as an inflexible rod that is free to move around a fixed point referred to as a **fulcrum.** The lever is used to transmit and modify force and motion.

A lever may be defined as an inflexible rod that is free to move around a fixed point referred to as a fulcrum.

Levers vary depending on the location of the fulcrum in relation to the applied force and the delivered force. In view A of Figure 2-31 the fulcrum is positioned between the applied force and the delivered force. In some cases such as view B of Figure 2-31, the fulcrum is at one end of the lever, and the applied force is at the opposite end of the lever with the delivered force between the fulcrum and the applied force. In view C of Figure 2-31 the fulcrum is at one end of the lever and the applied force is near the center of the lever, with delivered force at the opposite end of the lever from the fulcrum.

A law of levers states the applied force multiplied by the perpendicular distance between the line of force and the fulcrum is always equal to the delivered force multiplied by the perpendicular distance between the fulcrum and the line of force. In view A of Figure 2-31 the applied force of 100 lb (45.3 kg) is 2 ft (0.6 m) from the fulcrum and the delivered force is 200 lb (90.6 kg) at 1 ft (0.3 m) from the fulcrum.

The law of levers states the applied force multiplied by the perpendicular distance between the line of force and the fulcrum is always equal to the delivered force multiplied by the perpendicular distance between the fulcrum and the line of force.

In view B of Figure 2-31 the applied force of 100 lb (45.3 kg) 3 ft (0.9 m) from the fulcrum lifts 300 lb (136.08 kg) at a distance of 1 ft (0.3 m) from the fulcrum.

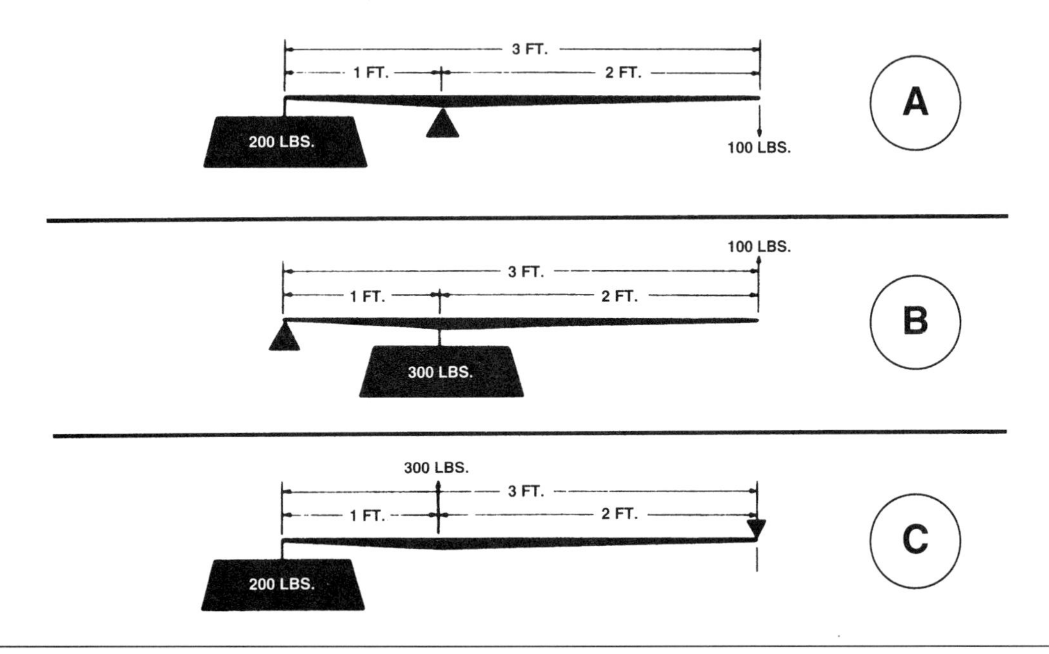

Figure 2-31 Lever principles. (Courtesy of Allied Signal Truck Brake Systems Co.)

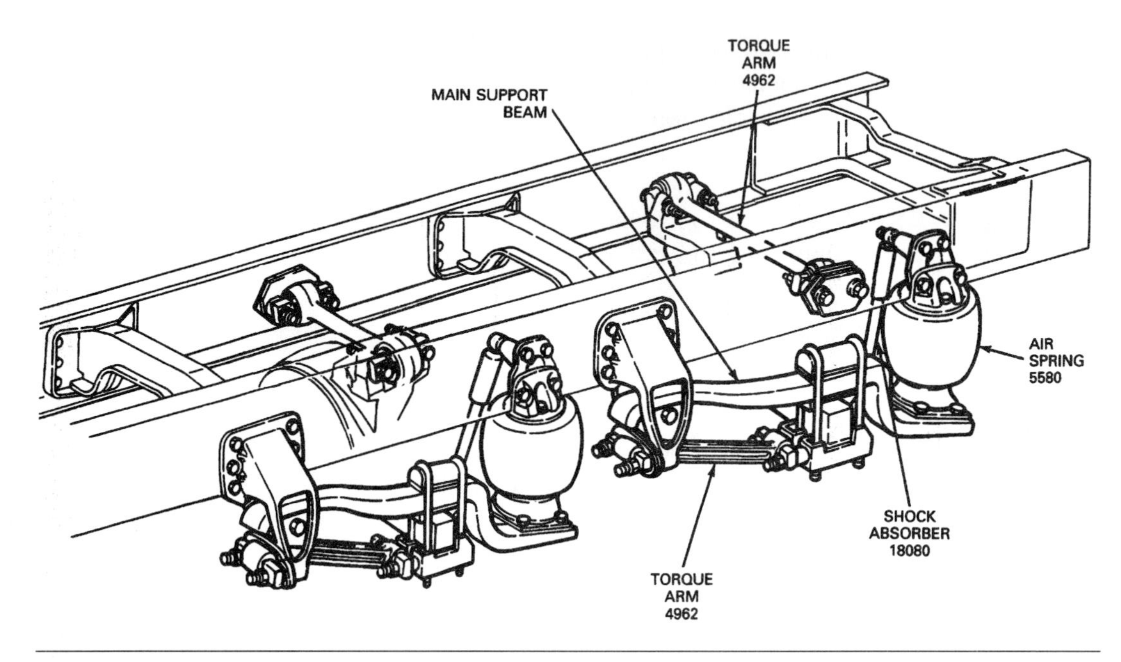

Figure 2-32 Rear axle air suspension system. (Courtesy of Sterling Truck Corporation.)

See Chapter 2 in the Shop Manual for safety practices, employee-employer obligations, and National Institute for Automotive Service Excellence (ASE) certification.

In view C of Figure 2-31 the delivered force is farthest from the fulcrum. Under this condition the delivered force is less than the applied force. The applied force is 300 lb (136.08 kg) 2 ft (0.6 m) from the fulcrum, and the delivered force is 200 lb (90.6 kg) at 3 ft (0.9 m) from the fulcrum.

Levers are used in truck suspension systems. For example, in a rear axle air suspension system the force of the vehicle weight is applied through the air spring to the rear end of the main support beam (Figure 2-32). Because the tires are resting on the road surface, the force of the vehicle weight is supplied through the air spring and the main support beam to the rear axle and tires. The spring hanger at the front of the main support beam acts as a fulcrum.

TRADE JARGON: Mechanical advantage is a term that may be used to describe the relationship between the force applied to a lever and the force delivered by the lever. When the difference between the applied force and delivered force is greater, the lever has more mechanical advantage.

Summary

Terms to Know

Atmospheric pressure
Compressible
Dynamic balance
Energy
Free air
Friction
Fulcrum
Highly incompressible
Inertia
Mass, weight, and volume
Momentum
Power

- ❑ Work is the result of a force that is allowed to act over a distance.
- ❑ Work is measured in pounds and distance.
- ❑ Energy is the ability to do work, and there are six basic types of energy.
- ❑ Inertia is the tendency of an object at rest to remain at rest or the tendency of an object in motion to remain in motion.
- ❑ An object gains momentum when force overcomes static friction and increases the speed of the object.
- ❑ Mass is a measurement of an object's inertia.
- ❑ Weight is a measurement of the earth's gravitational pull on an object.
- ❑ Volume is the length, width, and height of a space occupied by an object.
- ❑ Power is a measurement for the rate at which work is done.
- ❑ Torque is the twisting effort produced by a force.

- ❑ Friction is the resistance to relative motion between two bodies in contact.
- ❑ Atmospheric pressure is the total weight of the earth's atmosphere.
- ❑ Vacuum is the absence of atmospheric pressure.
- ❑ Gases are compressible.
- ❑ Liquids are almost incompressible.
- ❑ Pascal's law states, "Pressure on a confined fluid is transmitted equally in all directions and acts with equal force on equal areas."
- ❑ A rolling, tilted wheel tends to move in the direction of the tilt.
- ❑ If the pivot point at the tire center line is behind the center line where the vehicle weight is projected on the road surface, the wheel tends to remain in the straight ahead position.
- ❑ Weight must be distributed equally around the center of wheel rotation viewed from the side to obtain static balance.
- ❑ Weight must be distributed equally on both sides of the tire center line viewed from the front of the tire to maintain proper dynamic balance.
- ❑ Energy is stored in compressed air much like energy is stored in a compressed coil spring.

Terms to Know (Continued)

Static balance

Torque

Vacuum

Review Questions

Short Answer Essays

1. Describe Newton's three laws of motion.
2. Describe six different forms of energy.
3. Describe four different types of energy conversion.
4. Explain mass, weight, and volume.
5. Explain horsepower.
6. Define the coefficient of friction.
7. Explain the relationship between atmospheric pressure and altitude.
8. Describe the operation of a venturi.
9. Explain the difference in the compressibility of gases and liquids.
10. Describe the motion of a tilted, rolling wheel.

Fill-in-the-Blanks

1. Work is calculated by multiplying ____________________ × ____________________.
2. Energy may be defined as the ability to do ____________________.
3. When one object is moved over another object, the resistance to motion is called ____________________.
4. Torque is a force that does work with a ____________________ action.
5. Power is a measurement for the rate at which ____________________ is done.
6. Vacuum is defined as the absence of ____________________.
7. When the weight on a wheel and tire is not distributed equally on both sides of the tire center line viewed from the front, the wheel and tire are ____________________ unbalanced.

8. When a coil spring is compressed, it stores ____________________.

9. If two balls are positioned on a metal bar so their weight is equally distributed on the center line of the axis of rotation, the path of rotation remains at a ____________________ angle to the center line of the axis when the bar is rotated.

10. When the temperature of the air in a truck tire increases, the pressure in the tire ____________________.

ASE Style Review Questions

1. While discussing different types of energy:
   Technician A says when energy is released to do work, it is called kinetic energy.
   Technician B says stored energy is referred to as potential energy.
   Who is correct?
   A. A only **C.** Both A and B
   B. B only **D.** Neither A nor B

2. While moving a metal weight on a metal surface:
   A. Friction only occurs in solids.
   B. Friction does not affect motion.
   C. Friction creates heat.
   D. Friction is not affected by the weight of the moving metal.

3. While discussing torque:
   Technician A says torque is calculated by multiplying force × distance.
   Technician B says work is accomplished if a bolt turns when torque is applied.
   Who is correct?
   A. A only **C.** Both A and B
   B. B only **D.** Neither A nor B

4. While discussing horsepower:
   Technician A says 1 horsepower is equal to 400 ft-lb per second.
   Technician B says when horsepower increases, the time required to do work remains the same.
   Who is correct?
   A. A only **C.** Both A and B
   B. B only **D.** Neither A nor B

5. When applying principles of liquids and gases in truck systems:
   A. Gases are incompressible.
   B. Gases and liquids may be used to transmit force.
   C. Liquids are compressible.
   D. Compressed air does not store energy.

6. When applying principles of liquids and gases in truck systems:
   A. When air becomes colder, it contracts and becomes heavier.
   B. Atmospheric pressure increases as altitude increases.
   C. The earth's atmosphere contains 45% oxygen.
   D. Atmospheric pressure is highest at the equator.

7. While discussing venturi operation:
   Technician A says when the speed of the fluid flowing through a venturi increases, the pressure in the venturi decreases.
   Technician B says venturi diameter is less than the diameter of the pipe above and below the venturi.
   Who is correct?
   A. A only **C.** Both A and B
   B. B only **D.** Neither A nor B

8. When applying principles of truck tires and wheels in motion:
   A. A tilted, rolling tire and wheel tend to move away from the direction of the tilt.
   B. A bicycle front wheel tends to remain in the straight ahead position if the tire pivot point is ahead of the center line of the front forks.
   C. If the weight is not distributed equally around the center of wheel rotation, the tire and wheel may have a hopping action.
   D. If two balls on a metal bar have their weight equally distributed on the center line of the axis of rotation, the path of rotation remains at a 75-degree angle to the center line of the axis.

9. While discussing lever action with the fulcrum between the applied force and the delivered force:
   Technician A says if the distance is increased from the fulcrum to the applied force, the delivered force is increased.
   Technician B says if the fulcrum is moved toward the applied force, the delivered force is increased.
   Who is correct?
   A. A only **C.** Both A and B
   B. B only **D.** Neither A nor B

10. When applying principles of liquids and gases in truck systems:

A. Because liquids and gases both have flow capabilities, they may be classified as fluids.

B. Pressure on a confined fluid is not transmitted equally inside the fluid container.

C. If the temperature of a pressurized gas decreases, the pressure increases.

D. The fluid in a power steering system may be compressed so it takes up less space.

Tires, Wheels, and Wheel Bearings

Upon completion and review of this chapter, you should be able to:

- ❏ Explain the difference in the bias ply and radial ply tire design.
- ❏ Explain the advantages of radial ply tires compared with bias ply tires.
- ❏ Describe the advantages of low-profile radial ply tires compared with conventional radial ply tires.
- ❏ Explain why tires must match on drive axles.
- ❏ Describe three methods of retaining tires on different types of rims.
- ❏ Describe two different types of wheels.
- ❏ Describe two different ways of retaining disk wheels on the wheel hub.
- ❏ Explain the results of overinflation and underinflation on tire tread wear.
- ❏ Explain toe-in and toe-out, and describe the type of tire tread wear caused by improper toc adjustment.
- ❏ Explain positive camber and negative camber, and describe the type of tire tread wear caused by improper camber.
- ❏ Explain static wheel balance, and describe the result of improper static wheel balance.
- ❏ Describe dynamic wheel balance, and describe the result of improper dynamic wheel balance.
- ❏ Explain the purposes of a bearing.
- ❏ Describe three different types of bearing loads.
- ❏ Identify the basic parts in ball bearings or roller bearings and describe the purpose of each part.
- ❏ Describe the action between the balls and the race when a ball bearing is rotating.
- ❏ Explain the purposes of bearing snap rings, shields, and seals.
- ❏ Describe the load-carrying capabilities of ball bearings, roller bearings, tapered roller bearings, and needle roller bearings.
- ❏ Explain the advantage of tapered roller bearings compared with other types of bearings.
- ❏ Explain the purpose of the garter spring behind a lip seal.
- ❏ Describe the purpose of flutes on seal lips.
- ❏ Describe spoke and disk wheels.
- ❏ Explain single and dual rim mountings on spoke and disk wheels.
- ❏ Explain unitized seal design and purpose.
- ❏ Explain the purpose of grit guards and wiper rings.
- ❏ Describe barrier seal design.

Introduction

Students and technicians must understand different types of tire design, tire rims and wheels, the causes of tire tread wear, and tire wheel balance theory to diagnose the causes of various tire and wheel problems, including different types of tire tread wear. Students and technicians must also understand proper tire inflation and the results of underinflation and overinflation.

Many different types of bearings are used in a truck. A bearing may be defined as a component that supports and guides one of these parts:

1. Pivot
2. Wheel
3. Rotating shaft
4. Oscillating shaft
5. Sliding shaft

While a bearing is supporting and guiding one of these components, the bearing is designed to reduce friction and support the load applied by the component and related assemblies. Bearings are precision machined assemblies that provide smooth operation and long life. When bearings are properly installed and maintained, bearing failure is rare.

Tires

Bias Ply Tires

Bias ply tires have body ply cords that are positioned at an angle of 30 degrees to 40 degrees between the tire beads (Figure 3-1). These tires may have narrow plies under the tread, and the cords in these plies are mounted at an angle that is close to the angle of the body ply cords. The narrow plies under the tread may be called "breakers." Many bias ply tires contain inner tubes.

Radial Ply Tires

The rubber used in bias ply or **radial ply tires** may be natural or synthetic. Synthetic rubber is manufactured in a laboratory. Most tires manufactured at present are made from a blend of various synthetic rubbers. In radial ply tires the body ply cords are placed at a 90-degree angle from one tire bead to the opposite bead. Radial ply tires also have belt plies that are positioned circumferentially around the tire under the tread (Figure 3-2). These belts provide additional tire rigidity. Tire body plies and breakers or belts in bias ply and radial ply tires may be manufactured from fiberglass, nylon, rayon, polyester, steel, or aramids. In radial ply tires these materials may be used in different combinations such as nylon body cords and steel belts, polyester body plies and polyester belt, and steel body plies with a steel belt. Many radial ply tires are tubeless. The advantages of tubeless tires compared with tube-type tires are as follows:

Narrow plies under the tread of a bias ply tire may be called breakers.

1. Reduced weight, up to 136 lb (61.68 kg) per axle
2. Reduced number of punctures and road delays
3. Easier, quicker mounting
4. Elimination of lock ring hazard on the wheels because tubeless tires have one-piece wheels
5. Problems with tubes and tube flaps are eliminated

Tires with radial body plies are more flexible compared with bias ply tires. Therefore radial tires have less rolling resistance compared with bias ply tires. Tire rolling resistance can affect fuel economy by 20% on a tractor and trailer, depending on the vehicle load. The advantages of radial ply tires compared with bias ply tires are as follows:

1. Reduced rolling resistance
2. Less noise
3. Improved ride quality because of increased flexibility
4. Reduced tread wear by elimination of sidewall scrubbing as the tire rolls

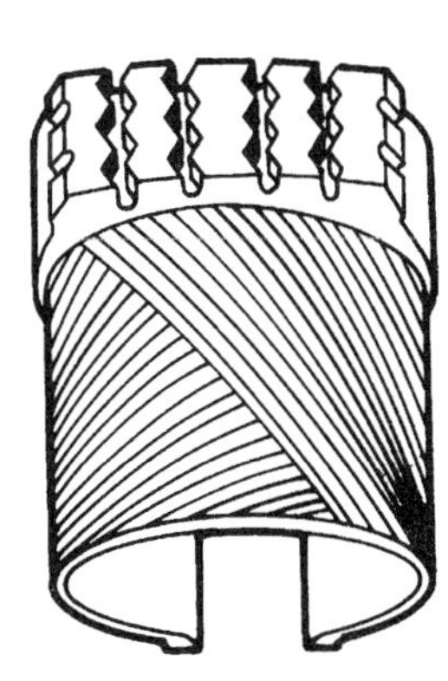

Figure 3-1 Bias ply tire design. (Courtesy of Volvo GM Heavy Truck Corporation.)

Figure 3-2 Radial ply tire design. (Courtesy of Volvo GM Heavy Truck Corporation.)

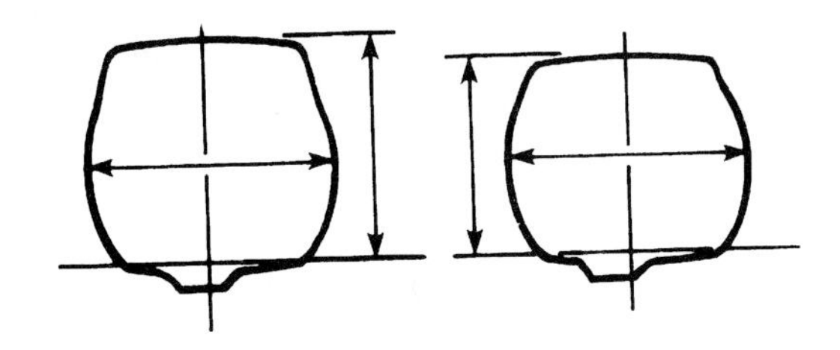

Figure 3-3 Aspect ratio on conventional and low-profile tires.

5. Improved traction because improved flexibility provides a more uniform tire pressure on the road surface
6. High load rating
7. Cooler running
8. Excellent recapability

Low-Profile Radial Ply Tires

Low-profile tires have a lower aspect ratio compared with conventional radial ply tires. The aspect ratio of a tire is the relationship between the height and width of the tire. The aspect ratio is calculated by dividing the tire height by the width. If a conventional radial ply tire has a height of 9.5 in (24.13 cm) and a width of 11.0 in (27.94 cm), the aspect ratio is $9.5 \div 11.0 = .86$. A low-profile radial ply tire may have a height of 8.4 in (21.33 cm) and a width of 10.9 in (27.68 cm), which provides an aspect ratio of $8.4 \div 10.9 = .77$ (Figure 3-3). Notice that the low-profile radial tire has an outside diameter of 41.3 in (104.90 cm), and the outside diameter of the conventional radial tire is 43.5 in (110.49 cm). Compared with conventional radial tires, low-profile radial tires are lighter in weight and provide longer life. The advantages of low-profile radial tires compared with conventional radial tires are as follows:

1. Reduced weight of approximately 12 lb (5.44 kg) per tire
2. Reduced rolling resistance
3. Cooler running
4. Longer tread life
5. Improved traction
6. Improved fuel economy
7. Reduced vehicle height with a lower center of gravity
8. Reduced deformation of tread contact area provides faster steering response

Tire Matching

WARNING: Mismatched tires on opposite sides of a drive axle may cause drive axle failure.

Radial ply and bias ply tires must not be mixed on the same axle. All the tires on an axle must be either radial ply or bias ply. On trucks with two or more drive axles, the **tire matching** on all the axles is necessary so all the tires are the same type. Because radial tires deflect more than bias ply tires under load, a mix of bias ply and radial ply tires overloads the bias ply tires. This action causes premature tire tread wear and possible tire blowout.

Tire Rims

Single-piece rims have a flange on each side of the rim to retain the tire on the rim. These flanges are an integral part of the rim (Figure 3-4). Tubeless tires are used on single-piece rims.

Split side rings are used on some rims. A fixed flange is positioned on one side of the rim on these rims. This flange is an integral part of the rim. A split side ring is mounted on the opposite side of the rim. This split side ring acts as a flange and self-contained lock ring (Figure 3-5).

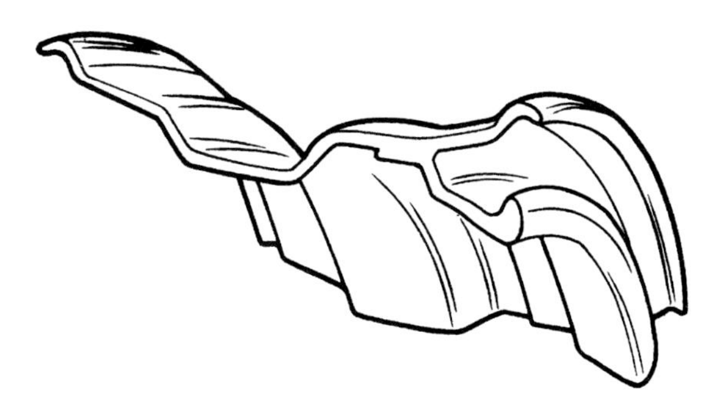

Figure 3-4 Single-piece rim. (Courtesy of Accuride Corporation.)

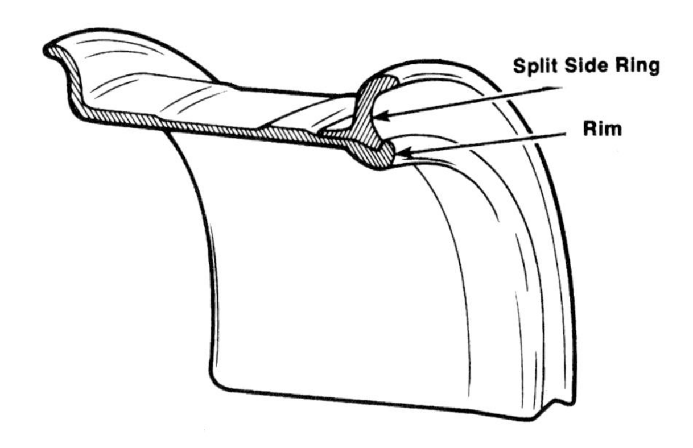

Figure 3-5 Rim with split side ring. (Courtesy of Accuride Corporation.)

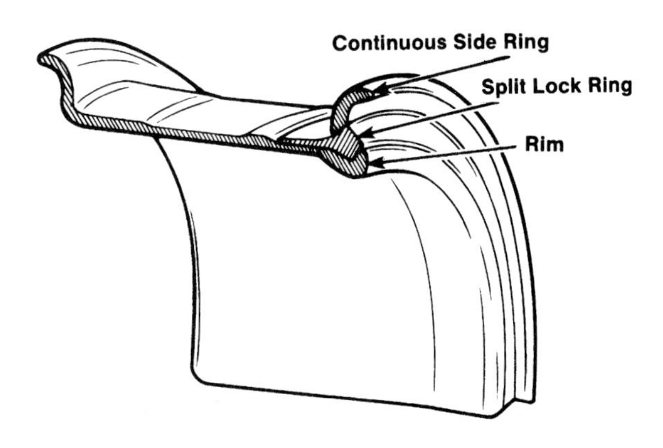

Figure 3-6 Rim with continuous side ring and separate split lock ring. (Courtesy of Accuride Corporation.)

Some rims have a continuous side ring that supports the tire on one side of the rim. The continuous side ring is retained on the rim by a separate split lock ring (Figure 3-6).

The most common medium- and heavy-duty truck rim diameters are 17.5 in (44.45 cm), 19.5 in (49.53 cm), 22.5 in (57.15 cm), and 24.5 in (62.23 cm).

Tire Standards and Identification

Federal Motor Vehicle Safety Standard (FMVSS-119) established standards for pneumatic tires on vehicles other than passenger cars. These standards established the following regulations:

1. No vehicle shall be operated when any tire on the vehicle has fabric exposed through the tread or sidewall.

2. Any tire on the front wheels of a bus, truck, or tractor shall have a minimum tread groove depth of $\frac{4}{32}$ in when measured at any point on a major tread groove. The tread depth measurements shall not be made where the tire bars, humps, or fillets are located.
3. Except as indicated in point 2, tires must have a minimum tread depth of $\frac{2}{32}$ in when measured in a major tread groove. The tread depth measurements shall not be made where the tire bars, humps, or fillets are located.
4. A bus shall not be operated with regrooved, recapped, or retreaded tires on the front wheels.
5. A truck or truck tractor shall not be operated with regrooved tires on the front wheels that have a load-carrying capacity equal to or greater than that of 8:25-20 8 ply rating tires.

All new and retreaded tires sold in the United States must have a Department of Transportation (DOT) number cured into the sidewall on one side of the tire. This DOT number contains a standard format for new and retread tires (Figure 3-7).

Tire Tread Designs

The rib-type tread design is the most common type of tire tread (Figure 3-8). This open groove tread design provides good skid resistance that provides improved steering control. The rib-type tread design may be used on all axles of a tractor and trailer or trailers.

Cross lug or cross rib and rib lug-type tire treads are designed for drive wheel service (Figure 3-9). This type of tire tread is suitable for over-the-road operation and provides im-

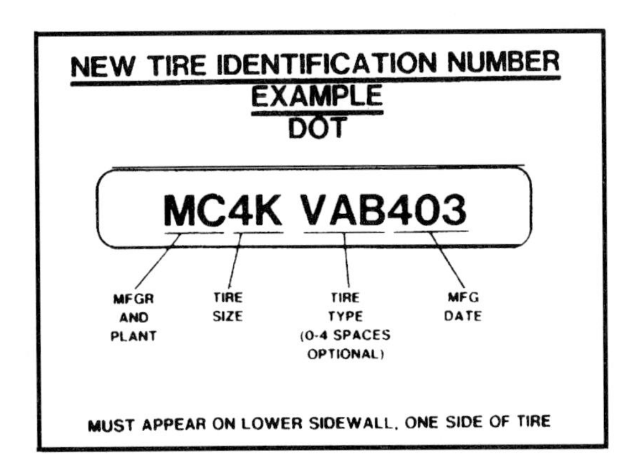

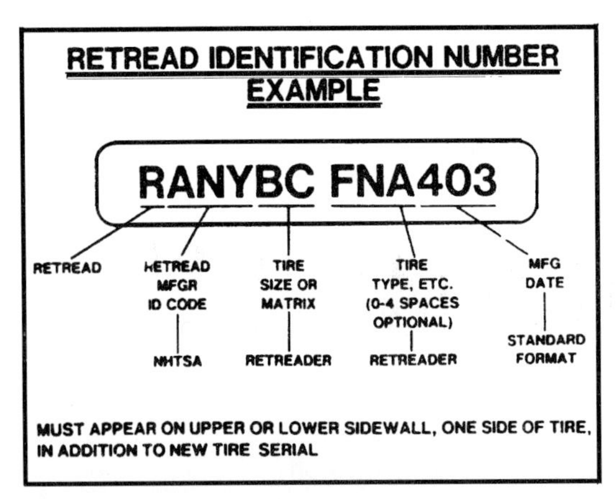

Figure 3-7 Department of Transportation (DOT) identification for new and retread tires.

Figure 3-8 Rib-type tire tread design. (Courtesy of Bridgestone/Firestone, Inc.)

Figure 3-9 Lug-type tire tread design. (Courtesy of Bridgestone/Firestone, Inc.)

Figure 3-10 Special service mud and snow lug-type tire tread design. (Courtesy of Bridgestone/Firestone, Inc.)

proved traction and excellent resistance to wear. This type of tire tread is suitable for some off-road service.

Special service mud and snow lug-type tires provide improved drive wheel traction for both off-road and over-the-road service (Figure 3-10). This type of tire is intended for short, low-speed hauling when mud or snow is often encountered. In winter conditions this type of tire may be selected for high-speed or long-distance hauling.

Wheels

Spoke Wheels

CAUTION: When servicing truck wheels, always follow the precautions and service procedures recommended by the truck manufacturer. Failure to observe these precautions and service procedures may cause personal injury.

WARNING: When removing and installing truck wheels, always follow the service procedures recommended by the truck manufacturer. Failure to follow these procedures may cause excessive wheel wobble, out-of-round, or wheel nuts coming loose. These conditions may cause a wheel to come off, resulting in extensive truck and property damage or personal injury.

Truck wheels may be spoke-type or disk-type. Front **spoke wheels** are a six-spoke design, and they may be used with disk or drum brakes (Figure 3-11). The spoke casting includes the hub and spokes, and this casting may be made of ductile iron, cast steel, or aluminum. The tire rim is

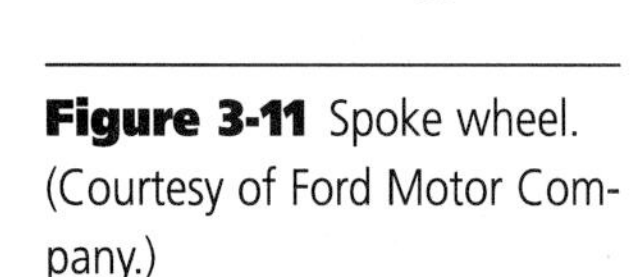

Figure 3-11 Spoke wheel. (Courtesy of Ford Motor Company.)

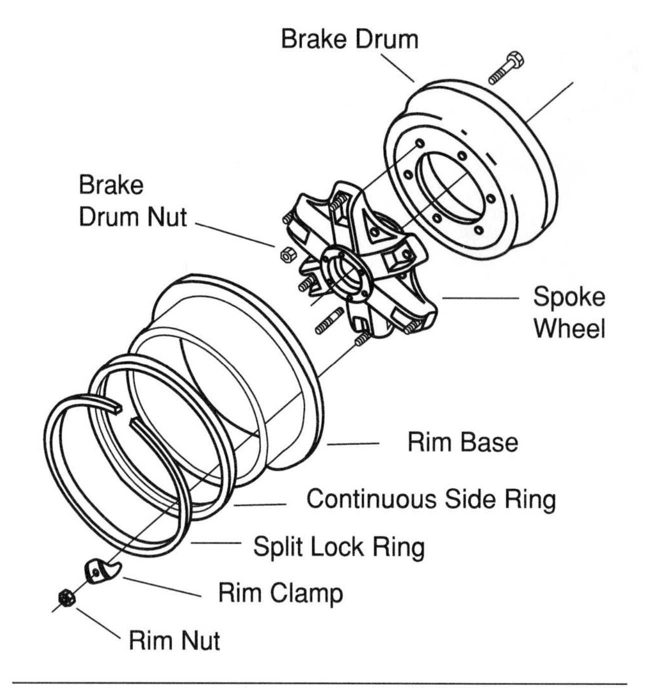

Figure 3-12 The tire rim is retained on the spoke wheel with rim nuts and clamps. (This information courtesy of Freightliner Corporation.)

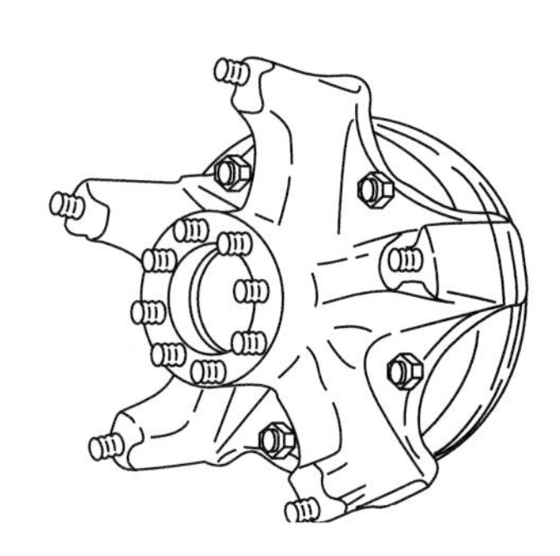

Figure 3-13 Rear spoke wheel. (Courtesy of Ford Motor Company.)

retained on the spoke wheel with rim nuts and clamps (Figure 3-12). Separate bolts retain the brake drum to the spoke wheel. Rear spoke wheels have five or six spokes depending on the axle weight rating (Figure 3-13). Spoke wheels with three spokes may be used on trailers. On dual wheel applications a spacer is positioned between the inner and outer rims (Figure 3-14). This spacer maintains proper spacing between the tires. The rims are retained on the spoke wheel with nuts and clamps, and separate bolts retain the brake drum to the wheel. Proper installation procedures are critical on spoke wheels. If the rim clamps are not properly installed, wheel out-of-round or wobble may be excessive. However, if spoke wheel rims are installed and tightened properly, they provide trouble-free service.

Disk Wheels

In a stud-piloted wheel mounting, the wheel rims are centered on the mounting studs by matching tapers on the cap nuts and wheels.

In a hub-piloted wheel mounting, the wheel rims are centered on mounting studs by the openings in the wheels, and the cap nuts provide a clamping force to retain the wheels.

Medium-duty trucks may be equipped with eight-hole **disk wheels,** and heavy-duty trucks may have ten-hole disk wheels (Figure 3-15). Studs in the hub extend through the brake drum and wheel or wheels, and nuts are threaded onto these studs to retain the wheels and brake drum. In some single disk wheels the retaining nuts have a ball seat that fits into a taper in the rim openings. This ball seat and taper center the rim on the wheel studs (Figure 3-16). On some dual disk wheel assemblies inner cap nuts are installed on the wheel studs to retain the inner rim, and the outer nuts with ball seats retain and center the outer wheel (Figure 3-17). This type of wheel mounting may be called **stud-piloted.**

On other disk wheels the hub acts as a pilot to position the rims properly, and the inner side of the nuts has a flat surface that clamps the rims and brake drum to the hub (Figure 3-18). This type of wheel mounting may be called a **hub-piloted** mounting.

Because the brake drum is bolted on the inside of a spoke wheel, the wheel and hub assembly must be removed to service the brake drum (Figure 3-19). When the wheel and hub assembly are removed, wheel bearing and seal service are required. On disk wheels the wheels and brake drum are mounted on the outside of the hub. Therefore the brake drum may be serviced without disturbing the hub and bearings.

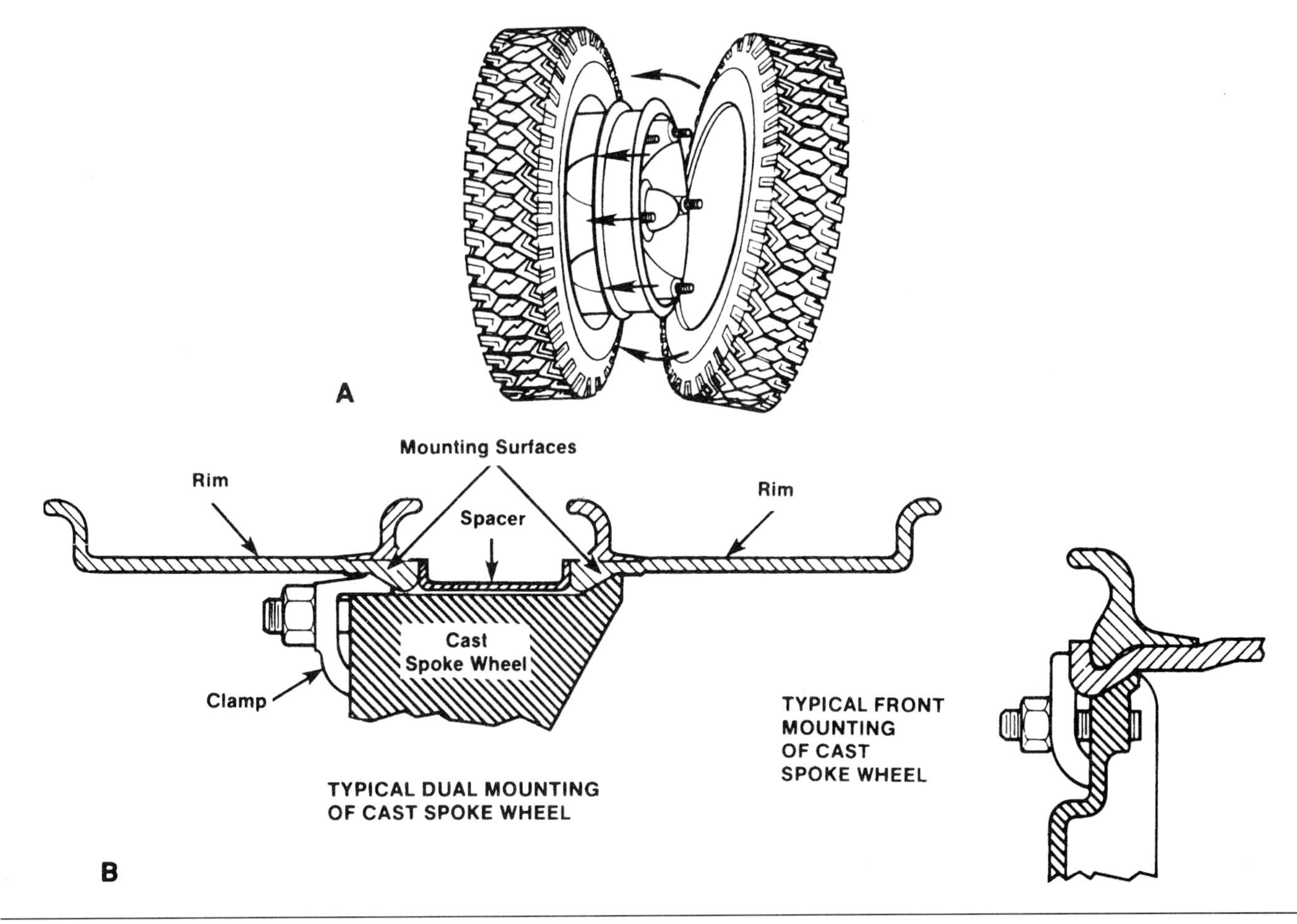

Figure 3-14 **(A)** Position of the spoke wheel dual mounting spacer and **(B)** cross-section view of mounted dual wheels. (Courtesy of Navistar International Transport Corp.)

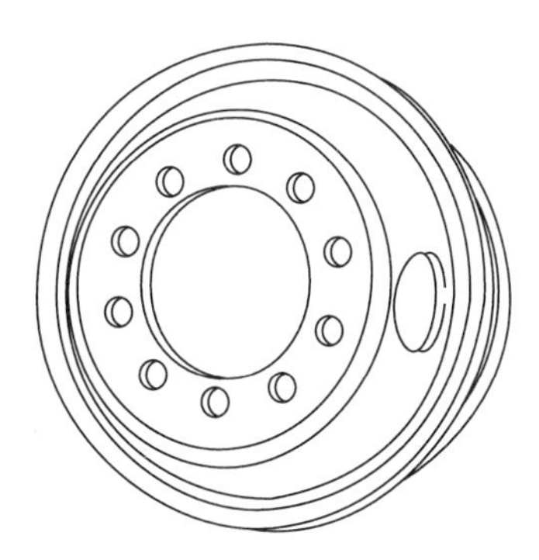

Figure 3-15 Ten-hole disk wheel for heavy-duty trucks. (Courtesy of Ford Motor Company.)

Tire Inflation and Wear

Tire labels are molded into the tire sidewall. A conventional-size tire may have an 11R22.5 label. This indicates the tire section width is 11 in, and the R indicates a radial tire. The 22.5 on the label is the inside diameter of the rim in inches. Some tires have a metric tire label. If the label is 295/75R22.5, the 295 indicates the tire section width in millimeters, and the 75 represents the aspect ratio. The R indicates a radial tire, and the 22.5 is the inside rim diameter in inches.

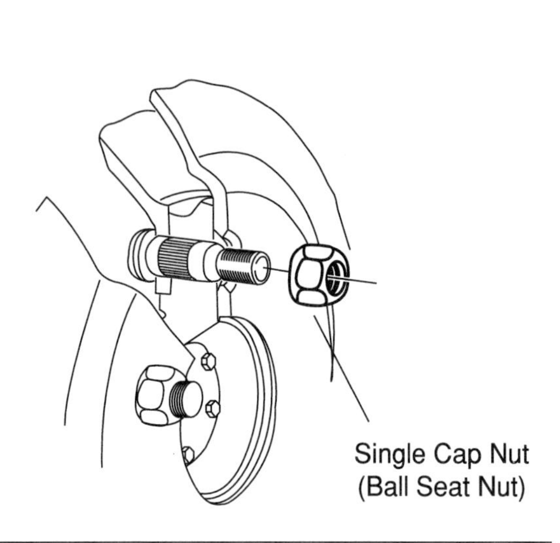

Figure 3-16 Single disk wheel with tapered stud openings and a ball seat on the wheel nuts. (Courtesy of Ford Motor Company.)

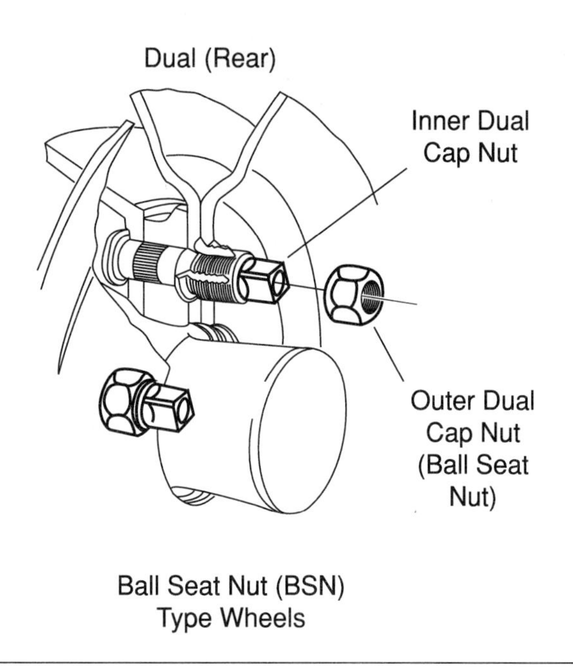

Figure 3-17 Dual disk wheels retained with inner and outer cap nuts. (Courtesy of Ford Motor Company.)

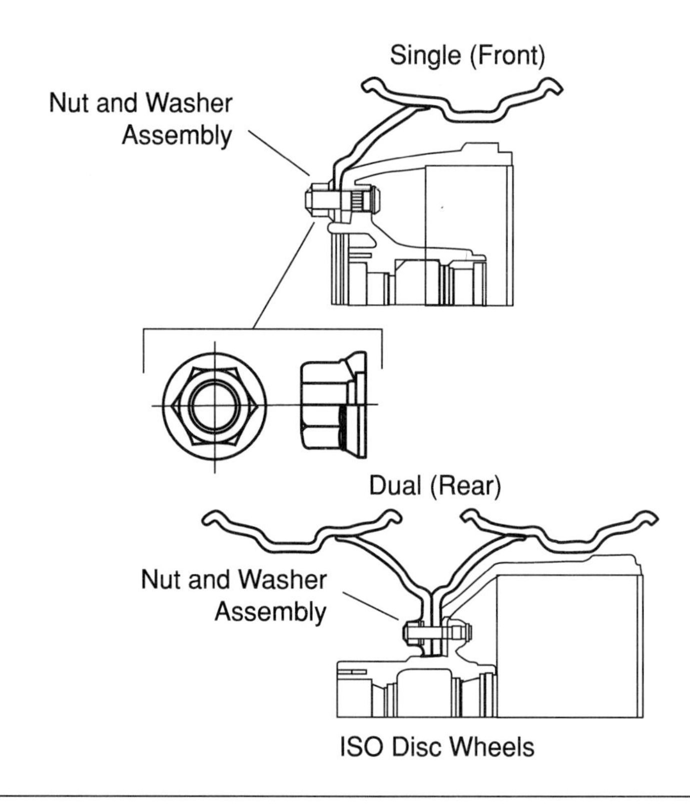

Figure 3-18 Disk wheels with pilot-type hub. (Courtesy of Ford Motor Company.)

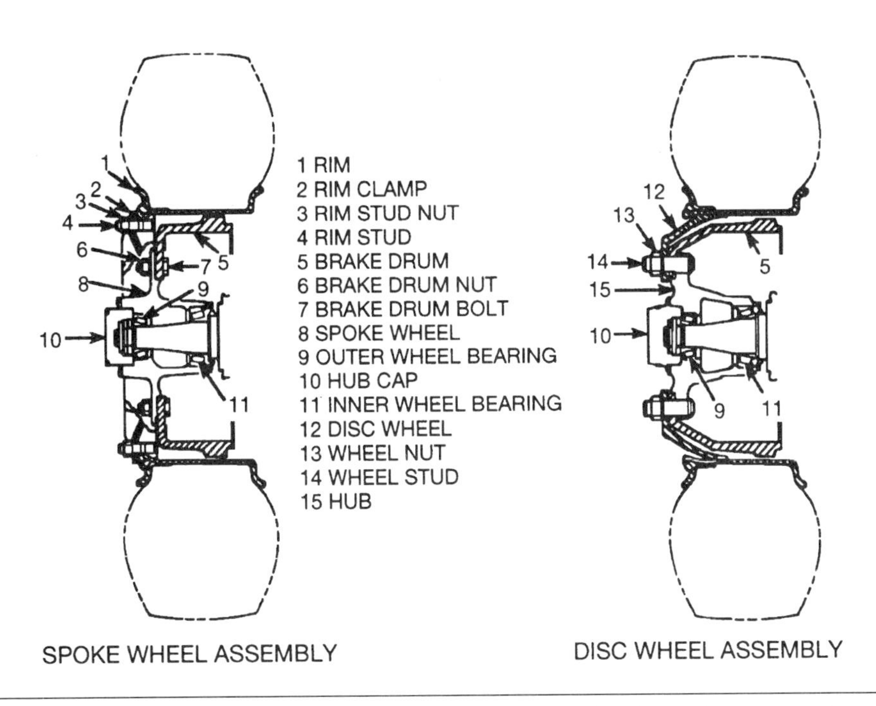

Figure 3-19 Brake drum and wheel mountings with spoke and disk wheels. (This information courtesy of Freightliner Corporation.)

Truck manufacturers provide tire pressure specifications for various tire types and sizes. These specifications include the tire load and ply rating. For example an H (16) in the rating indicates an H load rating and a sixteen-ply tire. Tire pressure specifications also include the maximum weight capacity of the tires on each axle and the inflation pressure. Inflation pressure specifications are made on the basis of cool temperatures (Figure 3-20).

Tire inflation pressures are extremely important to maintain proper tire life. If a tire has the specified inflation pressure, the entire tread surface is in contact with the road surface. When a tire is underinflated, the vehicle weight is concentrated on the outer edges of the tire, and the center of the tread may not even contact the road surface (Figure 3-21). Tire overinflation lifts the edges of the tire off the road surface, and the vehicle weight is concentrated on the center of the tread (Figure 3-21). Therefore tire underinflation causes wear on the edges of the tread (Figure 3-22), and overinflation results in wear to the center of the tread (Figure 3-23).

Front wheel **toe-in** occurs when the distance between the rear inside edges of the tires is greater than the distance between the front inside edges of the front tires (Figure 3-24). **Toe-out** is present when the distance between the front inside edges of the tires is greater than the distance between the rear inside edges of the tires (Figure 3-25). The toe must be adjusted to the truck manufacturer's specifications. Improper toe adjustment causes a feathered tire tread wear (Figure 3-26).

Negative camber is present when the tire and wheel center line tilts inward in relation to the true vertical centerline of the wheel. **Positive camber** occurs when the tire and wheel center line tilts outward in relation to the true vertical center line of the wheel (Figure 3-27). Improper camber causes wear on one side of the tire tread (Figure 3-28).

Wear bars are raised portions across the tire tread positioned at the bottom of the tread. When the tire tread is worn to the minimum tread depth, the wear bars appear as straight lines across the tread (Figure 3-29). When the wear bars appear on the tire tread, the tire should be replaced.

Other causes of excessive tire tread wear include overloading, which causes overheating; improper load distribution; and excessive speed. If the load is not distributed equally on both sides of a truck or trailer, the tires on the side with the heavier load wear faster than the tires with the lighter load. Overloading or excessive speed increases tire friction on the road surface, and this causes excessive tire heating and increased tread wear.

Tire Size, Load Range (Ply Rating)	Wheel Width (In.) and Type	Single Rear Axle: Max. Tire and Wheel Capacity (Lbs.) @ psi (Cold) by Axle		Tandem Rear Axle: Max. Tire and Wheel Capacity (Lbs.) @ psi (Cold) by Axle	
		Front	Rear	Front	Rear
Tube Type Bias Ply					
9.00-20 E (10)	7.5 CS	9,220 @ 80	16,160 @ 70	9,220 @ 80	32,320 @ 70
9.00-20 F (12)	7.5 CS	10,300 @ 95	18,080 @ 85	10,300 @ 95	36,160 @ 85
10.00-20 F (12)	7.5 CS 8.0 10H or CS	10,860 @ 85 10.860 @ 85	19,040 @ 75 19,040 @ 75	10,860 @ 85 10,860 @ 85	38,080 @ 75 38,080 @ 75
10.00-20 G (14)	7.5 CS 8.0 10H or CS	10,860 @ 85 12,080 @ 100	21,200 @ 90 21,200 @ 90	10,860 @ 85 12,080 @ 100	42,400 @ 90 42,400 @ 90
11.00-20 G (14)	8.0 10H or CS	13,180 @ 100	23,120 @ 90	13,180 @ 100	46,240 @ 90
Tube Type Radial Ply					
9.00R20 F (12)	7.5 CS	9,940 @ 95	18,080 @ 90	9,940 @ 95	36,160 @ 90
10.00R20 G (14)	7.5 CS 8.0 10H or CS	10,860 @ 90 12,080 @ 105	21,200 @ 95 21,200 @ 100	10,860 @ 90 12,080 @ 105	42,400 @ 95 42,400 @ 100
Tubeless Type Bias Ply					
11-22.5 F (12)	7.50 CS	10,860 @ 85	19,040 @ 75	10,860 @ 85	38,080 @ 75
	7.50 10H (ISO or BSN)	10,860 @ 85	19,040 @ 75	10,860 @ 85	38,080 @ 75
	8.25 CS	10,860 @ 85	19,040 @ 75	10,860 @ 85	38,080 @ 75
	8.25 10H or AL (ISO or BSN)	10,860 @ 85	19,040 @ 75	10,860 @ 85	38,080 @ 75
Tubeless Type Radial Ply					
9R22.5 F (12)	6.75 CS	8,100 @ 90	15,280 @ 90	—	—
	7.50 CS	9,000 @ 105	15,800 @ 95	—	—
	7.50 10H (ISO or BSN)	9,000 @ 105	15,800 @ 95	—	—
10R22.5 F (12)	7.50 CS	10,300 @ 100	18,080 @ 90	10,300 @ 100	36,160 @ 90
	7.50 10H (ISO or BSN)	10,300 @ 100	18,080 @ 90	10,300 @ 100	36,160 @ 90
10R22.5 G (14)	7.50 CS	11,340 @ 115	19,880 @ 105	11,340 @ 115	39,760 @ 105
	7.50 10H (ISO or BSN)	11,340 @ 115	19,880 @ 105	11,340 @ 115	39,760 @ 105
11R22.5 G (14)	7.50 CS	12,080 @ 105	21,200 @ 95	12,080 @ 105	42,400 @ 95
	7.50 10H (ISO or BSN)	12,080 @ 105	21,200 @ 95	12,080 @ 105	42,400 @ 95
	8.25 CS	12,080 @ 105	21,200 @ 95	12,080 @ 105	42,400 @ 95
	8.25 10H or AL (ISO or BSN)	12,080 @ 105	21,200 @ 95	12,080 @ 105	42,400 @ 95
11R22.5 H (16)	8.25 CS	13,220 @ 120	23,200 @ 110	13,220 @ 120	46,400 @ 110
	8.25 10H or AL (ISO or BSN)	13,220 @ 120	23,200 @ 110	13,220 @ 120	46,400 @ 110
12R22.5 H (16)	8.25 CS	14,400 @ 120	25,280 @ 110	14,400 @ 120	50,560 @ 110
	8.25 10H or AL (ISO or BSN)	14,400 @ 120	25,280 @ 110	14,400 @ 120	50,560 @ 110
	9.00 CS	14,400 @ 120	—	14,400 @ 120	—
	9.00 10H or AL (ISO or BSN)	14,400 @ 120	—	14,400 @ 120	—
255/80R22.5 G (14)	7.50 CS	10,410 @ 95	19,240 @ 95	10,410 @ 95	38,480 @ 95
	7.50 10H (ISO or BSN)	10,410 @ 95	19,240 @ 95	10,410 @ 95	38,480 @ 95
	8.25 CS	10,410 @ 95	19,240 @ 95	10,410 @ 95	38,480 @ 95
	8.25 10H or AL (ISO or BSN)	10,410 @ 95	19,240 @ 95	10,410 @ 95	38,480 @ 95

Figure 3-20 Tire inflation pressure specifications. (Courtesy of Sterling Truck Corporation.)

Automatic Tire Inflation System

Automatic tire inflation systems (ATIS) provide automatic tire inflation, and these systems supply a warning to the driver if an air leak is present in a tire or in the automatic inflation system. ATIS are still in the developmental stage, so they are not widely used at present.

In a trailer ATIS system an air supply line is connected to the air brake supply line on the trailer. A control box for the automatic tire inflation system is mounted on the trailer. This box contains an air supply valve, air regulator, unloader valve, flasher switch, and low-pressure switch (Figure 3-30).

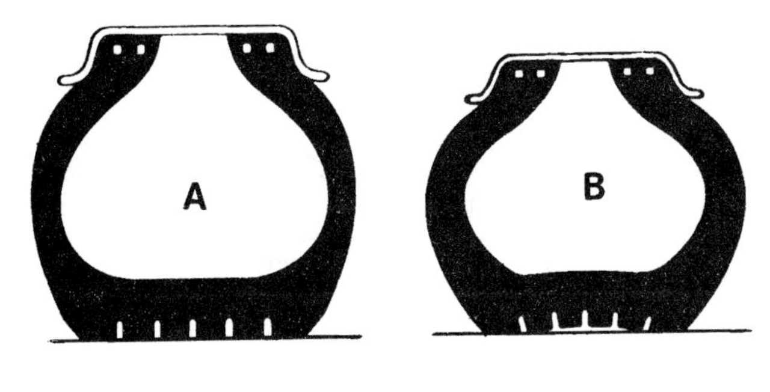

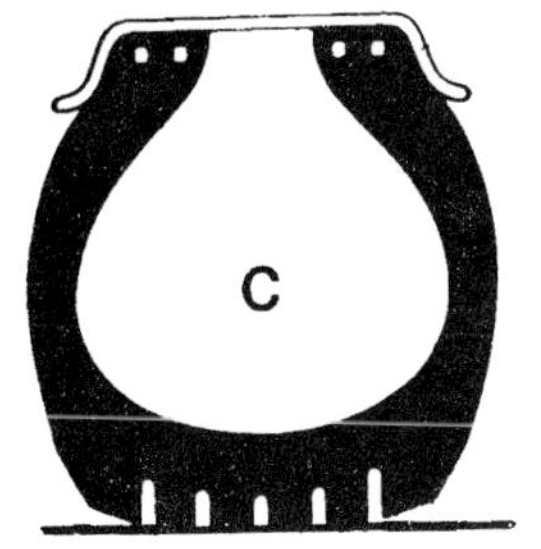

A. Proper Inflation
B. Underinflation
C. Overinflation

Figure 3-21 Results of proper tire inflation, underinflation, and overinflation. (Courtesy of General Motors Corporation, Service Technology Group.)

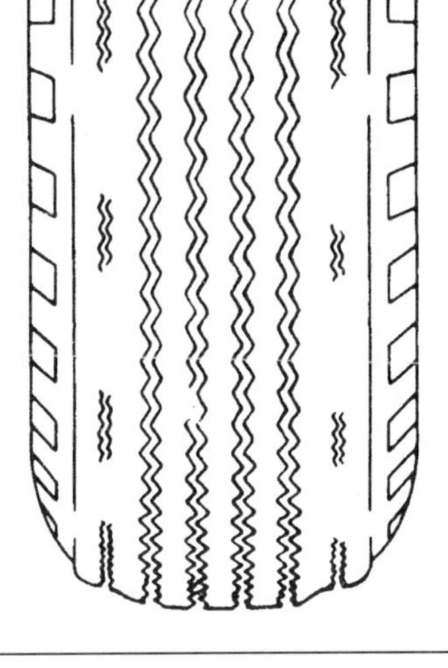

Figure 3-22 Underinflation causes wear on the edges of the tire tread. (Courtesy of General Motors Corporation, Service Technology Group.)

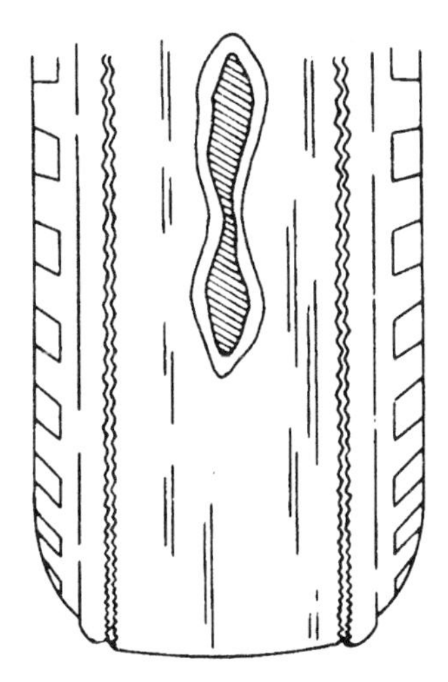

Figure 3-23 Overinflation causes wear on the center of the tire tread. (Courtesy of General Motors Corporation, Service Technology Group.)

Figure 3-24 Toe-in. (Courtesy of General Motors Corporation, Service Technology Group.)

Figure 3-25 Toe-out. (Courtesy of General Motors Corporation, Service Technology Group.)

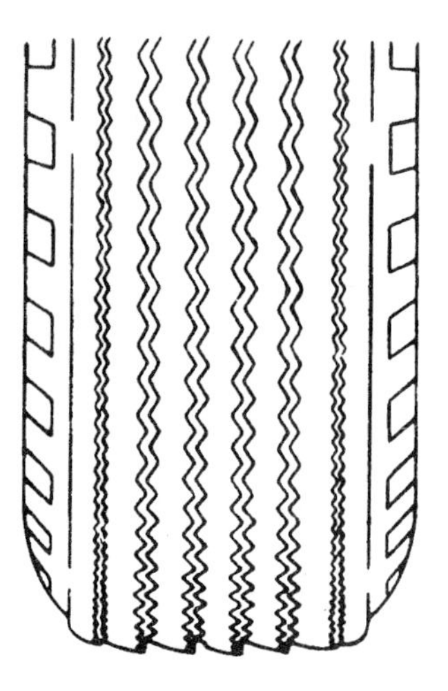

Figure 3-26 Feathered tire tread wear caused by improper toe adjustment. (Courtesy of General Motors Corporation, Service Technology Group.)

Figure 3-27 Excessive positive camber on the front wheels. (Courtesy of General Motors Corporation, Service Technology Group.)

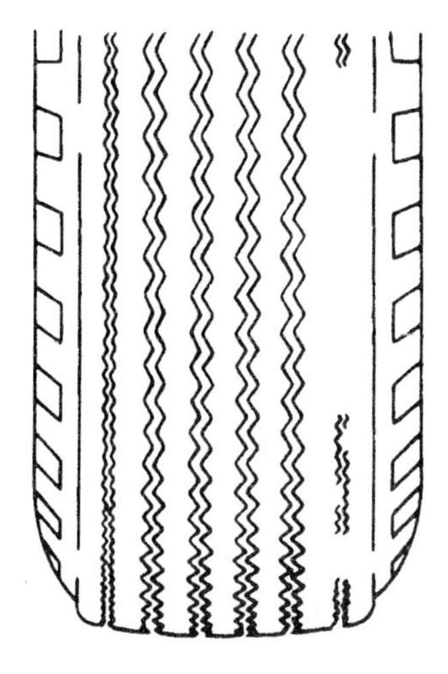

Figure 3-28 Wear on one edge of the tire tread caused by improper camber. (Courtesy of General Motors Corporation, Service Technology Group.)

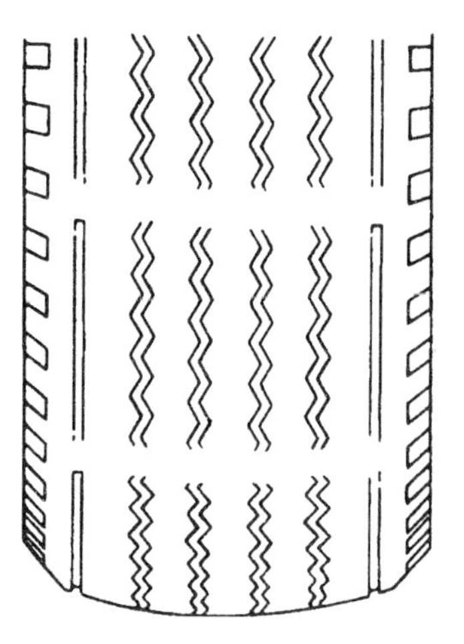

Figure 3-29 When wear bars appear across the tire tread, the tire should be replaced. (Courtesy of General Motors Corporation, Service Technology Group.)

Air pressure is supplied from the air brake supply line to the control box. Air pressure is supplied from the control box to a T fitting mounted on the chassis near the trailer axles. All tubing in the ATIS system is nylon tubing that meets SAE standard J844 for nonmetallic air brake system tubing. Coiled nylon tubing is connected from each side of the T fitting on the chassis to fittings in the rear axle housing. The coiled nylon tubing allows for suspension movement. Air pressure is supplied through the nylon tubing and fittings into the rear axle housing.

A special spindle nut replaces the conventional wheel bearing lock nut. This special spindle nut contains a rotary union. Air pressure is supplied from the rear axle housing through the rotary union to a T fitting on the outboard side of the special spindle nut. A seal prevents air leaks between the special spindle nut and the spindle, and another seal prevents air leaks between the rotary union and the special spindle nut (Figure 3-30). The rotary union can rotate inside the special spindle nut.

A hub cap extension is attached to the oil cap that is bolted to the wheel hub. The T fitting on the outboard side of the special spindle nut is connected to fittings in the hub cap extension.

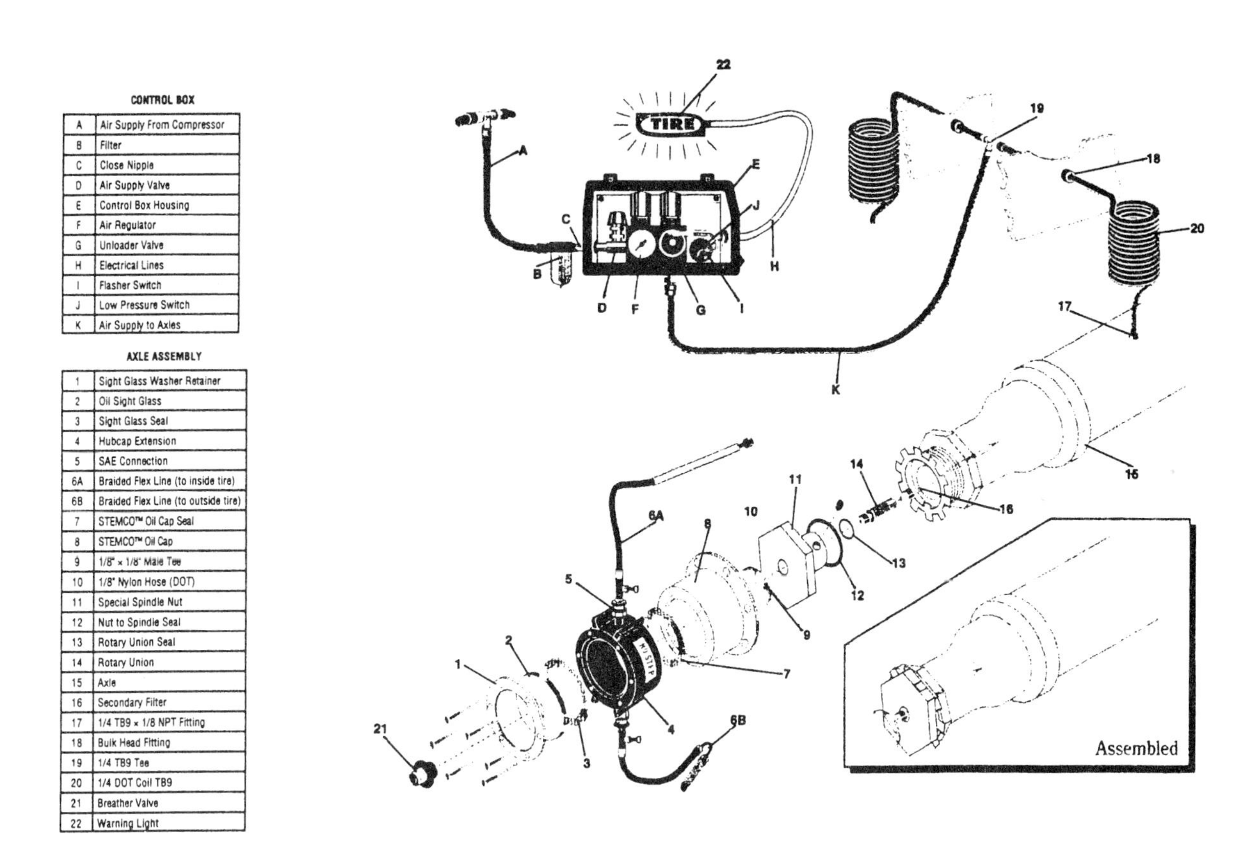

Figure 3-30 Automatic tire inflation system. (Reprinted with permission from SAE Paper 952591 ©1995 Society of Automotive Engineers, Inc.)

Air lines are connected externally from the hub cap extension to the valve stems on the inner and outer dual wheels. The hub cap extension, T fitting, and rotary union rotate with the wheel. Air pressure is supplied through the T fitting, fittings in the hub cap extension, and the air lines to each tire. The ATIS system maintains tire pressure at a predetermined pressure. If a leak occurs in a tire or in the system and air pressure drops to less than a preset value, the low-pressure switch activates the flasher, and the flasher flashes the tire warning light. This light is mounted on the trailer so it is visible to the driver in the tractor mirrors. When a tire develops a slow leak, the ATIS system will maintain tire pressure and allow the rig to be driven to a tire repair shop.

ATIS systems have also been designed for tractor drive wheels. In these applications the rotary unions are mounted externally on the center of the drive axles. The rotary unions support the nylon air lines on the outside of the tires.

Tire Rotation

WARNING: If tire sizes are not matched on tandem drive wheels, tire slippage and premature wear may occur.

When the front wheel tire tread depth is $\frac{4}{32}$ in (3.17 mm), the front wheels and tires should be moved to the inside rear wheels (Figure 3-31). When rotating tires to the rear drive wheels, each pair of tires on the dual wheels must be matched so they are the same size. The tire sizes on dual drive wheels must be matched to prevent excessive tire tread wear from slippage. The tire sizes on dual drive wheels may be matched by measuring them with a tape measure, string gauge, square, tire caliper, or straight edge.

The tire rotation procedure is the same on a single axle or tandem axle truck. When the front tire tread depth decreases to $\frac{4}{32}$ in (3.17 mm), the front wheels are moved to the inside dual wheels on the forward drive axle.

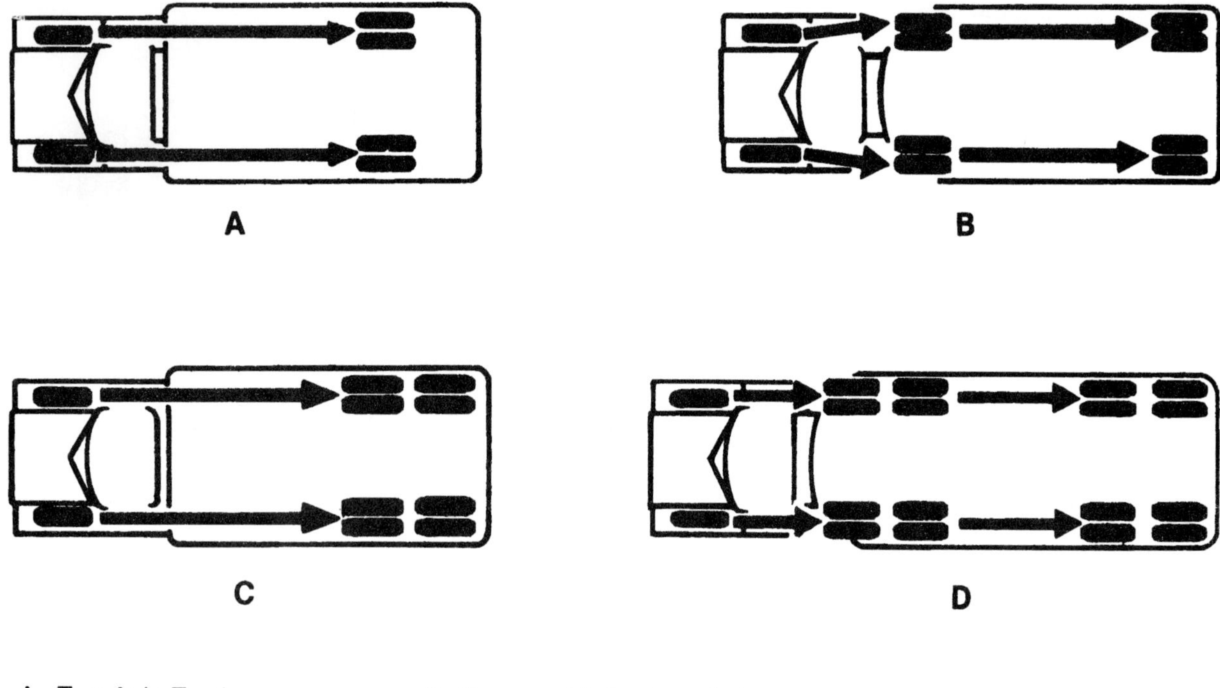

Figure 3-31 Tire rotation on trucks and tractor trailer combinations. (Courtesy of General Motors Corporation, Service Technology Group.)

On a tractor and trailer with single axles the front tires are moved to the inside or outside on the drive wheels. The tires removed from the drive wheels may be moved to the trailer axle. Trailer tires have a greater possibility of tire damage from cuts and bruises. Therefore this rotation plan with the tires ending their service on the trailer wheels provides maximum tire life.

On a tractor with a tandem drive axle and a tandem axle trailer the front tires may be moved to the forward drive axle, and tires removed from the tractor drive axle may be moved to one of the trailer axles. On tandem tractor drive axles the sum of the circumferences on the four tires on the forward axle must equal the sum of the circumferences on the four tires on the rearmost drive axle. If a third differential exists between these two drive axles, the tire circumferences do not require matching. Because trailer tandem axles have free-rolling wheels with no drive function, the tire circumferences do not need to be matched on these axles.

Static Wheel Balance Theory

When a wheel and tire have proper **static balance,** the weight is equally distributed around the axis of rotation, and gravity will not force it to rotate from its rest position. If a vehicle is raised off the floor and a wheel is rotated in 120-degree intervals, a statically balanced wheel will remain stationary at each interval. When wheel and tire are statically unbalanced, the tire has a heavy portion at one location. The force of gravity acting on this heavy portion will cause the wheel to rotate when the heavy portion is located near the top of the tire (Figure 3-32).

Results of Static Imbalance

Centrifugal force may be defined as the force that tends to move a rotating mass away from its axis of rotation. As explained previously, a tire and wheel are subjected to very strong acceleration and

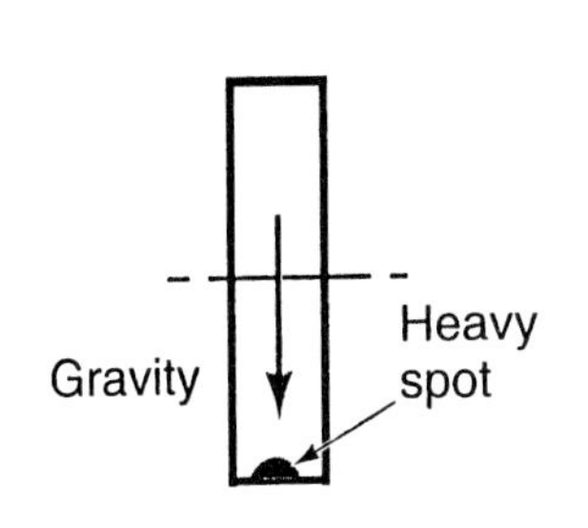

Figure 3-32 Static wheel imbalance. (Courtesy of Chevrolet Motor Division, General Motors Corporation.)

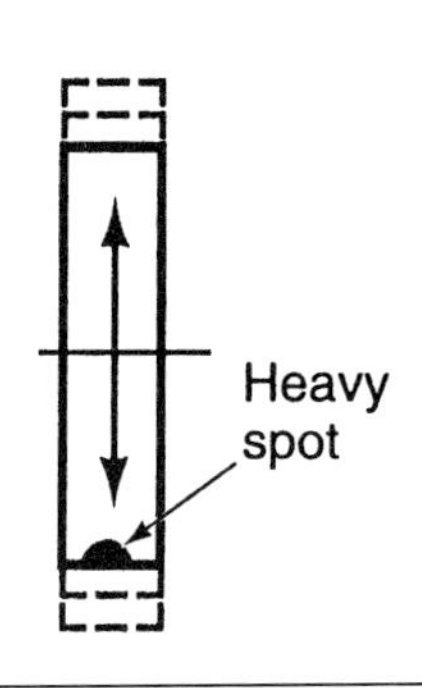

Figure 3-33 Effects of static wheel imbalance. (Courtesy of Chevrolet Motor Division, General Motors Corporation.)

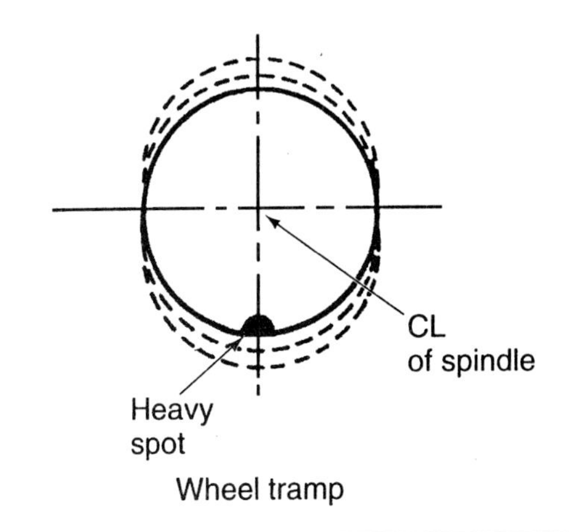

Figure 3-34 Wheel tramp. (Courtesy of Chevrolet Motor Division, General Motors Corporation.)

deceleration forces when a vehicle is in motion. The heavy portion of a statically unbalanced wheel is influenced by centrifugal force. This influence attempts to move the heavy spot on a tangent line away from the wheel axis. This action attempts to lift the wheel assembly off the road surface (Figure 3-33).

The wheel-lifting action caused by static imbalance may be referred to as wheel tramp (Figure 3-34). When the wheel and tire move downward as the heavy spot decelerates, the tire strikes the road surface with a pounding action. This repeated pounding action causes cupping (Figure 3-35).

The vertical wheel motion from static imbalance is transferred to the suspension system and then absorbed by the chassis and body. This action causes rapid wear on suspension and steering components. The wheel tramp action resulting from static imbalance is also transmitted to the passenger compartment, which causes driver discomfort and fatigue.

Dynamic Wheel Balance Theory

When a wheel and tire assembly have correct **dynamic balance,** the weight of the assembly is distributed equally on both sides of the wheel center viewed from the front. Dynamic wheel balance may be explained by dividing the tire into four sections (Figure 3-36).

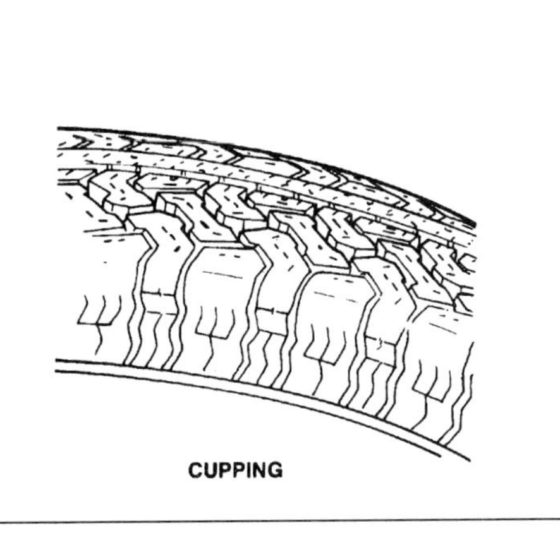

Figure 3-35 Tire cupping may be caused by static imbalance. (Courtesy of Sterling Truck Corporation.)

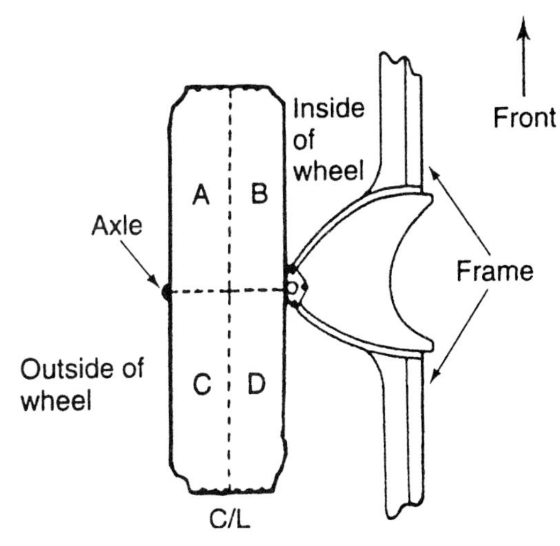

Figure 3-36 Dynamic wheel balance theory.

In Figure 3-36, if sections A and C have the same weight and sections B and D also have the same weight, the tire has proper dynamic balance. If a tire has dynamic imbalance, section D may have a heavy spot, and thus sections D and B have different weights (Figure 3-37).

From our discussion of dynamic balance we can understand that a tire and wheel assembly may be in static balance but have dynamic imbalance. Therefore wheels must be in balance statically and dynamically.

Results of Dynamic Wheel Imbalance

When a dynamically unbalanced wheel is rotating, centrifugal force moves the heavy spot toward the tire center line. The center line of the heavy spot arc is at a 90-degree angle to the spindle. This action turns the true center line of the left front wheel inward when the heavy spot is at the rear of the wheel (Figure 3-38).

When the wheel rotates until the heavy spot is at the front of the wheel, the heavy spot movement turns the left front wheel outward (Figure 3-39).

From the illustrations in Figures 3-38 and 3-39 we can understand that dynamic wheel imbalance causes lateral wheel shake or shimmy (Figure 3-40). This action causes steering wheel oscillations at medium and high speeds, with resultant driver fatigue and passenger discomfort. Wheel shimmy and steering wheel oscillations also cause unstable directional control of the vehicle.

A wheel with dynamic imbalance is forced to pivot on the tire area in contact with the road surface, which results in excessive tire scuffing and wear. Dynamic wheel imbalance causes premature wear on steering linkage and suspension components. Therefore dynamic wheel balance is important to provide normal tire life, reduce steering and suspension component wear, increase directional control, and decrease driver fatigue. The main purposes of proper wheel balance may be summarized as follows:

1. Maintains normal tire tread life
2. Provides extended life of suspension and steering components
3. Helps to provide directional control of the vehicle
4. Reduces driver fatigue
5. Increases passenger comfort
6. Helps to maintain the life of body and chassis components

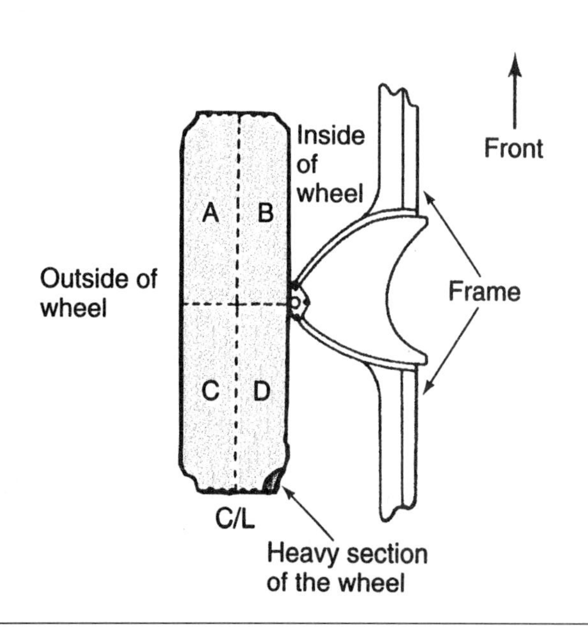

Figure 3-37 Dynamic imbalance.

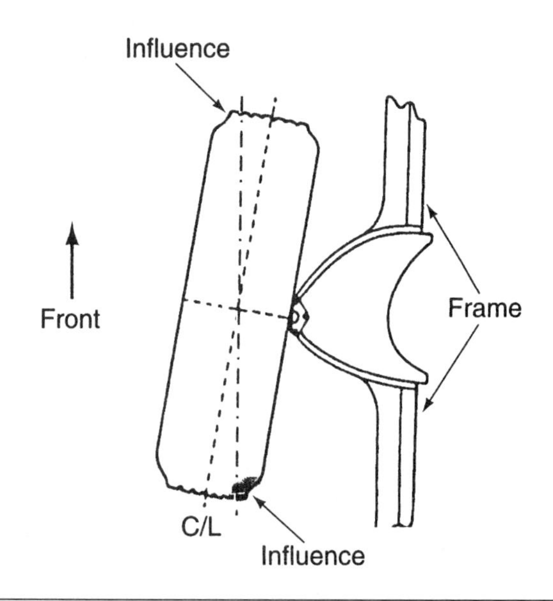

Figure 3-38 Dynamic imbalance with heavy spot at the rear of the left front wheel.

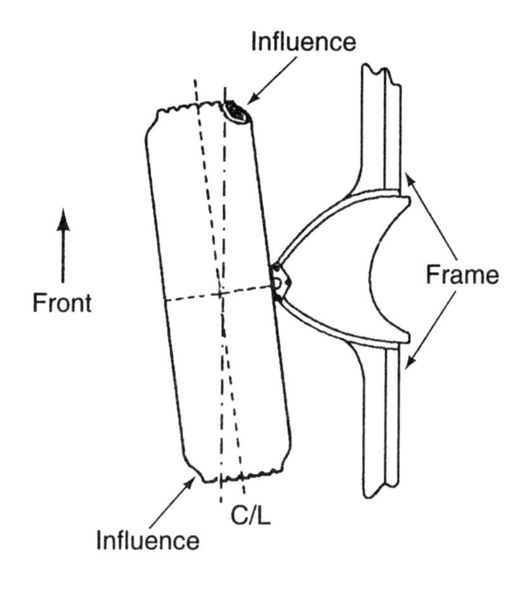

Figure 3-39 Dynamic wheel imbalance with heavy spot at the front of the left front wheel.

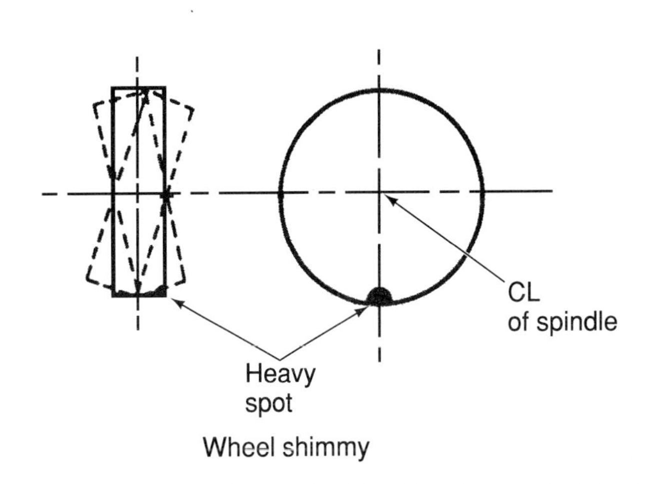

Figure 3-40 Wheel shimmy caused by dynamic wheel imbalance. (Courtesy of Chevrolet Motor Division, General Motors Corporation.)

Bearing Loads

TRADE JARGON: A thrust-bearing load may be referred to as an axial load.

A bearing is used to support and guide such components as rotating shafts, oscillating shafts, sliding shafts, pivots, and wheels. While a bearing is supporting and guiding one of these components, the bearing is designed to reduce friction and support the load applied by the component. Because the bearing reduces friction, it also decreases the power required to rotate or move the component. Bearings are precision machined assemblies that provide smooth operation and long life. When bearings are properly installed and maintained, bearing failure is rare.

When a bearing load is applied in a vertical direction, it is called a **radial load.** If the vehicle weight is applied straight downward on a bearing, this weight is a radial load on the bearing. A **thrust-bearing load** is applied in a horizontal direction (Figure 3-41). For example, while a vehicle is turning a corner, horizontal force is applied to the front wheel bearings. When **angular load**

A radial-bearing load is applied in a vertical direction.

A thrust-bearing load is applied in a horizontal direction.

An angular-bearing load is applied at an angle between the vertical and the horizontal.

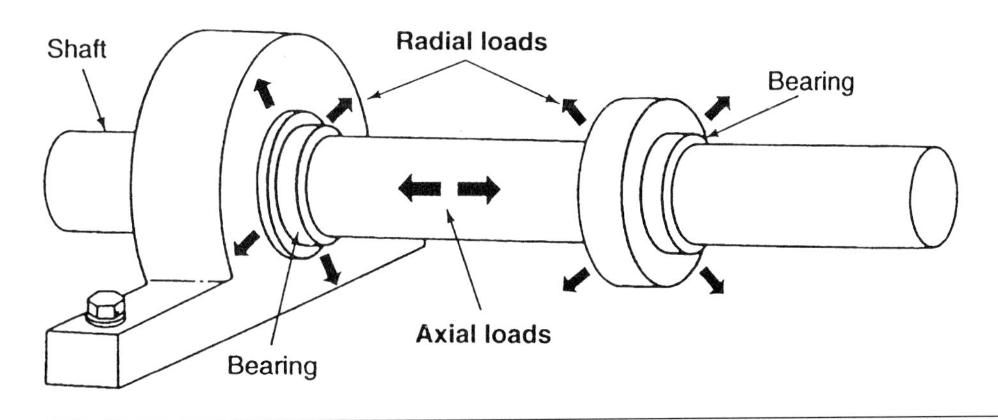

Figure 3-41 Types of bearing loads. (Courtesy of Deere and Company.)

is applied to a bearing, the angle of the applied load is somewhere between the horizontal and vertical positions.

Truck wheel bearings are mounted between the wheel hubs and the spindles or axle housing. The weight of the truck or trailer and cargo is transmitted from the spindles or axle housing through the wheel bearings to the hubs, wheels, and tires. Wheel bearings are subjected to very heavy vertical or radial loads on a loaded truck or tractor and trailer. While cornering, wheel bearings must also withstand very strong axial loads. Wheel bearings must also provide a very smooth rotational surface for the hubs, wheels, and tires. Wheel bearings must support and withstand these loads without overheating for many miles of operation.

Cylindrical Ball Bearings

Front and rear wheel bearings may be **cylindrical ball bearings** or roller bearings. Either type of bearing contains the following basic parts:

1. Inner race or cone
2. Separator, also called a cage or retainer
3. Rolling elements, balls or rollers
4. Outer race or cup

The **inner race** is an accurately machined component, and the inner surface of the race is mounted on the shaft with a precision fit. The **rolling elements** are mounted on a very smooth machined surface on the inner race. Positioned between the inner and outer races, the **separator** retains the rolling elements and keeps them evenly spaced. The rolling elements have precision machined surfaces, and these elements are mounted between the inner and outer races. The **outer race** is the bearing's exterior ring, and both sides of this component have precision machined surfaces. The outer surface of this race supports the bearing in the housing, and the inner surface is in contact with the rolling elements.

A **single-row ball bearing** has a crescent-shaped machined surface in the inner and outer races in which the balls are mounted (Figure 3-42). When a ball bearing is at rest, the load is distributed equally through the balls and races in the contact area. When one of the races and the balls begin to rotate, the bearing load causes the metal in the race to bulge out in front of the ball and flatten out behind the ball (Figure 3-43). This action creates a certain amount of friction within the bearing, and the same action is repeated for each ball while the bearing is rotating. If metal-to-metal contact is allowed between the balls and races, these components would experience very fast wear. Therefore bearing lubrication is extremely important to eliminate metal-to-metal contact in the bearing and reduce wear.

A cylindrical ball bearing is designed primarily to handle radial loads. However, this type of bearing can also withstand a considerable amount of thrust load in either direction even at high speeds. A maximum capacity ball bearing has extra balls for greater radial load-carrying capacity. Ball bearings are available in many different sizes for various applications.

Double-row ball bearings contain two rows of balls side by side. As in the single-row ball bearing, the balls in the double-row bearing are mounted in crescent-shaped grooves in the inner and outer races. The double-row ball bearing can support heavy radial loads, and this type of bearing can also withstand thrust loads in either direction.

Ball Bearing Seals, Shields, and Snap Rings

On some applications a ball bearing is held in place with a **snap ring.** A groove is cut around the outside surface of the outer race, and the snap ring is mounted in this groove. The snap ring may fit against a machined housing surface, or in other applications the outer circumference of the snap ring is mounted in a groove in the housing. Ball bearings retained with a snap ring are not used on wheels because they are not designed to withstand high-thrust loads encountered by wheel bearings. Thrust loads would damage the snap ring and groove.

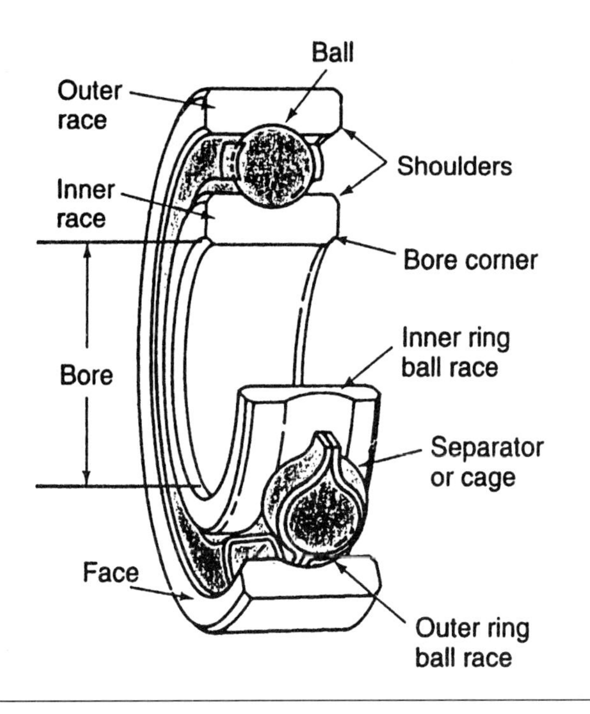

Figure 3-42 Parts of a cylindrical ball bearing. (Courtesy of Deere and Company.)

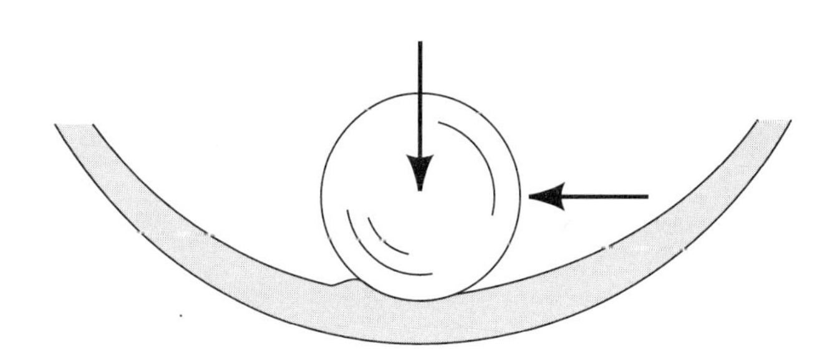

Figure 3-43 When a load is applied to a ball bearing, the metal in the race bulges out in front of the ball and flattens out behind the ball.

Bearing shields cover the space between the two bearing races on one or both sides of the bearing. These shields are usually attached to the outer race, but space is left between the shield and the inner race. Bearing shields prevent dirt from entering the bearing, but excess lubrication can still flow through the bearing.

A **bearing seal** is a circular metal ring with a sealing lip on the inner edge. The seals are usually attached to the outer bearing race on each side of the bearing, and the lip surface contacts the inner race. The seal lip may have single, double, or triple lips made of rubber, synthetic or nonsynthetic, or elastomers. Lubricant is retained in the bearing by the seal, and the seal also keeps moisture, dirt, and contaminants out of the bearing. The drive shaft center support bearing is an example of a bearing with seals on each side (Figure 3-44).

Roller Bearings

Cylindrical Roller Bearing

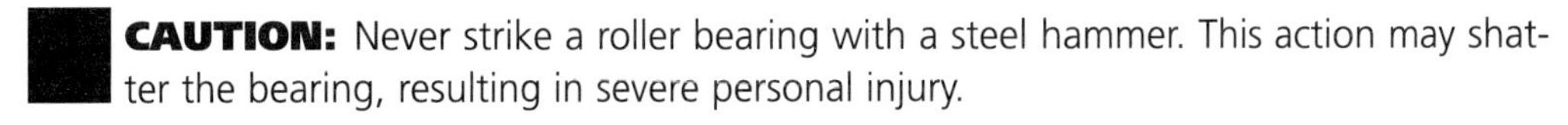

CAUTION: Never strike a roller bearing with a steel hammer. This action may shatter the bearing, resulting in severe personal injury.

CAUTION: Spinning a roller bearing with compressed air may rotate the bearing at extremely high speed and cause the bearing to disintegrate, resulting in serious personal injury.

A **cylindrical roller bearing** contains precision machined rollers that have the same diameter at both ends. These rollers are mounted in square-cut grooves in the outer and inner races (Figure 3-45). In the cylindrical roller bearing the races and rollers run parallel to one another. Cylindrical roller bearings are designed primarily to carry radial loads, but they can withstand some thrust load. Because cylindrical roller bearings do not withstand high thrust loads, they are not used in truck wheel bearings.

A cylindrical roller bearing may be called a nontapered roller bearing.

Figure 3-44 The drive shaft center support bearing has seals on both sides. (Courtesy of CR Services.)

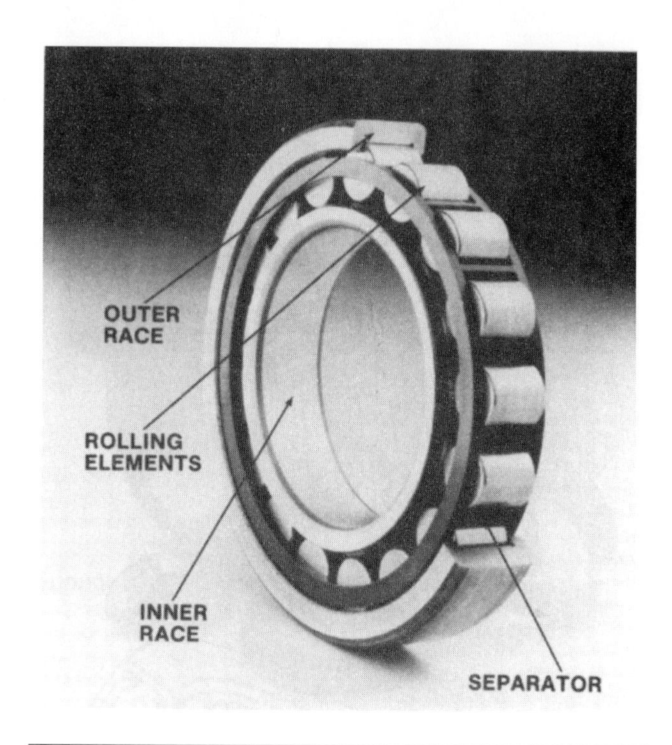

Figure 3-45 Parts of a cylindrical roller bearing. (Courtesy of SKF USA Inc.)

Tapered Roller Bearing

In a **tapered roller bearing** the inner and outer races are cone shaped. If imaginary lines extend through the inner and outer races, these lines taper and eventually meet at a point extended through the center of the bearing (Figure 3-46). The most important advantage of the tapered roller bearing compared with other bearings is an excellent capability to carry radial, thrust, and angular loads. Because tapered roller bearings withstand all these loads, they are commonly used in truck wheel bearings. In the tapered roller bearing the rollers are mounted on cone-shaped precision surfaces in the outer and inner races. The bearing separator has an open space over each roller (Figure 3-47). Grooves cut in the side of the separator roller openings match the curvature of the roller. This design allows the rollers to rotate evenly without interference between the rollers and the separator. Lubrication and proper end-play adjustment are critical on tapered roller bearings.

Needle Roller Bearings

A **needle roller bearing** contains many small-diameter steel rollers in a thin outer race. This type of bearing is very compact, and it may be used on sector shafts in steering gears where mounting space is limited. Most needle roller bearings do not have a separator, but the steel rollers push against each other and maintain the roller position. Rather than having an inner race, a machined

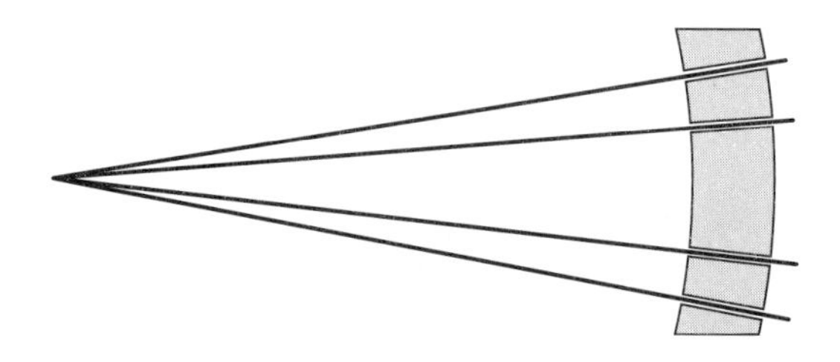

Figure 3-46 Imaginary lines extending from tapered roller bearing races eventually meet at a point extending from the bearing center.

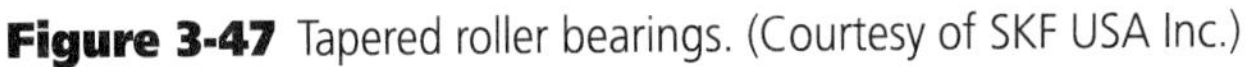

Figure 3-47 Tapered roller bearings. (Courtesy of SKF USA Inc.)

Figure 3-48 Needle roller bearing. (Courtesy of SKF USA Inc.)

surface on the mounting shaft contacts the inner surface of the rollers (Figure 3-48). The needle roller bearing is designed to carry radial loads, but it does not withstand thrust loads. Tapered needle bearings are used in some applications.

See Chapter 3 in the Shop Manual for seal service.

Seals

Seals are designed to keep lubricant in the bearing and prevent dirt particles and contaminants from entering the bearing. Wheel bearing seals are mounted in front and rear wheel hubs. The metal seal case has a surface coating that resists corrosion and rust and acts as a bonding agent for the seal material. Seals have many different designs, including single lip, double lip, and fluted. The seal material is usually made of a synthetic rubber compound such as nitrile, silicon, polyacrylate, or a fluoroelastomer such as Viton. The actual seal material depends on the lubricant and contaminants that the seal encounters. All seals may be divided into two groups, springless and spring loaded. **Springless seals** are used in some front wheel hubs, where they seal a heavy lubricant into the hub (Figure 3-49).

In a **spring-loaded seal** the **garter spring** behind the seal provides additional force on the seal lip to compensate for lip wear, shaft movement, and bore eccentricity (Figure 3-50). A **fluted lip seal** has angled serrations on the seal that direct oil back into a housing. This seal design provides a pumping action to redirect the oil back into the housing.

Some seals have a sealer painted on the outside surface of the metal seal housing. When the seal is installed, this sealer prevents leaks between the seal case and the housing (Figure 3-51).

Front Wheels, Hubs and Bearings

Front Hubs and Wheel Bearings

Many medium-duty and heavy-duty trucks have two tapered roller bearings mounted in the front wheel hubs. The outer bearing cups are pressed into the hub, and the inner bearing races are

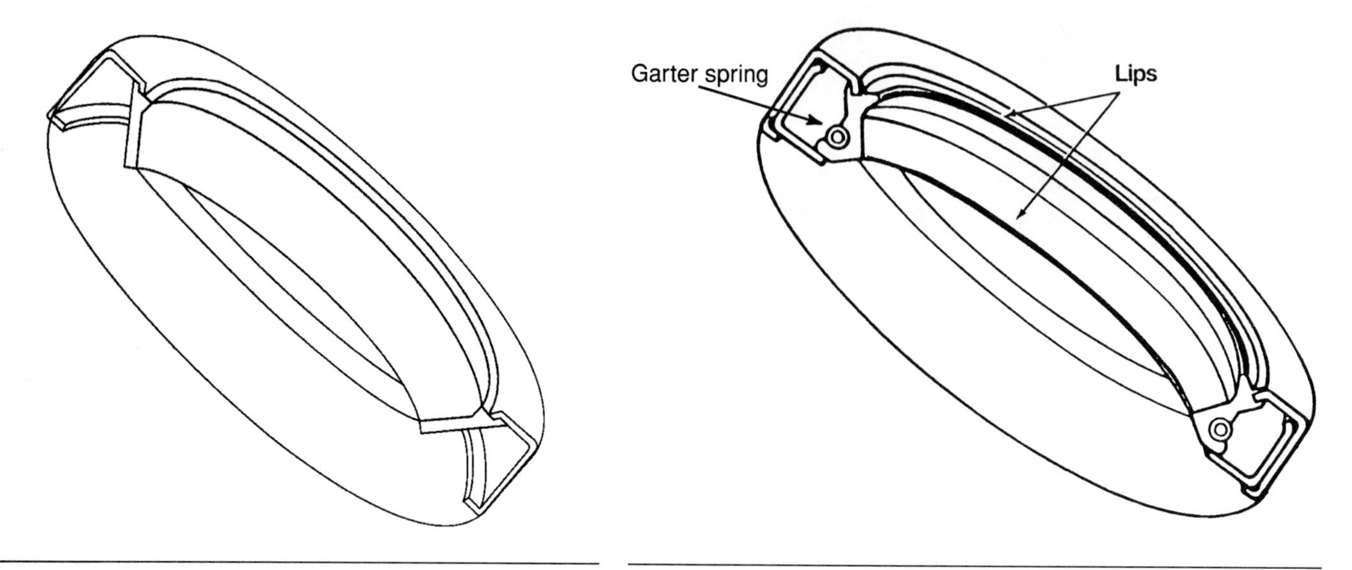

Figure 3-49 Springless seal. (Courtesy of Chrysler Corporation.)

Figure 3-50 Spring-loaded seal. (Courtesy of Chrysler Corporation.)

mounted on the spindle. An inner seal prevents lubricant leaks between the hub and spindle. A cap and gasket prevent leaks at the outer end of the hub. On heavy-duty trucks the hub and bearings are retained on the spindle with a wheel bearing adjusting nut, lock washer, lock ring, and jam nut (Figure 3-52). On some medium-duty trucks the hub and bearings are retained on the spindle with a castellated nut, washer, and cotter key (Figure 3-53). Front wheel hub and bearing design is very similar on trucks with front wheel disk brakes.

Some front wheel bearings must be packed with the truck manufacturer's specified wheel bearing grease. Many truck front wheel bearings have a transparent oil level window in the bearing hub cap and a filler plug in the hub. With the filler plug positioned at the top, this plug is removed and the specified lubricant is installed in the hub until the lubricant level is at the proper level in the transparent window (Figure 3-54). Many truck manufacturers recommend **Society of Automotive Engineers (SAE)** 90 rear axle lubricant in the front hubs.

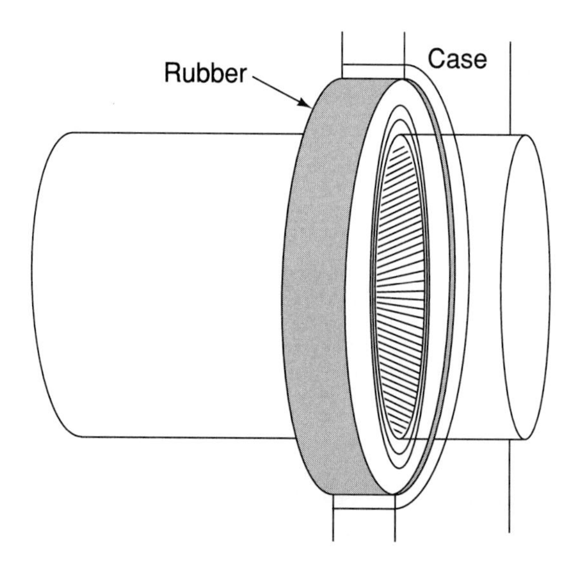

Figure 3-51 Sealer painted on seal case prevents leaks between the case and the housing. (Courtesy of Chrysler Corporation.)

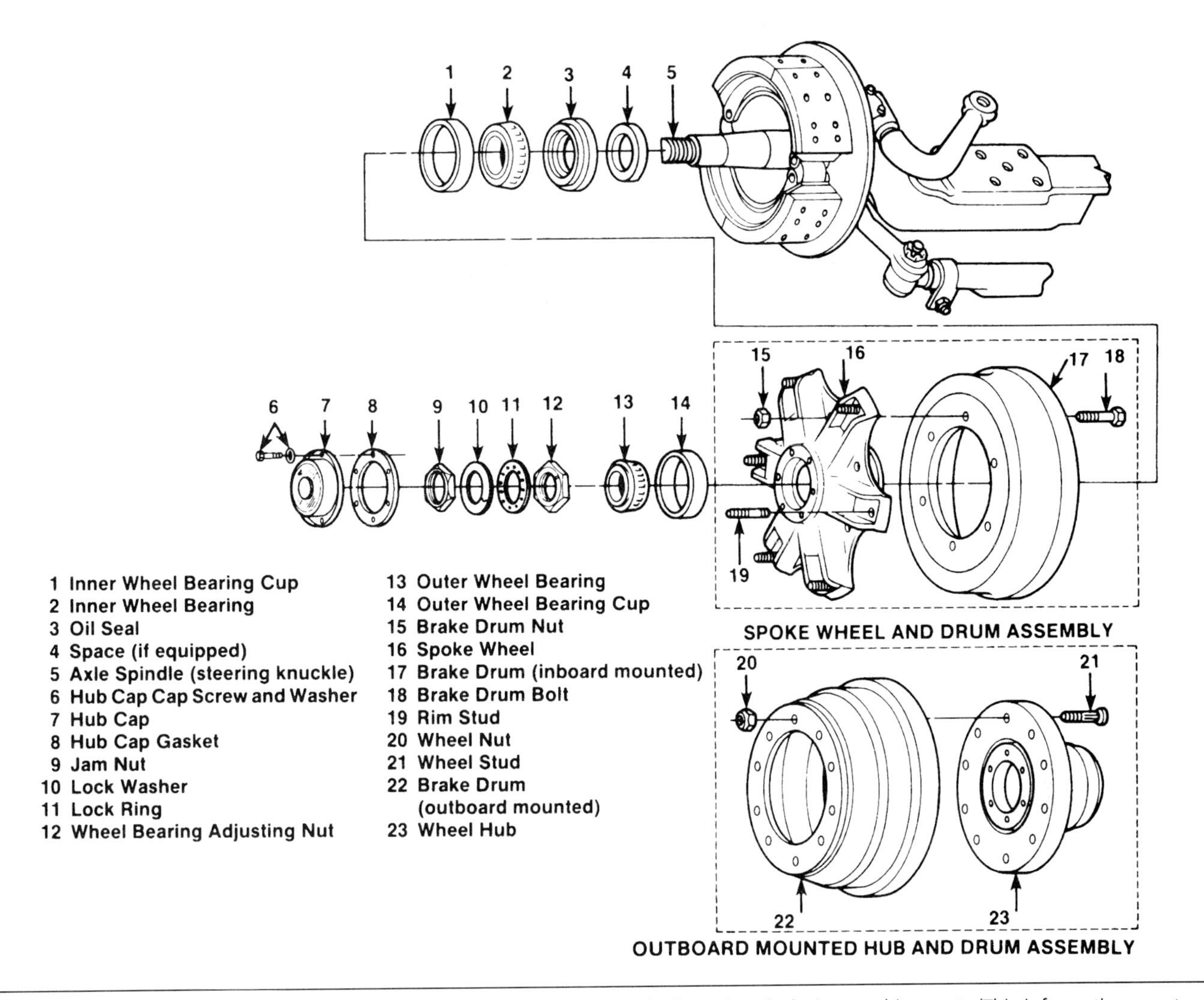

Figure 3-52 Front wheel bearings and hub retained with adjusting nut, lock washer, lock ring, and jam nut. (This information courtesy of Freightliner Corporation.)

Unitized Front Wheel Bearing Hubs

Some front axles are now available with **unitized hub assemblies** (Figure 3-55). These hub assemblies are a one-piece unit containing the front wheel bearings and seals. Each hub unit contains two tapered roller bearings. Because these hub assemblies are permanently lubricated and sealed, they do not require lubrication or adjustment. The bearings in these hub units are preloaded to eliminate endplay and reduce tire wear.

Rear Wheel Bearings

Many rear wheel hubs on medium-duty and heavy-duty trucks have two tapered roller bearings mounted in the hub. The outer bearing cups are pressed into the wheel hubs, and the inner bearing races are supported on the rear spindle (Figure 3-56). An inner seal prevents lubricant leaks between the hub and the spindle.

A row of studs is threaded into the outer edge of the rear hub. Holes in the rear axle flange fit over these studs, and a gasket is positioned between the axle flange and the hub. This gasket prevents lubricant leaks between the axle flange and the hub. Nuts are threaded onto the studs to retain the rear axle to the hub. The rear axle extends through the hub, and the splines on the inner end of the axle fit into the differential side gear. This type of rear axle may be called a full-floating axle.

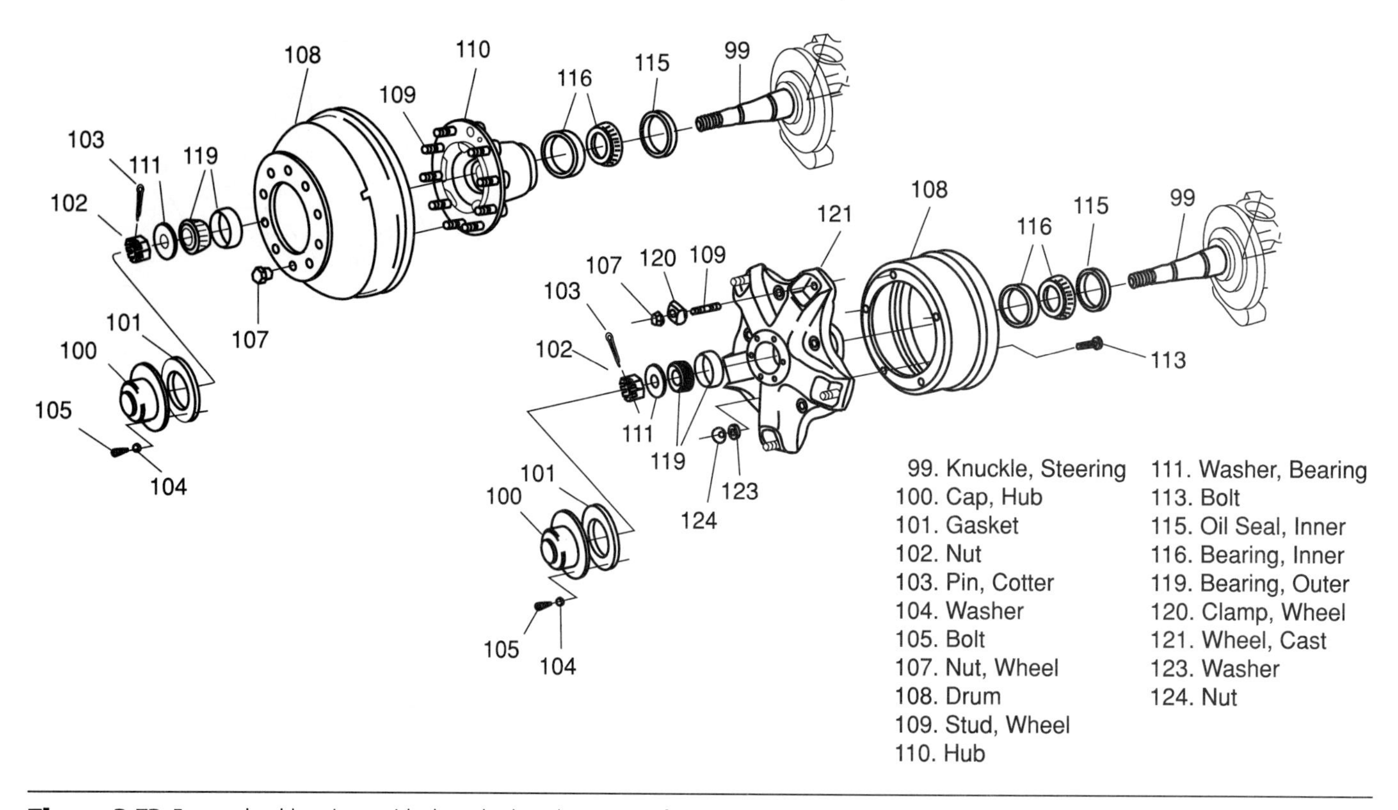

Figure 3-53 Front wheel bearings with drum brakes. (Courtesy of General Motors Corporation, Service Technology Group.)

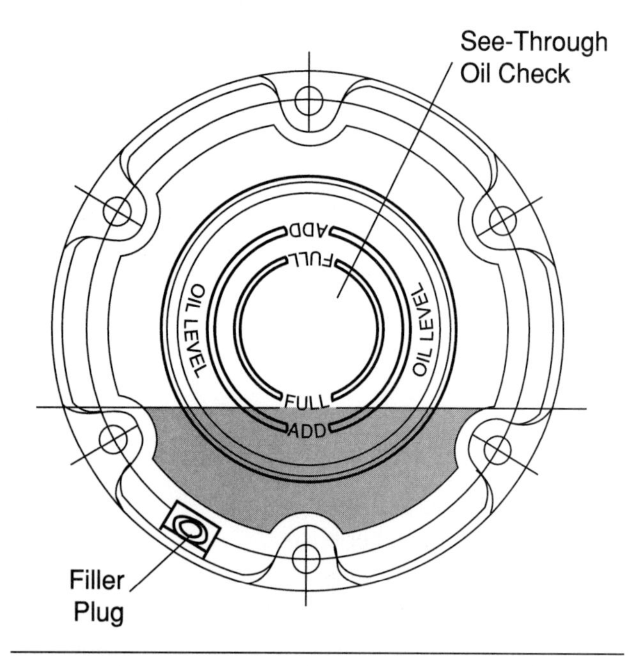

Figure 3-54 Transparent lubricant level window in a front hub. (Courtesy of General Motors Corporation, Service Technology Group.)

Figure 3-55 Unitized front wheel bearing hub assembly. (Reprinted with permission from SAE Publication Automotive Engineering November 96 ©1996 Society of Automotive Engineers, Inc.)

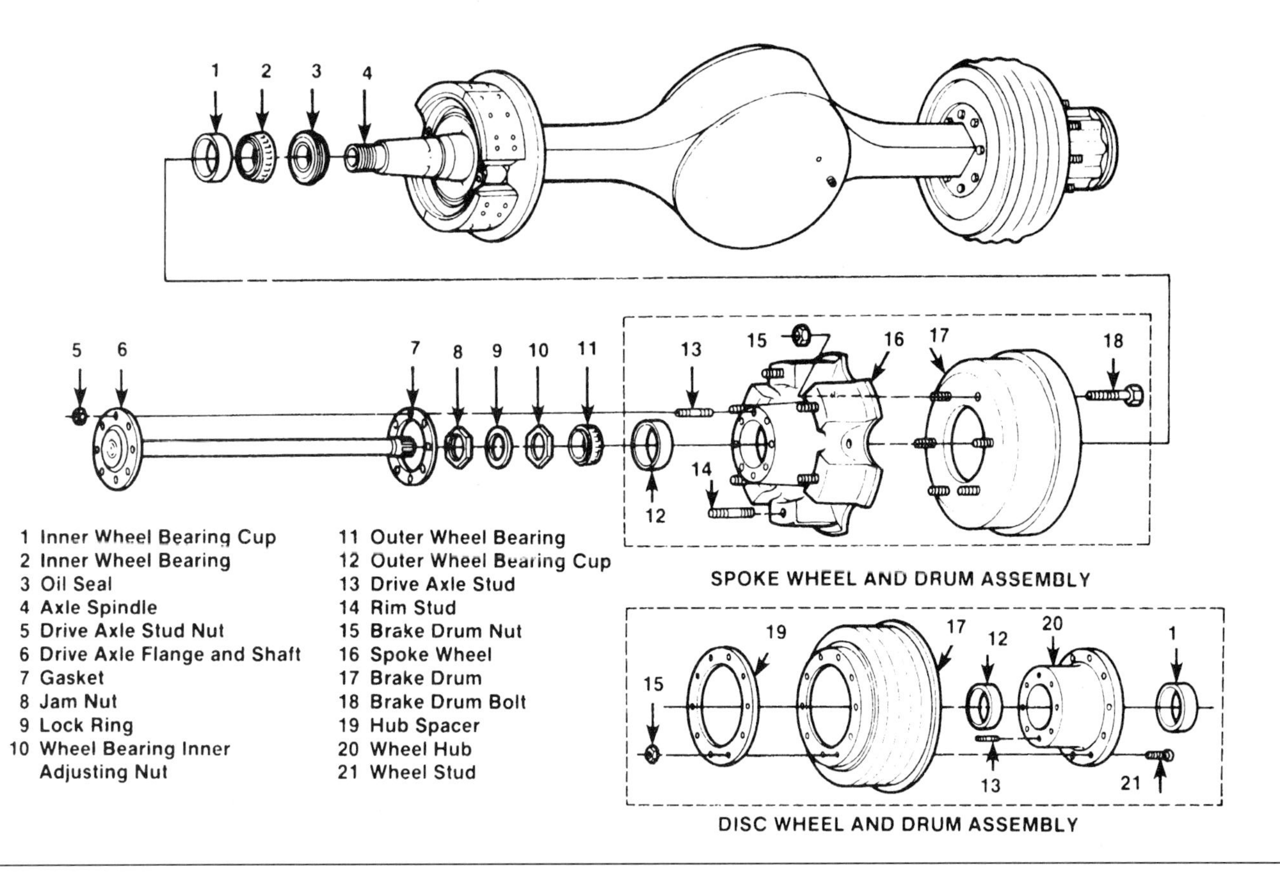

Figure 3-56 Rear wheel bearings, hub, and axle. (This information courtesy of Freightliner Corporation.)

Many hub and bearing assemblies are retained on the rear spindle with an adjusting nut, a lock washer, and a lock nut. (Figure 3-57). These lock washers may be dowel type or tang type (Figure 3-58).

Some rear wheel hubs are designed so they are lubricated from the differential sump (Figure 3-59). This type of rear wheel hub does not have an oil filler hole. Other rear wheel hubs have an oil filler hole and plug, and the specified lubricant must be installed in the rear wheel hub through the oil filler hole with the plug removed and the hole located at the top of the hub (Figure 3-60).

In a full-floating rear axle two tapered roller bearings in the rear wheel hub support the vehicle weight. The axle shaft is bolted to the outer side of the rear hub, and the axle shaft extends through the hub into the differential.

Hub Seals

Unitized Lip-Type Seal

Hub seals must keep lubricant in the hub and keep dirt and contaminants out of the hub to provide long bearing life. If a hub seal leak allows lubricant to escape from the hub, the wheel bearing will overheat very quickly because of the heavy loads on truck wheel bearings. When a hub seal does not provide a tight seal, dirt may enter the wheel bearings and greatly reduce bearing life. Proper wheel bearing adjustment is absolutely necessary to provide proper sealing action on the wheel bearing seals. Loose wheel bearings cause excessive hub and wheel lateral movement, which may cause improper contact between the hub seal lips and the spindle. This condition may cause a lubricant leak at the seal lips.

Some truck hubs have a **unitized one-piece seal.** The purpose of a hub seal is to keep contaminants out of the hub and keep lubricant in the hub. The outer shell of the seal is pressed into the hub, and the seal rotates with the hub. The seal contains a garter spring and a dirt lip (Figure 3-61). The seal lip contacts the sealing element surface on the spindle.

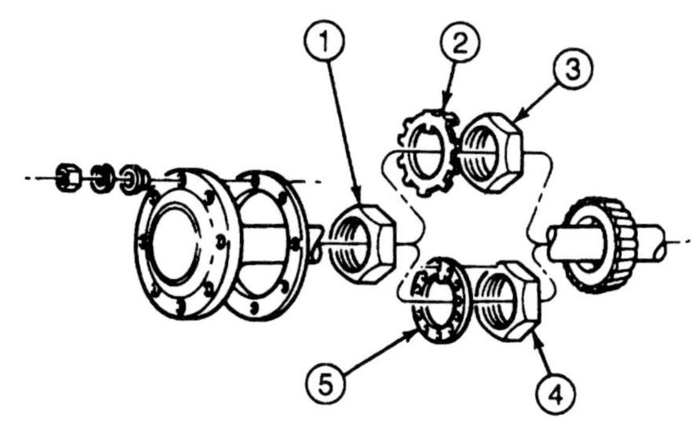

Item	Part Number	Description
1	4254	Locknut
2	1124	Lockwasher (Design for 21K, 23K, 40K and 46K Rear Axles)
3	4254	Adjusting Nut (for 13K, 15K, 17.5K, 19K and 34K Rear Axles)

Figure 3-57 Rear wheel bearing adjusting nut, lock washer, and lock nut. (Courtesy of Ford Motor Company.)

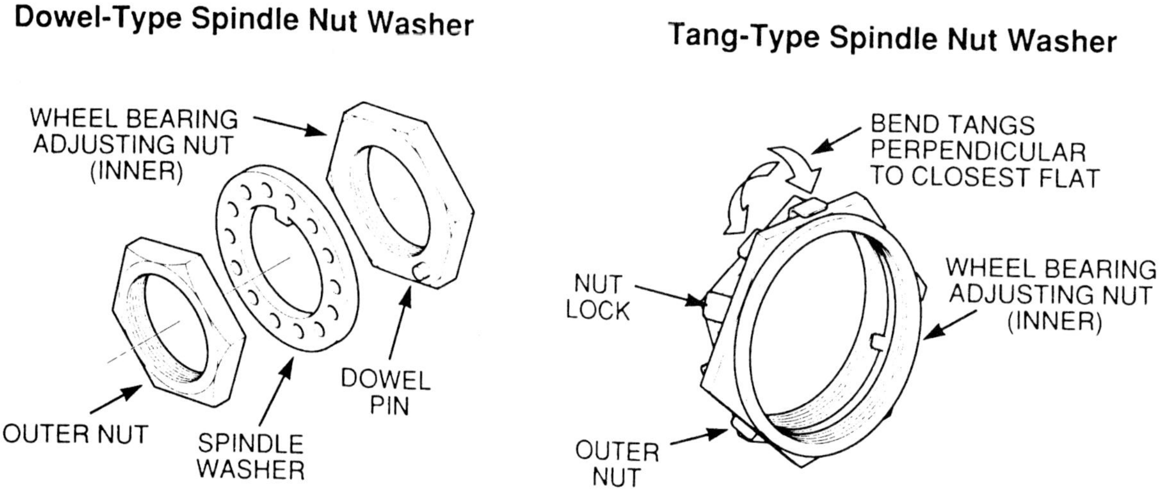

Dowel-Type Spindle Nut Washer
WHEEL BEARING ADJUSTING NUT (INNER)
OUTER NUT
SPINDLE WASHER
DOWEL PIN

Tang-Type Spindle Nut Washer
BEND TANGS PERPENDICULAR TO CLOSEST FLAT
NUT LOCK
WHEEL BEARING ADJUSTING NUT (INNER)
OUTER NUT

Figure 3-58 Rear wheel bearing lock washers may be dowel type or tang type. (Courtesy of Eaton Corporation.)

Lip-Type Seal, Wiper Rings, and Grit Guards

Some truck hubs have a lip-type seal with a wiper ring that is pressed on the axle or spindle (Figure 3-62). The **wiper ring** surface provides a very smooth contact surface for the seal lip. The wiper ring and seal are replaced as a set. If a spindle does not have a wiper ring, the seal lip may wear a groove in the spindle contact area that may require spindle replacement. Some wiper rings have a **grit guard** that is curled up over the seal to help prevent dirt from entering the seal lip area

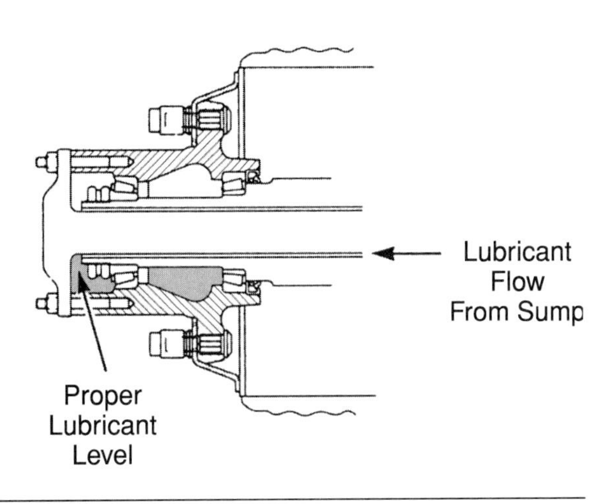

Figure 3-59 Some rear wheel bearings are lubricated from the differential sump. (Courtesy of Eaton Corporation.)

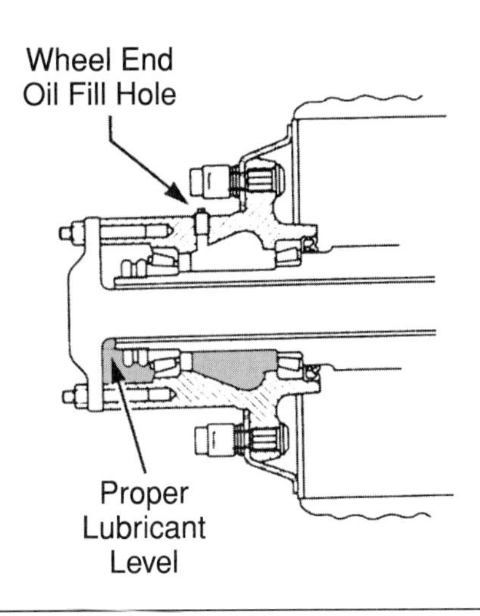

Figure 3-60 Some rear wheel hubs have an oil filler hole and plug. (Courtesy of Eaton Corporation.)

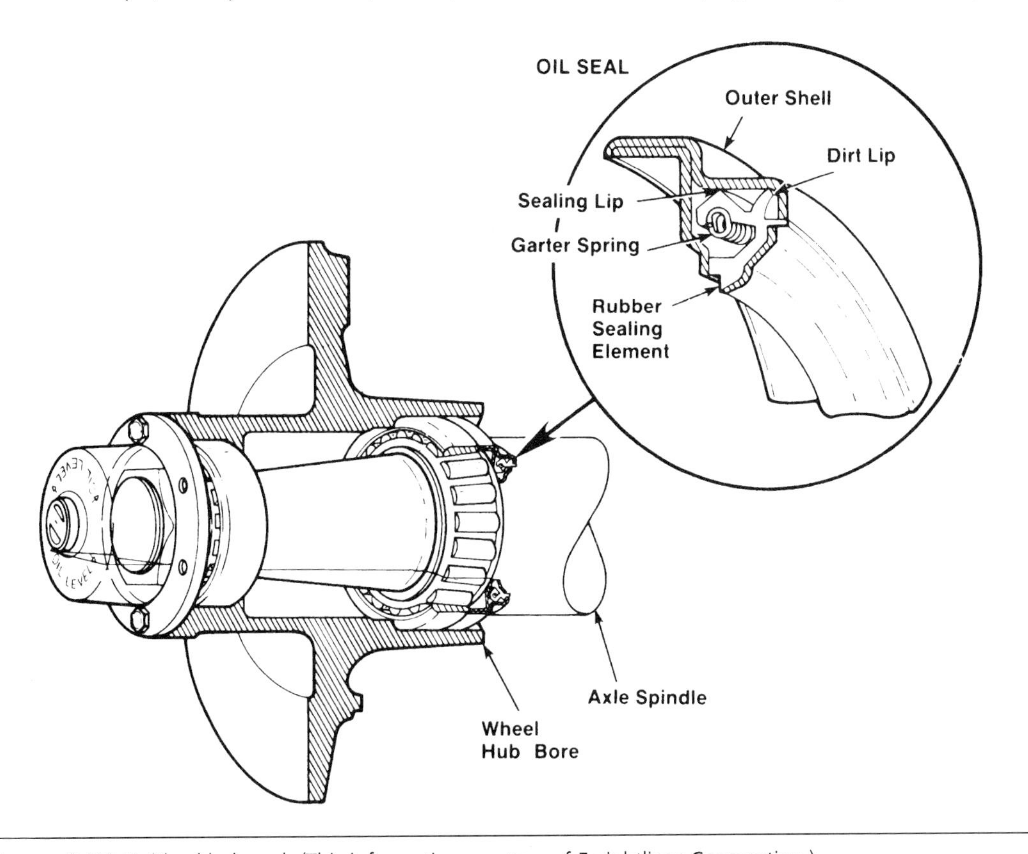

Figure 3-61 Unitized hub seal. (This information courtesy of Freightliner Corporation.)

(Figure 3-63). Because dirt in the seal lip area shortens seal life, wiper rings with a grit guard provide longer seal life.

A guardian-type seal is a variation of the unitized seal with a wiper ring and grit guard. In a guardian-type seal direct contact is made between the grit guard and the seal to provide improved protection against dirt entering the seal.

Barrier-Type Seal

A **barrier-type hub seal** may be used in some truck hubs. This type of seal has sharp, pointed ribs on the outer and inner mounting surfaces. These mounting surfaces are lubricated on a new

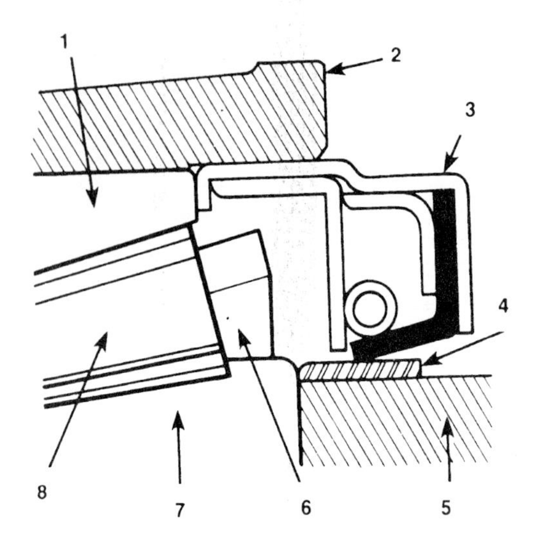

1 Bearing Cup
2 Hub
3 Wheel Seal
4 Wiper Ring with Grit Guard
5 Axle
6 Bearing Cage
7 Bearing Cone
8 Bearing Rollers

Figure 3-62 Unitized hub seal with wiper ring. (Courtesy of Navistar International Transportation Corp.)

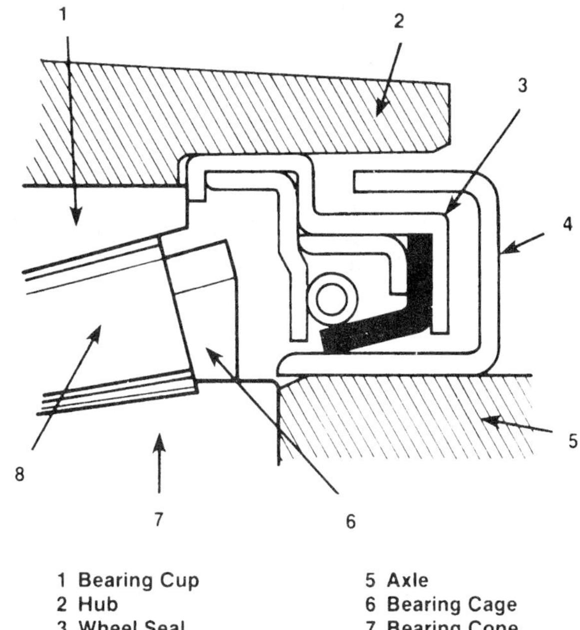

1 Bearing Cup
2 Hub
3 Wheel Seal
4 Wiper Ring
5 Axle
6 Bearing Cage
7 Bearing Cone
8 Bearing Rollers

Figure 3-63 Unitized hub seal with wiper ring and grit guard. (Courtesy of Navistar International Transportation Corp.)

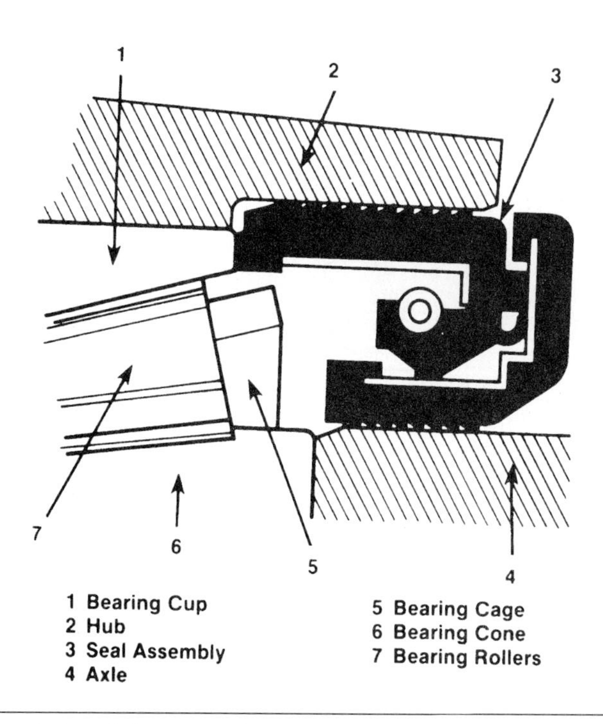

Figure 3-64 Barrier-type hub seal. (Courtesy of Navistar International Transportation Corp.)

seal, and the seal is pushed by hand into the hub. Do not add any other sealant or lubricant to the mounting surfaces. When the seal is installed, the outer part of the seal rotates with the hub and the inner part of the seal remains stationary on the spindle (Figure 3-64). The sealing surfaces are between the seal lips and the encased metal ring in the stationary part of the seal.

Bearing Lubrication

WARNING: If a bearing is operated without proper lubrication, bearing life will be very short.

Proper bearing lubrication is extremely important to maintain bearing life. Bearing lubricant reduces friction and wear, dissipates heat, and protects surfaces from dirt and corrosion. Sealed or shielded bearings are lubricated during the manufacturing process, *and no attempt should be made to wash these bearings or pack them with grease.*

Bearings that are not sealed or shielded require cleaning and repacking at intervals specified by the vehicle manufacturer. *Always use the bearing grease specified by the vehicle manufacturer.* Bearing lubricants may be classified as greases or oils. **Lithium-based** or **sodium-based grease** is used in many wheel bearings.

New bearings usually have a protective coating to prevent rust and corrosion. This coating should not be washed from the bearing. When rear wheel bearings are lubricated from the differential housing, the type and level of oil in the housing is important.

Truck manufacturers usually recommend an SAE No. 90, or SAE No. 140 hypoid gear oil in the differential. In very cold climates the manufacturer may recommend an SAE No. 80 differential gear oil. The **American Petroleum Institute (API)** classifies gear lubricants as GL-1, GL-2, GL-3, GL-4, and GL-5. The GL-4 lubricant is used for hypoid gears under normal conditions. The GL-5 lubricant is used in heavy-duty hypoid gears. Always use the truck manufacturer's specified differential gear oil.

The differential should be filled until the lubricant is level with the bottom of the filler plug opening in the differential housing. If the differential is overfilled, excessive lubricant may be present at the bearings and seals. Under this condition the lubricant may leak past the seal. When the lubricant level is low in the differential, the lubricant may not be available in the axle housings. When this condition exists the bearings do not receive enough lubrication, and bearing life is shortened.

See Chapter 3 in the Shop Manual for wheel bearing lubrication and service.

A BIT OF HISTORY

Like many other industries the trucking business has become more competitive in recent years. Profit margins vary between different trucking companies. Today some trucking companies are operating on profit margins as low as 2%. To maintain profit margins trucking companies must keep their trucks on the job. A truck or tractor in the shop is not earning any profit. Therefore truck or tractor down time is critical. To minimize down time tractor and truck service is very important. Tire, wheel, and wheel bearing service must be performed properly the first time to provide safe operation and minimum down time.

Summary

- ❑ In a bias ply tire the cords are positioned at an angle of 30 degrees to 40 degrees between the tire beads.
- ❑ In a radial ply tire the cords are placed at a 90-degree angle between the tire beads.
- ❑ Radial ply tires have less rolling resistance compared with bias ply tires.
- ❑ The aspect ratio of a tire is determined by dividing the tire height by the width.

Terms to Know

American Petroleum Institute (API)

Angular load

Barrier-type hub seal

Bearing seals

Bearing shields

Terms to Know (Continued)

Bias ply tire
Cylindrical ball bearing
Cylindrical roller bearing
Disk wheel
Double-row ball bearing
Dynamic balance
Fluted lip seal
Garter spring
Grit guard
Hub-piloted wheel
Inner race
Lithium-based grease
Low-profile tire
Needle roller bearing
Negative camber
Outer race
Positive camber
Radial load
Radial ply tire
Rolling elements
Separator
Single-row ball bearing
Snap rings
Society of Automotive Engineers (SAE)
Sodium-based grease
Spoke wheel
Spring-loaded seal
Springless seal
Static balance
Stud-piloted wheel
Tapered roller bearing
Thrust-bearing load
Tire matching
Toe-in
Toe-out
Unitized hub assemblies
Unitized one-piece seal
Wiper ring

- ❑ Low profile radial ply tires have a lower aspect ratio compared with conventional radial ply tires.
- ❑ On trucks with two or more drive axles all the tires must be the same type and be matched so they are the same circumference.
- ❑ Tire rims may be spoke type or disk type.
- ❑ Disk wheels may be stud piloted or hub piloted.
- ❑ Overinflation of tires causes wear on the center of the tread.
- ❑ Underinflation of tires causes wear on the edges of the tire tread.
- ❑ Positive camber occurs when the center line of the tire and wheel is tilted outward from the true vertical center line of the tire and wheel.
- ❑ Negative camber occurs when the center line of the tire and wheel is tilted inward from the true vertical center line of the tire and wheel.
- ❑ Improper wheel camber causes excessive wear on one edge of the tire tread.
- ❑ Toe-in occurs when the distance between the front inside edges of the tires is less than the distance between the rear inside edges of the tires.
- ❑ Toe-out occurs when the distance between the front inside edges of the tires is more than the distance between the rear inside edges of the tires.
- ❑ Improper toe setting causes feathered tire tread wear.
- ❑ When a wheel and tire have proper static balance, the weight is distributed equally around the axis of wheel and tire rotation.
- ❑ Improper static wheel balance causes wheel tramp.
- ❑ When a wheel and tire have proper dynamic wheel balance, the weight of the tire and wheel is distributed equally on both sides of the wheel center.
- ❑ Improper dynamic wheel balance causes wheel shimmy.
- ❑ A bearing reduces friction, carries a load, and guides certain components such as pivots, shafts, and wheels.
- ❑ Radial-bearing loads are applied in a vertical direction.
- ❑ Thrust-bearing loads are applied in a horizontal direction.
- ❑ Angular-bearing loads are applied at an angle between the vertical and horizontal.
- ❑ The inner bearing race is positioned at the center of the bearing and supports the rolling elements.
- ❑ The rolling elements in a bearing are positioned between the inner and outer races.
- ❑ The bearing separator keeps the rolling elements evenly spaced.
- ❑ The outer bearing race forms the outer ring on a bearing.
- ❑ A cylindrical ball bearing is designed primarily to withstand radial loads, but these bearings can handle a considerable thrust load.
- ❑ A snap ring may be mounted in a groove in the outer bearing race, and the snap ring retains the bearing in the housing.
- ❑ A bearing shield prevents dirt from entering the bearing, but it is not designed to keep lubricant in the bearing.
- ❑ Bearing seals keep lubricant in the bearing and prevent dirt from entering the bearing.

- ❑ Cylindrical roller bearings are designed primarily to carry radial loads, but they can handle some thrust loads.
- ❑ Tapered roller bearings have excellent radial, thrust, and angular load-carrying capabilities.
- ❑ Needle roller bearings are very compact, and they are designed to carry radial loads. They will not carry thrust loads.
- ❑ Springless seals are used for wheel bearing seals in some wheel hubs.
- ❑ The garter spring provides additional force on the seal lip to compensate for lip wear, shaft movement, and bore eccentricity.
- ❑ Flutes on seal lips provide a pumping action to direct oil back into a housing.

Review Questions

Short Answer Essays

1. Explain the advantages of radial ply tires over bias ply tires.
2. Describe the results of improper static wheel balance.
3. Explain the purposes of proper wheel balance.
4. Define a radial-bearing load.
5. Define a thrust-bearing load and give another term for this type of load.
6. Describe the main parts of a bearing, including the location and purpose of each part.
7. Explain the design and purpose of bearing seals.
8. Explain the types of loads that may be carried by a cylindrical roller bearing.
9. Explain how dual rims are retained on a spoke wheel.
10. Describe barrier seal design.

Fill-in-the-Blanks

1. Improper dynamic wheel balance causes wheel ____________________ .
2. Tubeless tires have ____________________ rims.
3. A bearing is designed to support a load and ____________________ .
4. An angular bearing load is applied at an angle between the ____________________ , and the ____________________ .
5. A cylindrical ball bearing is designed primarily to withstand ____________________ loads.
6. A bearing shield is attached to the ____________________ bearing race.
7. Lubrication and proper bearing adjustment are important on ____________________ bearings.
8. When a vehicle is turning a corner, the front wheel bearings must carry a ____________________ load.
9. The garter spring in a seal compensates for lip wear ____________________ and ____________________ .
10. In a barrier seal the inner part of the seal remains stationary on the ____________________ .

ASE Style Review Questions

1. While discussing types of tires:
   Technician A says bias ply tires may have narrow plies under the tread.
   Technician B says radial ply tires may have nylon body cords and steel belts.
   Who is correct?
   A. A only **C.** Both A and B
   B. B only **D.** Neither A nor B

2. During truck tire and wheel inspections:
   A. Tires on the front steering axle must be replaced if the tread depth is $\frac{4}{32}$ in (3.17 mm).
   B. Regrooved tires may be used on the front wheels of a bus.
   C. Recapped tires may be used on the front wheels of a bus.
   D. Tubeless tires are mounted on rims with split side rings.

3. While discussing bearing loads:
   Technician A says tapered roller bearings withstand high thrust loads.
   Technician B says radial-bearing load is applied in a horizontal direction.
   Who is correct?
   A. A only **C.** Both A and B
   B. B only **D.** Neither A nor B

4. While discussing bearing loads:
   Technician A says a wheel bearing encounters high thrust-bearing loads.
   Technician B says a ball bearing has improved axial load-carrying capabilities compared with a tapered roller bearing.
   Who is correct?
   A. A only **C.** Both A and B
   B. B only **D.** Neither A nor B

5. While discussing types of bearings:
   Technician A says a maximum capacity ball bearing has heavier races than an ordinary ball bearing.
   Technician B says a ball bearing cannot carry any thrust load.
   A. A only **C.** Both A and B
   B. B only **D.** Neither A nor B

6. Spoke truck wheels:
   A. Are made only from cast steel or ductile iron.
   B. Are mounted on the same bolts as the brake drums.
   C. On this type of wheel dual wheels are kept apart by individual spacers.
   D. Improper wheel nut torquing may cause wheel wobble.

7. All of these statements about disk wheels are true *except:*
   A. Disk wheels have a brake drum that may be removed before the hub and bearings.
   B. Disk wheels are mounted on the same bolts as the brake drums.
   C. May be retained with wheel nuts having a ball seat.
   D. May have retaining bolts that act as pilots to position the wheels.

8. While discussing truck front wheel bearings:
   Technician A says some front wheel hubs have a transparent oil level window in the bearing hub cap.
   Technician B says some truck manufacturers recommend SAE 90 rear axle lubricant in the front wheel hubs.
   Who is correct?
   A. A only **C.** Both A and B
   B. B only **D.** Neither A nor B

9. When diagnosing and servicing truck rear wheel bearings:
   A. Some rear hub and bearing assemblies are retained on the rear spindle with an adjusting nut, lock ring, and cotter key.
   B. A full-floating rear axle has **C** clips on the inner end of the axle that retain the axle in the differential side gear.
   C. Some rear wheel bearings may be damaged by low lubricant level in the differential.
   D. On some rear wheel hubs lubricant must be installed through one of the axle stud holes.

10. When diagnosing and servicing truck hub seals:
    A. A unitized seal and wiper ring may be replaced separately.
    B. A grit guard helps to seal the bearing lubricant in the hub.
    C. A grit guard is attached to the unitized seal shell.
    D. A unitized seal rotates with the hub.

Steering Columns, Manual Steering Gears, and Steering Linkages

Upon completion and review of this chapter, you should be able to:

- ❑ Explain the purpose of the steering column.
- ❑ Describe the purpose of the steering gear.
- ❑ Describe the advantage of a collapsible steering column.
- ❑ Explain the design features that allow a steering column to collapse.
- ❑ Describe how a steering column bracket may allow steering column movement if the driver is thrown against the steering wheel in a frontal collision.
- ❑ List four switches that may be mounted on the steering column.
- ❑ Describe how the tilting action occurs in a tilt steering column.
- ❑ Explain how the telescoping action occurs in a telescoping steering column.
- ❑ Describe the design of a worm-and-roller manual steering gear.
- ❑ Explain two methods of worm bearing preload adjustment in a manual worm-and-roller or recirculating ball steering gear.
- ❑ Explain the purpose of the worm bearing preload adjustment in a manual steering gear.
- ❑ Describe how the sector shaft is adjusted in a manual steering gear.
- ❑ Explain the purpose of proper sector shaft adjustment in a manual steering gear.
- ❑ Describe the difference in operation between a manual steering gear with a 13:1 ratio compared with a steering gear with a 16:1 ratio.
- ❑ Explain vehicle directional stability.
- ❑ Describe the effect of a crosswind on vehicle steering in relation to the vehicle center of gravity.
- ❑ Explain understeer and oversteer.
- ❑ Describe the effect of load distribution on understeer and oversteer.
- ❑ Describe the steering linkage components connected from the pitman arm to the front wheels.
- ❑ Explain wheel jounce and rebound.
- ❑ Describe the effect of improper pitman arm ball position on steering wheel rotation during wheel jounce and rebound.
- ❑ Explain the advantages of all-wheel steering on a truck.
- ❑ Describe the basic operation of an all-wheel steering system.

Introduction

The purpose of the steering column is to connect the steering wheel to the steering gear (Figure 4-1). Steering column condition is very important to provide proper steering quality and control. For example, loose or worn steering column components may cause excessive **steering wheel free-play,** reduced steering control, and a rattling noise. Worn steering column components may also cause a binding condition when the steering wheel is rotated and this reduces steering control.

The purpose of the steering gearbox is to provide a mechanical advantage that allows the driver to turn the front wheels with a reasonable amount of effort.

A BIT OF HISTORY

In the early 1900s steering gears were a worm-and-gear or worm-and-sector design. These gears gave the driver a mechanical advantage to turn the front wheels, but they created a lot of friction.

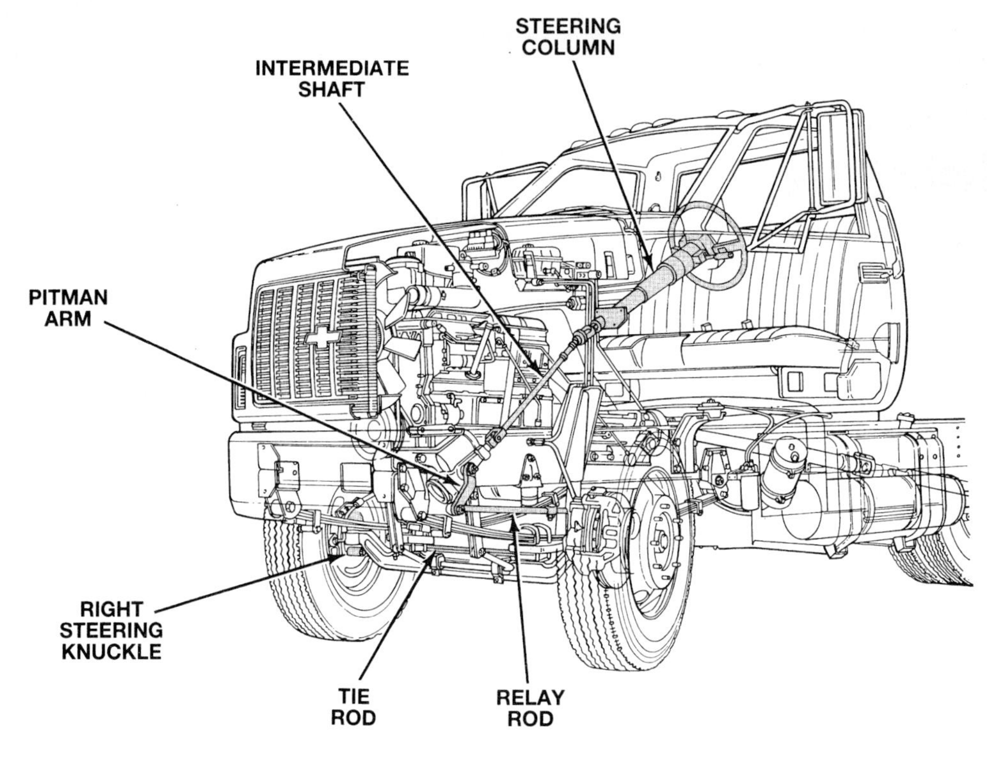

Figure 4-1 The steering column connects the steering wheel to the steering gear. (Courtesy of General Motors Corporation, Service Technology Group.)

The Ross cam and lever steering gear was introduced in 1923. The cam in this gear was a spiral groove machined into the end of the **steering shaft.** A pin on the pitman shaft was mounted in the spiral groove in the steering shaft. When the steering wheel and shaft were turned, the pin was forced to move, and this action rotated the pitman shaft. When a front wheel struck a road irregularity, this steering gear design prevented serious **steering wheel kickback.** However, this steering gear design still created a considerable amount of friction and required higher steering effort.

A sector may be defined as part of a gear.

Steering kickback refers to a strong and sudden movement of the steering wheel in the opposite direction to which the steering wheel is turned. This kickback action tends to occur if a front wheel strikes a road irregularity during a turn.

In the mid-1920s Saginaw Steering Division of General Motors Corporation developed the **worm-and-roller steering gear.** In this steering gear the sector became a roller that greatly reduced friction and steering effort.

Steering Columns

Fixed Steering Column with Floor Shift Transmission

CAUTION: Always disconnect the battery ground cable before servicing a steering column. If the battery terminals are connected and a wire is shorted to ground, severe wiring harness damage may occur. A technician may receive severe burns from an overheated wiring harness.

A **fixed steering column** does not provide any tilting or telescoping action. A steering shaft is mounted in the center of the steering column. Some trucks have collapsible steering columns that collapse a certain amount if the driver is thrown forward against the steering wheel in a front-end collision. This collapsing action helps to protect the driver. In some collapsible steering columns the steering shaft is a two-piece design that collapses in a collision (Figure 4-2).

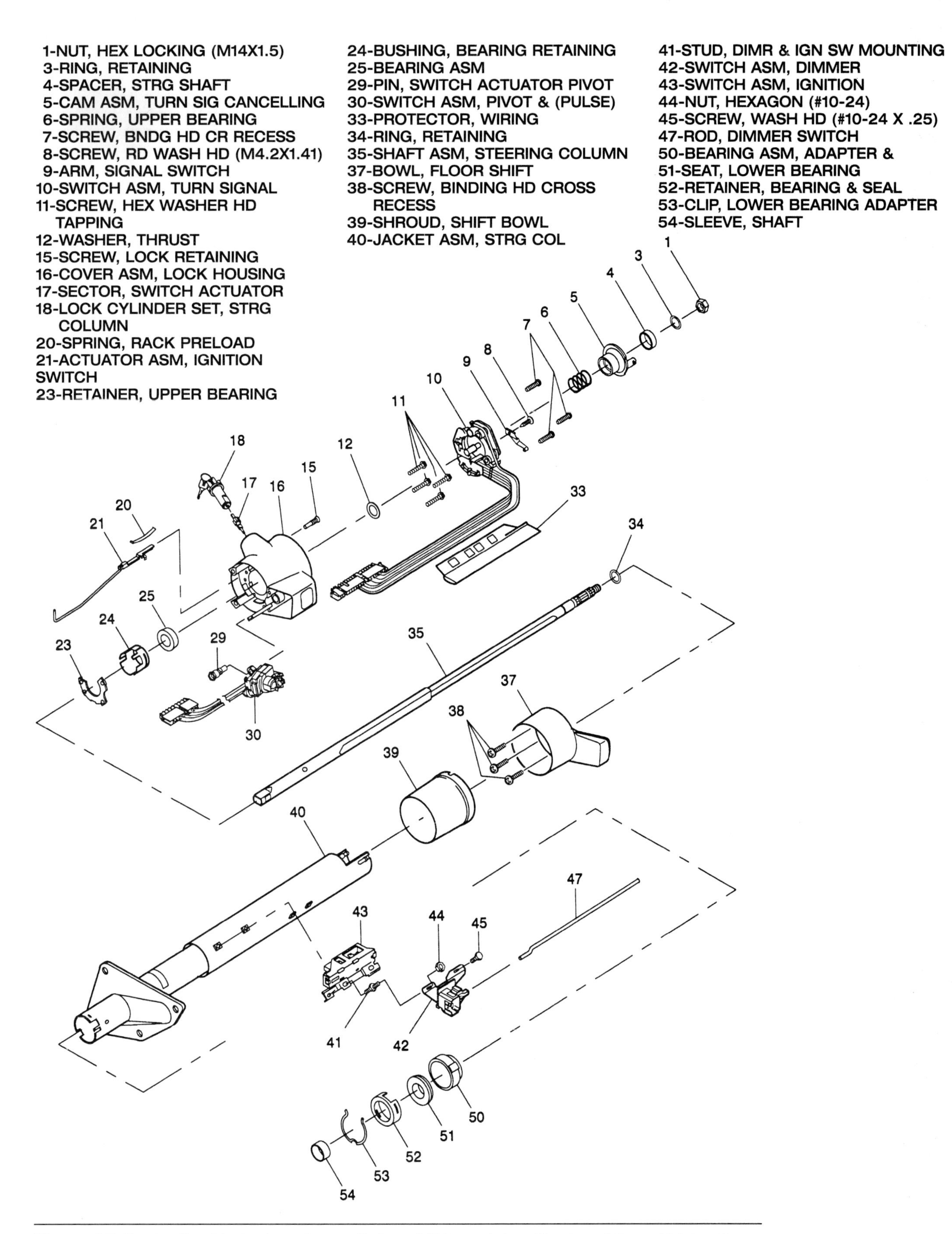

Figure 4-2 Fixed, collapsible steering column with floor shift transmission. (Courtesy of General Motors Corporation, Service Technology Group.)

A jacket assembly (item 40 Figure 4-2) surrounds the lower part of the steering shaft (item 35). This steering jacket is also a two-piece collapsible design. Plastic pins are injected between the two parts of the steering jacket, and these pins shear off, allowing the jacket to collapse if the driver is thrown against the steering wheel in a frontal collision. The steering shaft is supported on bearings (items 50 and 25) in the lower end of the steering jacket and in the ignition switch housing mounted on top of the steering jacket.

WARNING: After a collision, steering column measurements should be performed to determine whether the steering column is collapsed. Collapsed or partially collapsed steering column components must be replaced because they may cause improper steering control.

The ignition lock cylinder is retained in the ignition lock housing with a hard steel screw (item 15). This hard steel screw makes it more difficult for thieves to remove the ignition lock cylinder. An actuator rod (item 21) is connected from the ignition lock cylinder to the ignition switch (item 43), and this switch is bolted to the steering jacket. The turn signal switch (item 10) is bolted on top of the ignition lock housing. Wires are connected from this switch to the vehicle wiring harness. The windshield wiper and washer switch (item 30) are mounted on the lower side of the ignition lock housing. The dimmer switch (item 42) is mounted on the steering jacket with the ignition switch. A rod (item 47) is connected from the dimmer switch to the turn signal lever mechanism. Vertical movement of the turn signal lever operates the dimmer switch. On some tractors the trailer brake control is mounted on the steering column.

The steering wheel is mounted on splines on the upper end of the steering shaft. A locking nut (item 1) retains the steering wheel on the steering shaft. A floor seal on the lower end of the steering jacket seals the jacket to the truck cab floor. A steering column support bracket and mounting bolts retain the steering column to the dash (Figure 4-3). Some steering column support brackets are designed with capsules around the bolt openings. These capsules allow the steering column and bracket to move away from the driver if the driver's weight is thrown against the column in a

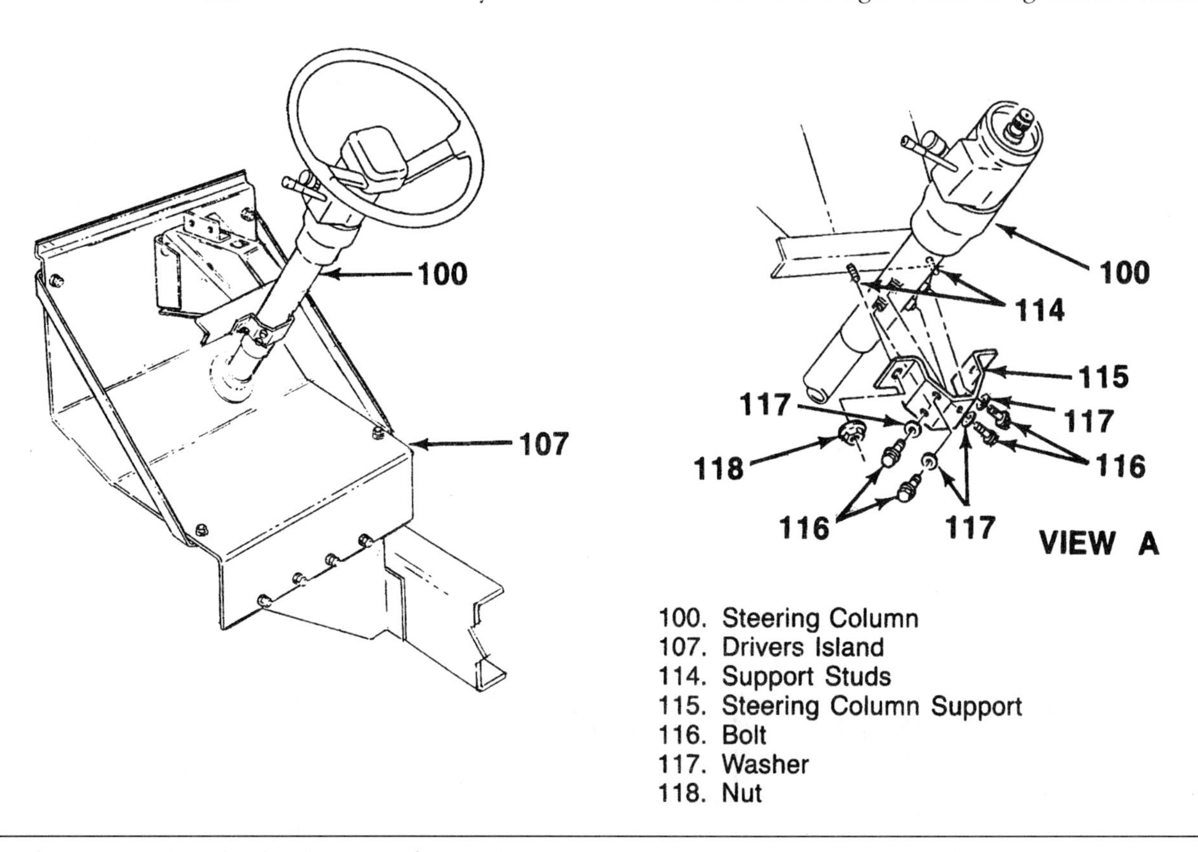

Figure 4-3 Steering column bracket. (Courtesy of General Motors Corporation, Service Technology Group.)

frontal collision. The capsules in the steering column bracket must be properly positioned to provide the proper column and bracket movement during a frontal collision. When properly torqued, the heads of the retaining bolts must not contact the capsules or steering column, and bracket movement is restricted during a frontal collision (Figure 4-4).

Fixed Steering Column with Column Shift Transmission

Some medium duty trucks have a gear shift selector mounted on the steering column. These steering columns contain a shift tube assembly (item 59 Figure 4-5). The gear selector lever is mounted in a gear shift lever bowl (item 47) under the ignition lock cylinder housing. The upper end of the shift tube is connected to the gear shift lever bowl, and the lower end of the shift tube is connected to the gear shift linkage. A linkage is connected from the shift tube to the transmission linkage.

Tilt and Telescoping Steering Column

A **tilt and telescoping steering column** allows the driver to tilt the steering wheel at various positions between his or her body and the dash. The telescoping action allows the driver to move the wheel vertically so it is higher or lower. The tilt and telescoping steering column allows the steering wheel to be positioned to suit various driver heights and sizes. This type of steering wheel also allows the driver to change the steering wheel position during a trip to reduce driver fatigue.

A collapsible steering column may be called an energy-absorbing column.

An actuation lever on the left side of the steering column releases the tilt and telescoping mechanisms. A universal joint assembly (item 22 Figure 4-6) in the steering shaft below the tilt and telescoping mechanism allows the steering shaft to tilt. The support bracket (item 23) is mounted inside

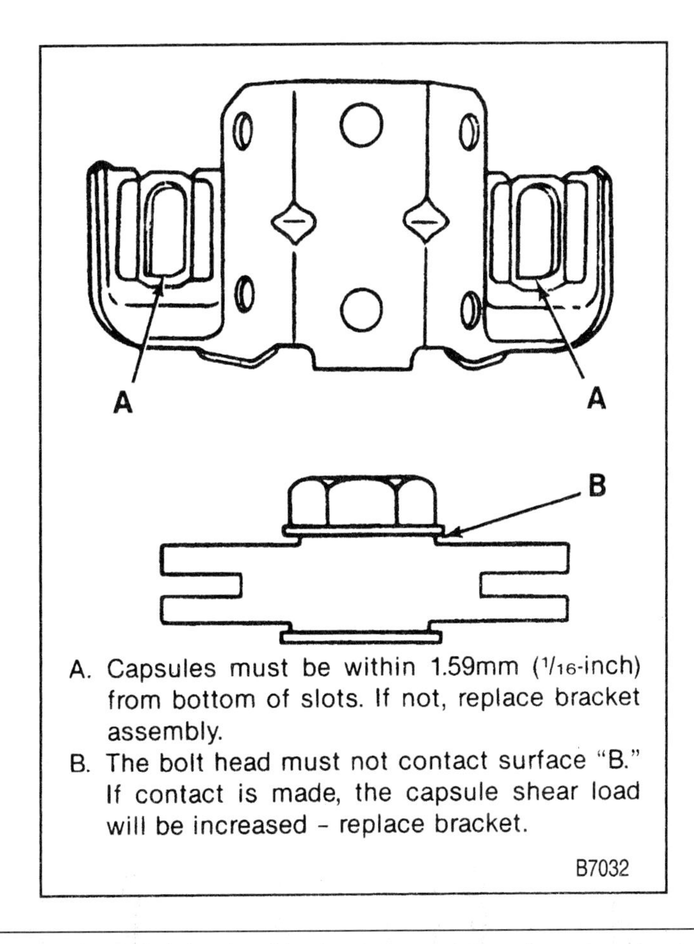

Figure 4-4 Proper capsule and bolt head position in relation to the steering column bracket. (Courtesy of General Motors Corporation, Service Technology Group.)

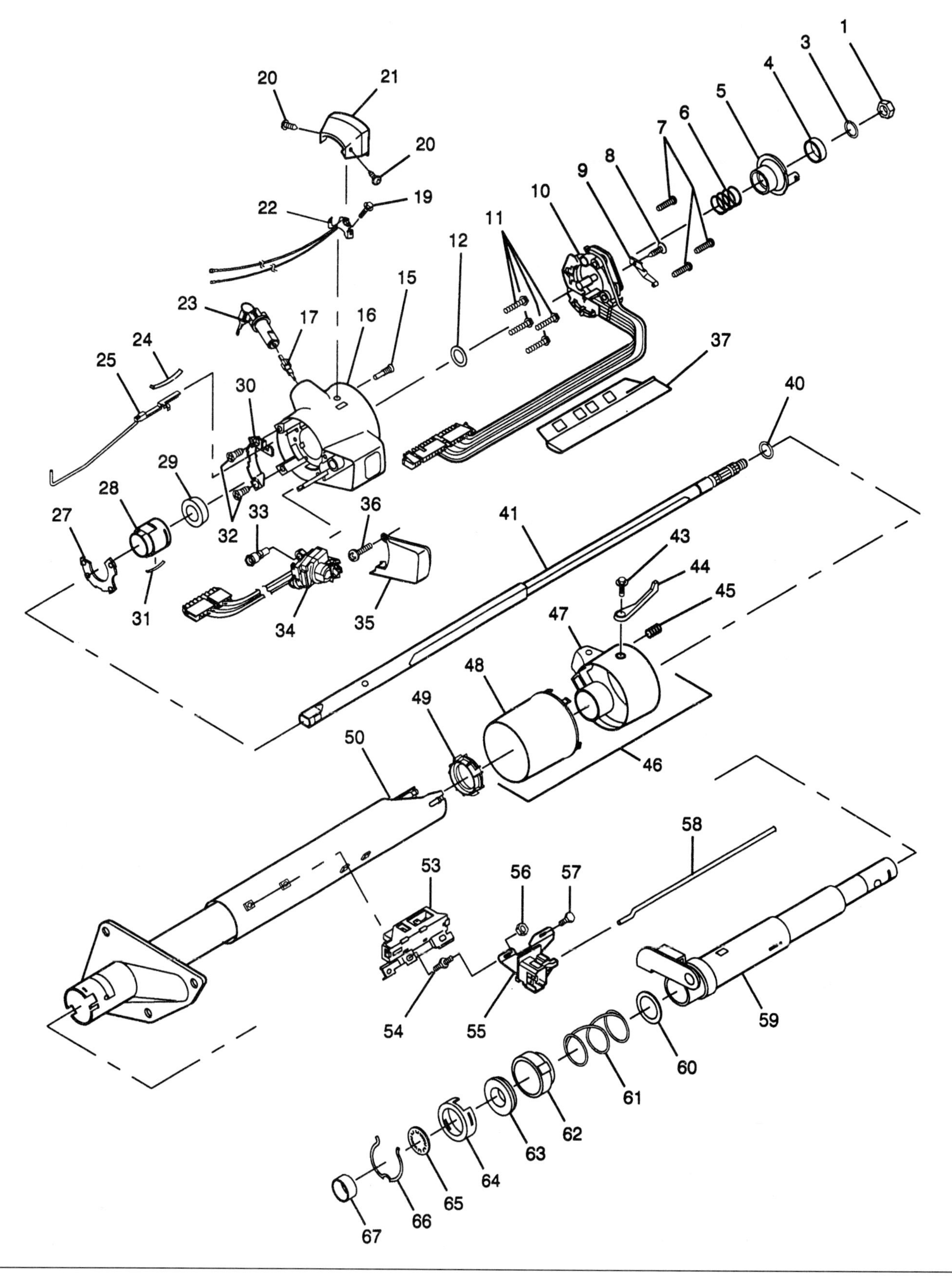

Figure 4-5 Fixed, collapsible steering column with column shift transmission. (Courtesy of General Motors Corporation, Service Technology Group.) See key on opposite page.

the main steering column housing (item 43). Pivot screws on each side of this main housing allow the main housing to pivot on the support bracket, and this action allows the main housing to tilt.

When the tilt lever is actuated downward, teeth on a wedge lock are disengaged from teeth on the support bracket (Figure 4-7). When these teeth are released, the driver can tilt the wheel to the desired position. Once the steering wheel is in the desired position, the driver releases the tilt lever, and the teeth on the wedge lock and support bracket re-engage to hold the steering column in position.

Upward actuation of the tilt and telescoping lever releases a lock pin from the steering jacket and tube (item 32 Figure 4-6). When this lock pin is released, the steering wheel may be pulled upward or pushed downward to the desired position. When the tilt and telescoping lever is released, the lock pin engages in a different hole in the steering jacket and tube to hold the steering wheel in the desired vertical position. When the steering wheel is moved to a new telescoping position, the splines on the wheel tube and sleeve (item 33 Figure 4-6) slide vertically on the matching splines on the shaft connected to the universal joint assembly to allow the steering shaft to be shortened or lengthened.

The lower shaft on the universal joint assembly is splined to the steering joint assembly (item 26 Figure 4-6). An intermediate steering shaft (item 27 Figure 4-6) is connected from the steering joint assembly to the steering gear. Another universal joint is positioned near the lower end of the intermediate steering shaft.

Some steering columns provide a tilting action without the telescoping action. Some of these steering columns have centering spheres between the upper and lower steering shafts to allow the steering shaft to pivot (Figure 4-8).

See Chapter 4 in the Shop Manual for steering column service and diagnosis.

Manual Steering Gears

Worm-and-Roller Steering Gears

End-play of the wormshaft refers to lateral movement of this shaft between the bearings on which the shaft is mounted.

Most medium and heavy duty trucks manufactured in recent years are equipped with power steering gears. Some older trucks may be equipped with manual steering gears. In a worm-and-roller steering gear a worm gear is mounted on bearings in the gear housing. The steering wheel is connected through the steering shaft in the steering column to the worm gear. The worm gear has a

1-NUT, HEXAGON JAM (M14x1.5)
3-RING, RETAINING
4-SPACER, STRG SHAFT
5-CAM ASM, TURN SIG CANCELLING
6-SPRING, UPPER BEARING
7-SCREW, BINDING HD CROSS RECESS
8-SCREW, RD WASH HD (M4.2x1.41)
9-ARM, SWITCH ACTUATOR
10-SWITCH ASM, TURN SIGNAL
11-SCREW, HEX WASHER HD TAPPING
12-WASHER, THRUST
15-SCREW, LOCK RETAINING
16-HOUSING ASM, STRG COLUMN
17-SECTOR, SWITCH ACTUATOR
19-SCREW, HEX WASHER HEAD
20-SCREW, OVAL HD (M4.2 X 1.41 X 14)
21-HOUSING ASM, TRANS IND DIAL &
22-BRACKET ASM, SOCKET &
23-LOCK CYLINDER SET, STRG COLUMN
24-SPRING, RACK PRELOAD
25-ACTUATOR ASM, IGNITION SWITCH
27-RETAINER, UPPER BEARING
28-BUSHING, BEARING RETAINING
29-BEARING ASM
30-GATE, SHIFT LEVER
31-CONTACT, HORN CIRCUIT
32-SCREW, FLT HD CROSS RECESS
33-PIN, SWITCH ACTUATOR PIVOT
34-SWITCH ASM, PIVOT & (PULSE)
35-COVER, HOUSING
36-SCREW, BINDING HD CROSS RECESS
37-PROTECTOR, WIRING
40-RING, RETAINING
41-SHAFT ASM, STEERING
43-SCREW, HEX WA HD (M4 X 0.7 X 8)
44-POINTER, TRANS CONT INDICATOR
45-SPRING, SHIFT LEVER
46-BOWL ASM, SHROUD &
47-BOWL, GEARSHIFT BOWL
48-SHROUD, GEARSHIFT LEVER
49-BEARING, BOWL LOWER
50-JACKET ASM, STRG COL
53-SWITCH ASM, IGNITION
54-STUD, DIMR & IGN SW MOUNTING
55-SWITCH ASM, DIMMER
56-NUT, HEXAGON (#10-24)
57-SCREW, WASH HD (#10-24 x .25)
58-ROD, DIMMER SWITCH ACTUATOR
59-TUBE ASM, SHIFT
60-WASHER, SPRING THRUST
61-SPRING, SHIFT TUBE RETURN
62-ADAPTER, LOWER BEARING
63-BEARING ASM
64-RETAINER, BEARING ADAPTER
65-RETAINER, LOWER SPRING
66-CLIP, LOWER BEARING ADAPTER
67-SLEEVE, SHAFT

Service Kits

201-HSG ASM SERV KIT, STRG COL
-INCLUDES: 16, 17, 27, 28, 29, 31
202-SECTOR SERV KIT, IGN SW ACTR
-INLUDES: 17
203-GREASE SERV KIT, (SYNTHETIC)

Figure 4-5 *continued* Key to fixed, collapsible steering column with column shift transmission. (Courtesy of General Motors Corporation, Service Technology Group.)

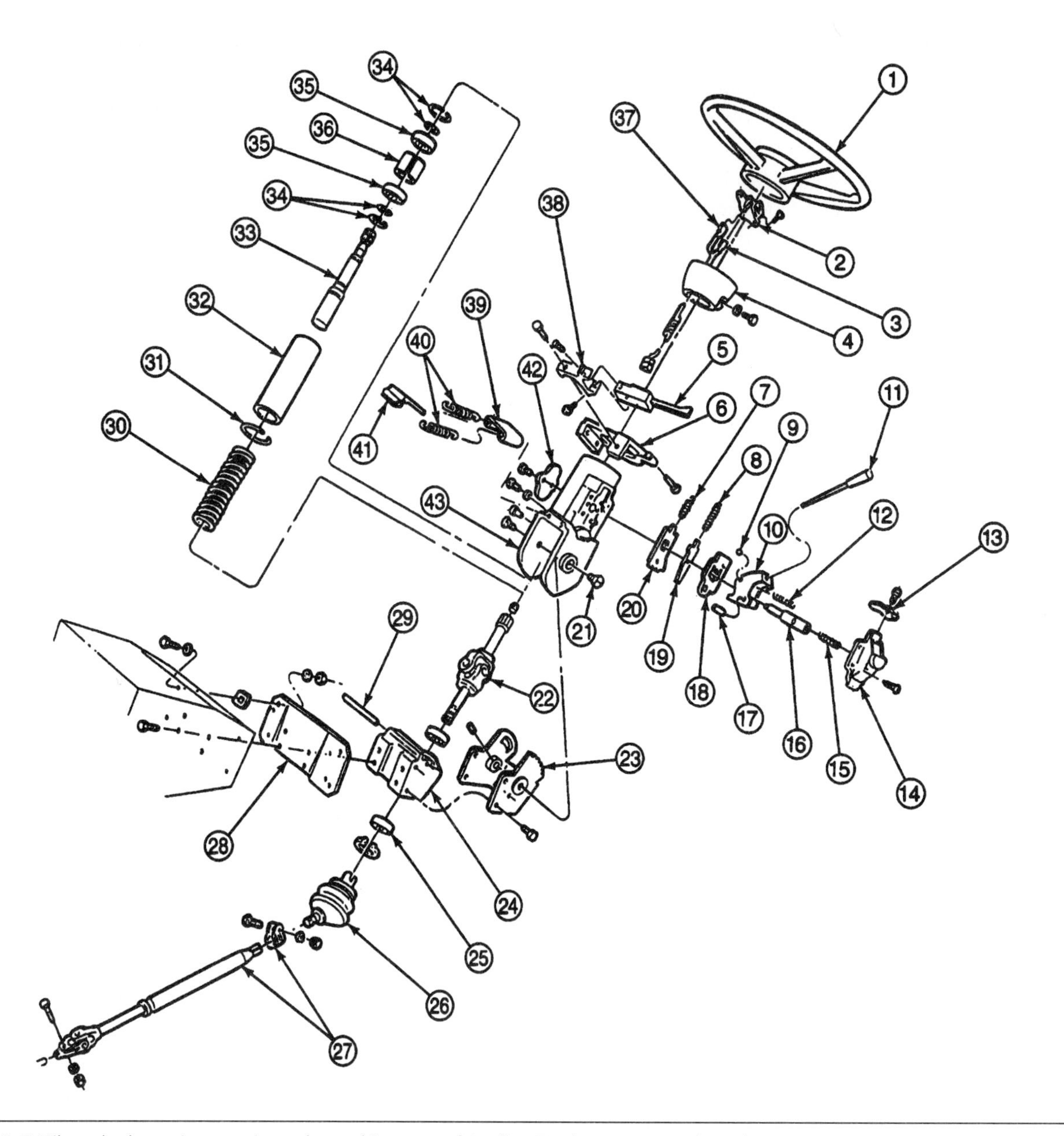

Figure 4-6 Tilt and telescoping steering column. (Courtesy of Sterling Truck Corporation.) See key on opposite page.

When a bearing has preload, all end-play is removed, and a slight tension is placed on the bearing.

Steering wheel free-play refers to the amount of steering wheel rotation before the front wheels begin to turn right or left.

slight hourglass shape (Figure 4-9). Proper preloading of the **wormshaft** bearings is necessary to eliminate wormshaft end-play and prevent steering gear free-play and **vehicle wander.** A shim adjustment between the lower steering gear cover and the gear housing provides the necessary wormshaft **bearing preload.**

Matching teeth on the **sector shaft** or roller are meshed with the worm gear (Figure 4-10). When the front wheels are straight ahead, an interference fit exists between the sector shaft and ball nut teeth. This interference fit eliminates **gear tooth lash** when the front wheels are straight ahead and provides the driver with a positive feel of the road. Proper adjustment of the sector shaft is necessary to obtain the necessary **interference fit** between the sector shaft and wormshaft teeth. A sector shaft adjuster screw is threaded into the sector shaft cover to provide sector shaft adjustment. The steering gear must be filled to the proper level with lubricant specified by the truck manufacturer.

A **pitman arm** is mounted on the outer end of the sector shaft, and the drag link connects the pitman arm to the steering arm on the left front wheel (Figure 4-11). The steering arm is

Item	Part Number	Description
1	3600	Steering Wheel
2	13836	Horn Contact Assembly
3	13N866	Horn Wire Assembly
4	13K314	Horn Contact Housing Assembly
5	13B302	Switch, Brake
6	3L774	Bracket, Air Brake
7	—	Spring, Lock Bar (Part of 3F777)
8	—	Spring, Wedge Lock (Part of 3F777)
9	—	Ball, Rubber (Part of 3F777)

Item	Part Number	Description
10	—	Cam Actuator (Part of 3F777)
11	3F527	Lever, Tilt / Telescoping
12	—	Spring, Anti-Rattle (Part of 3F777)
13	—	Plate, Spring Cover (Part of 3F777)
14	—	Housing, Actuator (Part of 3F777)
15	30570	Spring, Actuator Housing
16	—	Lock Pin, Telescope (Part of 3F777)
17	—	Pin, Pivot (Part of 3F777)

Item	Part Number	Description
18	—	Plate, Disengaging (Part of 3F777)
19	—	Wedgelock (Part of 3F777)
20	—	Lock Bar (Part of 3F777)
21	3F773	Screw, Pivot
22	3B734	Universal Joint Assembly
23	3B635	Bracket, Support
24	3L502	Housing, Lower Bearing
25	3517	Bearing
26	3B676	Steering Joint Assembly
27	3B676	Shaft, Steering Intermediate
28	3N720	Bracket, Column Mounting
29	3E517	Rod, Spring Retaining
30	3D655	Spring, Extension

Item	Part Number	Description
31	3499	Ring, Retaining
32	3A618	Tube, Jacket
33	3514	Wheel Tube / Sleeve Assembly
34	3510	Rings, Retaining
35	3517	Bearing
36	3D640	Spacer, Bearing
37	13N866	Wire, Horn
38	13K025	Bracket, Brake Switch Mounting
39	3T501	Cover, Lower, Spring
40	3E541	Springs, Tilt
41	3A731	Cover, Lower, Spring
42	3D505	Cover, Stop
43	3F774	Main Housing

Figure 4-6 *continued* Components for tilt and telescoping steering column. (Courtesy of Sterling Truck Corporation.)

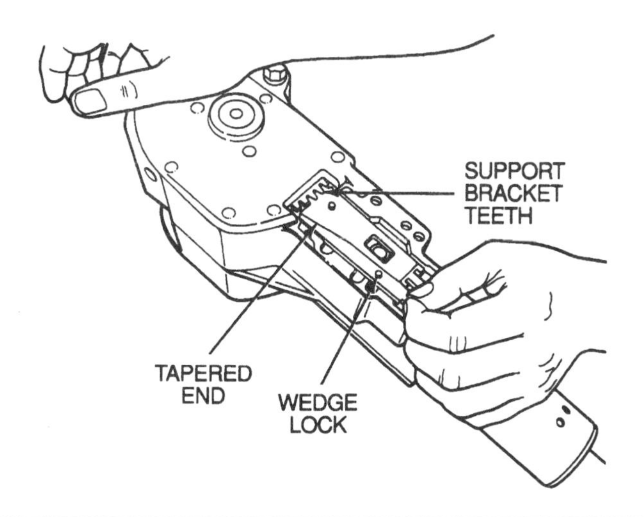

Figure 4-7 Teeth on the wedge lock and support bracket are disengaged when the tilt lever is actuated. (Courtesy of Sterling Truck Corporation.)

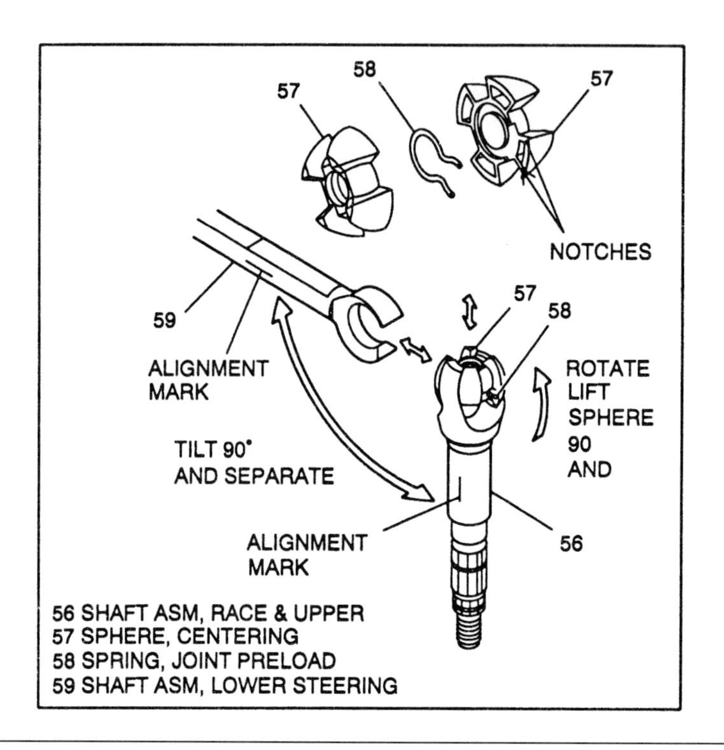

Figure 4-8 Tilt steering column with centering spheres between the upper and lower steering shafts. (Courtesy of General Motors Corporation, Service Technology Group.)

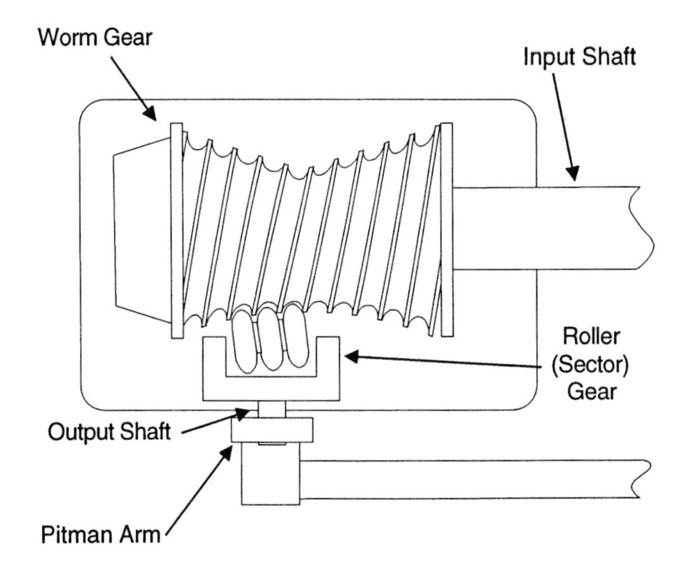

Figure 4-9 Worm-and-roller steering gear design.

Vehicle wander is the tendency of a vehicle to steer to the right or left as it is driven straight ahead.

When two gears have an interference fit, the gear teeth are meshed together so there is a slight tension or preload on the teeth.

Gear tooth lash refers to movement between gear teeth that are meshed with each other.

mounted in an opening near the top of the left front spindle. A spindle arm is mounted in an opening near the bottom of the left front spindle. A tie rod is connected from the spindle arm in the left front spindle to the steering arm in the right front spindle (Figure 4-12).

Steering wheel rotation turns the worm gear. Because the sector shaft teeth are meshed with the worm gear, rotation of this gear turns the sector shaft. Sector shaft movement is transferred through the pitman arm, steering arms, spindle arm, and tie rod to the front wheels. Therefore steering wheel rotation turns the front wheels to the right or left.

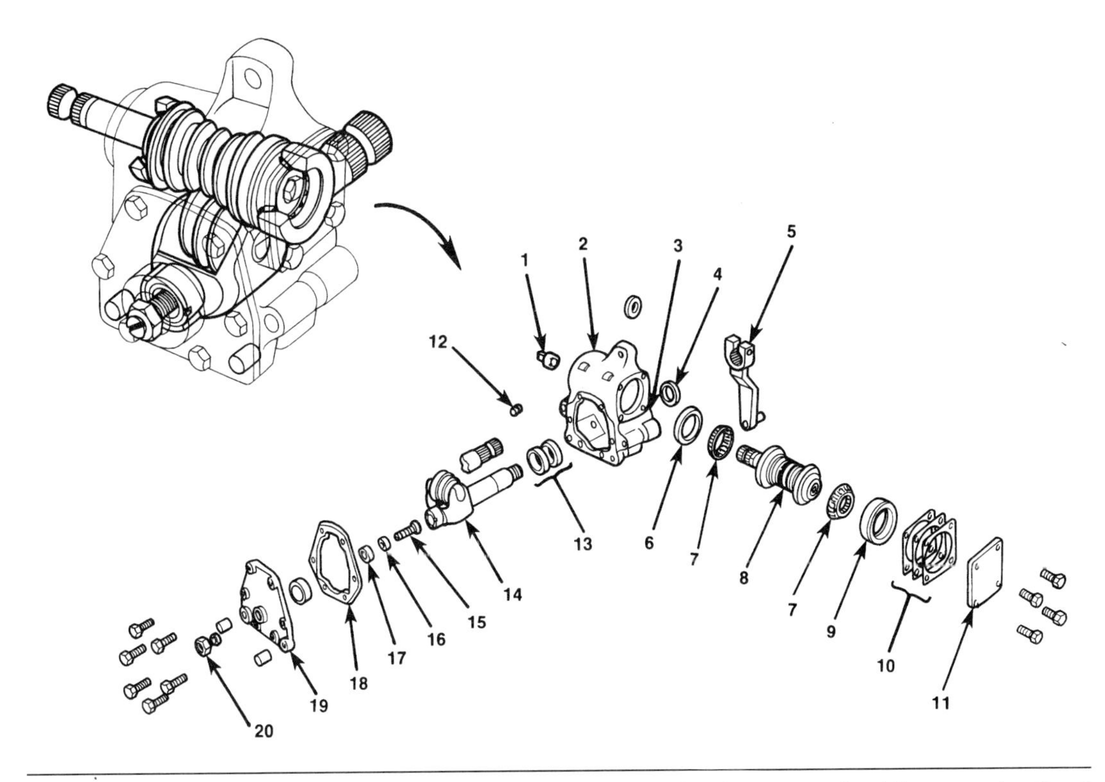

Figure 4-10 Complete worm-and-roller steering gear. (Courtesy of Navistar International Transportation Corp.)

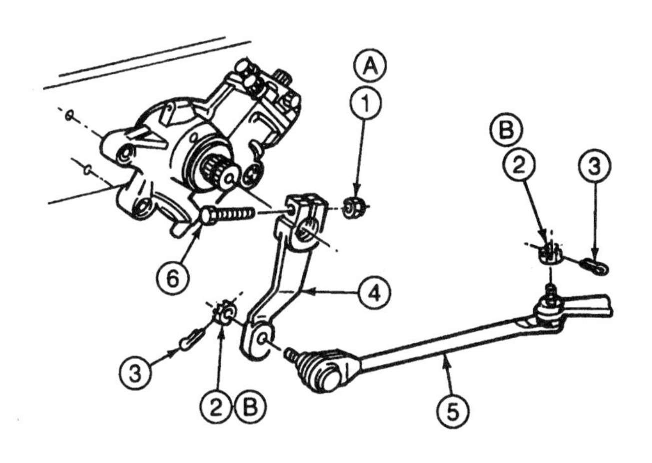

Item	Part Number	Description
1	34992-S2	Nut
2	383251-S	Nut
3	72107-S	Cotter Pin
4	3590	Pitman Arm
5	3304	Drag Link
6	383251-S	Bolt
A	—	Tighten to 298-400 N•m (220-300 Lb-Ft)
B	—	Tighten to 149-203 N•m (110-150 Lb-Ft)

Figure 4-11 The pitman arm and drag link connect the sector shaft to the steering arm on the left front wheel. (Courtesy of Sterling Truck Corporation.)

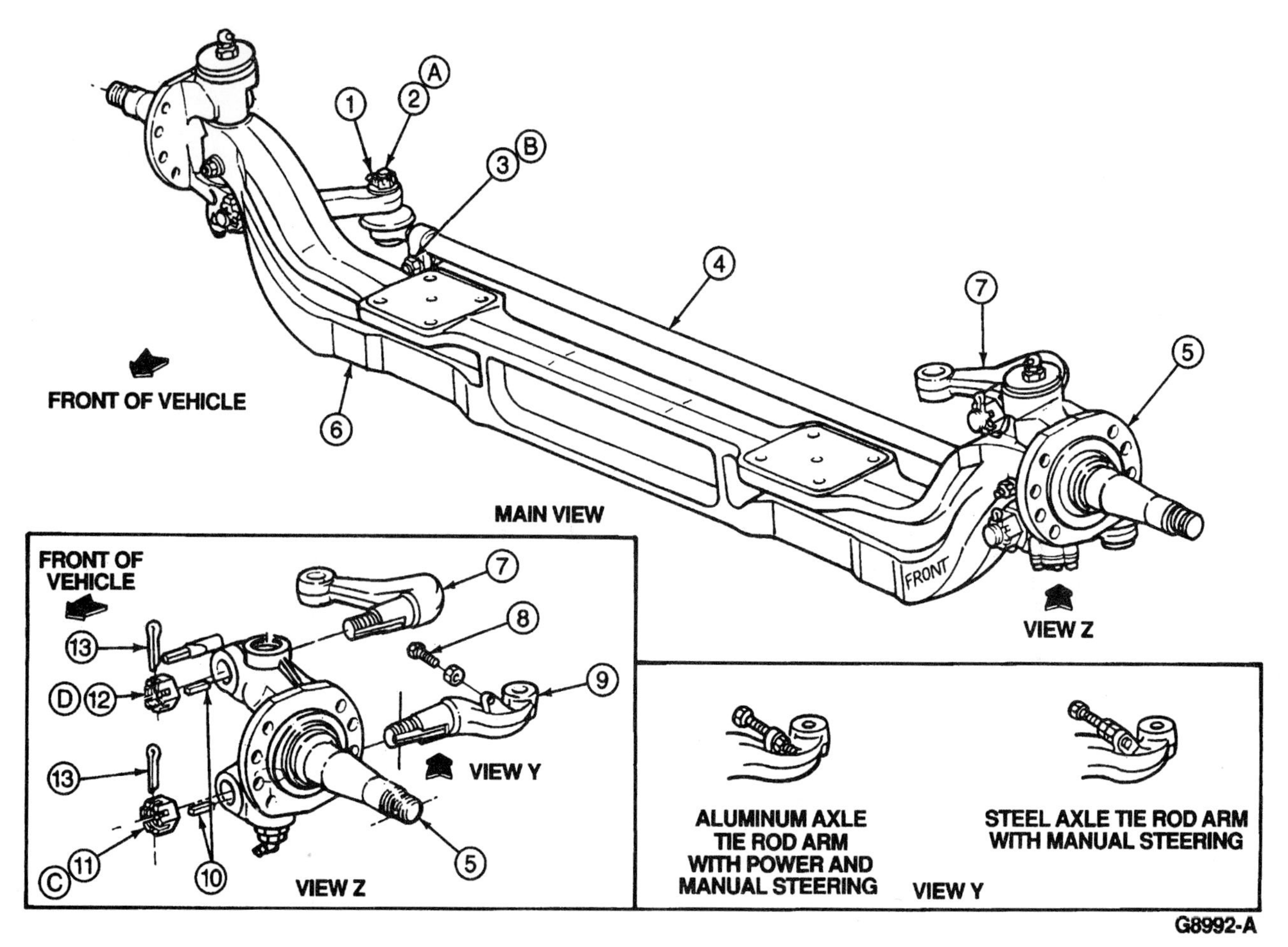

Item	Part Number	Description
1	72107-S36	Cotter Pin
2	383251-S100	Nut
3	33850-S	End Clamp
4	3280	Spindle Connecting Rod (Tie Rod)
5	3106	Spindle Assembly (LH)
6	3010	Front Axle, Steel; Aluminum
7	3146	Steering Arm
8	385123-S2	Bolt
9	3131	Spindle Arm (LH)
10	3138	Key
11	383252-S2	Nut
12	386180-S2	Nut

Item	Part Number	Description
13	72109-S36	Cotter Pin
A	—	Tighten to 150-203 N·m (110-150 Lb-Ft). Advance to Next Castellation if Necessary to Install Pin.
B	—	Tighten to 150-203 N·m (110-150 Lb-Ft) After Adjusting Toe-in.
C	—	Tighten to 488-650 N·m (360-480 Lb-Ft). Advance to Next Castellation if Necessary to Install Pin.
D	—	Tighten to 732-990 N·m (540-730 Lb-Ft)

Figure 4-12 The spindle arm on the left front wheel is connected through the tie rod to the steering arm on the right front wheel. (Courtesy of Sterling Truck Corporation.)

Recirculating Ball Steering Gear

In a **recirculating ball steering gear** the steering wheel and steering shaft are connected to the wormshaft. Roller bearings support both ends of the wormshaft in the steering gear housing. A seal above the upper wormshaft bearing prevents oil leaks, and an adjusting plug is provided on the upper wormshaft bearing to adjust wormshaft bearing preload. A ball nut is mounted over the wormshaft, and internal threads, or grooves, on the ball nut match the grooves on the wormshaft. Ball bearings run in the ball nut and wormshaft grooves (Figure 4-13).

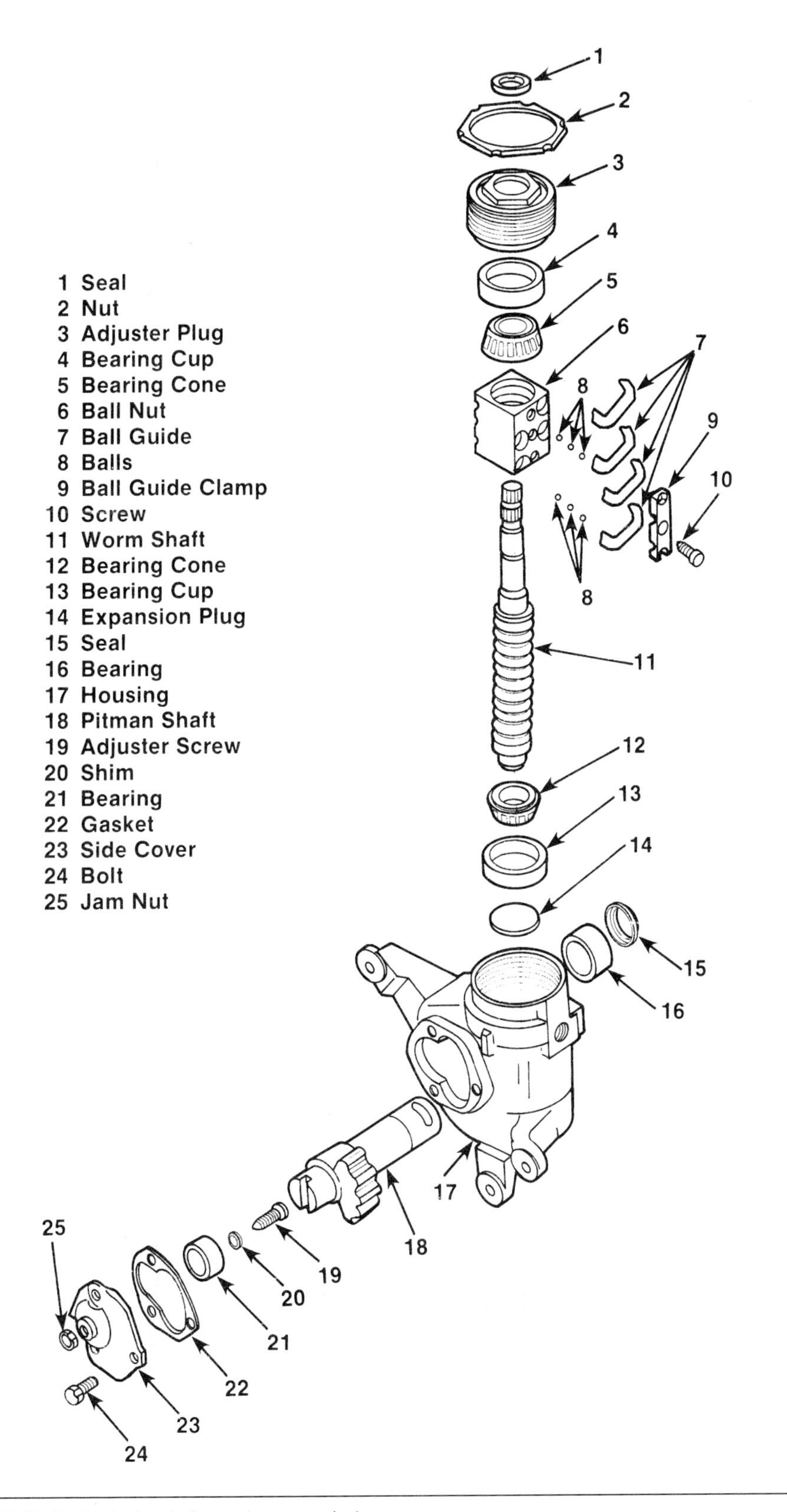

Figure 4-13 Recirculating ball steering gear design.

When the wormshaft is rotated by the steering wheel, the ball nut is moved up or down on the wormshaft. The gear teeth on the ball nut are meshed with matching gear teeth on the pitman shaft sector. Therefore ball nut movement causes pitman shaft sector rotation. Because the pitman shaft sector is connected through the pitman arm and steering linkage to the front wheels, the front

Gear ratio refers to the relationship between the rotation of the drive and driven gears. If 13 turns of the drive gear are necessary to obtain one turn of the driven gear, the gear ratio is 13:1.

See Chapter 4 in the Shop Manual for manual steering gear service and diagnosis.

wheels are turned by the pitman shaft sector. The lower end of the pitman shaft sector is usually supported by a bushing or a needle bearing in the steering gear housing, and a bushing in the sector cover supports the upper end of this shaft.

Steering **gear ratio** refers to the number of times the steering wheel is rotated to obtain one revolution of the sector shaft. The steering gear and linkage design only allow the sector shaft to move a specific amount in either direction. Steering gear ratio may vary from 13:1 to 16:1. When the same types of steering gears are compared, a higher numerical ratio provides reduced steering effort and increased steering wheel movement in relation to the amount of front wheel movement.

If a steering gear has a 16:1 ratio, 16 rotations of the steering wheel are required to obtain one revolution of the sector shaft. Because the sector shaft can only move a certain amount in either direction, on the average truck 5 or 6 steering wheel turns will move the front wheels from fully left to fully right. When the design engineers calculate the proper steering gear ratio for a truck or tractor, many factors are used, such as the maximum weight on the steering axle and the fifth wheel position on tractors. Manual steering gears usually have a higher numerical ratio compared with power steering gears.

TRADE JARGON: The term *faster steering* refers to a steering gear with a lower numerical ratio compared with a steering gear with a higher numerical gear ratio. With the faster steering gear, less steering wheel movement is required to obtain front wheel movement to the right or left.

Steering Linkages

Steering Linkage Purpose

Directional stability is the tendency of the vehicle steering to remain in the straight ahead position when the vehicle is driven straight ahead on a smooth level road surface.

Steering linkages connect the steering gear to the front wheels. Steering linkages and the front suspension must provide vehicle **directional stability** under all driving conditions. Many variables affect directional stability, including road condition, vehicle load, and side winds. The steering linkages and suspension system must be designed and maintained to provide directional stability under these varying conditions. Steering linkages must also be designed and maintained to minimize tire wear and to provide a reasonable steering effort for the driver. Steering linkages connect the steering gear to the front wheels. The steering linkages must position the front wheels properly in relation to each other. Steering linkages must also allow the front wheels to turn to the right or left without binding; at the same time these linkages must not have any looseness that would cause improper front wheel position and movement. Loose steering linkages also result in excessive steering wheel free-play and reduced steering control.

Directional Stability, Center of Gravity, and Load Distribution

The directional stability of a truck or tractor in a crosswind is determined partly by the vehicle center of gravity and the pressure center on the vehicle. The center of gravity on a vehicle is the point on the vehicle from which the load is equally distributed front-to-rear and side-to-side. Engineers determine the vehicle pressure center with a wind tunnel. Scale models may be used in the wind tunnel. If the pressure center is located ahead of the center of gravity, the wind force tends to turn the vehicle away from the wind (Figure 4-14). Under this condition, the driver must turn the steering wheel into the wind to keep the vehicle moving straight ahead. When the pressure center is behind the center of gravity, the wind force tends to turn the vehicle into the wind (Figure 4-15). This action causes the driver to turn the front wheels away from the wind to steer the vehicle straight ahead. Among truck and tractor engineers it is generally thought that turning the steering wheel into the wind to correct for wind force is a more natural reaction than turning the steering wheel into the wind.

The center of gravity and the pressure center are determined by the user requirements of wheelbase, load distribution, and body type. Load distribution may have a significant effect on the

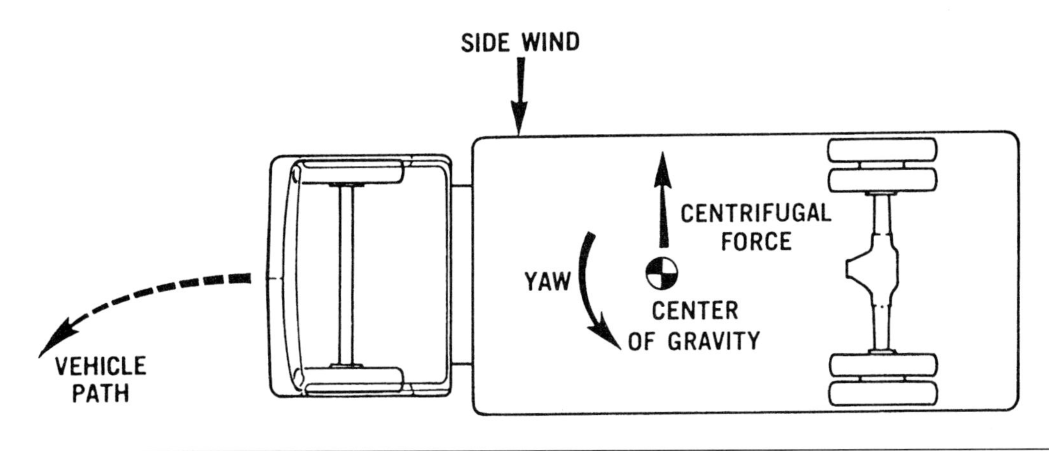

Figure 4-14 The effect of crosswind on vehicle direction when pressure is applied ahead of the center of gravity. (Reprinted with permission from SAE Publication The Truck Steering System Hand Wheel to Road Wheel ©1973 Society of Automotive Engineers, Inc.)

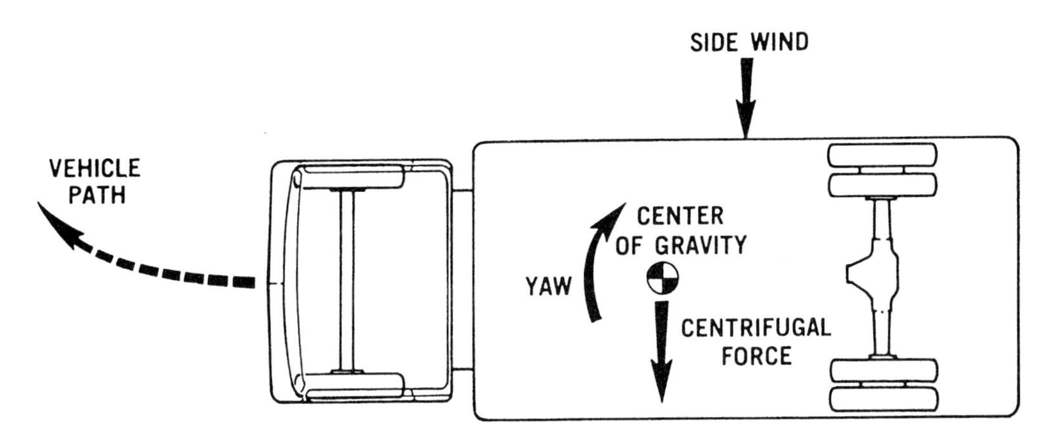

Figure 4-15 Effect of crosswind on vehicle direction when pressure is applied behind the center of gravity. (Reprinted with permission from SAE Publication The Truck Steering System Hand Wheel to Road Wheel © 1973 Society of Automotive Engineers, Inc.)

center of gravity and directional stability. For example in Figure 4-14 if the load is placed near the rear of the truck, the center of gravity is moved rearward. Because the crosswind force tends to turn the truck around the center of gravity, the crosswind force on the truck is increased when the center of gravity is moved rearward by excessive load near the rear of the truck. Under this condition the driver must supply more force to the steering wheel to turn the truck into the wind and keep the vehicle moving straight ahead. The technician must understand the center of gravity and the pressure center to diagnose the effect of improper load distribution on vehicle directional stability.

Understeer and Oversteer

Technicians must understand understeer and oversteer in relation to the vehicle center of gravity and load distribution to diagnose understeer and oversteer complaints. Weight distribution on a truck and tire cornering force characteristics affect the directional control of the vehicle while steering around a curve if the lateral acceleration is changed. While driving around a curve, the cornering force tends to force the vehicle toward the outer circumference of the curve. The truck design and load distribution should position the vehicle center of gravity so the cornering force is applied at the center of gravity (Figure 4-16). Because the center of gravity is positioned closer to the front of the vehicle, the front tires must supply more cornering force than the rear tires if the cornering force is applied at the center of gravity.

Understeer occurs when the vehicle tends to steer toward the outside of a curve when accelerating around the curve. The driver must turn the steering wheel toward the inside of the curve to maintain the original vehicle direction.

Oversteer occurs when the vehicle tends to steer toward the inside of a curve when accelerating around the curve. The driver must turn the steering wheel toward the outside of the curve to maintain the original vehicle direction.

When the driver accelerates the vehicle while turning around a curve, the increased forward velocity of the vehicle causes more lateral acceleration. This increase in lateral acceleration must be offset by additional cornering force supplied by the tires to achieve a steady state condition. However, the additional cornering force supplied by the tires does not occur instantly. A time lag is present between the increase in lateral acceleration and the additional cornering force supplied by the tires. Because the center of gravity is closer to the front of the vehicle, during this time lag more deficiency of force occurs at the front tires than the rear tires. Under this condition the lateral acceleration forces the vehicle toward the outside of the curve, and the driver must turn the steering wheel slightly toward the inside of the curve to maintain the original path of vehicle travel. Refer to Figure 4-16. This condition is called **understeer.**

If the center of gravity is positioned more toward the rear of the vehicle and the vehicle is accelerated during a curve, the lateral acceleration causes the vehicle to steer toward the inside of the curve. Therefore the driver must turn the steering wheel slightly toward the outside of the curve to maintain the original vehicle direction. This condition is referred to as **oversteer.** Trucks are designed to provide a small amount of understeer because the natural driver reaction is to expect to turn the steering wheel and steer the vehicle toward the inside of the curve when accelerating during a curve. Load distribution is important to provide the proper amount of understeer. If the load is placed near the rear of the truck, the center of gravity is moved toward the rear. This results in an oversteer condition while accelerating during a curve. When the load is placed near the front of the vehicle, the center of gravity is moved toward the front. This condition causes excessive understeer.

Steering Linkage with I Beam Front Suspension

Drag Links

WARNING: Never attempt to straighten steering linkages. This action may weaken the linkage and cause it to break suddenly, causing loss of steering, expensive truck damage, or personal injury.

WARNING: Never heat steering linkages with an oxyacetylene torch. This action may weaken the linkage and cause it to break suddenly, causing loss of steering, expensive truck damage, or personal injury.

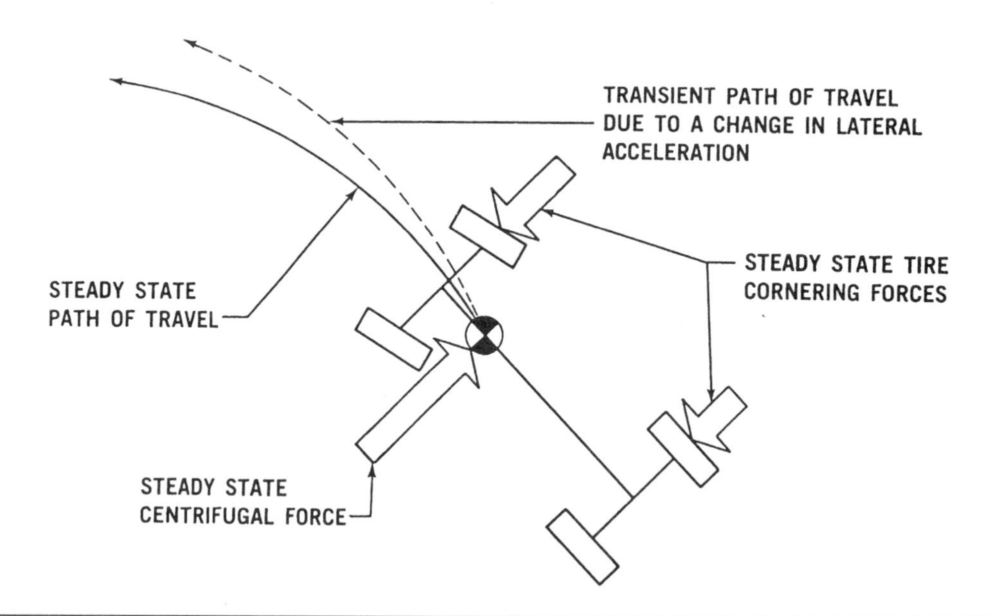

Figure 4-16 Change in vehicle direction caused by lateral acceleration while turning around a curve. (Reprinted with permission from SAE Publication The Truck Steering System Hand Wheel to Road Wheel © 1973 Society of Automotive Engineers, Inc.)

Many medium and heavy duty trucks have a solid I beam front axle. This type of front suspension system may be classified as nonindependent. In this type of front suspension vertical movement of one front wheel affects the position of the other front wheel to some extent.

A nonindependent front suspension system is one in which vertical movement of one front wheel affects the position of the opposite front wheel.

A **drag link** or relay rod is connected from the pitman arm on the steering gear to the **steering arm** on the left front wheel. A nonremovable tie rod end is mounted on each end of the drag link (Figure 4-17). Some drag links have removable tie rod ends. The inner end of each tie rod end has a ball that is mounted in a matching socket in the drag link. These tie rod ends allow drag link movement without looseness. Many drag links contain grease fittings to allow lubrication of the tie rod ends.

Drag Link and Pitman Arm Ball Position

As the front wheels of a truck or tractor are driven over road irregularities, the front wheels are forced to move upward and downward. When a tire strikes a hump in the road surface, the resulting upward wheel movement is called **wheel jounce.** Downward wheel and tire movement is referred to as **wheel rebound.** During wheel jounce and rebound as the axle moves upward and downward, the drag link moves through an arc as it pivots on the pitman arm ball (Figure 4-18).

Wheel jounce refers to upward wheel movement when the tire strikes a hump in the road surface.

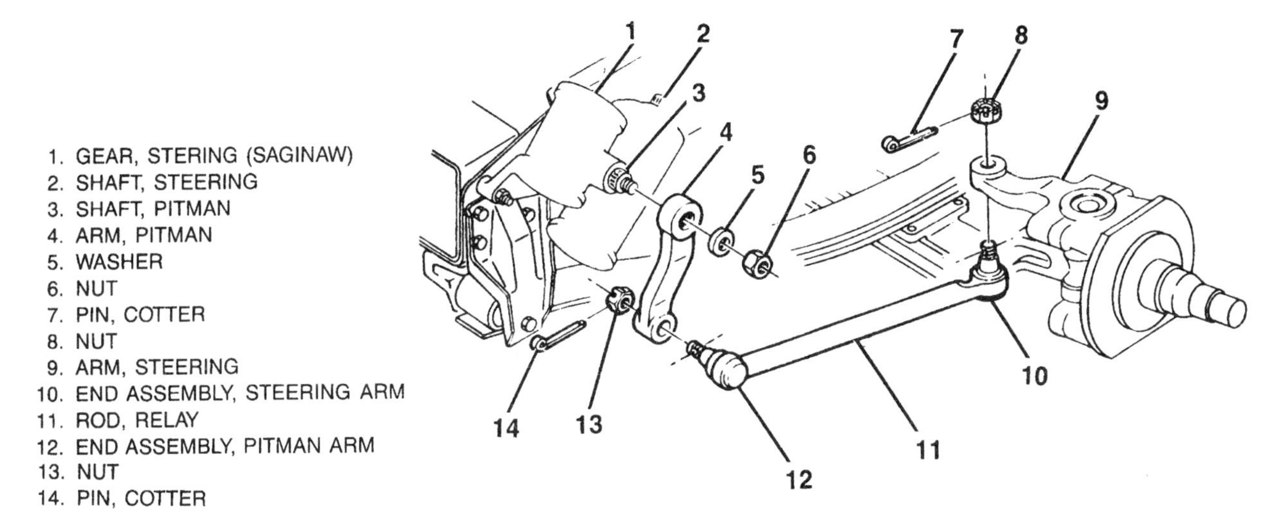

Figure 4-17 Drag link or relay rod and related components. (Courtesy of General Motors Corporation, Service Technology Group.)

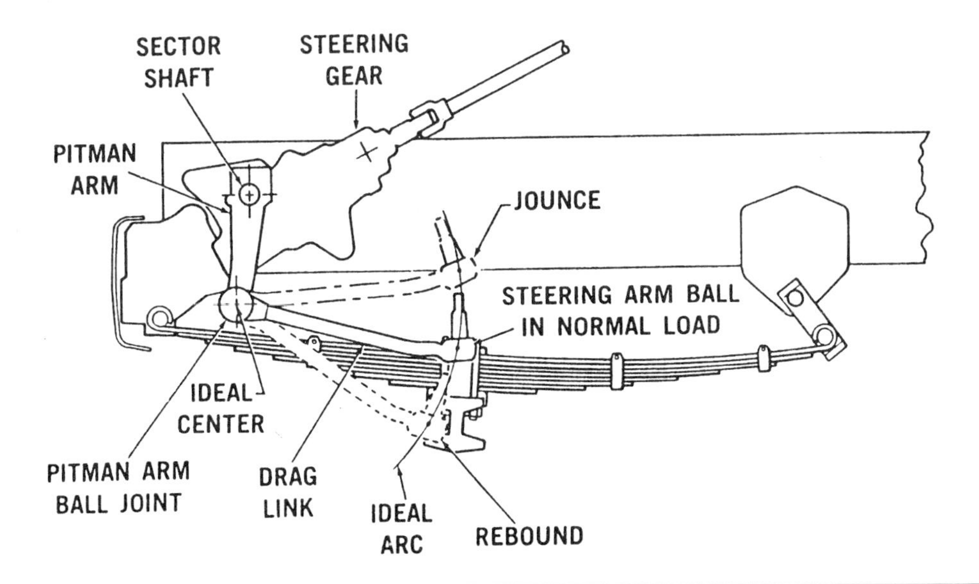

Figure 4-18 Drag link arc during wheel jounce and rebound. (Reprinted with permission from SAE Publication The Truck Steering System Hand Wheel to Road Wheel © 1973 Society of Automotive Engineers, Inc.)

Wheel rebound is downward wheel movement.

When the drag link moves through this arc during wheel jounce and rebound, it tends to pull the pitman arm and rotate the steering wheel back and forth. This steering wheel rotation as a result of wheel jounce and rebound must be minimized to allow the driver to maintain directional control of the vehicle. To minimize steering wheel rotation during wheel jounce and rebound the center of the pitman arm ball must be positioned on the center of the arc of steering arm ball travel (Figure 4-19). When the pitman arm ball is properly positioned, steering wheel rotation is minimized during wheel jounce and rebound.

An improperly positioned or loose steering gear, sagging or broken springs, or worn spring bushings may cause the pitman arm ball to be located at a lower position that is not at the center of the arc of steering arm ball travel. Under this condition the drag link arc is changed during wheel jounce and rebound. When the drag link arc is changed by improper pitman arm ball position, steering wheel rotation is increased during wheel jounce and rebound (Figure 4-20). This excessive steering wheel rotation makes it more difficult for the driver to maintain directional control of the vehicle. Students and technicians must understand the effect of improper pitman arm ball position to diagnose excessive steering wheel rotation during wheel jounce and rebound.

Spring windup is the bending of the spring as the axle attempts to rotate with the wheels during vehicle braking. This spring windup bends the front half of the spring downward and distorts the rear half of the spring upward (Figure 4-21). The center of the drag link tie rod end connected to the steering arm must be positioned at the center of rotation on the spring. If the center of the rear drag tie rod end is in this position, this tie rod end does not move during spring windup. However if the center of the rear drag link ball is not in this position, the drag link is moved during spring windup, and this action rotates the steering wheel. This action makes it more difficult for the driver to maintain directional control of the vehicle.

Tie Rods and Tie Rod Ends

A **tie rod** with removable tie rod ends is connected from the lower steering arm on the left front wheel to the steering arm on the right front wheel. These tie rod ends are threaded into the ends of

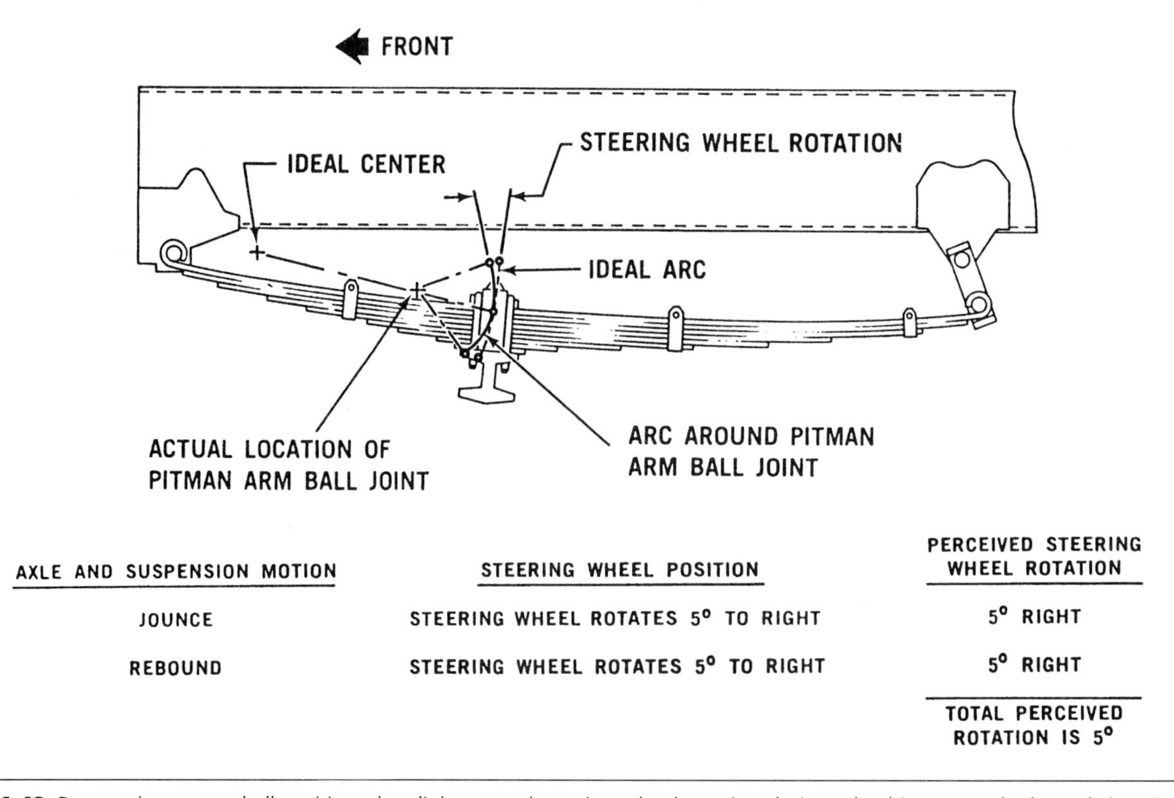

AXLE AND SUSPENSION MOTION	STEERING WHEEL POSITION	PERCEIVED STEERING WHEEL ROTATION
JOUNCE	STEERING WHEEL ROTATES 5° TO RIGHT	5° RIGHT
REBOUND	STEERING WHEEL ROTATES 5° TO RIGHT	5° RIGHT
		TOTAL PERCEIVED ROTATION IS 5°

Figure 4-19 Proper pitman arm ball position, drag link arc, and steering wheel rotation during wheel jounce and rebound. (Reprinted with permission from SAE Publication The Truck Steering System Hand Wheel to Road Wheel © 1973 Society of Automotive Engineers, Inc.)

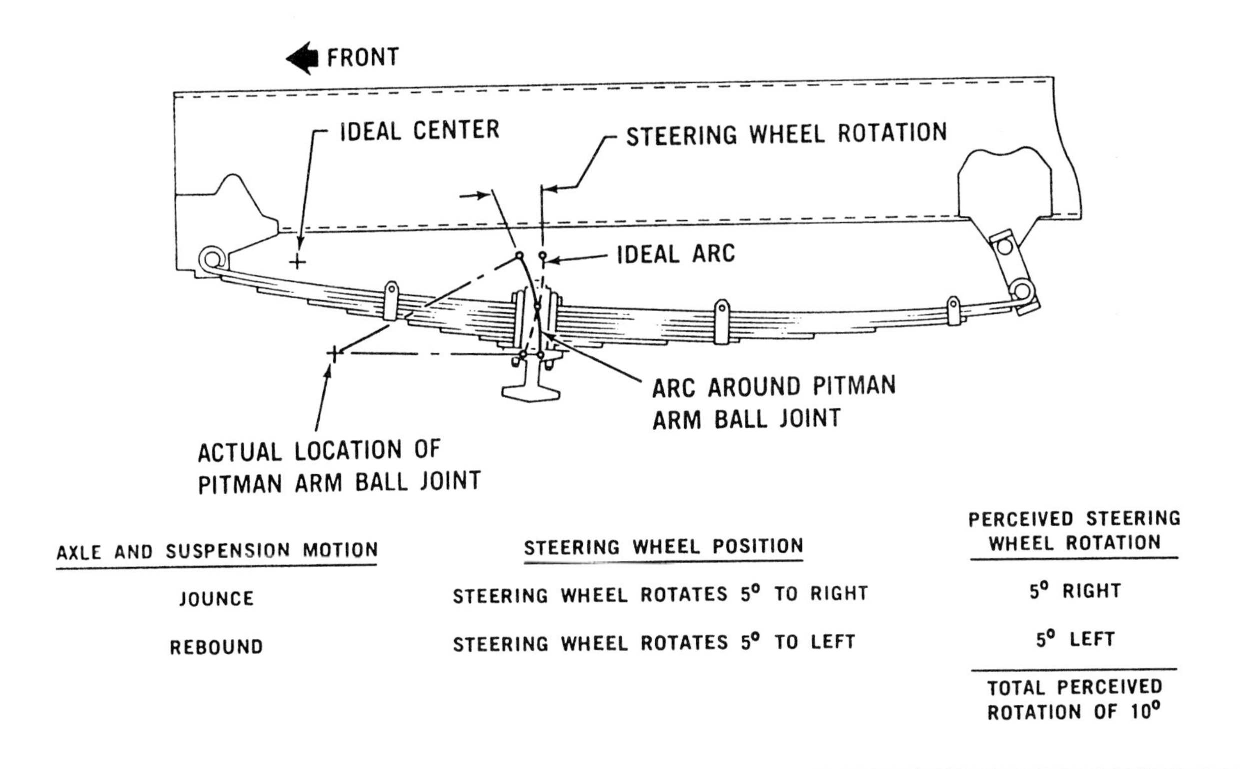

AXLE AND SUSPENSION MOTION	STEERING WHEEL POSITION	PERCEIVED STEERING WHEEL ROTATION
JOUNCE	STEERING WHEEL ROTATES 5° TO RIGHT	5° RIGHT
REBOUND	STEERING WHEEL ROTATES 5° TO LEFT	5° LEFT
		TOTAL PERCEIVED ROTATION OF 10°

Figure 4-20 Improper pitman ball position and drag link arc cause excessive steering wheel rotation during wheel jounce and rebound. (Reprinted with permission from SAE Publication The Truck Steering System Hand Wheel to Road Wheel © 1973 Society of Automotive Engineers, Inc.)

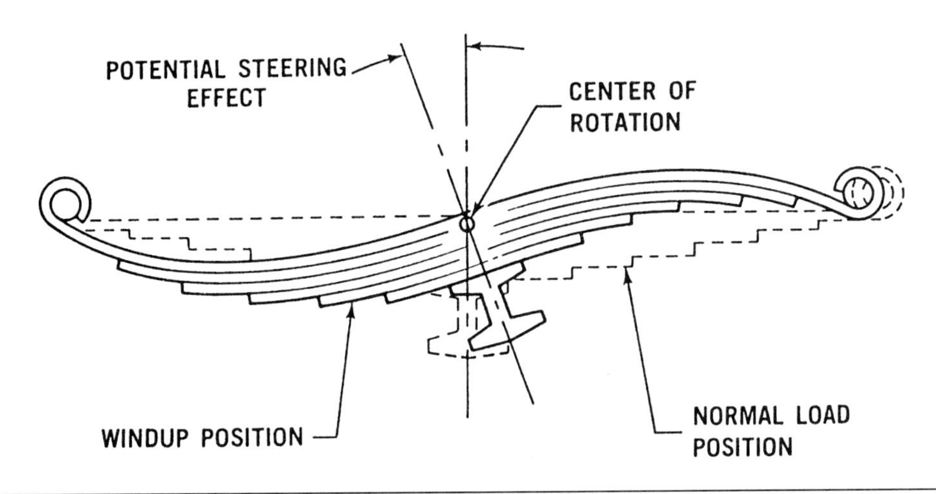

Figure 4-21 The center of the rear drag link tie rod end must be positioned at the center of spring rotation during spring windup. (Reprinted with permission from SAE Publication The Truck Steering System Hand Wheel to Road Wheel © 1973 Society of Automotive Engineers, Inc.)

the tie rod, and a clamp is tightened on the outer ends of the tie rod to prevent tie rod rotation. The slots in the tie rod clamps must be positioned away from the slots in the ends of the tie rod (Figure 4-22). Tie rod ends contain a ball stud enclosed in a metal housing. A spherical bearing at the top of the housing contacts the upper side of the ball stud. A spring seat and spring are mounted in the lower end of the housing. The spring maintains upward pressure on the spring seat to prevent ball stud vertical movement (Figure 4-23). Many tie rod ends contain grease fittings to allow lubrication of the ball joints in these ends. In some tie rod ends the lower end of the ball stud is surrounded by a polyethylene insert (Figure 4-24). Other tie rod ends have the lower end of the ball stud encapsulated in rubber (Figure 4-25). The length of the tie rod determines the front wheel toe.

Front wheel toe is the distance between the inside front edges of the front tires compared with the distance between the inside rear edges of the front tires.

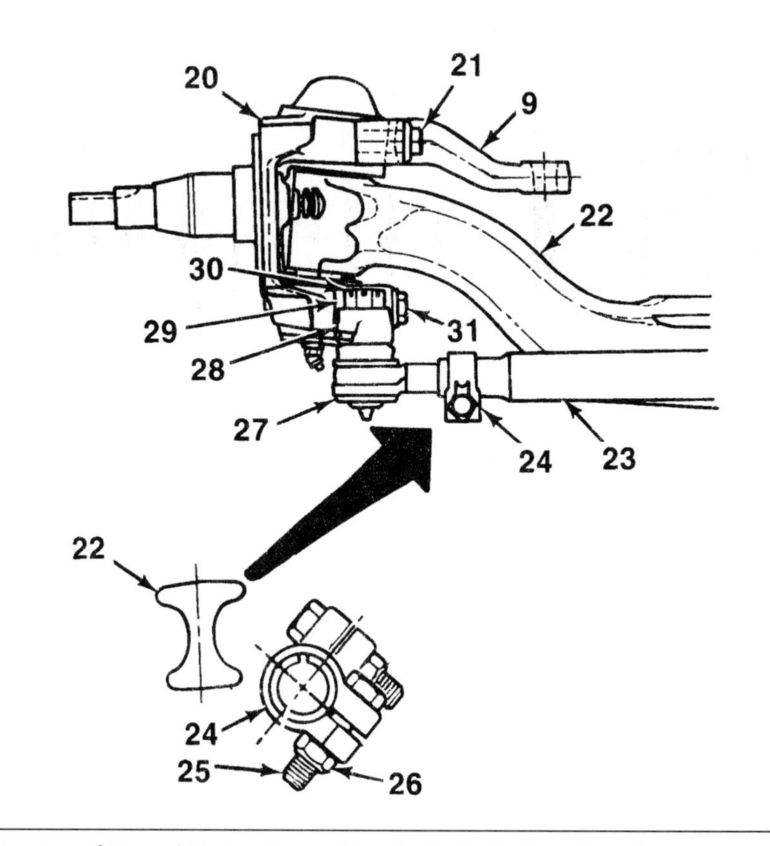

9. Steering Arm
20. Knuckle
21. Steering Arm Bolt
22. Axle
23. Tie Rod Tube
24. Tie Rod Clamp
25. Tie Rod Clamp Bolt
26. Tie Rod Clamp Nut
27. Tie Rod Ball Joint
28. Tie Rod Arm
29. Ball Joint Nut
30. Cotter Pin
31. Tie Rod Arm Bolt

Figure 4-22 Tie rod and removable tie rod end. (Courtesy of General Motors Corporation, Service Technology Group.)

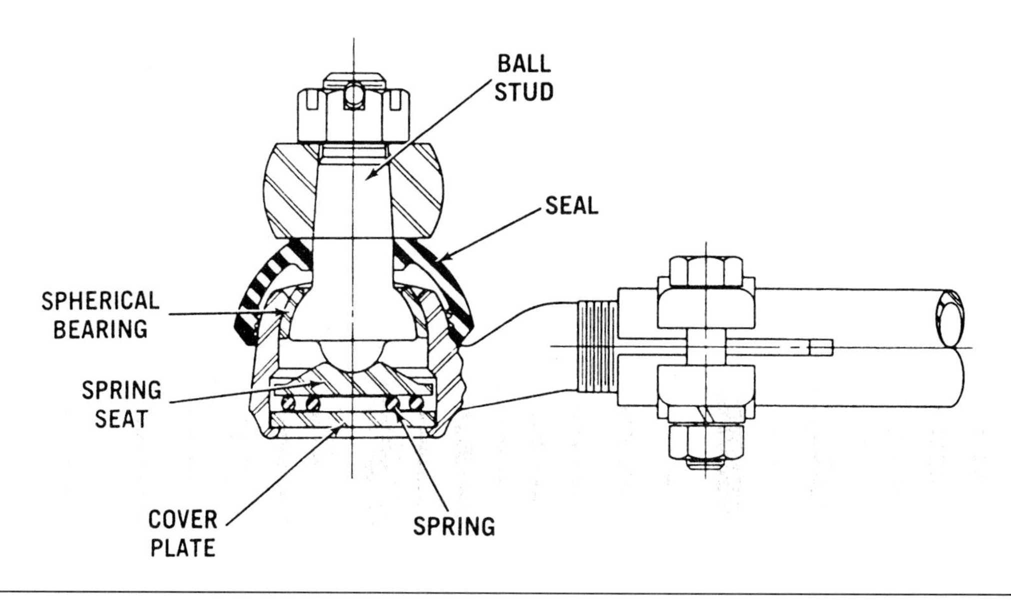

Figure 4-23 Tie rod end with upper spherical bearing and lower spring seat. (Reprinted with permission from SAE Publication The Truck Steering System Hand Wheel to Road Wheel © 1973 Society of Automotive Engineers, Inc.)

Parallelogram Steering Linkage with Independent Front Suspension

WARNING: Always remember that a customer's life may depend on the condition of the steering linkages on his or her vehicle. During under-truck service always make a quick check of steering linkage condition.

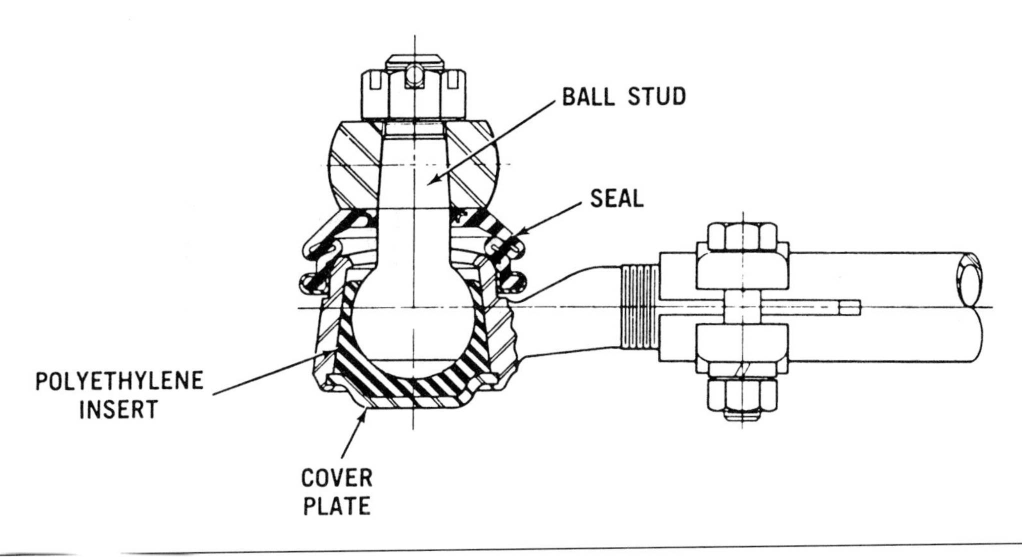

Figure 4-24 Tie rod end with polyethylene insert. (Reprinted with permission from SAE Publication The Truck Steering System Hand Wheel to Road Wheel © 1973 Society of Automotive Engineers, Inc.)

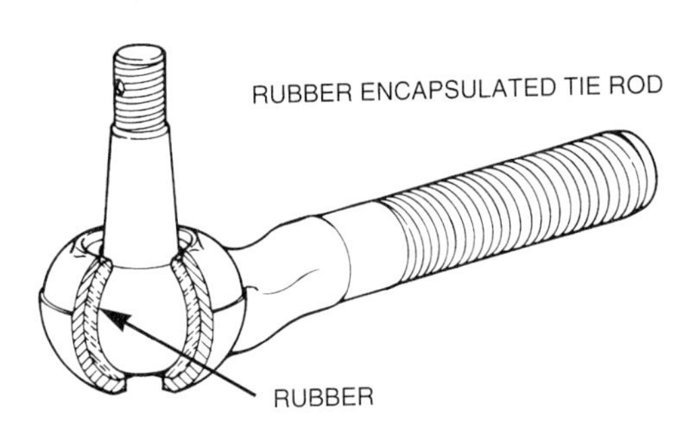

Figure 4-25 Tie rod end with lower end of ball stud encapsulated in rubber. (Courtesy of Moog Automotive.)

Some medium duty trucks have an independent front suspension system with a parallelogram steering linkage. Independent front suspension systems do not have enough load-carrying capabilities for use in heavy duty trucks. In an independent front suspension system each front wheel can move vertically without affecting the position of the opposite front wheel.

An independent front suspension system is one in which vertical movement of one front wheel does not affect the position of the opposite front wheel.

Steering linkage mechanisms are used to connect the steering gear to the front wheels. Parallelogram steering linkages may be mounted behind the front suspension or ahead of the front suspension (Figure 4-26). The **parallelogram steering linkage** must not interfere with the engine oil pan or chassis components.

In a parallelogram steering linkage the tie rods are positioned so they are parallel to the lower control arms.

Regardless of the parallelogram steering linkage mounting position, this type of steering linkage contains the same components. The main components in this steering linkage mechanism are as follows:

1. Pitman arm
2. Center link
3. **Idler arm** assembly
4. Tie rods with sockets
5. Tie rod ends

In a parallelogram steering linkage the tie rods are connected parallel to the lower control arms. Road vibration and shock are transmitted from the tires and wheels to the steering linkage, and these forces tend to wear the linkages and cause steering looseness. If the steering linkage components are worn, steering control is reduced. Because loose steering linkage components

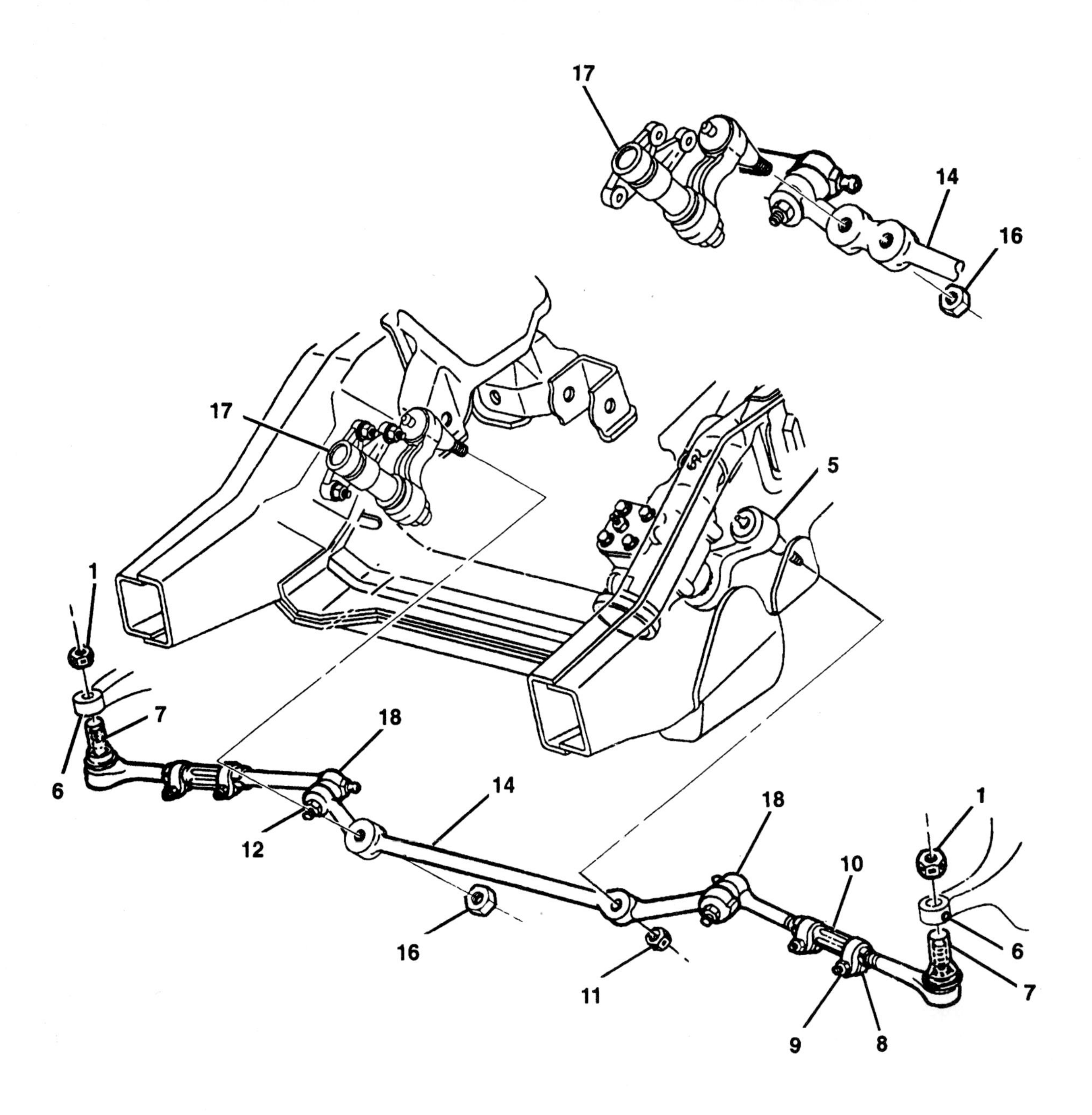

1. NUT, OUTER TIE ROD, 54 N·m (40 FT. LBS.)
5. ARM, PITMAN
6. KNUCKLE
7. TIE ROD BALL STUD
8. CLAMP ADJUSTER
9. NUT, 19 N·m (14 FT. LBS.)
10. SLEEVE, ADJUSTER
11. NUT, PITMAN ARM, 70 N·m (52 FT. LBS.)
12. NUT, INNER TIE ROD, 54 N·m (40 FT. LBS.)
14. ROD, RELAY
16. NUT, IDLER ARM, 70 N·m (52 FT. LBS.)
17. ARM, IDLER
18. ROD, INNER TIE

Figure 4-26 Parallelogram steering linkage. (Courtesy of General Motors Corporation, Service Technology Group.)

cause intermittent toe changes, this problem increases tire wear. The wear points in a parallelogram steering linkage are the tie rod sockets and ends, idler arm, and center link end.

Tie Rods

The tie rod assemblies connect the center link to the steering arms, which are bolted to the front steering knuckles. In some front suspensions the steering arms are part of the steering knuckle, whereas in other front suspension systems the steering arm is bolted to the knuckle. A ball socket is mounted on the inner end to each tie rod, and a tapered stud on this socket is mounted in a center link opening. A castellated nut and cotter key retain the tie rods to the center link. A threaded sleeve is mounted on the outer end of each tie rod, and a tie rod end is threaded into the outer end of this sleeve.

Each tie rod sleeve contains a left hand and a right hand thread where they are threaded onto the tie rod end and the tie rod. Therefore sleeve rotation changes the tie rod length and provides a toe adjustment. Clamps are used to tighten the tie rod sleeves. The clamp opening must be positioned away from the slot in the tie rod sleeve. The design of the steering linkage mechanism allows multiaxial movement, because the front suspension moves vertically and horizontally. Ball-and-socket–type pivots are used on the tie rod assemblies and center link.

If the front wheels hit a bump, the wheels move up and down, and the control arms move through their respective arcs. Because the tie rods are connected to the steering arms, these rods must move upward with the wheel. Under this condition the inner end of the tie rod acts as a pivot, and the tie rod also moves through an arc. This arc is almost the same as the lower control arm arc because the tie rod is parallel to the lower control arm. Maintaining the same arc between the lower control arm and the tie rod minimizes toe change on the front wheels during upward and downward wheel movement. This action improves the directional stability of the vehicle and reduces tread wear on the front tires.

Pitman Arm

The pitman arm connects the steering gear to the center link. This arm also supports the left side of the center link. Motion from the steering wheel and steering gear is transmitted to the pitman arm, and this arm transfers the movement to the steering linkage. This pitman arm movement forces the steering linkage to move to the right or left, and the linkage moves the front wheels in the desired direction. The pitman arm also positions the center link at the proper height to maintain the parallel relationship between the tie rods and the lower control arms. The opening in the inner end of both types of pitman arms has serrations that fit over matching serrations on the steering gear shaft. A nut and lock washer retain the pitman arm to the steering gear shaft.

Idler Arm

An idler arm support is bolted to the frame or chassis on the opposite end of the center link from the pitman arm. The idler arm is connected from the support bracket to the center link. Two bolts retain the idler arm bracket to the frame or chassis. In some idler arms a ball stud on the outer end of the arm fits into a tapered opening in the center link, whereas in other idler arms a ball stud in the center link fits into a tapered opening in the idler arm.

The idler arm supports the right side of the center link and helps to maintain the parallel relationship between the tie rods and the lower control arms. The outer end of the idler arm is designed to swivel on the idler arm bracket, and this swivel is subject to wear. A worn idler arm swivel causes excessive vertical steering linkage movement and erratic toe. This action results in excessive steering wheel free-play, with reduced steering control and front tire wear.

Center Link

The center link controls the sideways steering linkage and wheel movement. The center link together with the pitman arm and idler arm provides the proper height for the tie rods, which is im-

See Chapter 4 in the Shop Manual for steering linkage service.

portant to minimize toe change on road irregularities. Some center links have tapered openings in each end, and the studs on the pitman arm and idler arm fit into these openings. This type of center link may be called a taper end or nonwear link. Other wear-type center links have ball sockets in each end with tapered studs extending from the sockets. These tapered studs fit into openings in the pitman arm and idler arm, and they are retained with castellated nuts and cotter keys.

All-Wheel Steering

Purpose

All-wheel steering is not commonly used on trucks at the present time. However, at least one truck manufacturer in the United States has developed an all-wheel computer-controlled steering system. These systems may be more widely used in the future. Because the all-wheel steering system is not widely used, the discussion is brief. The all-wheel steering system may be used on single or tandem rear axle trucks (Figure 4-27). The all-wheel steering system improves truck maneuverability. In the coordinated steering mode the rear wheels are steered in the opposite direction to the front wheels, and this reduces the turning radius of the truck by 30% (Figure 4-28). When the all-wheel steering system is in the crab mode, the rear wheels are steered in the same direction as the front wheels. This action provides diagonal movement of the truck for parking and repositioning of the vehicle in areas with limited space.

All-Wheel Steering System Components

The all-wheel steering system has a conventional front axle steering system. On some applications the same power steering pump supplies hydraulic pressure to both the front and rear axle steering systems. Other trucks have a separate power steering pump to supply hydraulic pressure to the rear

Figure 4-27 Tandem rear axle truck with all-wheel steering. (Reprinted with permission from SAE Publication All Wheel Steering for Heavy Truck Application © 1995 Society of Automotive Engineers, Inc.)

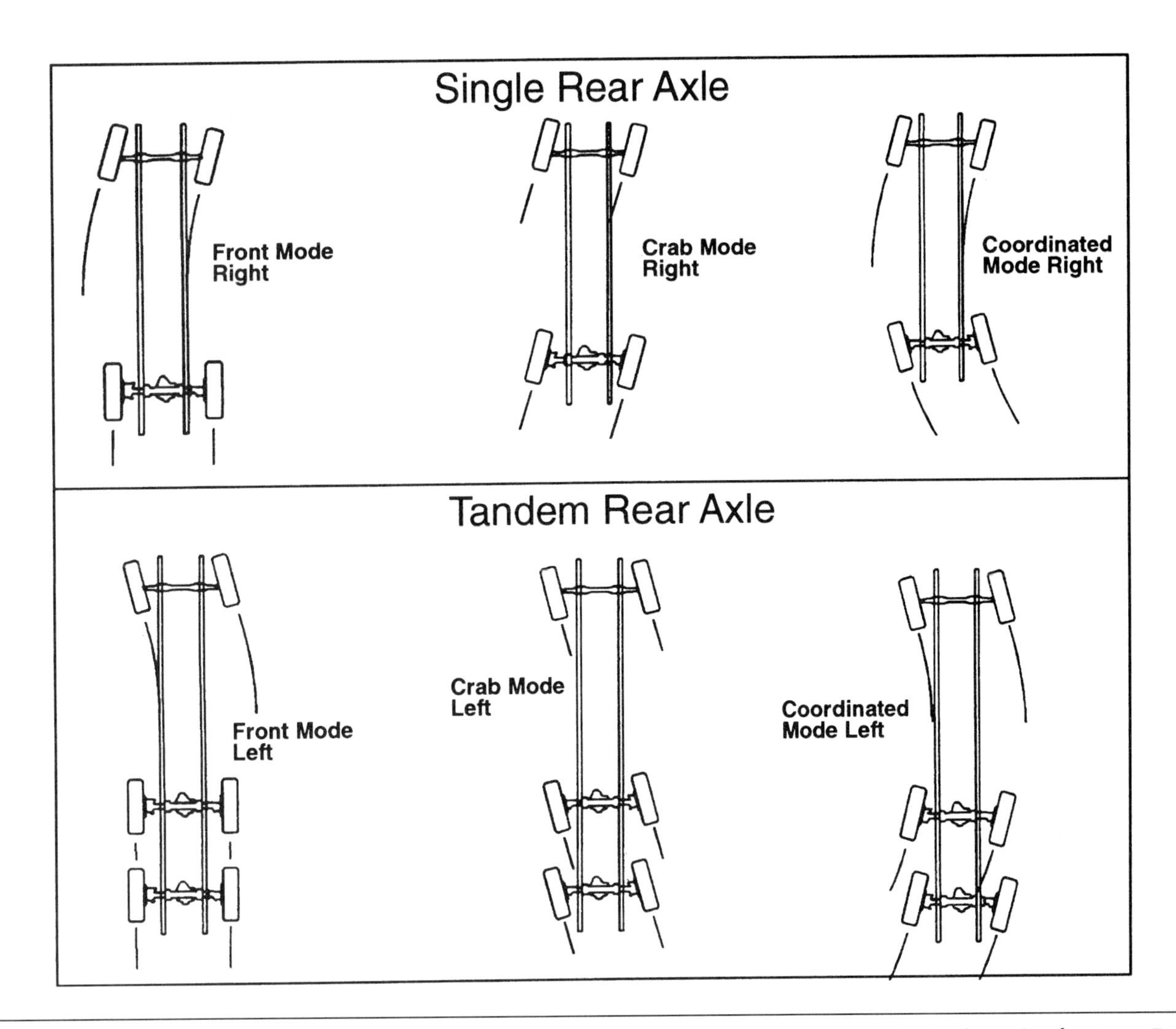

Figure 4-28 All-wheel steering operating modes. (Reprinted with permission from SAE Publication All Wheel Steering for Heavy Truck Application © 1995 Society of Automotive Engineers, Inc.)

axle steering system. Steering sensors are mounted on each axle. These sensors contain rotary potentiometers that send analog voltage signals to the steering computer in relation to the steering angle on each axle. On the front axle the steering sensor senses steering spindle rotation.

The steering computer receives input signals from the steering sensors and other inputs such as vehicle speed. The steering computer operates valves in the hydraulic manifold assembly to direct hydraulic fluid from the power steering pump to the appropriate side of the rear axle steering cylinder (Figure 4-29).

Each rear axle has a steering cylinder to provide the proper rear wheel steering angle (Figure 4-30). When the system is operating only in the front wheel steering mode or faster than a specific vehicle speed, the rear axle steering cylinders are locked in the straight-ahead position by check valves that lock hydraulic pressure on both sides of the steering cylinder piston. On some applications the rear steering cylinders are locked by a lock valve that is operated by voltage from the steering computer and air pressure. If the computer senses a defect in the all-wheel steering system, the rear wheels are locked in the straight-ahead position.

All-Wheel Steering System Operation

The driver may select front steer, coordinated steer, or low speed on the instrument panel steering switches (Figure 4-31). The front steer mode is used for normal highway driving. When increased maneuverability is required, the coordinated steer mode may be selected. The steering computer

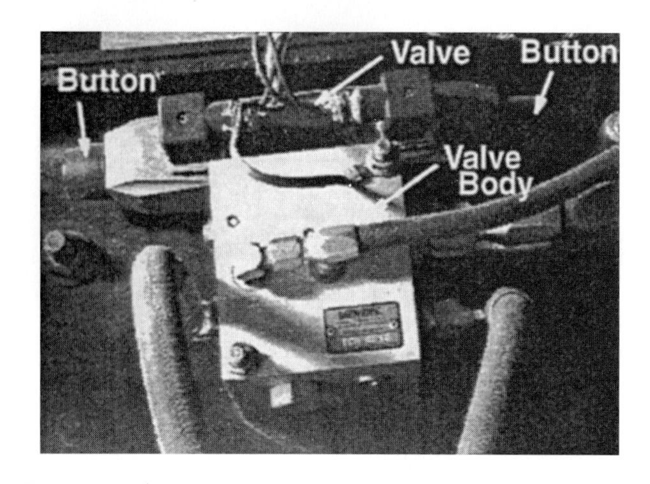

Figure 4-29 Hydraulic manifold, all-wheel steering system. (Reprinted with permission from SAE Publication All Wheel Steering for Heavy Truck Application © 1995 Society of Automotive Engineers, Inc.)

Figure 4-30 Rear wheel steering cylinder. (Reprinted with permission from SAE Publication All Wheel Steering for Heavy Truck Application © 1995 Society of Automotive Engineers, Inc.)

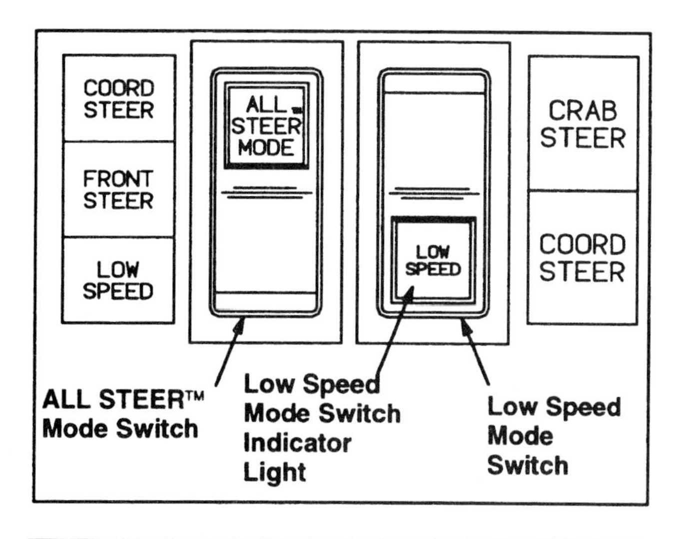

Figure 4-31 All-wheel steer selector switches. (Reprinted with permission from SAE Publication All Wheel Steering for Heavy Truck Application © 1995 Society of Automotive Engineers, Inc.)

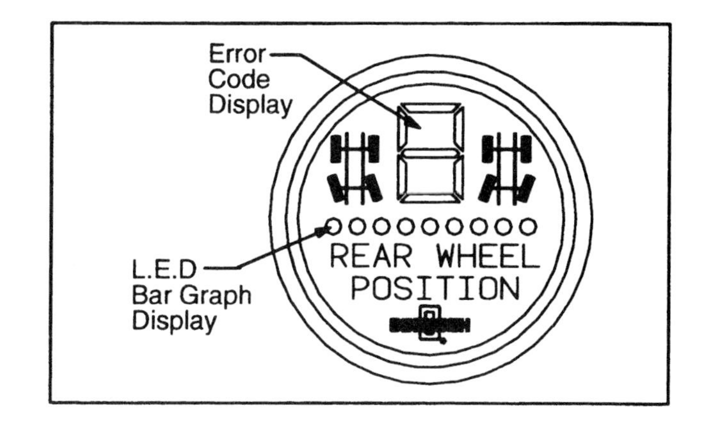

Figure 4-32 Wheel position gauge. (Reprinted with permission from SAE Publication All Wheel Steering for Heavy Truck Application © 1995 Society of Automotive Engineers, Inc.)

does not allow the use of this mode in speeds greater than 35 mph (56 kmh). When the low-speed button is pressed, a second low-speed button may be touched followed by the low-speed crab steer or coordinated steer buttons. These modes may be used when parking or maneuvering in small spaces traveling less than 10 mph (16 kmh). The all-wheel system also has a joystick within reach of the driver. The driver may use this joystick to manually operate the rear wheel steering in the low-speed mode. The joystick function may be useful when coupling a trailer behind the truck. Rear wheel steering angle is limited by stops on the rear axles.

A wheel position gauge is mounted in the instrument panel. Graphic displays of each wheel indicate the steering position of each wheel (Figure 4-32). A light-emitting diode (LED) bar graph also indicates the position of the rear wheels. If the rear wheels are straight ahead, the center LED is illuminated. When the rear wheels are turned to the right, the LED bar graph sweeps from the center to the right. If the steering computer senses a defect in the system, the computer illuminates the error code display.

Summary

- ❑ The steering gear provides a mechanical advantage that allows the driver to turn the front wheels with a reasonable amount of effort.
- ❑ A fixed steering column does not provide any tilting or telescoping action.
- ❑ In a collapsible steering column the steering shaft and column jacket are designed to collapse a certain amount if the driver is thrown against the steering wheel in a frontal collision. This collapsing action helps to prevent driver injury.
- ❑ Some collapsible steering columns have a mounting bracket that is designed to allow some column movement if the driver is thrown against the steering wheel in a frontal collision.
- ❑ On a tilt steering column the steering wheel may be tilted upward or downward when a release lever on the column is actuated. This action provides various steering wheel positions to suit the driver's size. Changing steering wheel positions may help to reduce driver fatigue.
- ❑ On a telescoping steering column the steering column may be lengthened or shortened so the steering wheel is moved closer to or farther from the driver. This action allows the driver to position the steering wheel in the most comfortable position, and the telescoping action also allows the driver to change the steering wheel position to reduce driver fatigue.
- ❑ In a manual steering gear proper preloading of the wormshaft is necessary to eliminate wormshaft end-play and prevent steering gear free-play and vehicle wander.
- ❑ In a manual steering gear proper adjustment of the sector shaft is necessary to obtain the proper interference fit between the sector shaft and wormshaft teeth.
- ❑ Manual steering gears usually have a higher numerical ratio compared with power steering gears.
- ❑ On many truck steering linkages a drag link is connected from the pitman arm on the steering gear to the steering arm on the left front wheel, and a tie rod is connected from the lower steering arm on the left front wheel to the steering arm on the right front wheel.

Terms to Know

Bearing preload
Directional stability
Drag link
Fixed steering column
Gear ratio
Gear tooth lash
Interference fit
Idler arm
Oversteer
Parallelogram steering linkage
Pitman arm
Recirculating ball steering gear
Sector shaft
Spring windup
Steering arm
Steering shaft
Steering wheel free-play
Steering wheel kickback
Tie rod
Tilt and telescoping steering column
Understeer
Vehicle wander
Wheel jounce
Wheel rebound
Worm-and-roller steering gear
Wormshaft

Review Questions

Short Answer Essays

1. Explain the purposes of the steering column.
2. Describe the purpose of the steering gear.
3. Describe two components that provide a collapsible action in a collapsible steering column.
4. Explain the advantages of a tilt/telescoping steering column.
5. List the two components that may be mounted in the steering shaft to provide the tilting action in a tilt steering column.
6. Explain why wormshaft bearing preloading is required in a manual steering gear, and describe how this preloading is obtained.
7. Describe why proper sector shaft adjustment is necessary in a manual steering gear, and explain how this adjustment is provided.
8. Explain the differences in steering action between a steering gear with a 13:1 ratio compared with a steering gear with a 16:1 ratio.
9. Describe some of the factors that engineers must consider when determining the steering gear ratio.
10. Explain the purposes of front-wheel steering linkages.

Fill-in-the-Blanks

1. Steering kickback is a strong, sudden movement of the steering wheel in the ____________________ to which the steering wheel is turned.
2. Lateral movement of the wormshaft between the bearings on which this shaft is mounted is referred to as ____________________.
3. Steering wheel free-play is the amount of steering wheel rotation before the ____________________ begin to turn.
4. A steering gear ratio of 16:1 requires ____________________ effort to rotate the steering wheel compared with a steering gear ratio of 13:1.
5. A steering column contains a ____________________ on a truck with a column shift transmission.
6. Tilting action in a tilt steering column may be provided by a universal joint or ____________________.
7. On some manual steering gears a ____________________ adjustment on the lower steering gear cover provides the wormshaft preload adjustment.
8. When the front wheels are straight ahead in a manual recirculating ball steering gear, a ____________________ fit occurs between the sector shaft teeth and the ball nut teeth.
9. In a recirculating ball steering gear ____________________ are mounted between the ball nut and the wormshaft.
10. In many truck steering linkages a ____________________ is connected between the pitman arm and the left front steering arm.

ASE Style Review Questions

1. While discussing collapsible steering columns:
   Technician A says on some collapsible steering columns, the column to instrument panel bracket is designed to allow column movement if the driver is thrown against the steering wheel in a frontal collision.
   Technician B says in some collapsible steering columns steel pins shear off in the two-piece jacket and steering shaft to allow column collapse if the driver is thrown against the steering wheel in a frontal collision.
   Who is correct?
   A. A only **C.** Both A and B
   B. B only **D.** Neither A nor B
2. While discussing tilt steering columns:
   Technician A says in some tilt steering columns the upper steering shaft is connected to the lower steering shaft by a universal joint.
   Technician B says two pivot bolts connect the main housing and the support bracket.
   Who is correct?
   A. A only **C.** Both A and B
   B. B only **D.** Neither A nor B
3. While discussing drag link steering linkages:
   Technician A says a long tie rod is connected between the steering arms on the front wheels.
   Technician B says the drag link usually has removable tie rod ends.
   Who is correct?
   A. A only **C.** Both A and B
   B. B only **D.** Neither A nor B
4. The component that is mounted in a steering column with a column shift transmission that is not required in a steering column with a floor shift transmission is the:
   A. Dimmer switch
   B. Wiper/washer switch
   C. Gear shift tube
   D. Trailer brake control
5. In a fixed steering column:
   A. The ignition lock cylinder is connected through a rod to the ignition switch.
   B. The ignition lock cylinder is retained in the ignition lock housing with a cotter key.
   C. The turn signal switch is bolted to the bottom of the ignition lock housing.
   D. The dimmer switch is connected directly to the turn signal arm.

6. While discussing manual steering gears:
Technician A says in a recirculating ball steering gear teeth on the sector shaft are meshed with the wormshaft.
Technician B says the wormshaft bearings do not require a preload.
Who is correct?
A. A only **C.** Both A and B
B. B only **D.** Neither A nor B

7. Worn or bent steering column components may cause all of these problems *except:*
A. A rattling noise
B. Excessive steering free-play
C. A binding condition when turning the steering wheel
D. Improper front suspension alignment

8. In manual steering gears:
A. With the front wheels straight ahead there should be a specific amount of free-play between the sector shaft teeth and the wormshaft.
B. In a worm-and-roller steering gear the wormshaft has a slight hourglass shape.
C. In many steering gears a shim adjustment is provided between the sector shaft cover and the gear housing.
D. In a recirculating ball steering gear the ball nut inner grooves are in direct contact with the wormshaft.

9. Compared with a steering gear with a 16:1 ratio a steering gear with a 13:1 ratio provides:
A. Reduced effort to turn the front wheels
B. Increased front wheel movement in relation to steering wheel movement
C. More revolutions of the steering wheel to move the front wheels from fully left to fully right
D. A longer drag link from the pitman arm to the left front steering arm

10. While discussing steering linkages:
Technician A says the tie rod clamps must be positioned so the slot in the clamp is aligned with the slot in the tie rod.
Technician B says the length of the tie rod determines the front wheel toe.
Who is correct?
A. A only **C.** Both A and B
B. B only **D.** Neither A nor B

Power Steering Pumps and Power Steering Gears

Upon completion and review of this chapter, you should be able to:

- ❑ Explain two different types of power steering pump reservoirs.
- ❑ List two different types of power steering pump rotor designs.
- ❑ Describe the power steering pump operation while driving with the front wheels straight ahead.
- ❑ Describe the power steering pump operation while the vehicle is turning a corner.
- ❑ Describe the operation of a Saginaw power steering pump pressure relief operation, and explain when this operation occurs.
- ❑ Explain the operation of a Saginaw power steering gear during a right turn.
- ❑ Explain the operation of a Saginaw power steering gear during a left turn.
- ❑ Describe the operation of a TRW/Ross steering gear when the steering wheel is turned so the steering linkage is one third of a turn from the stops.
- ❑ Explain the operation of the torque valve, power cylinder, and safety valve in an air-assisted steering system.
- ❑ Describe the operation of a variable-assist power steering system.

Introduction

Power steering systems have contributed to reduced driver fatigue and made driving a more pleasant experience. At present many power steering systems use fluid pressure to assist the driver in turning the front wheels. Because driver effort required to turn the front wheels is reduced, driver fatigue is decreased. The advantages of power steering have been made available on most trucks today, and safety has been maintained in these systems.

Basic Hydraulic Principles

Pressure and Compressibility

Because liquids and gases are both substances that flow, they may be classified as fluids. If a nail punctures an automotive tire, the air escapes until the pressure in the tire is equal to atmospheric pressure outside the tire. When the tire is repaired and inflated, air pressure is forced into the tire. If the tire is inflated to 32 psi (220.64 kPa), this pressure is applied to every square inch on the inner tire surface. Pressure is always supplied equally to the entire surface of a container. Because air is a gas, the molecules have plenty of space between them. When the tire is inflated, the pressure in the tire increases, and the air molecules are squeezed closer together or compressed. Thus the air molecules cannot move as freely, but extra molecules of air can still be forced into the tire. Therefore gases such as air are said to be compressible.

The air in the tire may be compared with a few balls on a billiard table without pockets. If a few more balls are placed on the table, the balls are closer together, but they can still move freely.

If the vehicle is driven at high speed, friction between the road surface and the tires heats the tires and the air in the tires. The tire is flexing continually at the road surface, and this action produces heat. When air temperature increases, the pressure in the tire also increases. Conversely, a temperature decrease reduces pressure.

If 100 ft^3 (2.8 m^3) of air is forced into a large truck tire and the same amount of air is forced into a much smaller car tire, the pressure in the car tire is much greater.

Molecules in a liquid may be compared with a billiard table without pockets that is completely filled with balls. These balls can roll around, but no additional balls can be placed on the table because the balls cannot be compressed. Similarly liquid molecules are nearly incompressible.

Liquid Flow

If a tube is filled with billiard balls and the outlet is open, more balls may be added to the inlet. When each ball is moved into the inlet, a ball is forced from the outlet. If the outlet is closed, no more balls can be forced into the inlet.

The billiard balls in the tube may be compared with molecules of power steering fluid in the line between the power steering pump and the power steering gear. Because fluid is almost noncompressible, and this fluid fills the line from the pump to the gear, the force developed by the power steering pump pressure is transmitted through the line to the steering gear (Figure 5-1).

This pressure is applied equally to every square inch in the power steering gear pistons. Pascal's law states "Pressure on a confined fluid is transmitted equally in all directions and acts with equal force on equal areas." This statement means the pressure is the same throughout the entire power steering system.

If the diameter of each power steering gear piston is 3 in (7.62 cm), the area of each piston is $3 \times 3.142 = 9.42$ in^2 (60.81 cm^2) (Figure 5-2). If the pressure supplied by the power steering pump is 500 psi (3,447.5 kPa), the force on each power steering gear piston is $500 \times 9.62 = 4{,}810$

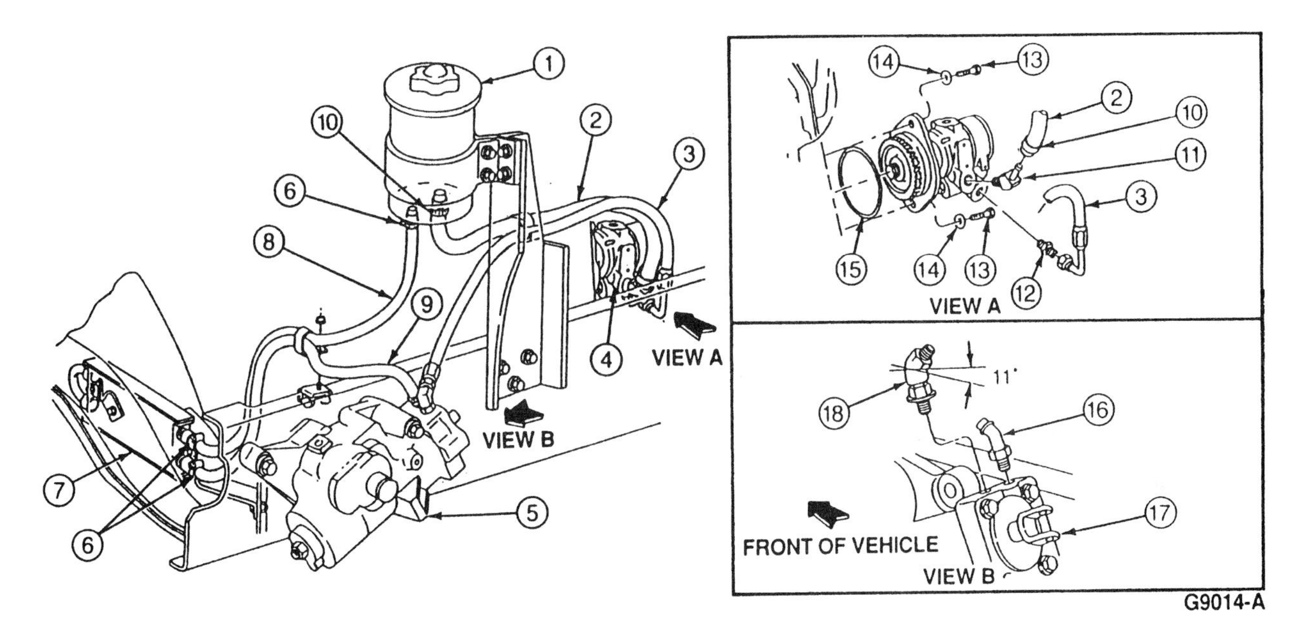

Item	Part Number	Description
1	3A697	Reservoir
2	381172-S160A	Hose, Reservoir to Pump
3	3A719	Hose, Pressure Hose — Pump to Steering Gear
4	3A674	Pump
5	3N503	Steering Gear
6	388651-S	Clamp
7	3D746	Oil Cooler — See Oil Cooler Installation Views
8	3881170-S250A	Hose, Oil Cooler to Reservoir

Item	Part Number	Description
9	3881170-S250A	Hose, Steering Gear to Cooler
10	388653-S	Clamp
11	3E599	Elbow
12	384074-S36	Adapter
13	58721-S2	Bolt
14	44881-S2	Washer
15	87093-S95	O-Ring
16	391068-S36	Nipple
17	3N503	Power Steering Gear
18	384285-S36	Elbow

Figure 5-1 Fluid pressure from the power steering pump is supplied through the high-pressure hose to the steering gear. (Courtesy of Sterling Truck Corporation.)

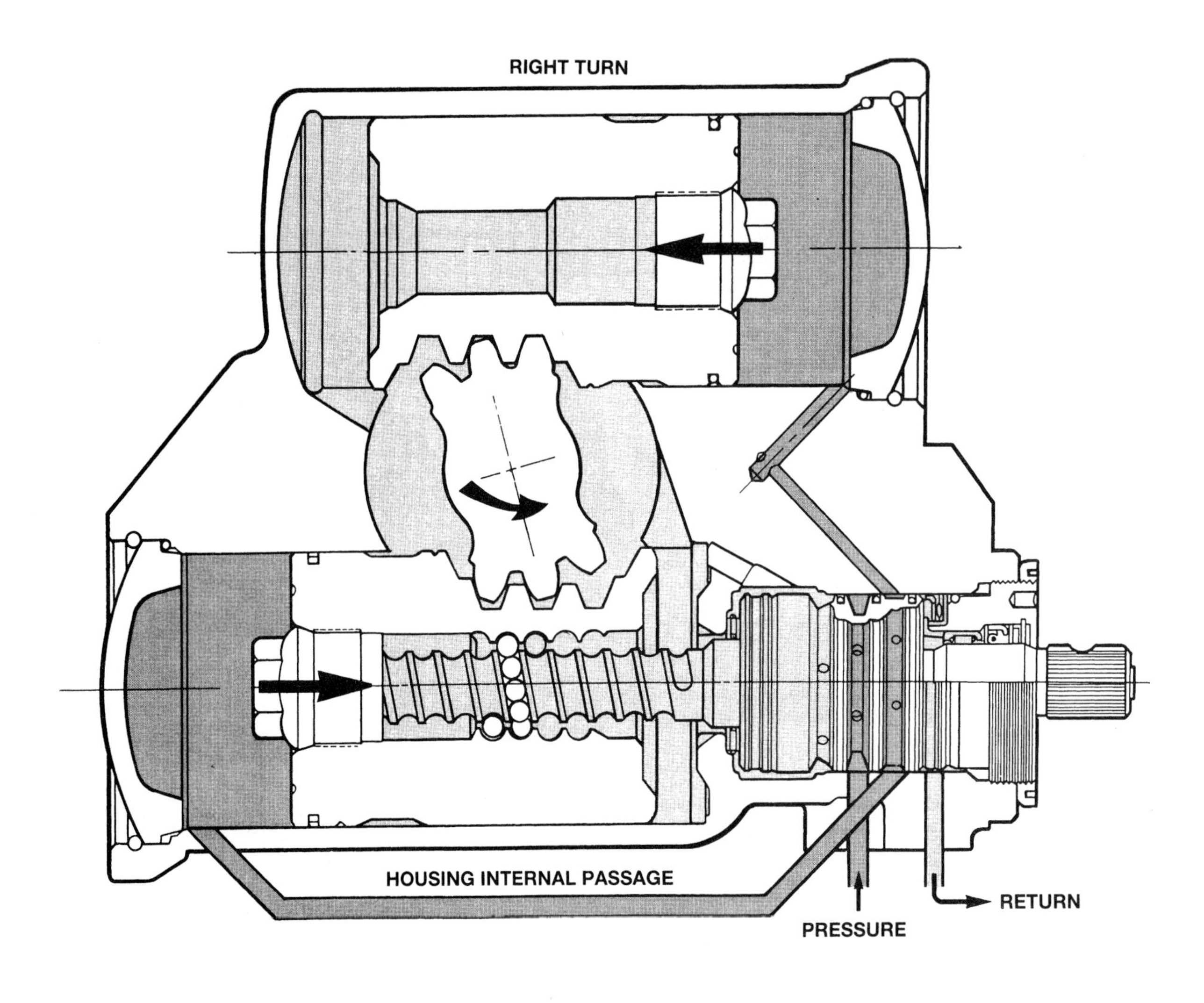

Figure 5-2 The pressure supplied to the steering gear pistons helps to move these pistons, and this action helps the driver to turn the steering wheel. (Courtesy of General Motors Corporation, Service Technology Group.)

psi (33,164.95 kPa). The combined force on the two power steering gear pistons is 4,810 × 2 = 9,620 psi (66,329 kPa). This pressure supplied to the power steering gear pistons helps to move these pistons, and this action helps the driver to turn the steering wheel. The front wheels are turned by the combined forces of the driver's rotational force on the steering wheel and the hydraulic force supplied by the power steering pump to the power steering gear pistons.

Power Steering Pump Drive Belts

CAUTION: Always keep hands, tools, and equipment away from rotating belts and pulleys. If any of these items become entangled in rotating belts, personal injury and equipment damage may result.

CAUTION: Always keep long hair tied back while working in the truck shop. If long hair is entangled in rotating belts and pulleys, personal injury *will* result.

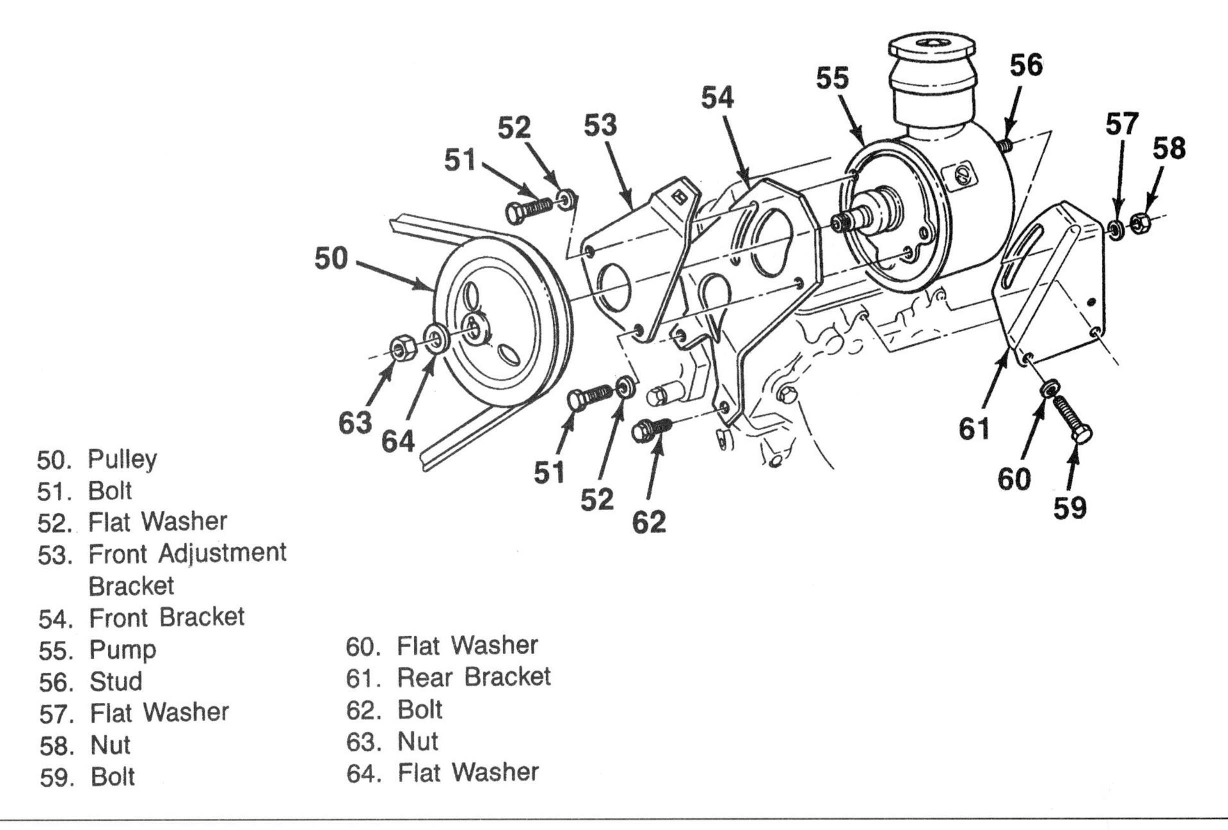

Figure 5-3 Conventional V-belt and power steering pump. (Courtesy of General Motors Corporation, Service Technology Group.)

Power steering pumps on trucks with gasoline engines are usually driven by a **V-belt** that surrounds the crankshaft pulley and the power steering pump pulley. The V-belt may drive other components such as the water pump. The sides of a V-belt are the friction surfaces that drive the power steering pump (Figure 5-3). If the sides of the belt are worn and the lower edge of the belt is contacting the bottom of the pulley, the belt will slip. The power steering pump pulley, crankshaft pulley, and any other pulleys driven by the V-belt must be properly aligned. If these pulleys are misaligned, excessive belt wear occurs. Because a power steering pump will never develop full pressure if the belt is slipping, the belt tension is critical. On many diesel engines the power steering pump is gear driven from the engine (Figure 5-4).

WARNING: Never hammer on the end of a power steering pump pulley to loosen or remove the pulley. This action will damage internal pump components.

Saginaw Power Steering Pump with Blade Vanes

Various types of power steering pumps have been used by truck manufacturers. Many vane-type power steering pumps have blade vanes, which seal the pump rotor to the elliptical pump cam ring (Figure 5-5).

WARNING: Prying on the reservoir must be avoided because this action may damage or puncture the reservoir.

A balanced pulley is pressed on the power steering pump drive shaft, and this pulley and shaft are belt driven by the engine. The pump reservoir is made from steel or plastic. The power steering pump may have an **integral fluid reservoir** or a **remote fluid reservoir.** If the pump has an integral reservoir, a large O-ring (item 8 Figure 5-6) seals the front of the reservoir to the

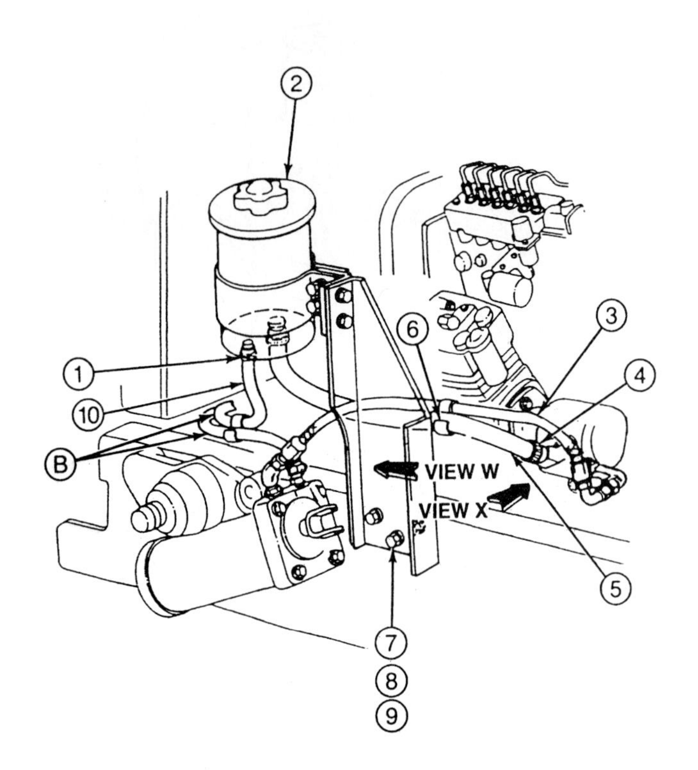

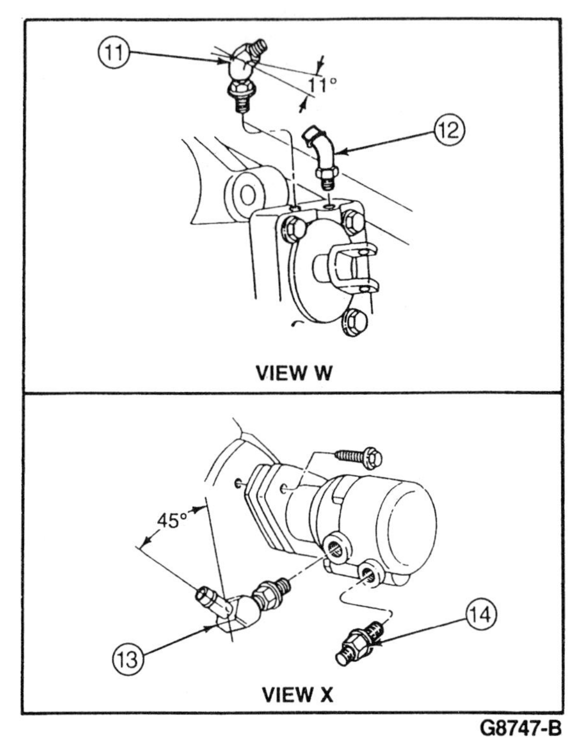

Item	Part Number	Description
1	388651-S	Clamp (2 Req'd)
2	3A697	Reservoir
3	3A719	Hose
4	388653-S	Clamp (2 Req'd)
5	381172-S190A	Hose
6	3E523	Clip
7	34987-S2	Nut (3 Req'd)
8	44877-S2	Washer (3 Req'd)

Item	Part Number	Description
9	58636-S2	Bolt (3 Req'd)
10	381170-S075A	Hose
11	384285-S36	Fitting
12	391068-S36	Nipple
13	3E599	45° Elbow
14	384074-S36	Connector
A	—	Oriented Inboard
B	—	To Oil Cooler — See Oil Cooler Installation Views

Figure 5-4 Gear-driven power steering pump on a diesel engine. (Courtesy of Sterling Truck Corporation.)

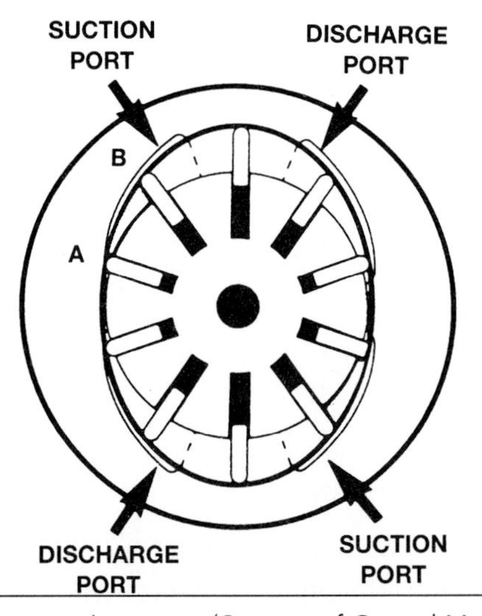

Figure 5-5 Blade vanes in a power steering pump. (Courtesy of General Motors Corporation, Service Technology Group.)

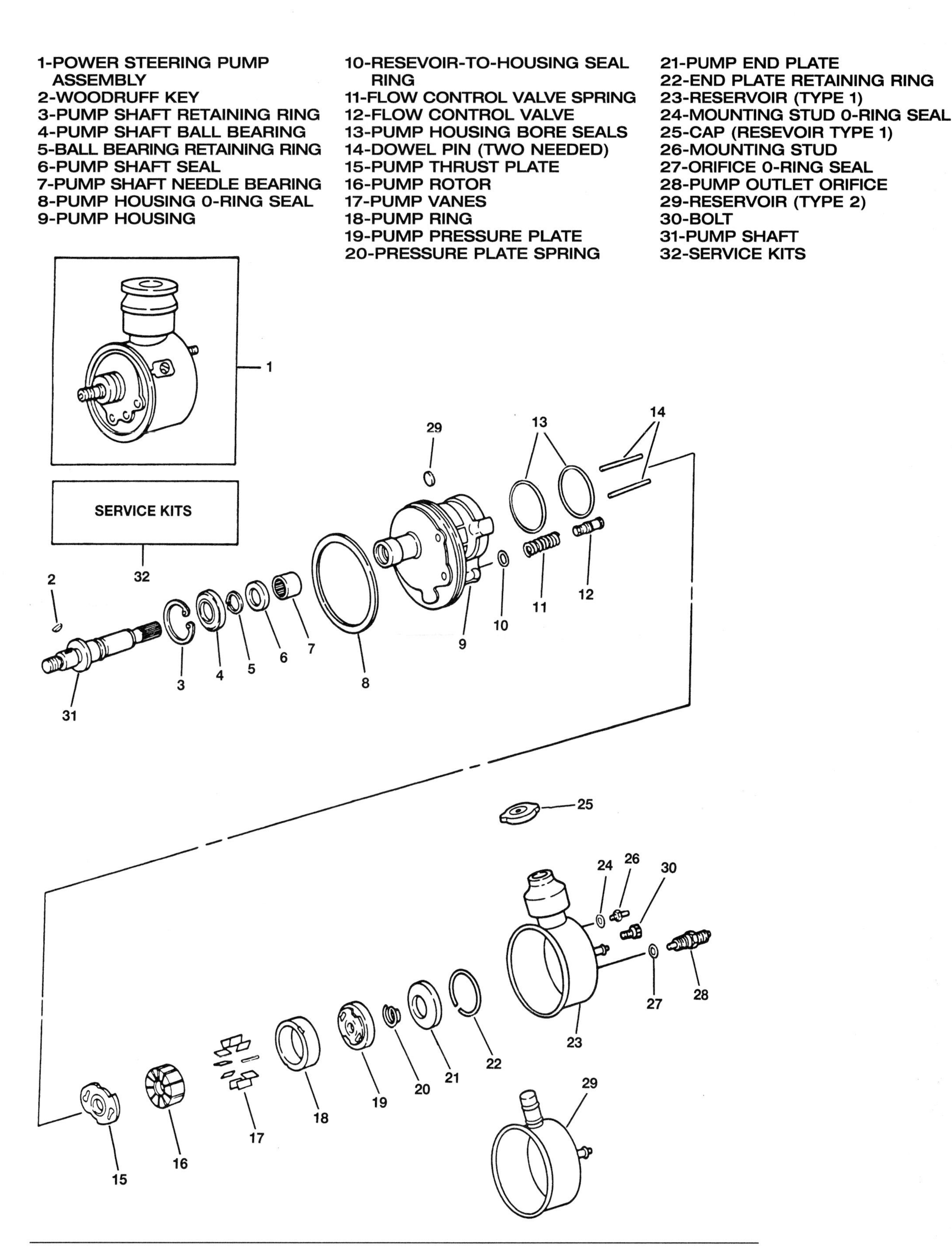

Figure 5-6 Components in a power steering pump with blade vanes. (Courtesy of General Motors Corporation, Service Technology Group.)

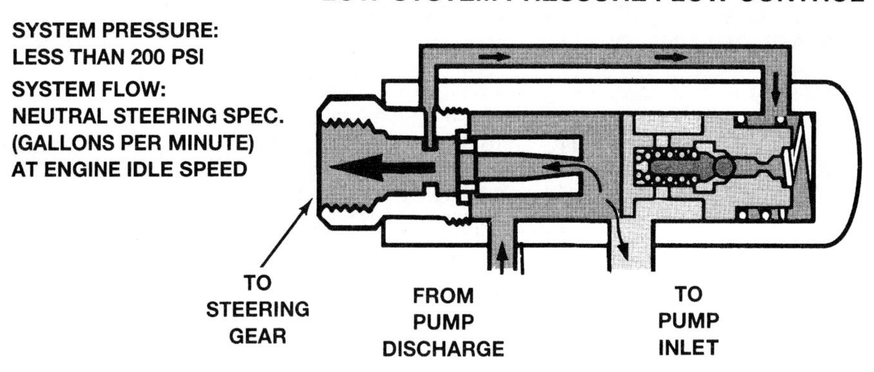

Figure 5-7 Power steering pump pressure relief valve. (Courtesy of General Motors Corporation, Service Technology Group.)

pump housing. Smaller O-rings (items 24 and 27) seal the bolt and fittings on the back of the reservoir. The reservoir cap keeps the fluid reserve in the pump and vents the reservoir to the atmosphere.

The rotating components inside the pump housing include the shaft and rotor (item 16), with the vanes mounted in the rotor slots. The vanes seal the rotor inside the **cam ring.** A seal (item 6) between the pump shaft and the housing prevents oil leaks around the shaft. As the pulley drives the pump shaft, the vanes rotate inside an elliptical-shaped opening in the cam ring. The cam ring remains in a fixed position inside the pump housing. A pressure plate (item 19) is installed in the housing behind the cam ring.

A spring (item 20) is positioned between the pressure plate and the end cover, and a retaining ring holds the end cover in the pump housing. The flow control valve (item 12) is mounted in the pump housing, and a magnet is positioned on the pump housing to pick up metal filings rather than allowing them to circulate through the power steering system.

The **flow control valve** is a precision fit valve controlled by spring pressure and fluid pressure. Any dirt or roughness on the valve results in erratic pump pressure. The flow control valve contains a **pressure relief valve** (Figure 5-7). High-pressure fluid is forced through an **outlet fitting venturi** and a high-pressure hose to the steering gear. When fluid flow in the venturi increases, the pressure in this area decreases. A low pressure hose returns the fluid from the steering gear to the inlet fitting in the pump reservoir.

A venturi is a narrow area in a pipe through which a liquid or gas is flowing. When liquid or gas flow increases in the venturi, the pressure drops proportionally. If liquid or gas flow decreases, the pressure in the venturi increases.

Saginaw Power Steering Pump Operation

Vane and Cam Action

As the belt rotates the rotor and vanes inside the cam ring, centrifugal force causes the vanes to slide out of the rotor slots. The vanes follow the elliptical surface of the cam. When the area between the vanes expands, a low-pressure area occurs between the vanes, and fluid flows from the reservoir through two suction ports into the space between the vanes. As the vanes approach the higher portion of the cam at the two outlet ports, the cam ring pushes the blades inward and the area between the vane becomes smaller. This action pressurizes the power steering fluid, and this fluid is forced out the two discharge ports (Figure 5-8).

High-pressure fluid is forced into two passages on the thrust plate to the flow control valve and outlet fitting. Fluid pressure from the pump discharge ports is also supplied to the flow control valve. Although most of the fluid is discharged from the pump, some fluid flows past the flow control valve and returns to the pump suction ports. Fluid pressure is supplied from the venturi area through a passage to the spring side of the flow control valve.

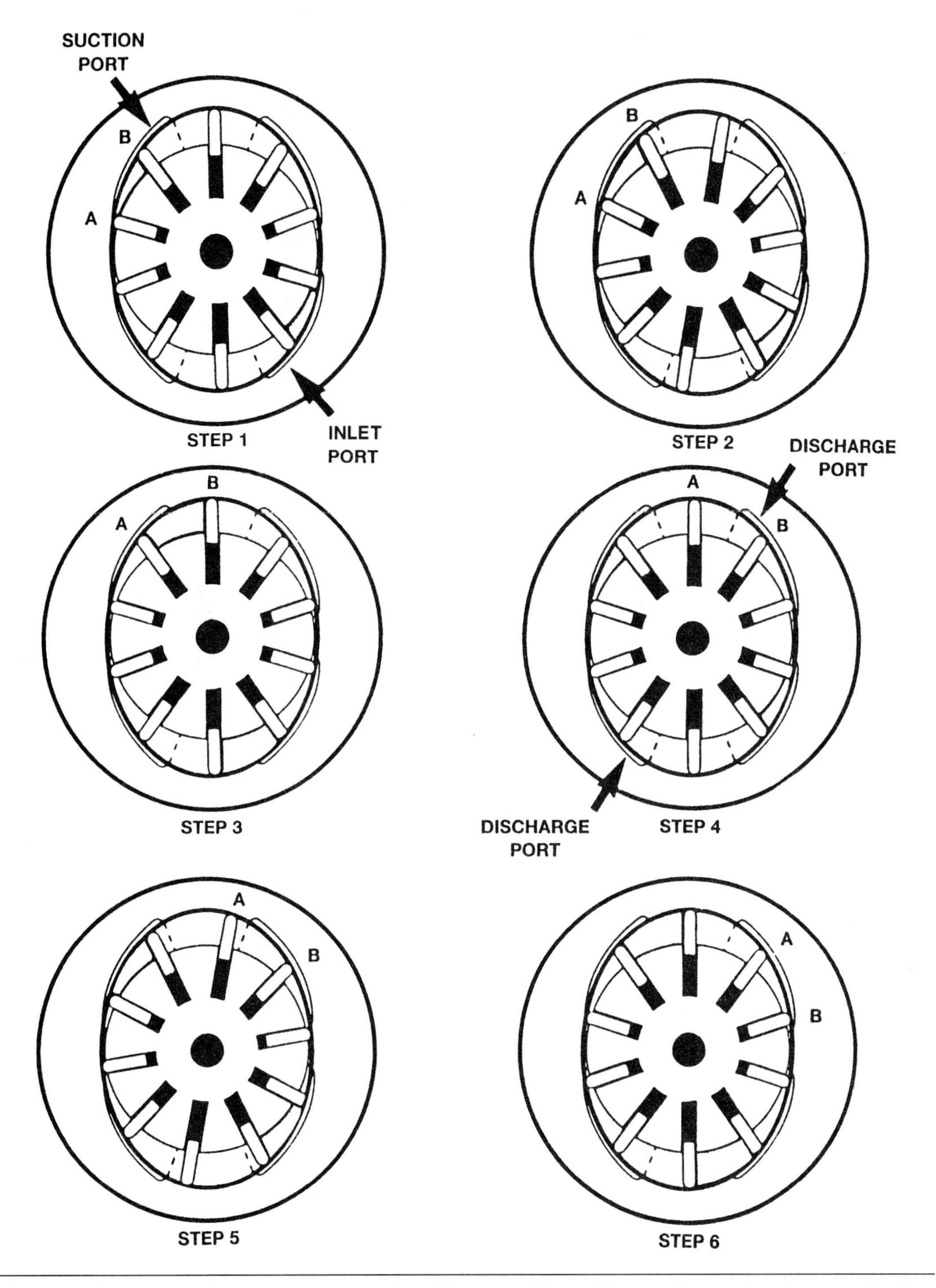

Figure 5-8 Power steering pump vane operation. (Courtesy of General Motors Corporation, Service Technology Group.)

Power Steering Pump Operation at Low System Pressure

At idle speed with the wheels straight ahead, the flow of fluid from the pump to the steering gear is relatively low. Under this condition a valve in the steering gear is positioned so the fluid discharged from the pump is directed through this valve and the low-pressure hose to the pump reservoir. With low fluid flow, the pressure is higher in the venturi. This higher pressure is applied to the spring side of the flow control valve and helps the valve spring to keep this flow control valve nearly closed (refer to Figure 5-7). Under this condition pump pressure remains low, and all the fluid is discharged from the pump through the steering gear valve and back to the pump inlet.

Power Steering Pump Operation at Higher Engine Speeds and Moderate System Pressure

When engine speed is increased, the power steering pump delivers more fluid than the system requires, and this increase in fluid flow creates a pressure decrease in the outlet fitting venturi. This pressure reduction is sensed at the spring side of the flow control valve, which allows the pump discharge pressure to force the flow control valve partially open. Under this condition the excess fluid from the pump is routed past the flow control valve to the pump inlet (Figure 5-9). If the steering wheel is turned, the fluid discharged from the pump rushes into the pressure chamber in the steering gear.

Power Steering Pump Operation with High Pump Pressure

If the steering wheel is turned fully in either direction until the steering linkage contacts the steering stops, the fluid discharged from the pump rushes into the pressure chamber in the steering gear. Under this condition pump pressure could become extremely high and damage hoses or other components. When this pressure chamber in the steering gear is filled with fluid, the flow from the pump decreases, but the high pump pressure is still supplied to the steering gear pressure chamber. Because the flow of fluid through the venturi and control orifice is now reduced, a higher pressure is present in this area. This extremely high pressure is also supplied through the passage to the spring end of the flow control valve.

See Chapter 5 in the Shop Manual for power steering gear adjustments.

At a predetermined pressure the pressure relief ball in the center of the flow control valve is unseated. This action allows the fluid pressure to flow through the pressure relief valve into the pump inlet ports. This action relieves the pressure on the spring end of the flow control valve, and the flow control valve moves to the wide open position. This action allows some of the high pressure in the steering system to return to the pump inlet, which limits pump pressure to a maximum safe value (Figure 5-10).

Hobourn Power Steering Pump with Roller Vanes

Pump Design and Operation

CAUTION: Power steering components may be extremely hot if the engine has been running. Use caution and wear gloves to avoid burns when servicing power steering components.

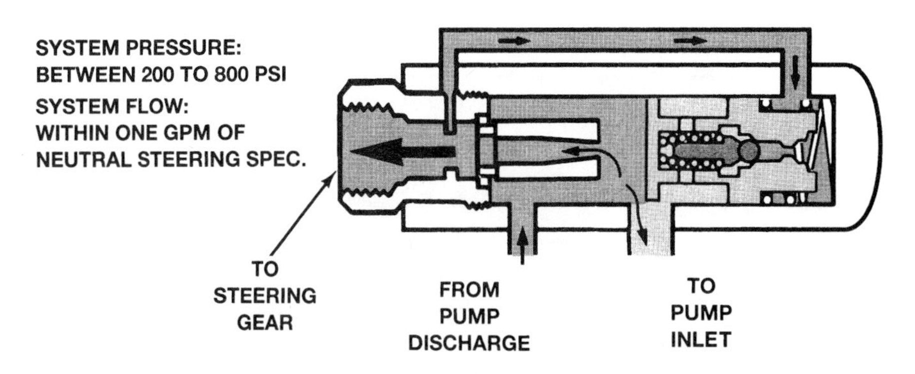

Figure 5-9 Flow control valve position with moderate power steering system pressure. (Courtesy of General Motors Corporation, Service Technology Group.)

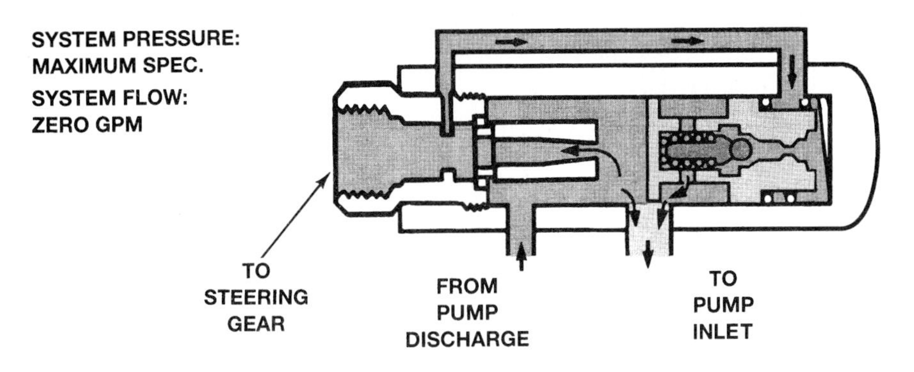

Figure 5-10 Flow control valve position with high power steering system pressure. (Courtesy of General Motors Corporation, Service Technology Group.)

Some power steering pumps have roller vanes in place of blade vanes. Power steering pumps with blade vanes or roller vanes operate in a similar manner. In a roller vane pump the roller vanes are mounted in carrier slots. The carrier is attached to the pump shaft and must rotate with the shaft. The power steering pump shaft is gear driven from the engine. An elliptical cam ring is mounted over the carrier and roller blades (Figure 5-11).

When the roller blades rotate past the two opposite suction ports, centrifugal force moves the roller blades outward against the cam ring. Under this condition a larger space is located between the roller blades and the carrier slots, and a low-pressure area is created in these slots. Atmospheric pressure on the power steering fluid in the pump reservoir and the low pressure between the pump carrier slots and roller blades cause fluid to flow from the pump reservoir into the area between the roller blades and slots.

As the roller blades rotate past the inlet ports, the fluid is trapped between the roller blades and slots. When the roller blades approach the two opposite discharge ports, the contour of the cam ring forces the roller blades back into the carrier slots. This action pressurizes the fluid between the roller blades and slots. While the roller blades are rotating past the discharge ports, the cam contour continues to force these blades farther back into the carrier slots. Under this condition

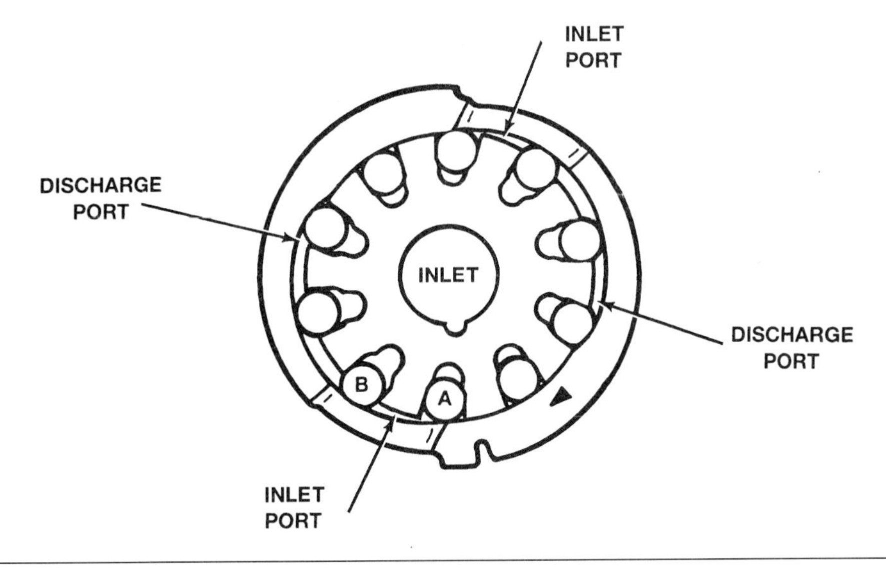

Figure 5-11 Power steering pump with roller blades. (Courtesy of General Motors Corporation, Service Technology Group.)

more pressure is exerted on the fluid, and the pressurized fluid is forced through the discharge ports, past the pressure relief valve, and through the restriction in the outlet fitting and high-pressure hose to the power steering gear (Figure 5-12). The fluid flows through the low-pressure return hose to the pump reservoir.

Flow Control Valve Operation

The flow control valve is mounted in the power steering pump end cover (Figure 5-13). A pressure relief valve is positioned in the center of the flow control valve (Figure 5-14). When driving the ve-

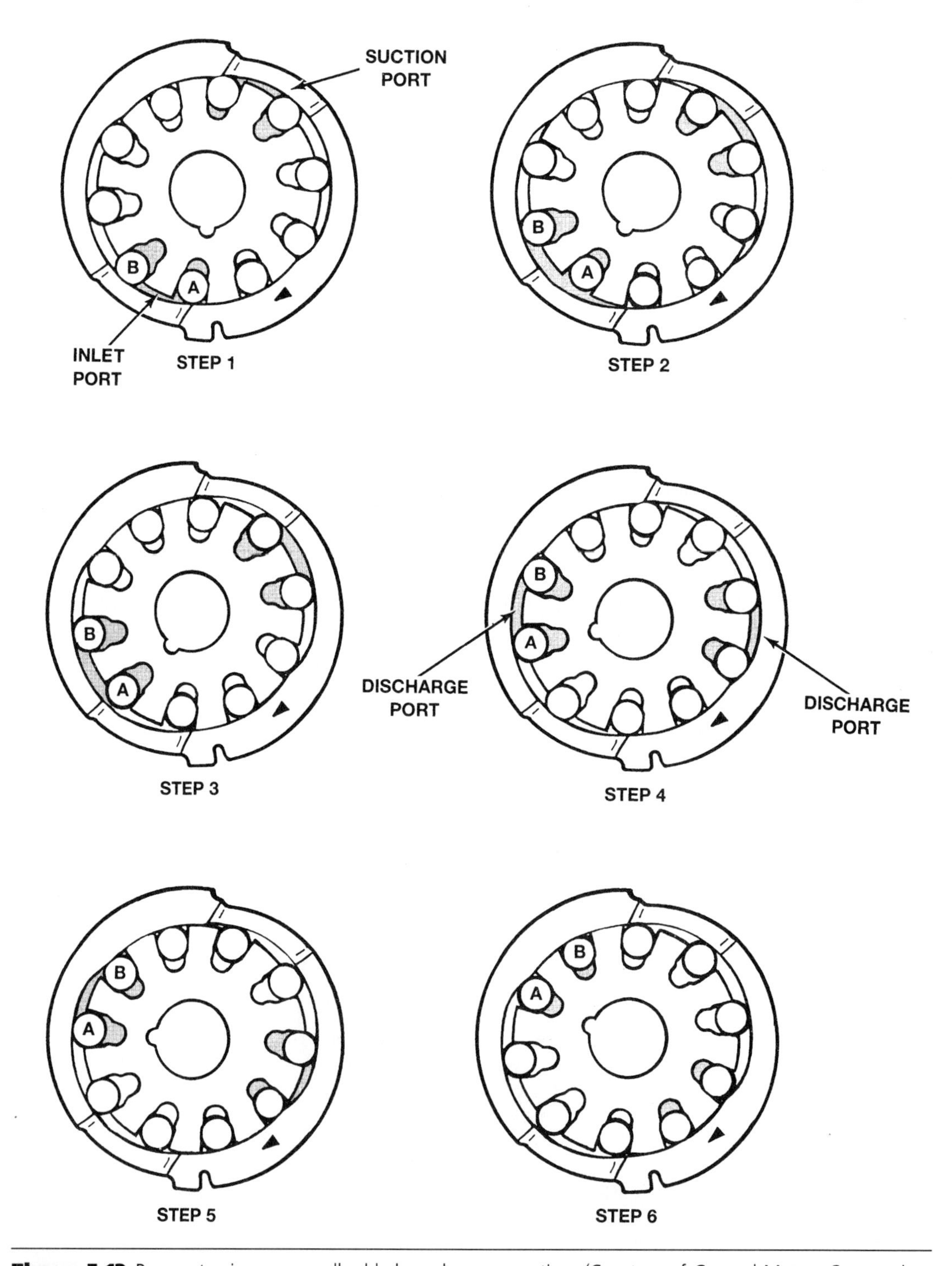

Figure 5-12 Power steering pump roller blade and cam operation. (Courtesy of General Motors Corporation, Service Technology Group.)

1-POWER STEERING PUMP ASSEMBLY
2-INLET FITTING BOLT (TWO NEEDED)
3-PUMP INLET TUBE
4-PUMP INLET TUBE CLAMP
5-PUMP INLET TUBE O-RING SEAL
6-PUMP MOUNTING STUD NUT
7-PUMP MOUNTING SPACER
8-OULET FITTING CAP
9-FLOW CONTROL VALVE CAP
10-CAP O-RING SEAL
11-FLOW CONTROL VALVE SPRING
12-FLOW CONTROL VALVE
13-PUMP COVER PASSAGE SEAL RING
14-PUMP COVER
15-PUMP BODY SEAL RING
16-PUMP ROLLER VANES
17-PUMP ELEMENT CARRIER
18-ELEMENT CARRIER DRIVE PIN
19-PUMP ELEMENT CAM
20-PUMP ELEMENT PORT PLATE
21-PORT PLATE SEAL (TWO NEEDED)
22-PORT PLATE O-RING SEALS (TWO NEEDED)
23-PUMP END PLATE
24-PUMP ELEMENT LOCATING PIN
25-MOUNTING FLANGE BOLT (AS NEEDED)
26-MOUNTING FLANGE BOLT (AS NEEDED)
27-PUMP MOUNTING FLANGE
28-PUMP SHAFT COUPLING BOLT
29-COUPLING BOLT WASHER
30-PUMP SHAFT RETAINING RING
31-PUMP SHAFT BALL BEARING
32-WOODRUFF KEY
33-PUMP SHAFT
34-PUMP MOUNTING STUD (AS NEEDED)
35-MOUNTING STUD NUT (AS NEEDED)
36-PUMP BODY
37-SERVICE KITS

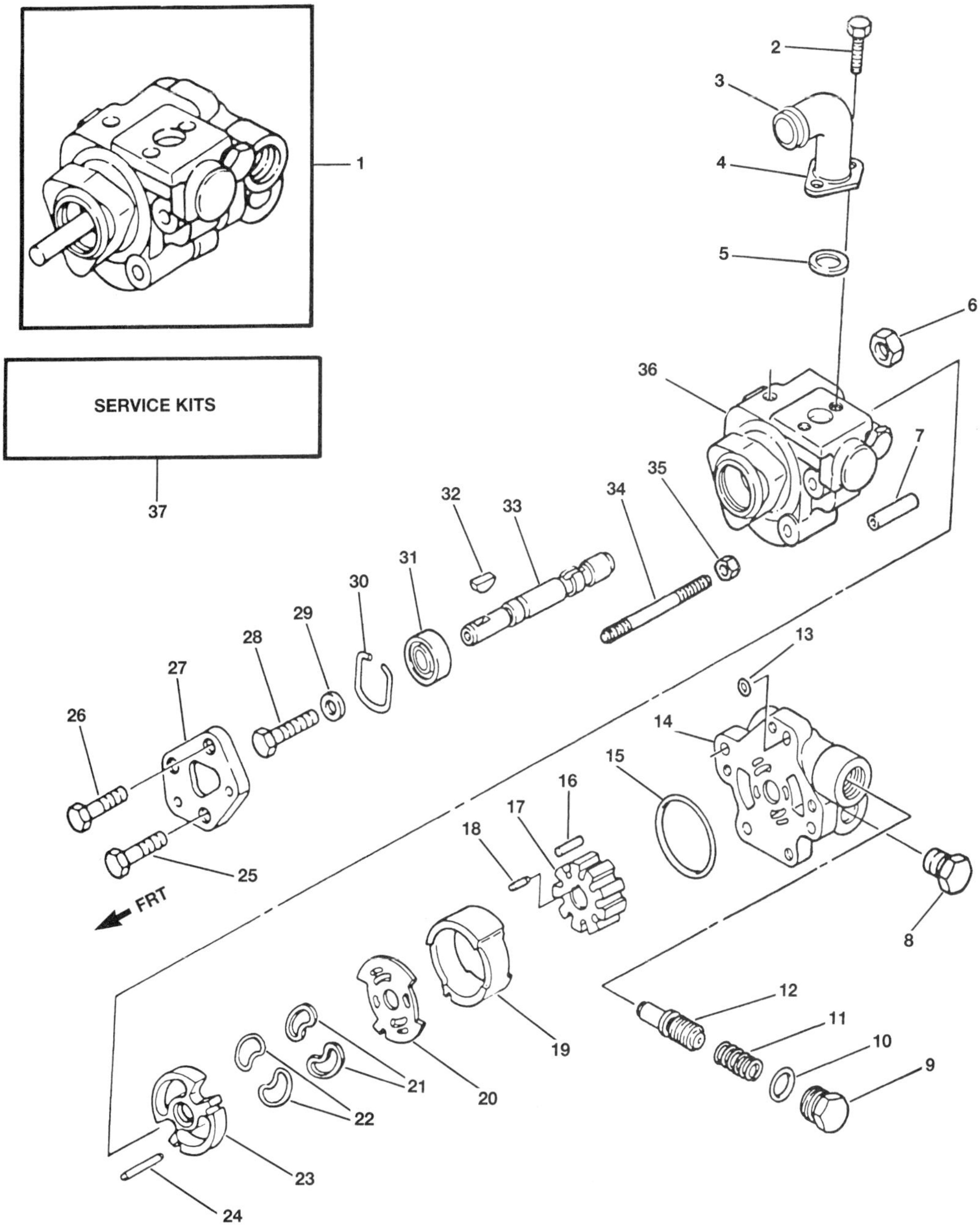

SERVICE KITS

FRT

Figure 5-13 Components in a power steering pump with roller blades. (Courtesy of General Motors Corporation, Service Technology Group.)

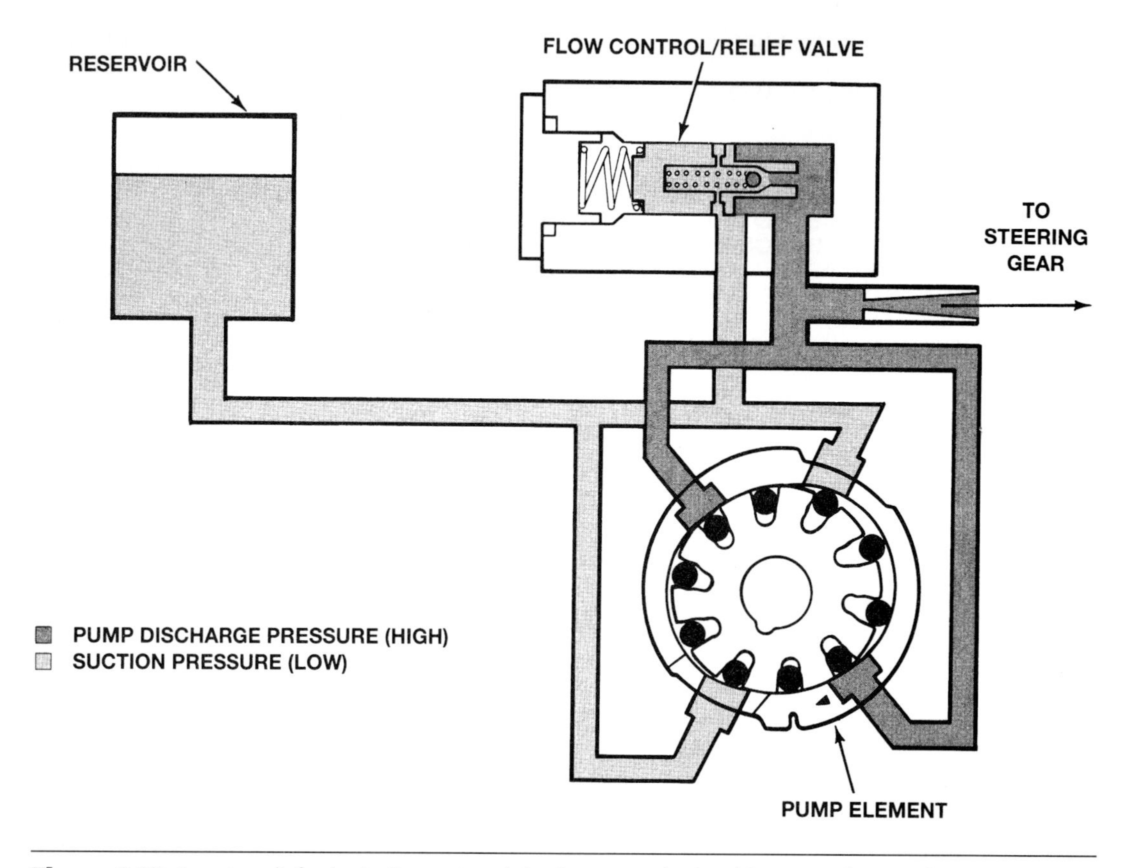

Figure 5-14 Pressure relief valve in the center of the flow control valve. (Courtesy of General Motors Corporation, Service Technology Group.)

hicle with the front wheels straight ahead, the power steering fluid is directed from the pump through the steering gear and back to the pump reservoir. Under this condition the pump pressure remains low, and this pressure moves the flow control valve a small amount against the valve spring. Under this condition a small amount of fluid flows past the flow control valve and back to the pump inlet ports (Figure 5-15).

When the front wheels are turned, the power steering pump pressure is supplied to the piston chamber in the steering gear. Under this condition the pump pressure increases to supply steering assist, and this higher pressure moves the flow control valve against the spring. This action allows more fluid to return past the flow control valve to the pump inlet ports (Figure 5-16).

LOW-SYSTEM PRESSURE FLOW CONTROL

SYSTEM PRESSURE:
LESS THAN 200 PSI
SYSTEM FLOW:
NEUTRAL STEERING SPEC.
(GALLONS PER MINUTE)
AT ENGINE IDLE SPEED
FLOW CONTROL/RELIEF VALVE
TO
STEERING
GEAR
TO
PUMP
INLET
FROM
PUMP
DISCHARGE

Figure 5-15 Flow control valve position with the front wheels straight ahead and low power steering pump pressure. (Courtesy of General Motors Corporation, Service Technology Group.)

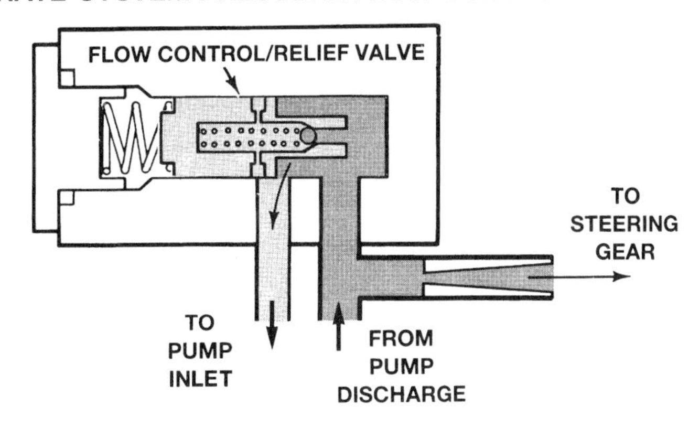

Figure 5-16 Flow control valve position with the front wheels turned and moderate power steering pump pressure. (Courtesy of General Motors Corporation, Service Technology Group.)

If the steering wheel is turned fully to the left or right, the power steering fluid is directed into the closed steering gear piston chamber and the front wheels cannot turn any further. This action stops the piston movement in the steering gear, and under this condition the pump pressure becomes very high. This high pressure unseats the pressure relief valve in the center of the flow control valve. Under this condition fluid flows past the pressure relief valve to the pump suction ports, and fluid continues flowing past the flow control valve to the pump inlet ports (Figure 5-17). The pressure relief valve action limits power steering pump pressure to a safe maximum value to protect power steering hoses and components.

See Chapter 5 in the Shop Manual for power steering belt diagnosis and adjustment.

Saginaw Power Steering Gear with Dual Pistons

Design and Operation

TRADE JARGON: A power steering gear in which all the hydraulic components and gears are mounted inside the steering gear may be called an integral steering gear.

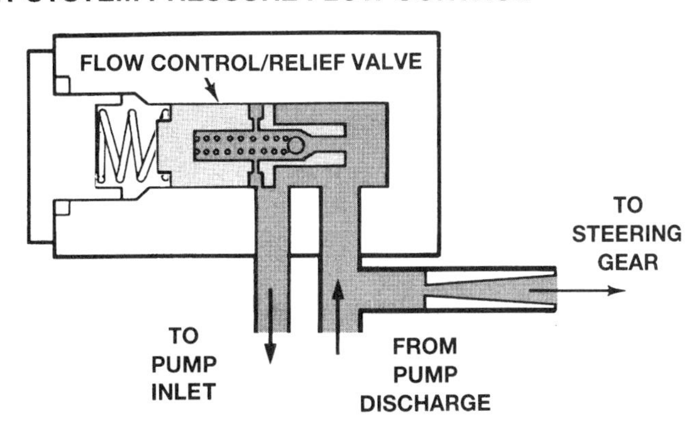

Figure 5-17 Flow control valve and pressure relief valve position with the front wheels turned fully to the right or left. (Courtesy of General Motors Corporation, Service Technology Group.)

In the power steering gear (see Figure 5-18), the lower piston (item 1) is mounted over the wormshaft. The wormshaft is supported on bearings in the gear housing, and an adjuster plug (item 2) in the upper end of the housing is used to adjust wormshaft bearing preload. Ball bearings are mounted in grooves in the **wormshaft** and the lower piston. Teeth on the lower piston are meshed with the sector teeth. The sector shaft is supported by a needle bearing in the gear housing, and an adjusting bolt in the sector shaft cover provides sector lash adjustment. Power steering fluid may be directed to chambers at either end of the lower piston. An upper piston (item 3) is mounted in a cylinder on the opposite side of the **sector shaft** (item 4), and teeth on this piston are meshed with the sector teeth. Power steering fluid may be directed to chambers at either end of the upper piston.

A **torsion bar** is connected between the steering shaft and the wormshaft. Because the front wheels are resting on the road surface, they resist turning, and the parts attached to the wormshaft also resist turning. This turning resistance causes torsion bar deflection when the wheels are turned, and this deflection is limited to a predetermined amount.

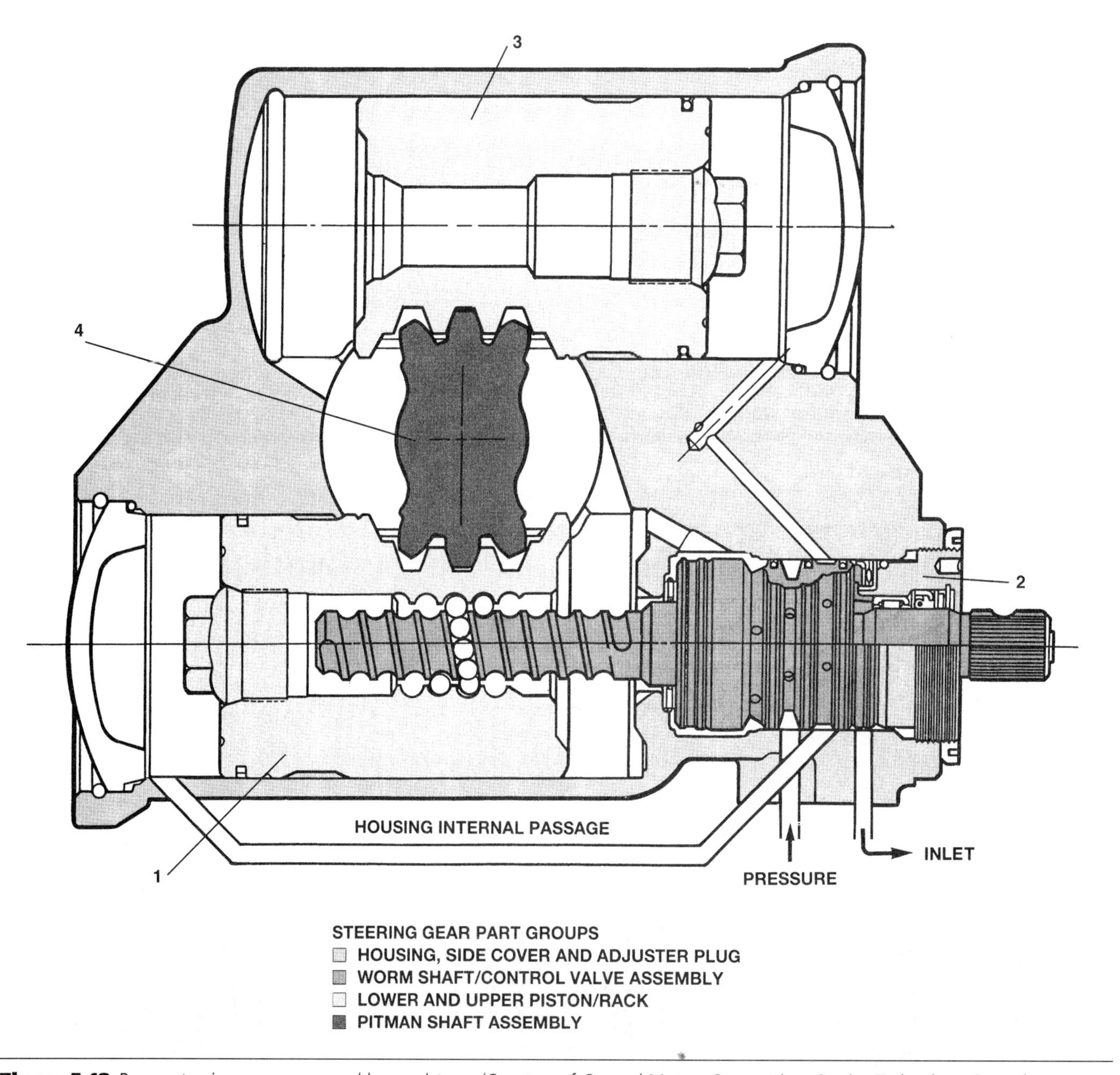

Figure 5-18 Power steering gear upper and lower pistons. (Courtesy of General Motors Corporation, Service Technology Group.)

The wormshaft is connected to the rotary valve body, and the torsion bar pin also connects the torsion bar to the wormshaft. The upper end of the torsion bar is attached to the steering shaft and wheel. A stub shaft is mounted inside the **rotary valve,** and a pin connects the outer end of this shaft to the torsion bar. The pin on the inner end of the stub shaft is connected to the **spool valve** in the center of the rotary valve body (Figure 5-19). The rotary valve body, torsion bar, and spool valve may be called a control valve. The control valve and wormshaft assembly provides a mechanical connection from the steering wheel and steering shaft to the lower piston in the steering gear. The control valve and wormshaft also provide a hydraulic connection to the ends of both pistons and also to the pressure and return lines.

CAUTION: Many power steering gears are mounted near the exhaust manifold, which may be extremely hot. Use caution and wear protective gloves when inspecting or servicing this type of steering gear.

When the truck is driven with the front wheels straight ahead, oil flows from the power steering pump through the spool valve, rotary valve, and low-pressure return line to the pump inlet (Figure 5-20). In the straight ahead steering gear position, power steering fluid is also directed to both ends of the lower and upper pistons. This condition may be called neutral steer. Fluid pressure is equal on both sides of these pistons, and the fluid acts as a cushion that prevents road shocks from causing **kick-back** on the steering wheel.

Neutral steer refers to the condition when the truck is driven with the front wheels straight ahead and equal fluid pressure is on both sides of the pistons in the steering gear.

If a driver makes a left turn, torsion bar deflection moves the valve spool inside the rotary valve body so that oil flow is directed through the rotary valve to the left turn passages in the rotary valve body (Figure 5-21). Power steering pump pressure is now directed through the rotary valve body to passages connected to the end of the lower piston next the rotary valve body. Fluid pressure is also directed to the opposite end of the upper piston (Figure 5-22). Fluid is directed into the return passage from the opposite ends of the lower and upper pistons to which pressure is supplied. This hydraulic pressure on the pistons assists the driver in turning the wheels to the left.

When the driver makes a right turn, torsion bar deflection moves the spool valve so that oil flows through the spool valve and rotary valve body to the right turn passages in the rotary valve

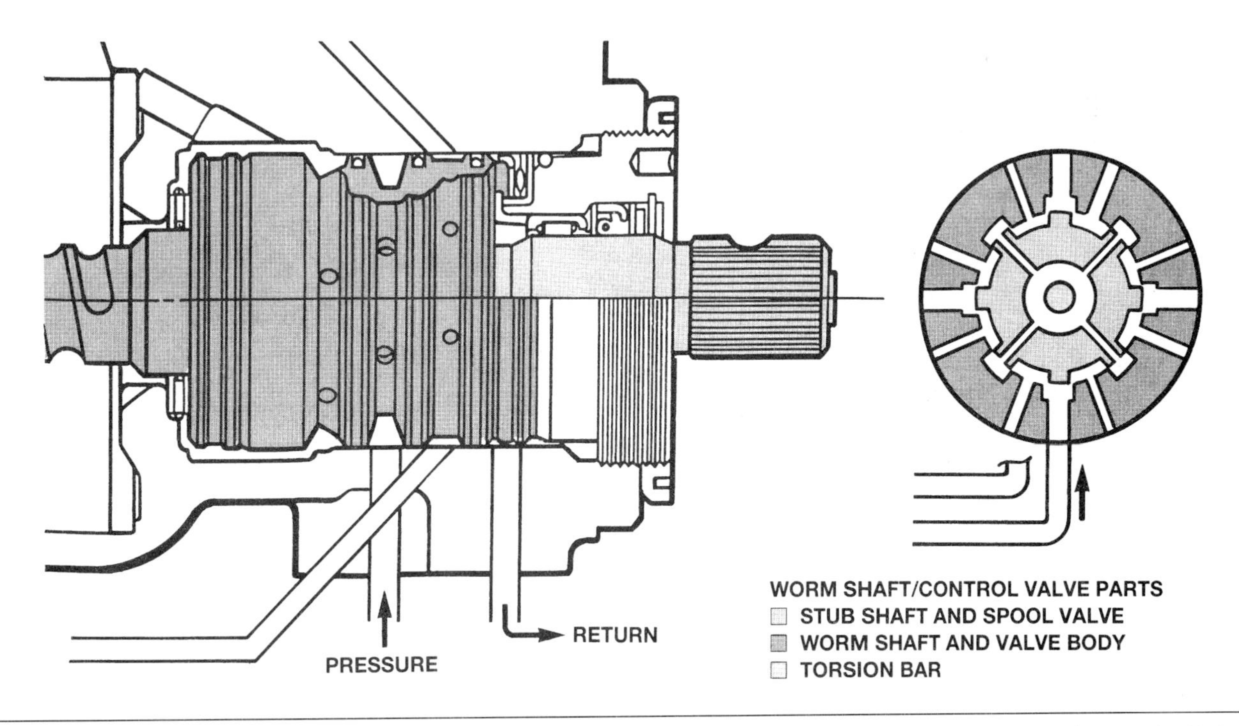

Figure 5-19 Stub shaft, spool valve, wormshaft, valve body, and torsion bar. (Courtesy of General Motors Corporation, Service Technology Group.)

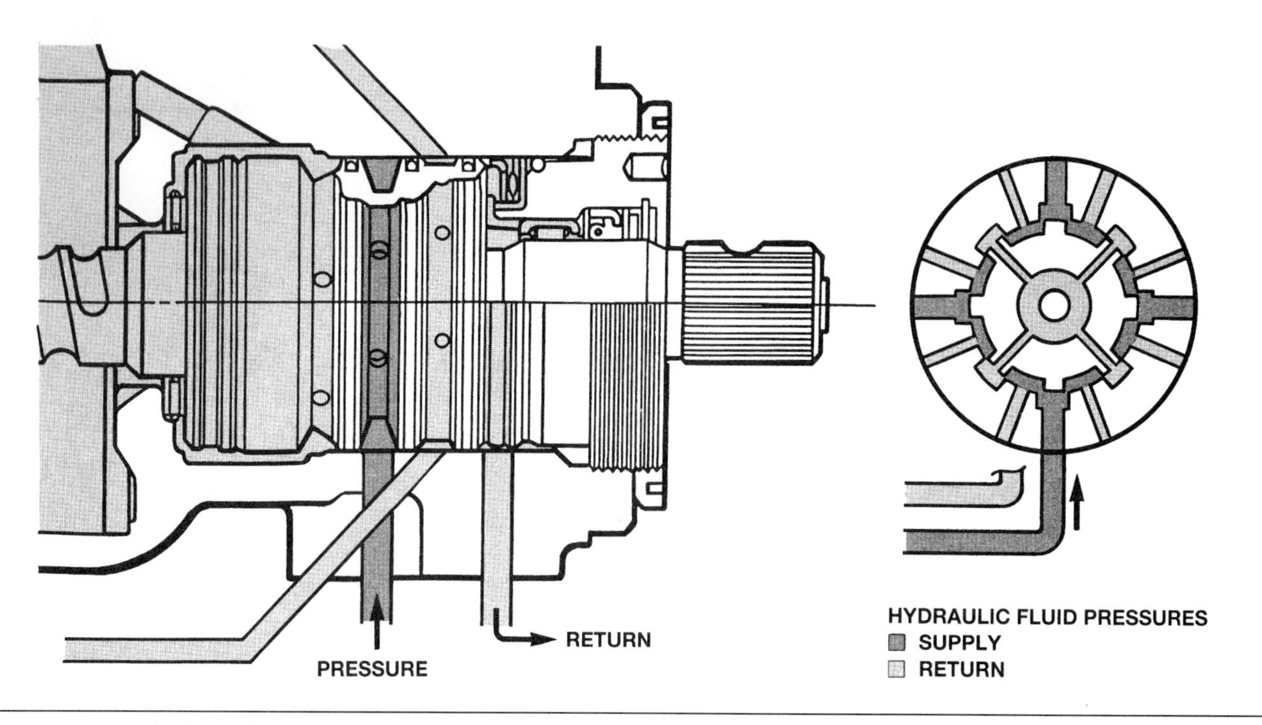

Figure 5-20 Spool valve and rotary valve position with the front wheels straight ahead. (Courtesy of General Motors Corporation, Service Technology Group.)

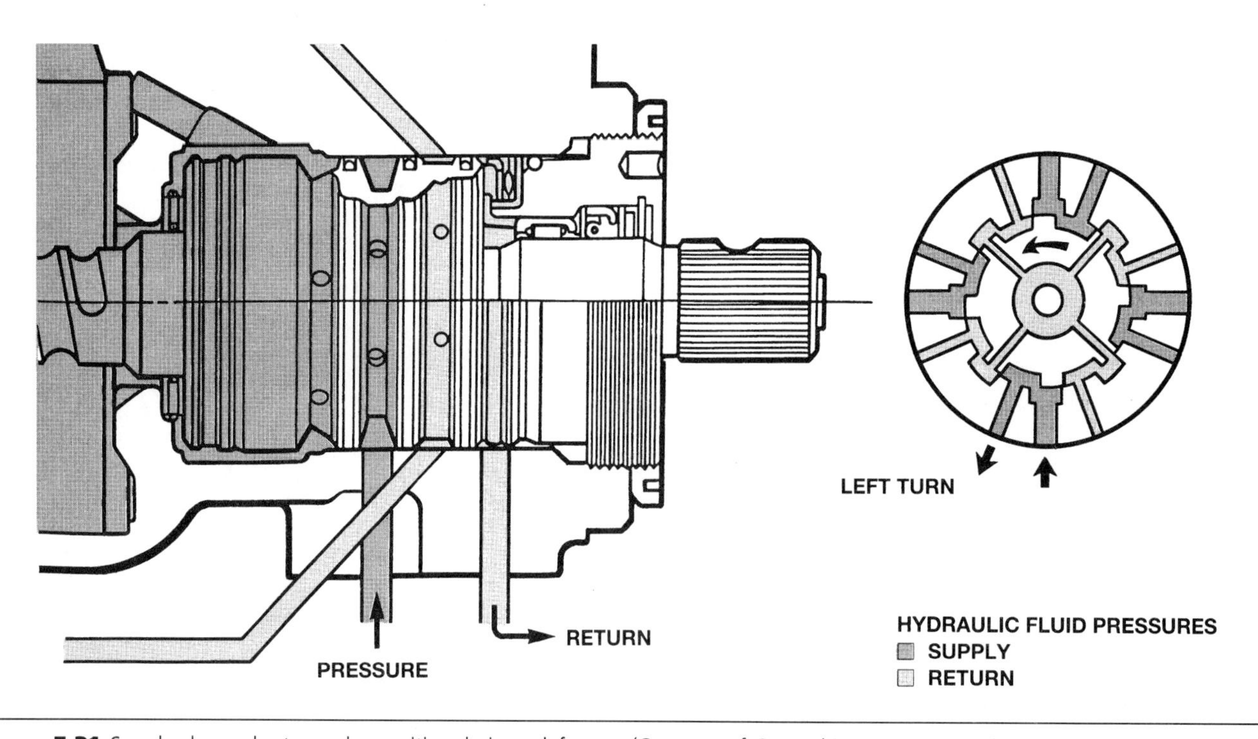

Figure 5-21 Spool valve and rotary valve position during a left turn. (Courtesy of General Motors Corporation, Service Technology Group.)

body (Figure 5-23). Power steering fluid pressure is now supplied through passages in the housing to the pressure chamber at the lower end of the lower piston and the opposite end of the upper piston (Figure 5-24). During a right turn hydraulic pressure applied to the lower end of the lower piston and the opposite end of the upper piston helps the driver to turn the wheels.

A spring-loaded pressure relief valve in the steering gear housing is normally closed. If the power steering pump pressure exceeds 1,550 psi (10,700 kPa), the relief valve opens to protect

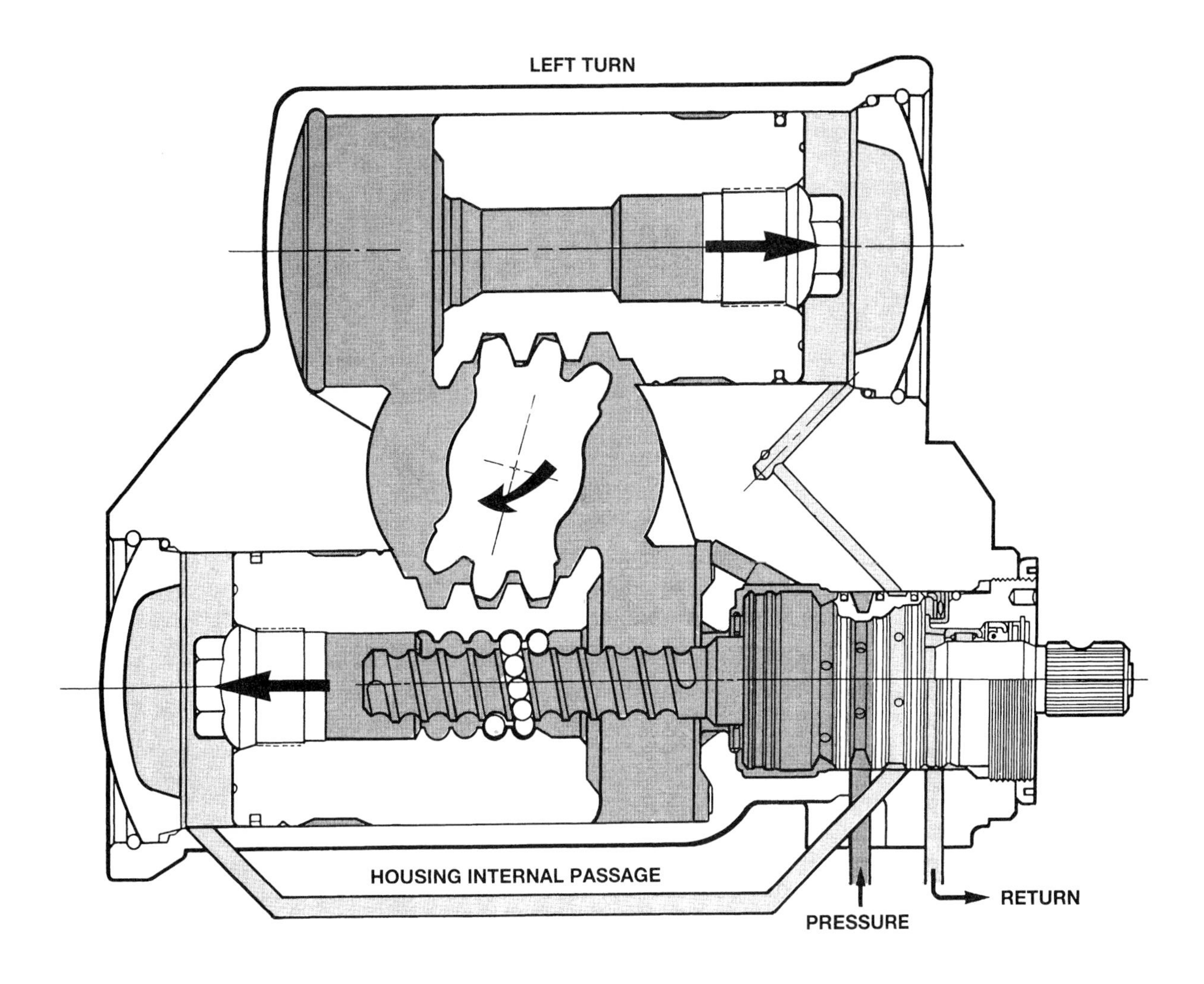

Figure 5-22 During a left turn power steering pump pressure is supplied to the end of the lower piston next to the rotary valve body and also to the opposite end of the upper piston. (Courtesy of General Motors Corporation, Service Technology Group.)

the system components (Figure 5-25). When the pressure relief valve opens, the power steering pump pressure drops to the point where no power steering assist is present. However, the driver may continue to turn the wheels without power assist until the pressure relief valve closes again.

Steering wheel kick-back is a force transmitted from the front wheels through the steering linkage, steering gear, and steering column to the steering wheel when a front wheel strikes an irregularity in the road surface.

TRW/Ross Power Steering Gear with Poppet Valves

Design and Operation

The TRW/Ross power steering gear contains a single piston with ball bearings mounted between the inner piston grooves and the wormshaft. The piston teeth are meshed with the sector shaft teeth. Bearings support the wormshaft on both ends, and a needle bearing supports the sector shaft

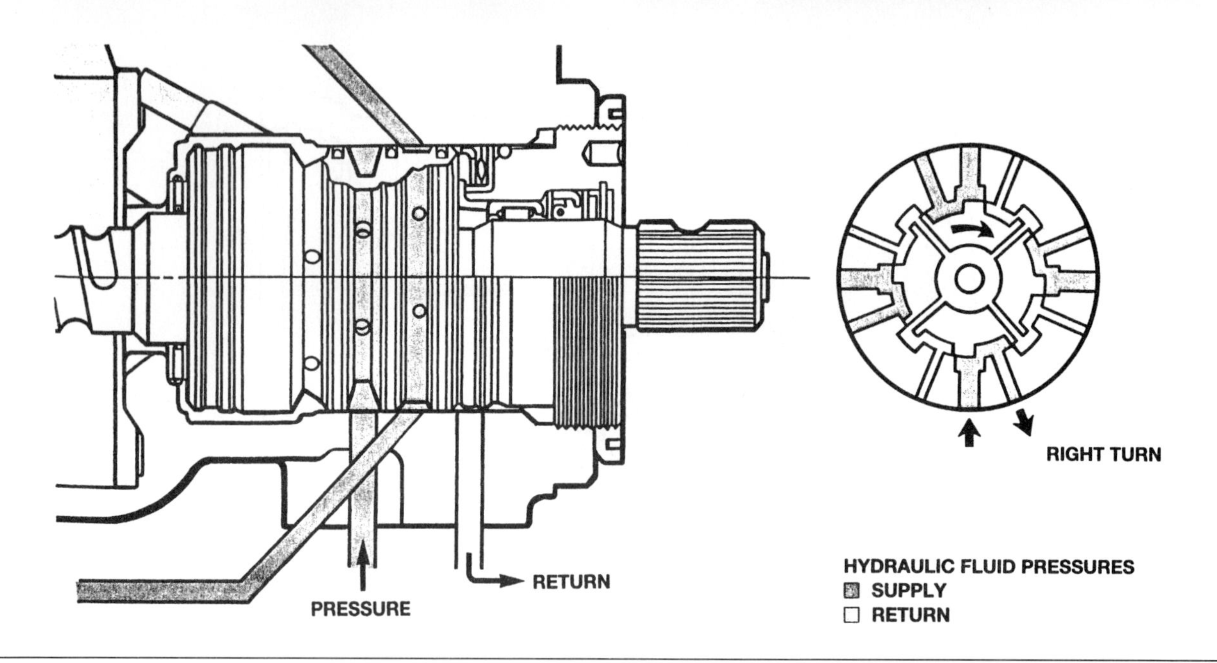

Figure 5-23 Spool valve and rotary valve body position during a right turn. (Courtesy of General Motors Corporation, Service Technology Group.)

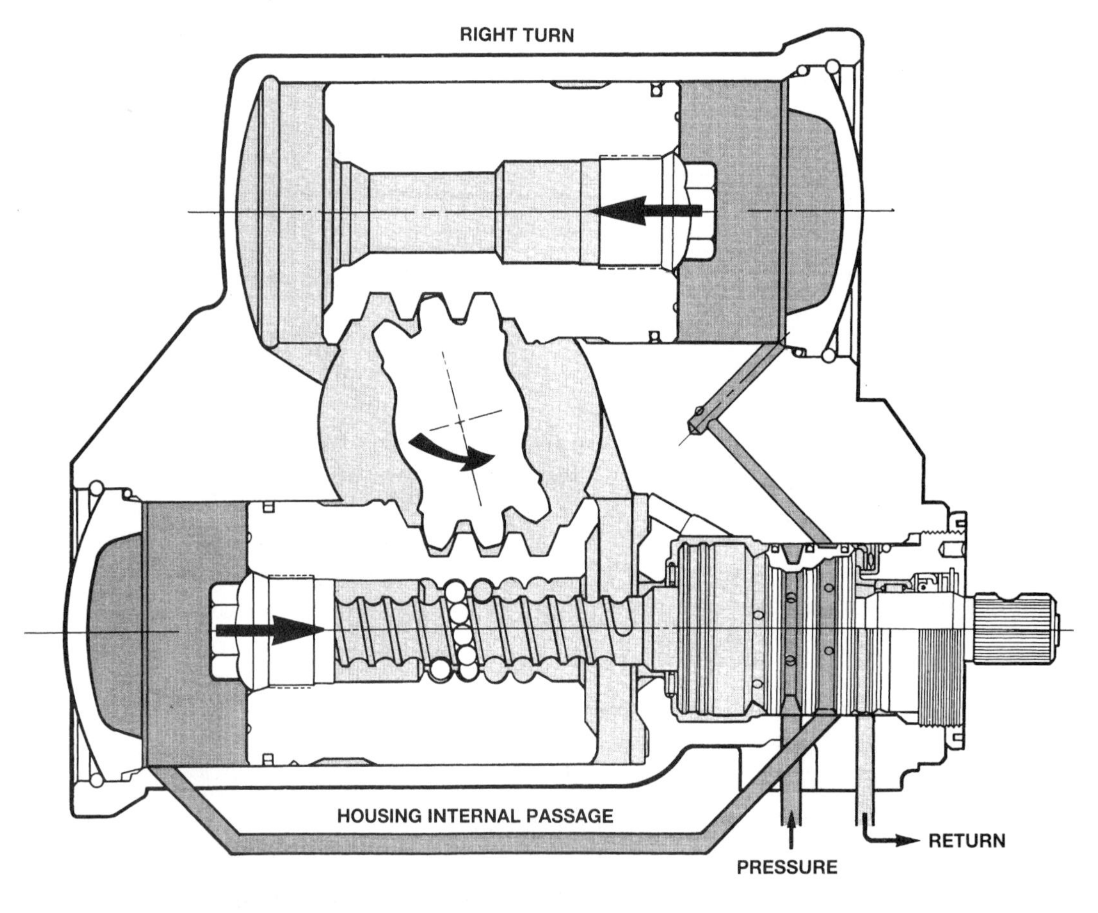

Figure 5-24 During a right turn power steering pump pressure is supplied to the lower end of the lower piston and the opposite end of the upper piston. (Courtesy of General Motors Corporation, Service Technology Group.)

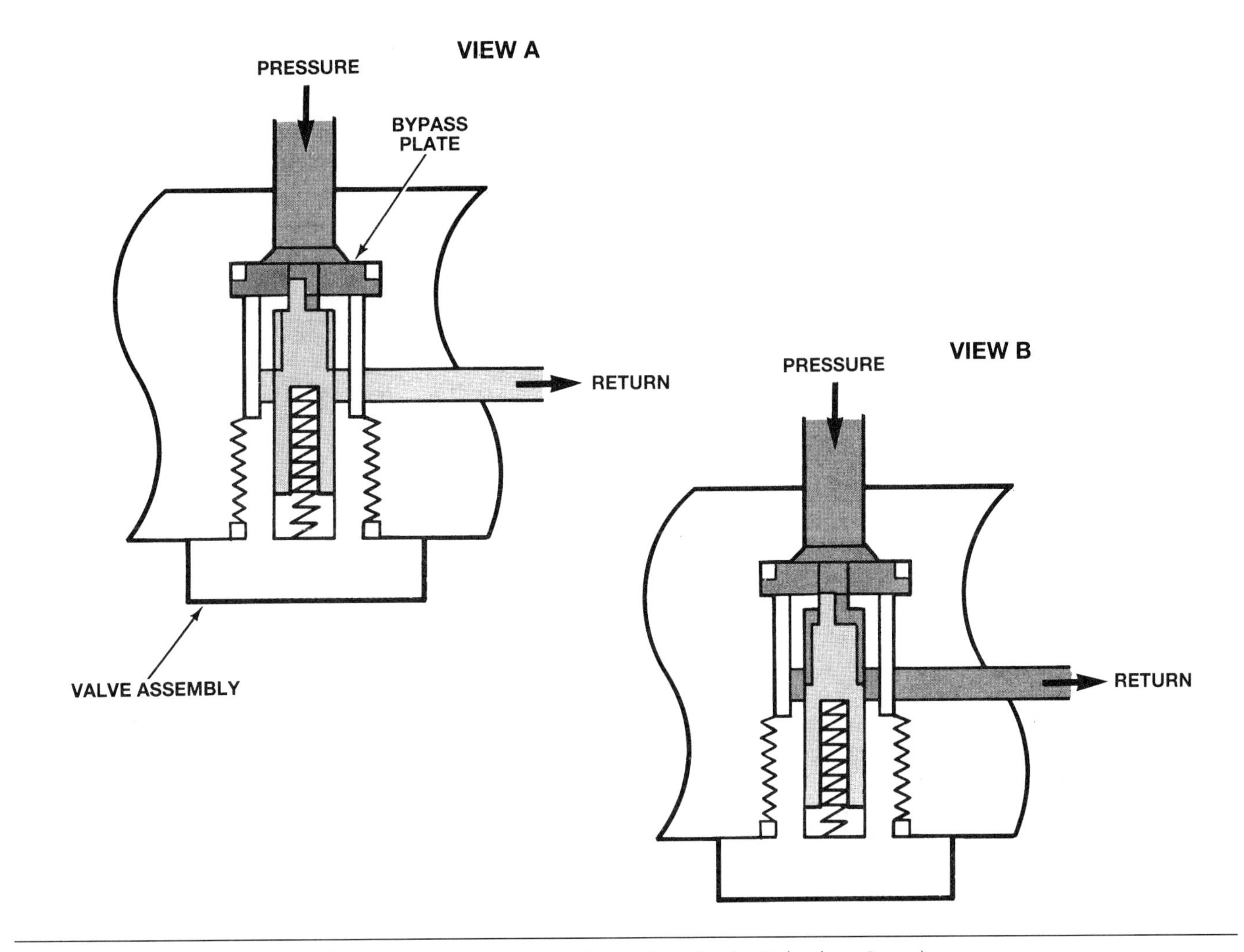

Figure 5-25 Pressure relief valve. (Courtesy of General Motors Corporation, Service Technology Group.)

in the gear housing. Two **poppet valves** are mounted in the piston, and a poppet valve stem extends from each end of the piston (Figure 5-26).

The upper end of the power steering gear input shaft is connected through a universal joint to the steering shaft and steering wheel. A torsion bar is connected from the input shaft to the wormshaft, and a rotary valve is attached to the wormshaft. A spool valve attached to the input shaft is mounted inside the rotary valve on the wormshaft (Figure 5-27). When the steering wheel is turned in either direction, the torsion bar deflects, and this action allows the spool valve to rotate a small amount inside the rotary valve.

When driving the truck with the front wheels straight ahead, power steering fluid is directed from the steering gear supply passage to the return passage. Fluid pressure is also supplied to both ends of the piston in the steering gear (Figure 5-28). This fluid pressure on both sides of the steering gear piston reduces steering wheel kick-back when one of the front wheels strikes an irregularity in the road surface.

If the steering wheel is turned to the left, power steering fluid pressure is supplied from the power steering gear supply passage to the lower end of the piston (Figure 5-29). Under this condition the fluid from the top of the piston flows into the gear return passage. The fluid pressure supplied to the lower end of the piston helps the driver to turn the front wheels to the left.

When the steering wheel is turned to the right, power steering pump pressure is supplied through the spool valve to the upper side of the piston (Figure 5-30). In this mode fluid at the lower end of the piston flows into the gear return passage. The power steering pump pressure supplied to the upper side of the piston helps the driver to turn the front wheels to the right.

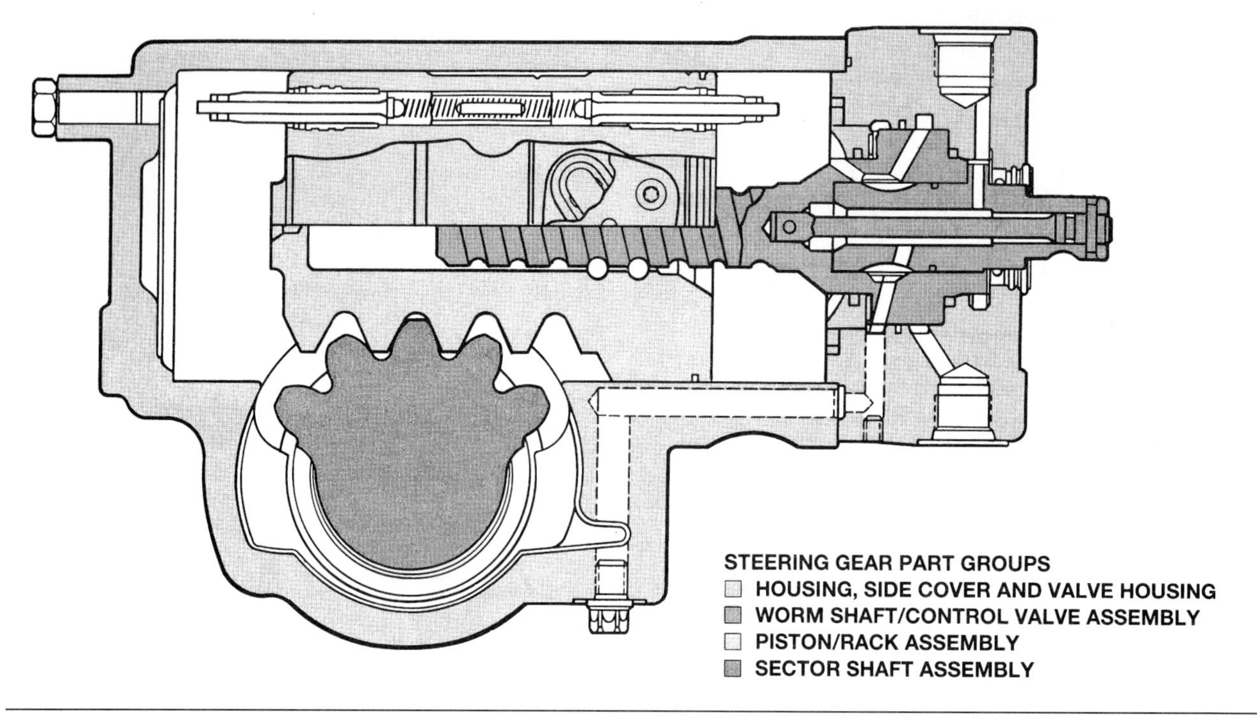

Figure 5-26 Power steering gear piston with poppet valves. (Courtesy of General Motors Corporation, Service Technology Group.)

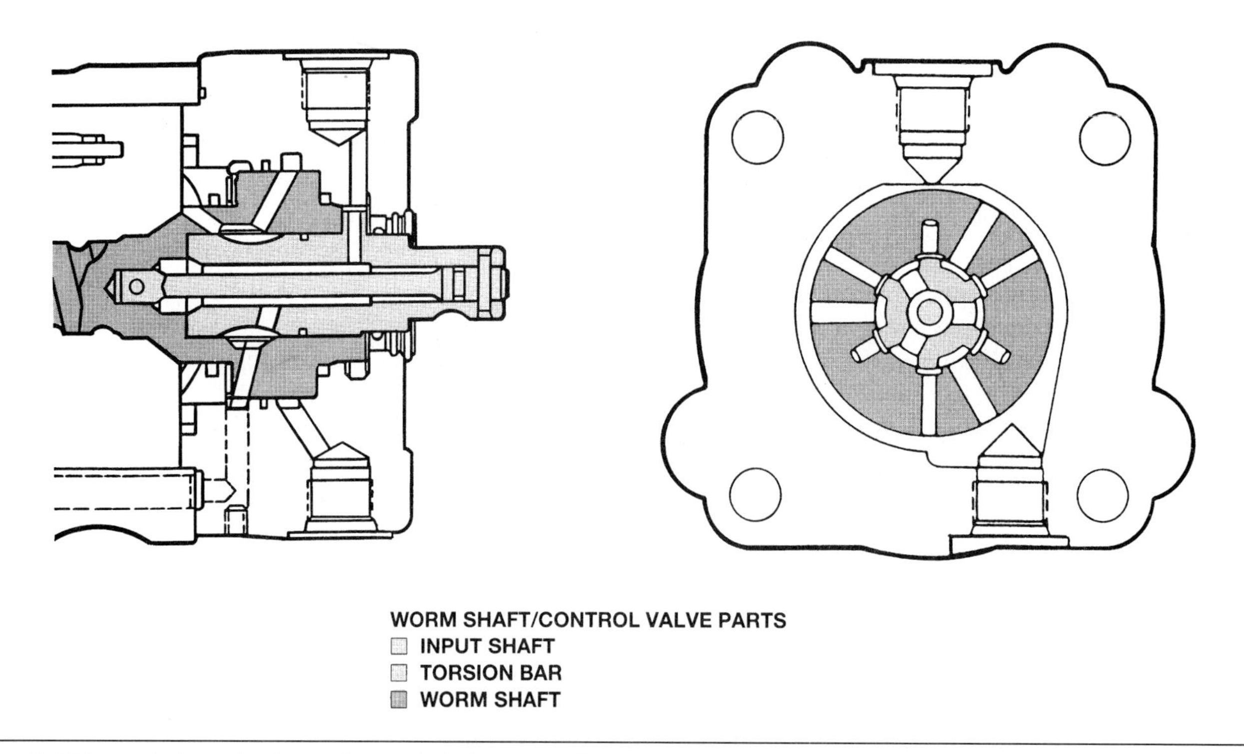

Figure 5-27 Input shaft, torsion bar, and wormshaft. (Courtesy of General Motors Corporation, Service Technology Group.)

During a right turn the power steering pump pressure on the upper side of the piston opens the poppet valve in the upper side of the piston. If the steering wheel is turned until it is one third of a turn from the fully right position, the poppet valve in the lower side of the piston contacts a bolt in the gear housing (Figure 5-31). This action opens this poppet valve, and the high fluid pressure on the upper side of the piston is released to the lower

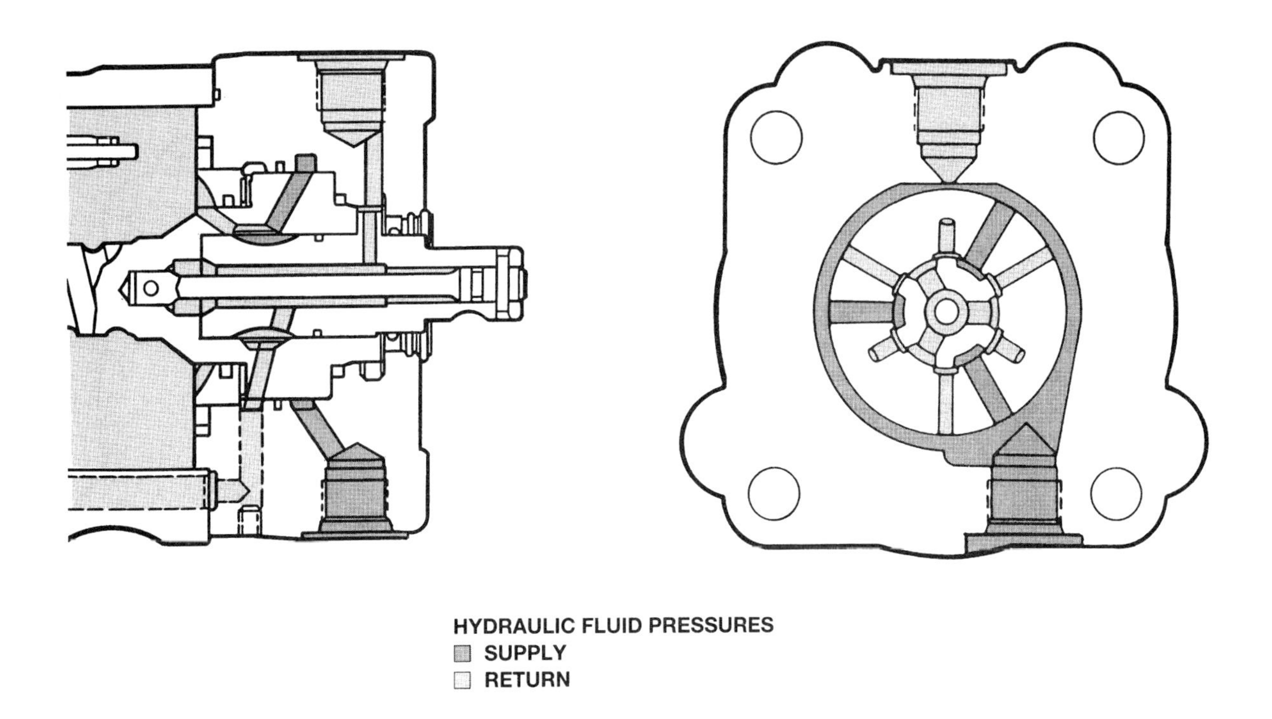

Figure 5-28 Steering gear operation with the front wheels straight ahead. (Courtesy of General Motors Corporation, Service Technology Group.)

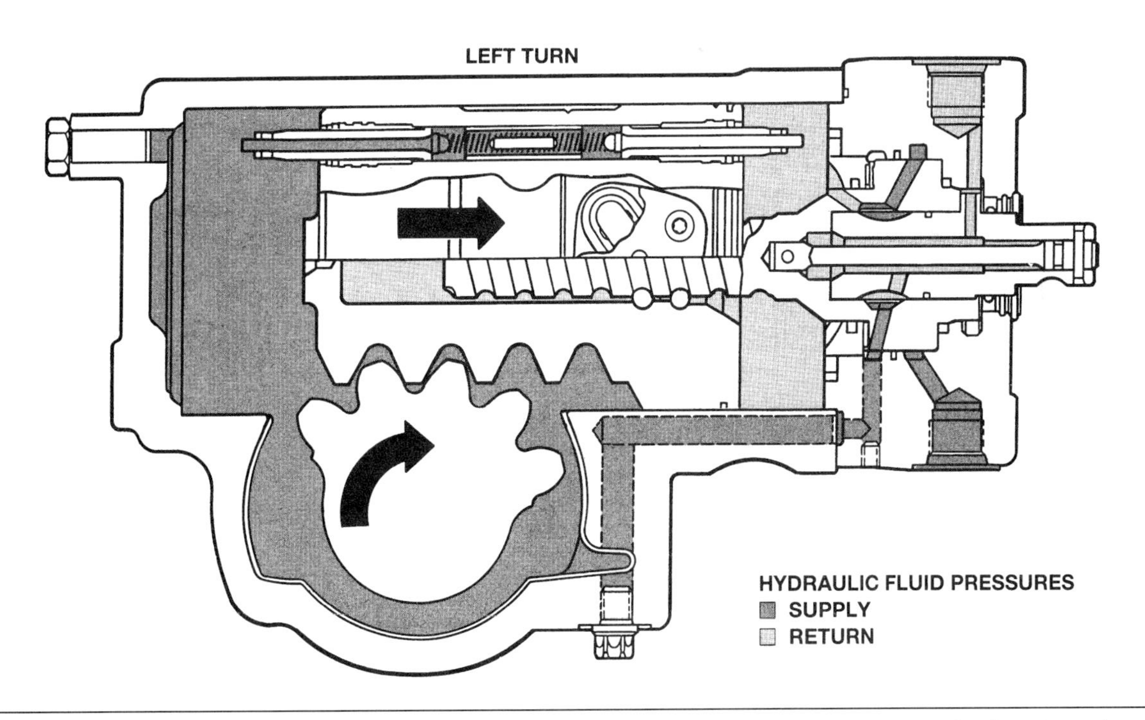

Figure 5-29 Steering gear operation during a left turn. (Courtesy of General Motors Corporation, Service Technology Group.)

side of the piston. Under this condition power steering assist is no longer available, but the driver may continue turning the steering wheel to the right without power assist until the steering arms contact the axle stops.

During a left turn the power steering pump pressure on the lower side of the piston opens the poppet valve in the lower side of the piston. When the steering wheel is turned until it is one

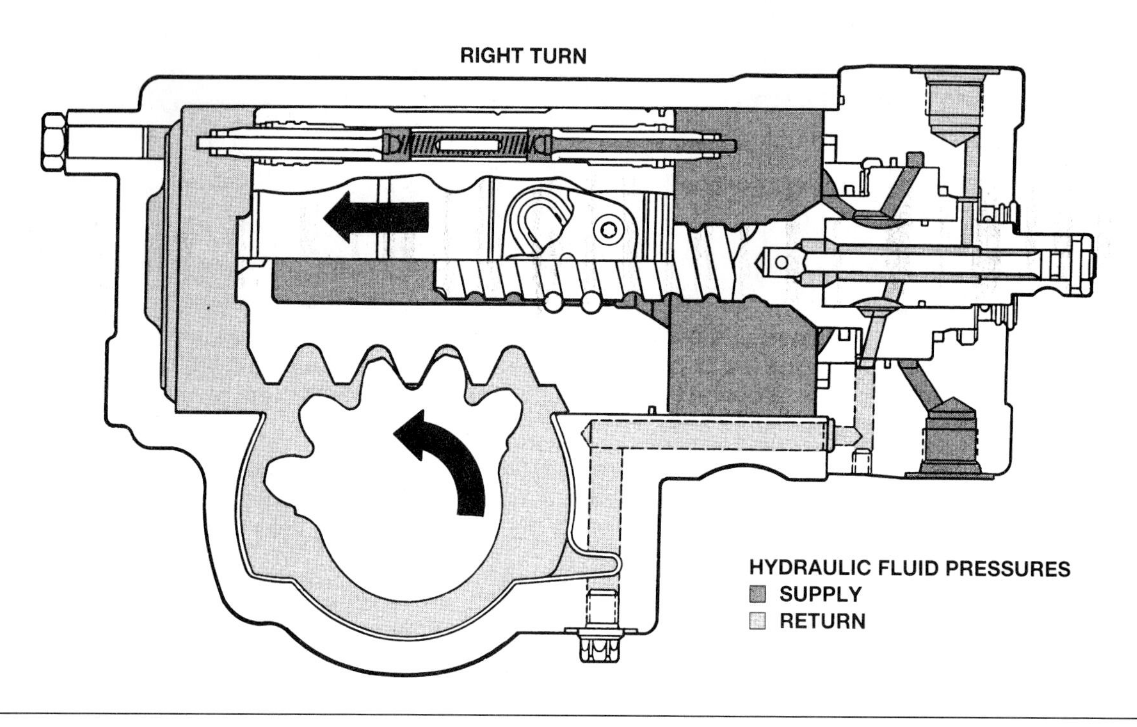

Figure 5-30 Steering gear operation during a right turn. (Courtesy of General Motors Corporation, Service Technology Group.)

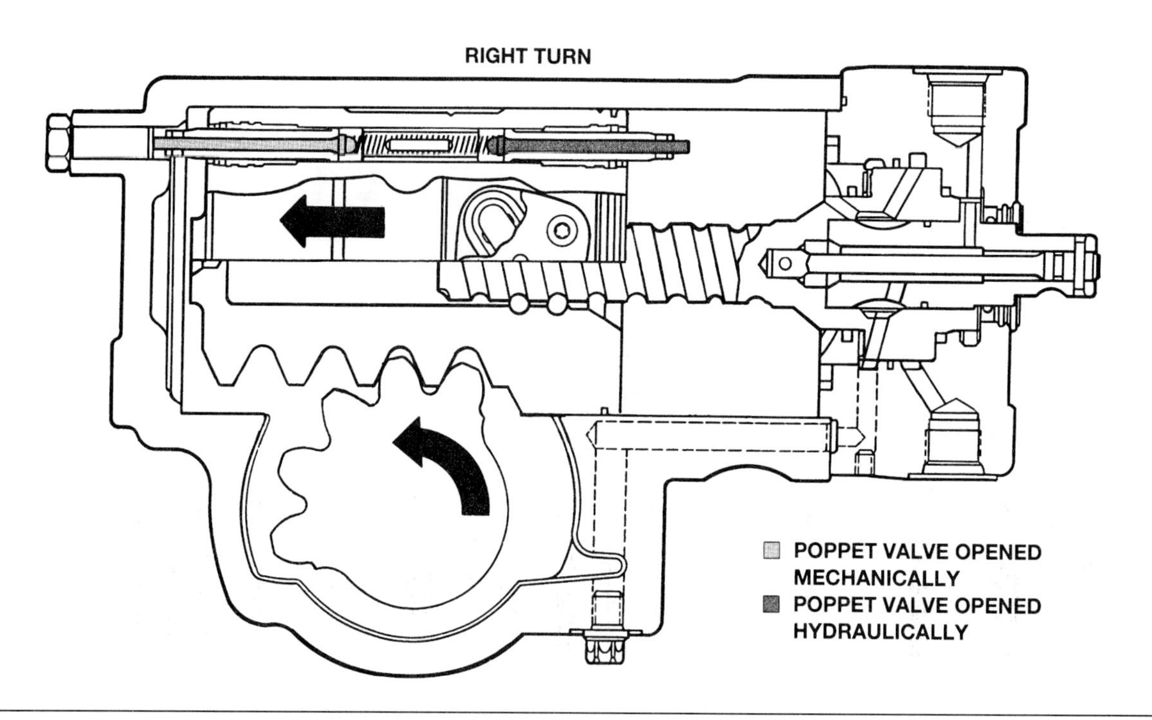

Figure 5-31 Poppet valve operation during a right turn. (Courtesy of General Motors Corporation, Service Technology Group.)

third of a turn from the fully left position, the poppet valve in the upper end of the piston contacts a bolt in the gear housing (Figure 5-32). This action opens the poppet valve, and the high fluid pressure on the lower side of the piston is released to the upper side of the piston. Under this condition power steering assist is no longer available, but the driver may continue turning the steering wheel to the right without power assist until the steering arms contact the axle stops.

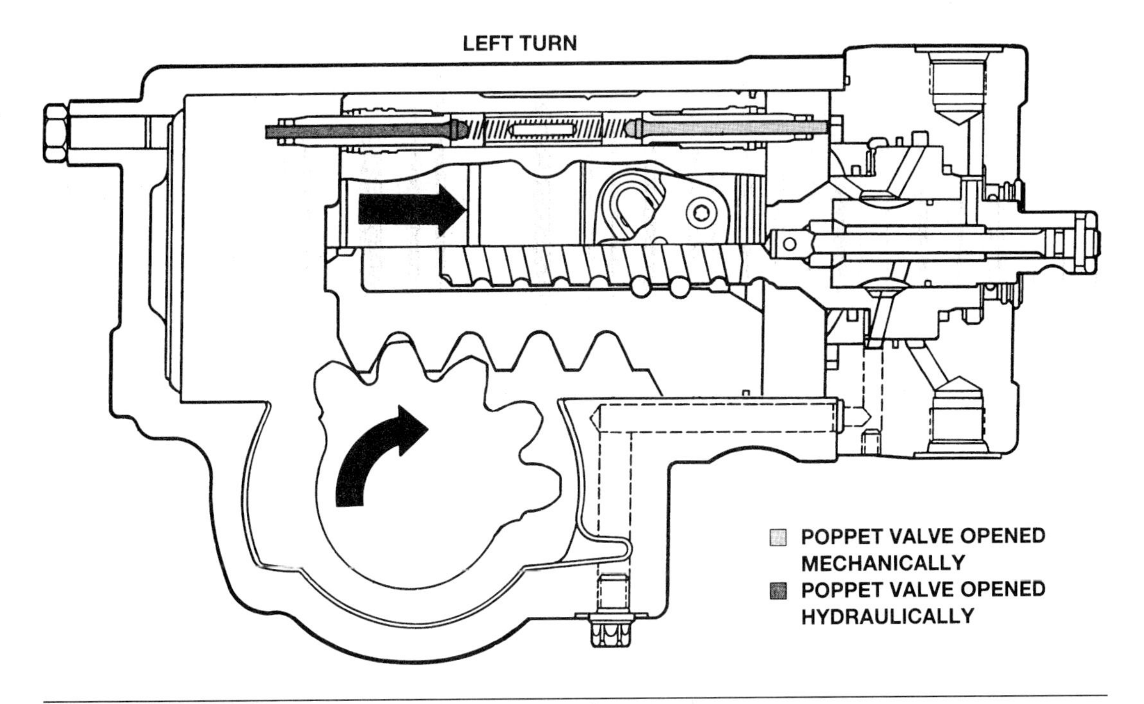

Figure 5-32 Poppet valve operation during a left turn. (Courtesy of General Motors Corporation, Service Technology Group.)

Bendix C-500N Power Steering Gear

See Chapter 5 in the Shop Manual for power steering poppet valve adjustments.

The Bendix C-500N power recirculating ball steering gear may be used on heavy duty trucks with a front (steer) axle weight rating up to 15,000 lb (Figure 5-33). The components in this steering gear are heavier and stronger than the components in a steering gear used on light or medium duty

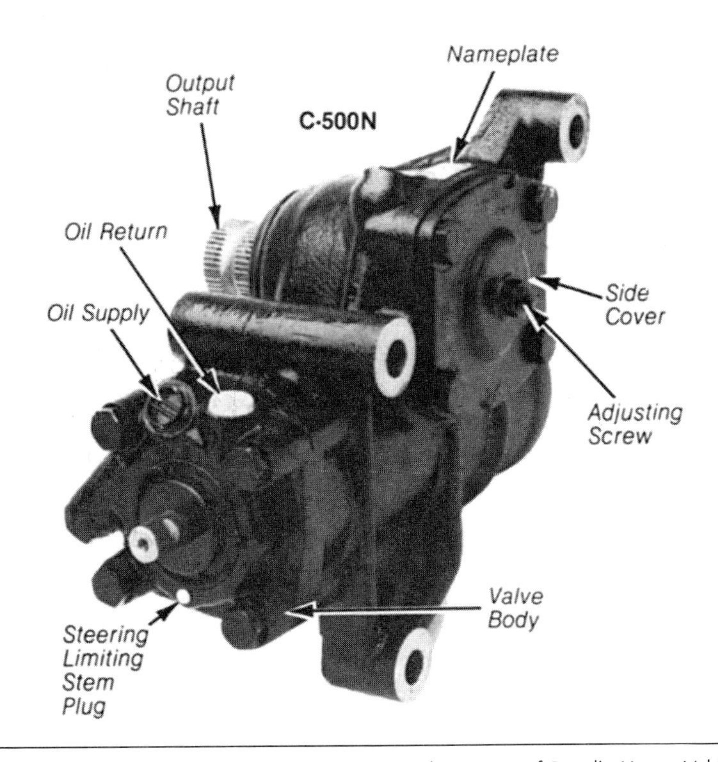

Figure 5-33 C-500N power recirculating ball steering gear. (Courtesy of Bendix Heavy Vehicle Systems, Allied Signal Automotive.)

trucks. In this steering gear the pitman shaft is supported on heavy roller bearings to withstand the additional load encountered in heavy duty trucks. The operation of this steering gear is similar to the steering gears explained previously in this chapter.

The Bendix C-500N steering gear may have a right-hand or left-hand thread on the wormshaft, depending on whether the ball screw and piston are mounted above or below the sector shaft (Figure 5-34). The position of the ball screw and piston in relation to the sector shaft is determined by the mounting requirements on the tractor. The rotary valve is mounted inside the upper end of the ball screw (Figure 5-35). The operation of the rotary valve during a right or left turn is similar to the operation of the rotary valves in other steering gears explained previously in this chapter (Figure 5-36). Poppet (limiting) valves are mounted in the ball nut piston, and poppet valve stems that contact these valves are mounted in the steering gear housing (Figure 5-37). The operation of these poppet valves is explained previously in this chapter. The action of the poppet valves reduces power steering assist when the steering wheel is turned almost fully to the right or left. This action limits the power assist force transmitted to the steering linkages when the steering wheel is turned fully to the right or left.

The steering gear may have an optional pressure relief valve or bypass valve. The pressure relief valve limits the maximum pressure in the power steering system to protect system components such as the high-pressure hose. If the power steering pump pressure is not available, the bypass valve allows fluid to flow from the return to the supply passages in the gear so fluid can be bypassed

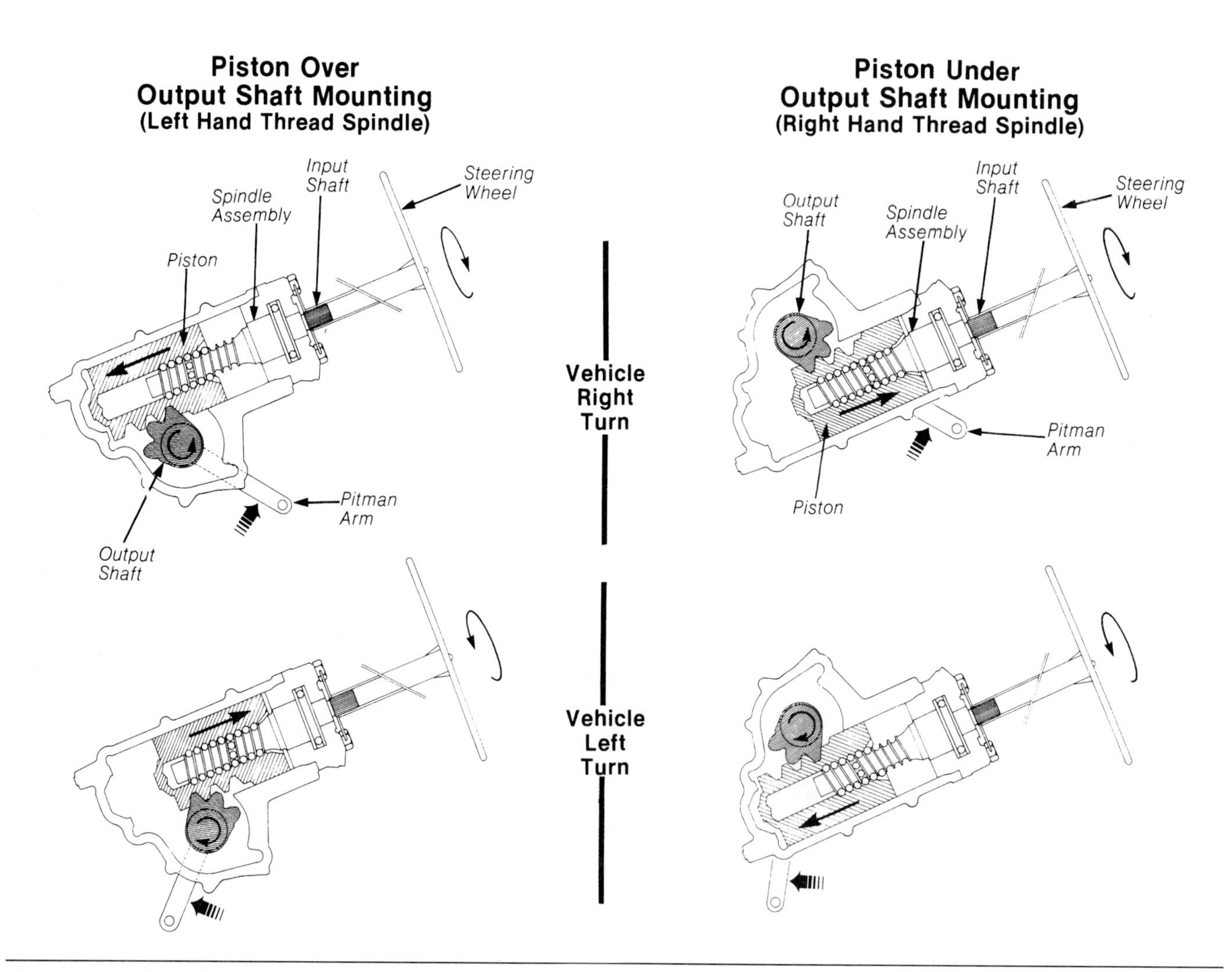

Figure 5-34 The ball screw may have a right-hand or a left-hand thread, depending on the ball nut position in relation to the sector shaft. (Courtesy of Bendix Heavy Vehicle Systems, Allied Signal Automotive.)

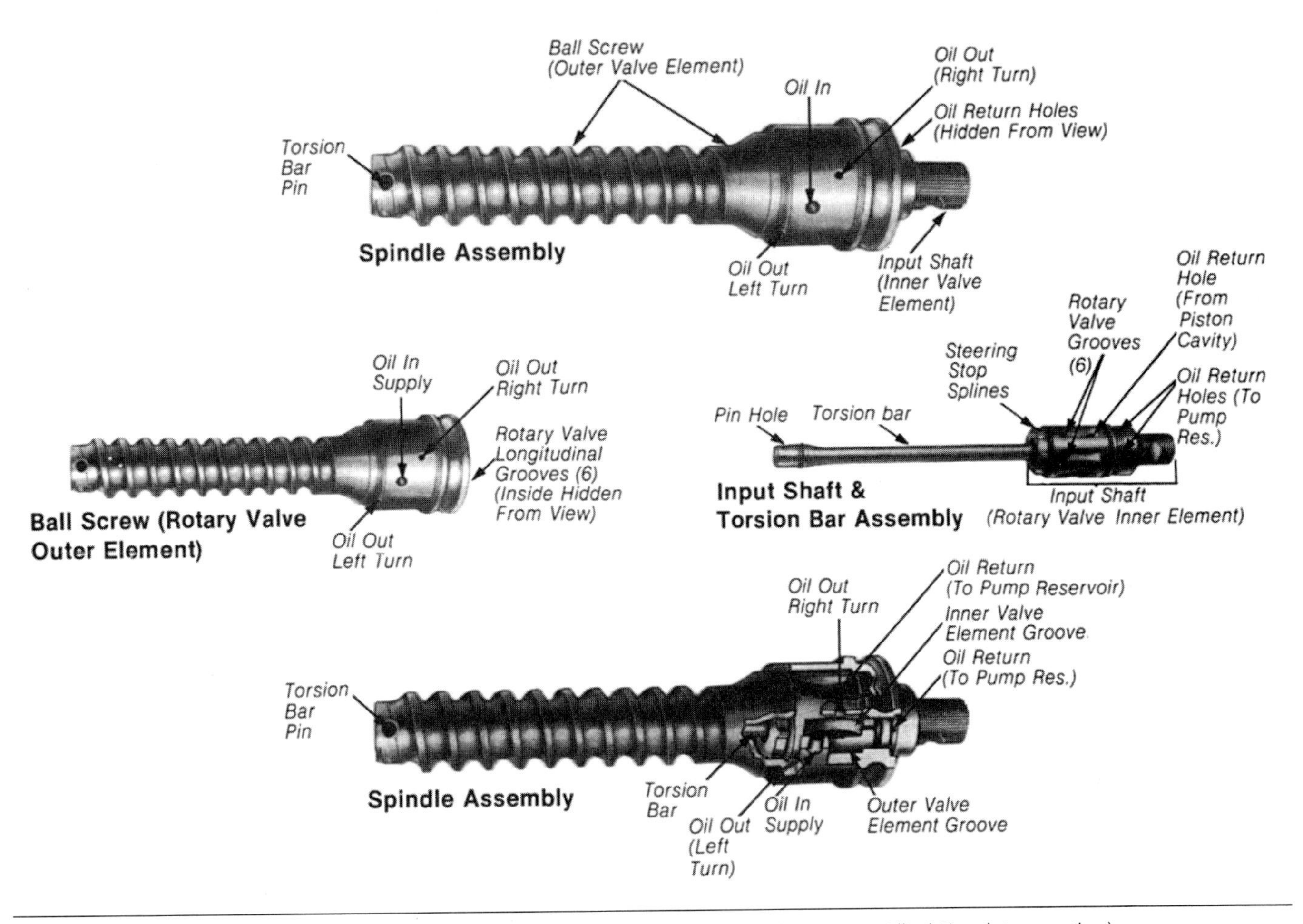

Figure 5-35 Ball screw and rotary valve assembly. (Courtesy of Bendix Heavy Vehicle Systems, Allied Signal Automotive.)

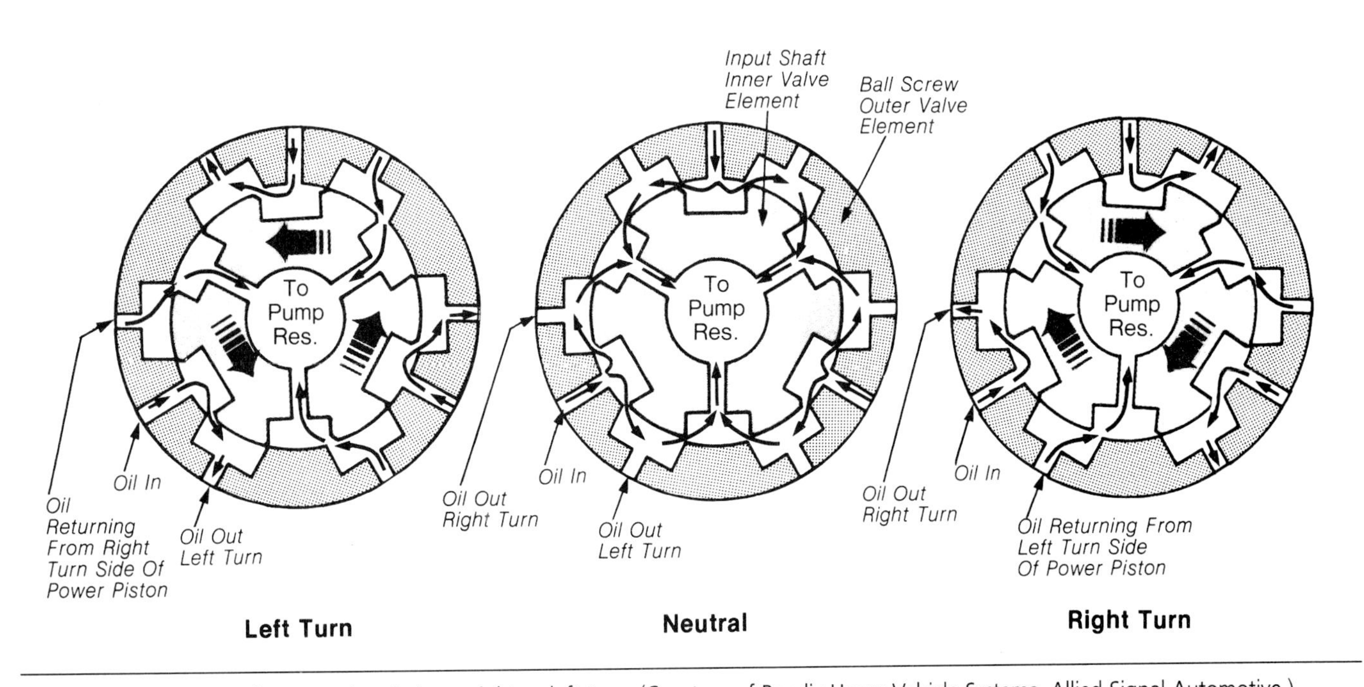

Figure 5-36 Rotary valve operation during a right or left turn. (Courtesy of Bendix Heavy Vehicle Systems, Allied Signal Automotive.)

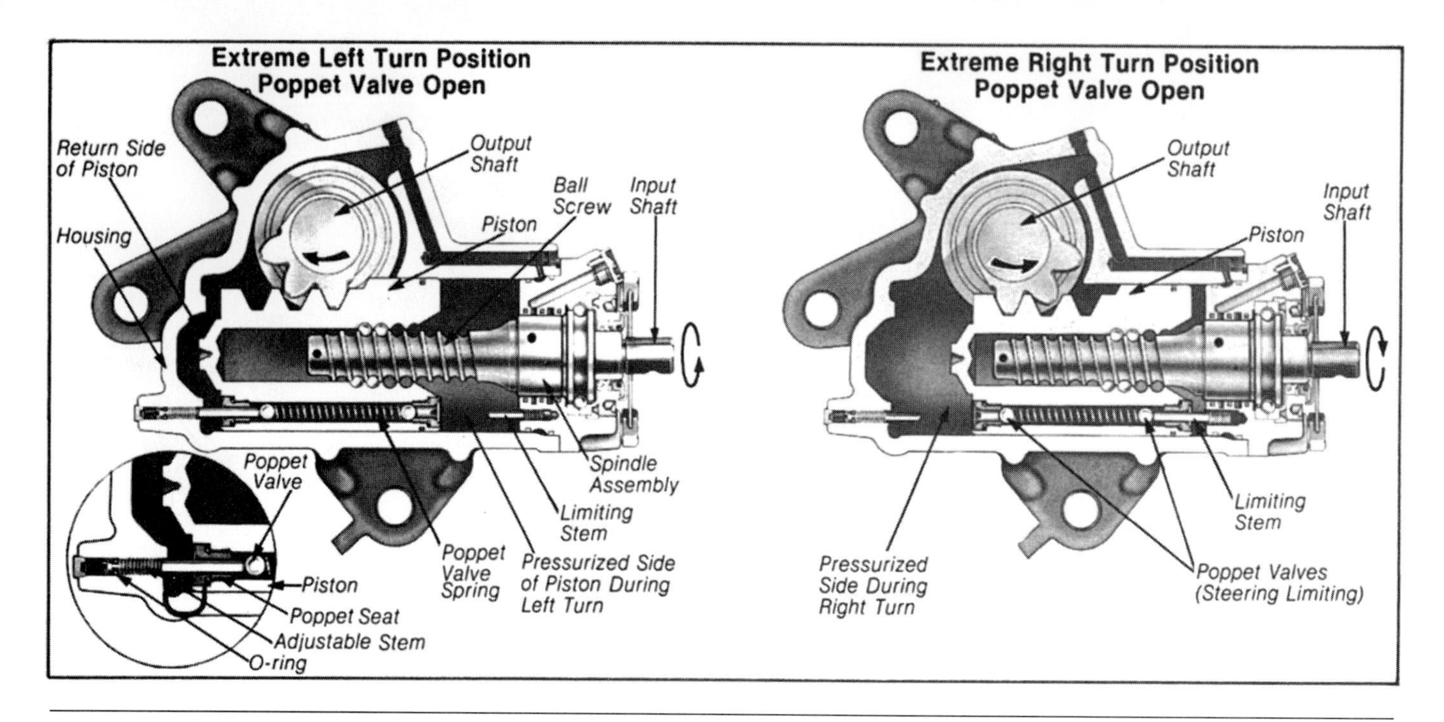

Figure 5-37 Poppet valve operation during a right or left turn. (Courtesy of Bendix Heavy Vehicle Systems, Allied Signal Automotive.)

See Chapter 5 in the Shop Manual for pBendix C-500N power steering gear adjustments.

from one side of the ball nut piston to the other side of this piston. This action provides easier steering wheel rotation and prevents reservoir overflowing and fluid cavitation in the high-pressure hose.

Air-Assisted Steering Systems

CAUTION: Never loosen an air line with air pressure in the line. If an air line fitting is disconnected with air pressure in the line, the line may whip around and cause personal injury.

Some heavy duty trucks are equipped with an **air-assisted steering system.** This system uses a manual steering gear. Air is supplied from the air brake system to the air-assisted steering system. A **torque valve** is mounted in the drag link connecting the pitman arm to the left front steering arm (Figure 5-38). Air pressure is supplied from the air brake system to the torque valve. The torque valve senses the direction of steering wheel rotation and steering linkage movement.

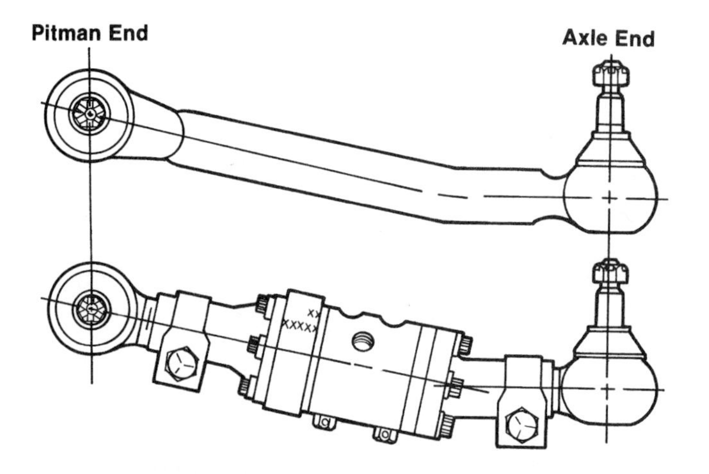

Figure 5-38 In an air-assisted steering system the conventional drag link is replaced with a drag link containing a torque valve. (Courtesy of Air-O-Matic Power Steering, Div. of Sycon Corp.)

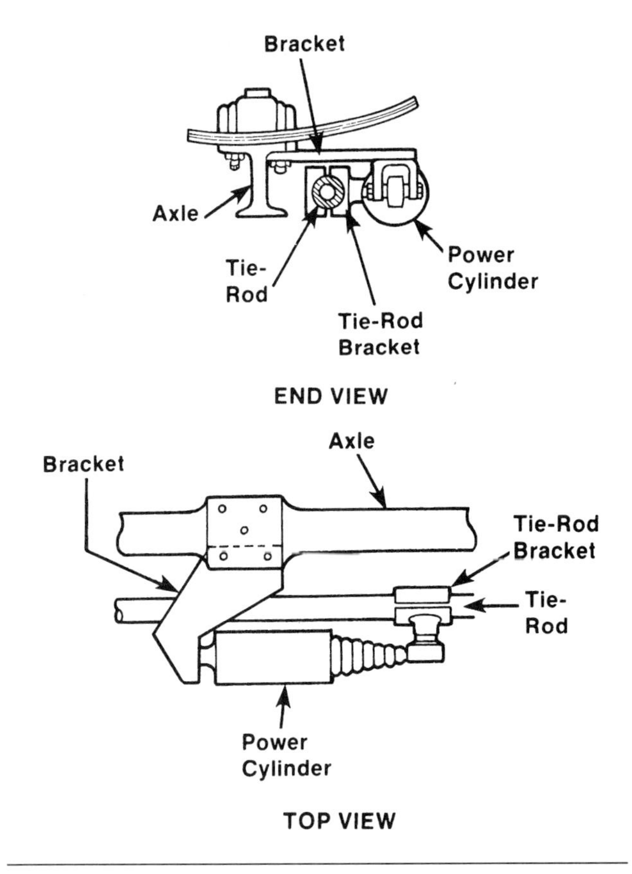

Figure 5-39 The power cylinder is mounted securely to the spring U-bolts, and the power cylinder piston rod is connected to the tie rod. (Courtesy of Air-O-Matic Power Steering, Div. of Sycon Corp.)

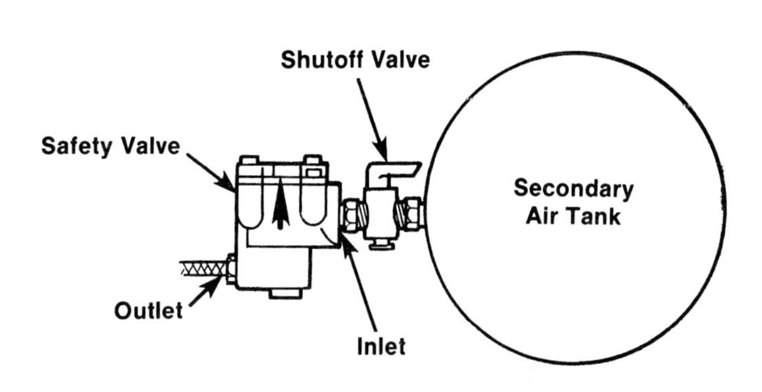

Figure 5-40 Safety valve in an air-assisted steering system. (Courtesy of Air-O-Matic Power Steering, Div. of Sycon Corp.)

Outlet air hoses are connected from the torque valve to the **power cylinder.** One end of the power cylinder is attached securely to the spring U-bolts on one side of the front axle (Figure 5-39). A piston is mounted in the center of the power cylinder, and a rod is connected from this piston to the tie rod. Air may be supplied from the torque valve to either side of the piston in the power cylinder.

During a right turn air pressure is supplied from the torque valve to the right side of the power cylinder piston. This air pressure helps to move the power cylinder piston and tie rod to the left, and this movement turns the front wheels to the right. During a left turn the torque valve supplies air pressure to the left side of the power cylinder piston, and this pressure helps to move the piston and tie rod to the right, which turns the front wheels to the left.

A **safety valve** is connected in the air hose between the air brake reservoir and the torque valve (Figure 5-40). This safety valve automatically closes and shuts off the air supply to the air-assisted steering system if the reservoir pressure drops to less than 65 psi (448.17 kPa). This action protects the air supply in the air brake system. When the safety valve closes and shuts off the air supply to the air-assisted steering system, the steering system reverts to manual steering. When the air pressure in the air brake reservoir increases to greater than 65 psi (448.17 kPa), the safety valve opens, and normal air-assisted steering operation is restored.

See Chapter 5 in the Shop Manual for air assisted power steering service and adjustments.

Load-Sensing Power Steering Systems

The load-sensing power steering system is hydraulically operated. The power steering pump in this system contains a special flow control and pressure relief valve that senses the load on the steer-

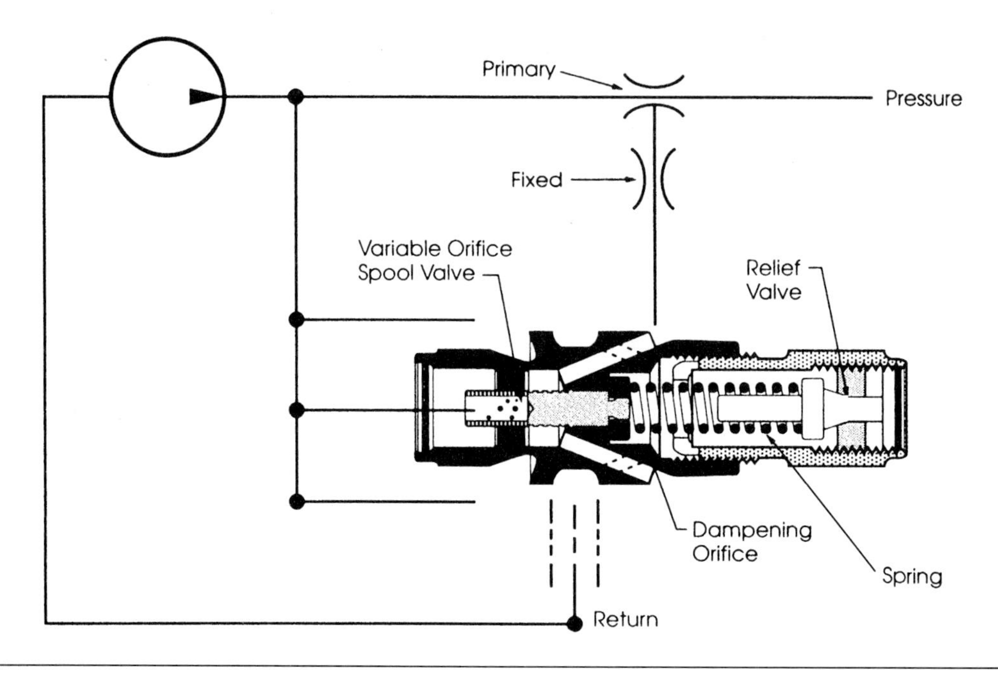

Figure 5-41 Combined flow control, pressure relief, and variable orifice spool valve assembly.

ing system. This valve contains a variable orifice spool valve (Figure 5-41). Fluid from the power steering pump may flow through a primary orifice to the steering gear. Some of the fluid from the pump is bypassed through the variable orifice spool valve and a fixed restriction to the power steering gear. If the restriction in the variable orifice spool valve increases, the pump flow through this spool valve decreases and the flow through the primary orifice to the steering gear increases to provide more steering assist.

When the wheels are straight ahead, the pump pressure and flow remain low. Under this condition fluid flow to the steering gear is divided between the variable orifice spool valve and the primary orifice, and the power steering system provides normal power steering assist. In this mode the variable orifice spool valve is partly open, and fluid flows through this valve and the fixed orifice to the steering gear. Return fluid also flows past the shoulder of this valve into the spring chamber. To escape from this chamber the return fluid must flow through passages in the valve housing to the pump inlet or through the orifice in the center of the valve and back into the outlet passage to the fixed orifice.

If the steering wheel is turned, higher pump pressure and flow are demanded by the power steering system. Under this condition the increased pump pressure and flow tend to force the variable orifice spool valve farther open. Under this condition flow past the variable orifice spool valve to the spring side of the variable orifice spool valve is increased. The orifice in the center of the variable orifice spool valve is designed to temporarily trap the fluid on the spring side of this valve. This action forces the variable orifice spool valve toward the closed position. Under this condition power steering pump pressure and flow increase through the primary orifice to the steering gear to provide improved steering assist. Therefore the load-sensing power steering system provides improved steering assist if the steering wheel is turned and higher power steering pump pressure and flow are demanded by the system.

Variable-Assist Power Steering

System Advantages and Design

Conventional power steering systems are designed to provide a reasonable amount of pressure and flow to meet the power assist demands during sharp turns or parking. These systems also provide a light steering effort when driving straight ahead, which decreases road feel. When the truck is driven straight ahead, the power steering pump flow is continually being circulated from the pump through the steering gear and back to the pump. Pump flow is actually much higher than required during straight ahead driving, and the engine power to drive the power steering pump is largely being wasted.

Some heavy duty trucks are now available with computer-controlled variable-assist power steering. In these systems the power steering pump flow is controlled by a computer and an electric solenoid. This computer and solenoid control the power steering pump flow more precisely to provide equal or more flow for increased power steering assist during sharp turns or when parking and reduced flow for better road feel when driving straight ahead. This action decreases the amount of engine power required to drive the power steering pump when driving straight ahead, and this provides a small savings in fuel consumption.

The variable-assist power steering system contains a driver-operated switch in the instrument panel. The driver may select normal or economy power steering modes with this switch (Figure 5-42). The driver-operated switch sends an electrical signal to the variable-assist power steering computer. This computer also receives voltage input signals from the vehicle speed sensor. These voltage signals inform the computer regarding the truck road speed. An optional two-speed axle shift sensor may be used in the system. In response to the input signals from the vehicle speed sensor and the switch, the computer controls the solenoid in the system.

The computer controls this solenoid with a pulse width–modulated (PWM) signal. This type of signal is pulsed on and off by the computer, and the computer can vary the solenoid on and off time. The solenoid contains a movable plunger surrounded by an electric winding. The computer uses the PWM signal to turn the current flow on and off in the solenoid winding. This action causes the solenoid plunger to be continually oscillated up and down. Power steering fluid from the pump must flow through a valve on the end of the solenoid plunger. By precisely controlling the solenoid plunger movement the computer can control the power steering pump flow. The steering gear in the variable-assist steering system is similar to the steering gear in a conventional power steering system, but the gear in the variable-assist system has a modified rotary valve. The vane-type pump in the variable-assist power steering system is also similar to the pump in a conventional power steering system except for the addition of the electric solenoid.

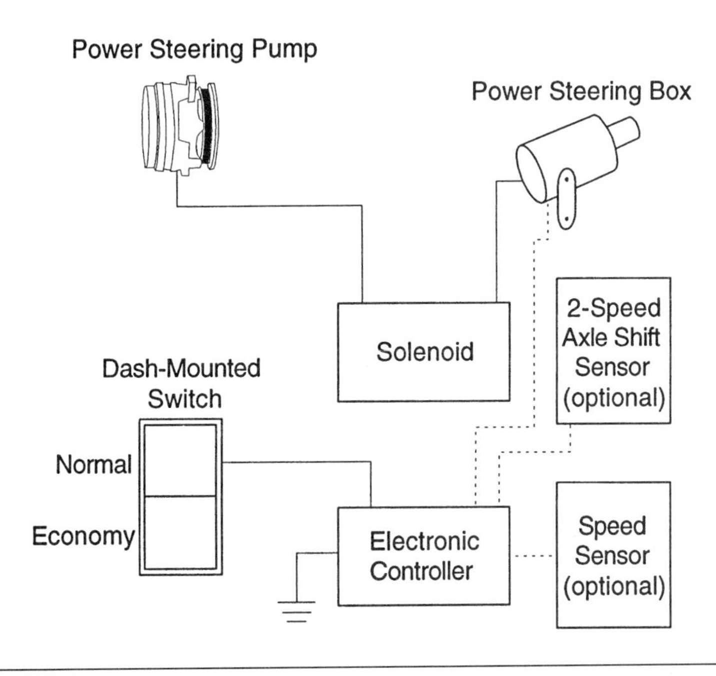

Figure 5-42 Variable-assist power steering system.

System Operation

When the dash switch is in the economy position and the truck is operating at less than 15 mph (24 kmh), the computer operates the solenoid to allow maximum power steering pump flow, and this provides maximum power steering assist. When the truck speed is increasing between 15 mph (24 kmh) and 50 mph (80 kmh), the computer operates the solenoid to gradually reduce power steering pump flow and this reduces power steering assist to provide improved road feel and better economy. If the truck speed is greater than 50 mph (80 kmh), the computer operates the solenoid to provide minimum power steering assist. Under this condition the power steering pump flow is 13.6 liters per minute (L/min).

If the truck is parked with the engine running for more than 10 minutes, the variable-assist power steering system reverts to a flow of 13.6 L/min. If a defect occurs in the system, variable-assist power steering function is inoperative and the system operation is similar to a conventional power steering system with a constant steering assist.

Linkage-Type Hydraulic Power Steering with Dual Power Cylinders

Some heavy-duty trucks are equipped with a hydraulic-assist power steering system. This system uses a manual steering gear and a conventional power steering pump. Pressure and return hoses are connected from the power steering pump to the integral valve in the main power cylinder (Figure 5-43). Two hydraulic hoses are connected from the integral valve to the auxiliary power cylinder. The drag link is connected from the pitman arm on the steering gear to the integral valve. The front end of the main power cylinder is connected to a pivoted arm attached to the frame. A linkage on the lower end of this pivoted arm is connected to the left upper steering arm. The rear end

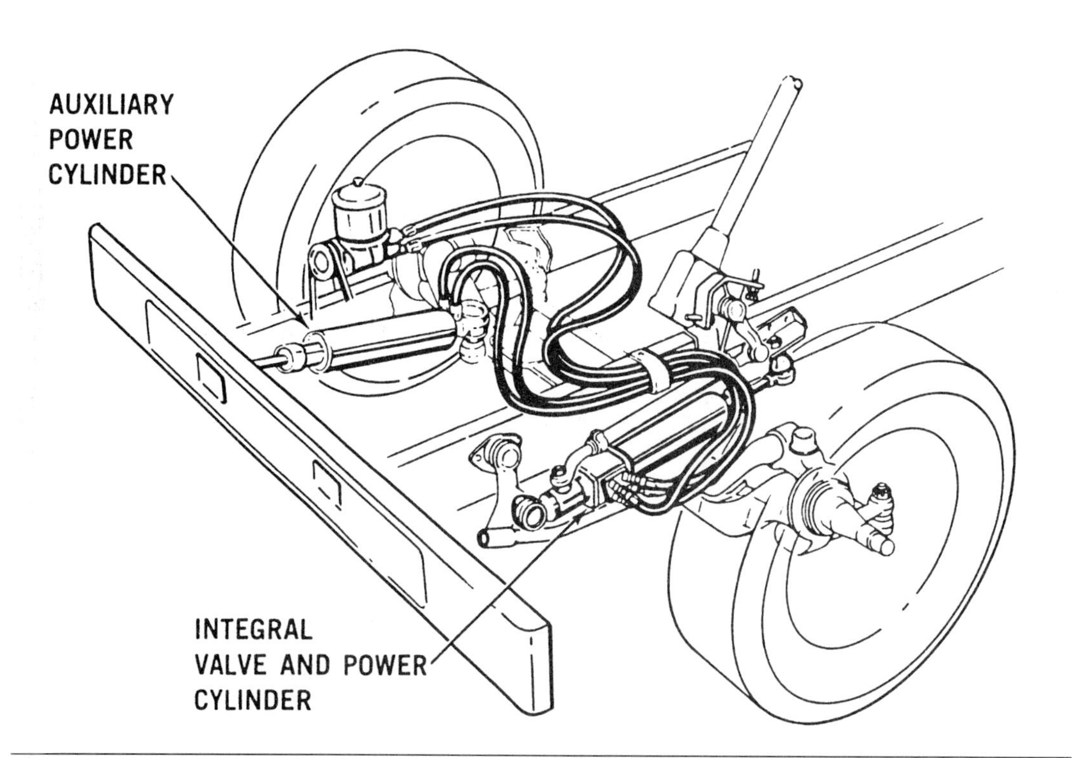

Figure 5-43 Linkage-type power steering system with dual power cylinders. (Reprinted with permission from SAE Publication All Wheel Steering for Heavy Truck Application ©1995 Society of Automotive Engineers, Inc.)

of the main power cylinder is attached to the frame at a point behind the steering gear. A piston and rod are mounted in each power cylinder.

When the steering wheel is turned with the engine running, the integral valve senses the direction of drag link movement. The integral valve supplies fluid pressure to the appropriate side of the main power cylinder piston. When this action occurs, the main power cylinder and the drag link push or pull on the pivoted arm to turn the front wheels. When the steering wheel is turned, the integral valve also supplies fluid pressure to the appropriate side of the auxiliary power cylinder piston, and this auxiliary power cylinder also supplies force to the steering linkage to help turn the front wheels.

See Chapter 5 in the Shop Manual for linkage type power steering with dual power cylinder diagnosis and service.

A BIT OF HISTORY

More electronically controlled systems are being installed on trucks each year. Antilock brake systems (ABS) have just become mandatory on heavy duty trucks according to Department of Transportation (DOT) regulations. On many heavy duty trucks traction control systems are available as an option with the ABS systems. Many engine functions such as injection control are now computer controlled. Some trucks are now available with navigation systems and collision avoidance systems. In the future electronic systems on trucks will be expanded and integrated so one computer controls more functions.

Summary

- A V-belt may be used to drive the power steering pump.
- The friction surfaces are on the sides of a V-belt.
- Proper belt tension is extremely important for adequate power steering pump operation.
- A power steering pump may have an integral or a remote fluid reservoir.
- A power steering pump may have a vane or roller-type rotor assembly, but both types of pumps operate on the same basic principle.
- The flow control valve in a power steering pump is moved toward the closed position by spring pressure and fluid pressure from the venturi in the pump outlet fitting.
- When a vehicle is driven at low speeds with the front wheels straight ahead, the power steering pump fluid flow is lower. Under this condition the pressure in the outlet fitting venturi is higher, and this higher pressure together with the spring tension keeps the flow control valve nearly closed. In this valve position a small amount of fluid is routed past the flow control valve to the pump inlet.
- If engine speed is increased, the pump delivers more fluid than the system requires. This increase in pump flow reduces pressure in the outlet fitting venturi, and this pressure decrease is applied to the spring side of the flow control valve. This action allows the flow control valve to move toward the open position, and excessive pump flow is returned past the flow control valve to the pump inlet.
- If the driver turns the steering wheel, fluid rushes from the pump outlet into the steering gear pressure chamber. Under this condition the flow control valve moves momentarily toward the closed position to maintain pump pressure and flow to the steering gear.
- If the front wheels are turned all the way against the stops, the power steering pump pressure could become high enough to damage hoses or other components. Under this condition the

Terms to Know

Air-assisted steering system
Cam ring
Flow control valve
Integral power steering pump reservoir
Kick-back
Outlet fitting venturi
Poppet valves
Pressure relief valve
Power cylinder
Remote power steering pump reservoir
Rotary valve
Safety valve
Sector shaft
Spool valve
Torque valve
Torsion bar
V-belt
Wormshaft

high pump pressure unseats a pressure relief ball in the center of the flow control valve, and this action allows some pump flow to move past the pressure relief ball to the pump inlet, which limits pump pressure.

- ❑ In a power steering gear with the front wheels straight ahead, equal pressure is applied to both sides of the piston or pistons, and the fluid from the pump is directed through the spool valve and rotary valve to the return hose and pump inlet.
- ❑ In a power steering gear the spool valve movement inside the rotary valve is controlled by torsion bar deflection.
- ❑ When the front wheels are turned in a power steering gear, torsion bar deflection moves the spool valve inside the rotary valve, and this valve movement directs the power steering fluid to the appropriate side of the piston to provide steering assist.
- ❑ In a TRW/Ross power steering gear when the steering wheel is turned in either direction so the steering linkage is one third of a turn from the stops, a poppet valve in the piston is opened and power steering fluid pressure on the piston is released past the poppet valve. This action stops the power steering assist.
- ❑ In an air-assisted steering system the torque valve in the drag link senses steering wheel and steering linkage movement and directs air pressure to the appropriate side of the power cylinder piston.
- ❑ In an air-assisted steering system the power cylinder is mounted on the spring U-bolts, and the power cylinder piston and rod are connected to the drag link. Air pressure supplied to the appropriate side of the power cylinder piston provides steering assist in the proper direction.
- ❑ In an air-assisted steering system the safety valve is connected in the air hose between the air brake system reservoir and the torque valve. The safety valve closes at 65 psi (448.17 kPa) and protects the air supply in the air brake system.
- ❑ A variable-assist power steering system provides improved road feel when driving straight ahead and increased power assist when turning the front wheels.

Review Questions

Short Answer Essays

1. Describe the proper position of a V-belt in a pulley.
2. Explain how the fluid pressure is produced in a vane-type power steering pump.
3. List the forces that move the flow control valve toward the closed position in a power steering pump.
4. Describe the operation of a venturi.
5. Explain the operation of the flow control valve when the steering wheel is turned.
6. Describe the operation of a Saginaw power steering pump when the steering wheel is rotated all the way to the right or left until the front wheels contact the stops.
7. Describe the flow of power steering fluid with the engine running and the front wheels straight ahead in a power steering gear.
8. Describe how torsion bar deflection occurs in a power steering gear.
9. Explain the purpose of torsion bar deflection in a power steering gear.
10. Describe the operation of the torque valve in an air-assisted steering system.

Fill-in-the-Blanks

1. The lower edge of a V-belt should never contact the ____________________.
2. Excessive belt wear occurs if pulleys are ____________________.
3. The purpose of the vanes in the power steering pump rotor is to ____________________ the rotor in the elliptical cam ring.
4. As the power steering pump shaft and rotor turn, ____________________ causes the vanes to move outward against the cam ring.
5. A machined venturi is located in the power steering pump ____________________.
6. When fluid movement through the venturi increases, the pressure in the venturi ____________________.
7. With the engine running and the front wheels straight ahead, the power steering pump pressure is ____________________.
8. The pressure relief ball in a Saginaw power steering pump is forced open when the front wheels are turned against the ____________________.
9. In a power steering gear with the front wheels straight ahead and the engine running, the fluid on each side of the recirculating ball piston helps to prevent ____________________ from reaching the steering wheel.
10. In a power steering gear during a turn the ____________________ movement directs the power steering fluid pressure to the appropriate side of the piston.

ASE Style Review Questions

1. While discussing power steering pump drive belts:
   Technician A says a slipping V-belt may be caused by the belt contacting the bottom of the pulley.
   Technician B says a power steering pump will not develop full pressure if the belt is slipping.
   Who is correct?
   A. A only
   B. B only
   C. Both A and B
   D. Neither A nor B
2. While discussing air-assisted steering operation:
   Technician A says the torque valve supplies air to the appropriate side of the power cylinder piston.
   Technician B says the safety valve limits the amount of steering linkage movement.
   Who is correct?
   A. A only
   B. B only
   C. Both A and B
   D. Neither A nor B
3. During normal power steering pump operation:
   A. An increase in fluid movement through the outlet fitting venturi causes a pressure increase in the venturi.
   B. An increase in flow control valve opening allows more fluid to return to the pump inlet.
   C. The flow control valve spring moves this valve toward the open position.
   D. A decrease in fluid pressure in the outlet fitting venturi moves the flow control valve toward the closed position.
4. While discussing power steering pump operation:
   Technician A says if the engine is idling and the front wheels are straight ahead, the flow control valve is near the closed position.
   Technician B says when the steering wheel is turned, the flow control valve moves toward the open position.
   Who is correct?
   A. A only
   B. B only
   C. Both A and B
   D. Neither A nor B
5. In a TRW/Ross power steering gear:
   A. Two poppet valves are located in the piston.
   B. The torsion bar connects the wormshaft and the outer part of the control valve.
   C. When the steering wheel is rotated from the center position, one of the poppet valves opens.
   D. When one of the poppet valves opens, fluid pressure on the piston increases.

6. While discussing TRW/Ross power steering gears:
Technician A says power steering assist is available when the steering wheel is turned so the steering linkage is against the stops.
Technician B says one poppet valve is open and the other poppet valve is closed when the steering linkage is against the stops.
Who is correct?
   A. A only **C.** Both A and B
   B. B only **D.** Neither A nor B

7. All of these statements about power steering pumps are true *except:*
   A. The cam ring rotates with the rotor and shaft.
   B. The rotor shaft and vanes rotate inside the cam ring.
   C. The pressure relief valve is in the center of the flow control valve.
   D. The pressure relief valve remains closed at idle speed.

8. When the steering wheel is turned fully to the right against the stops with a Saginaw power steering pump and gear:
   A. The power steering pump flow to the steering gear increases.
   B. The pressure relief ball in the flow control valve opens to limit pump pressure.
   C. The flow control valve moves toward the closed position.
   D. The flow control valve opens and supplies fluid pressure to the pressure relief valve.

9. While discussing steering kick-back:
Technician A says that steering kick-back is caused by a defective steering gear.
Technician B says that steering kick-back is a strong and sudden movement of the steering wheel in the opposite direction to which the steering wheel is turned.
Who is correct?
   A. A only **C.** Both A and B
   B. B only **D.** Neither A nor B

10. During normal Saginaw power steering gear operation:
   A. During a turn torsion bar twisting positions the spool valve inside the rotary valve so fluid pressure is directed to the appropriate side of the pistons to provide steering assist.
   B. When the truck is driven straight ahead, unequal pressure exists on each side of the pistons.
   C. During a turn when fluid pressure is supplied to one side of the pistons, fluid pressure is trapped on the opposite side of the pistons.
   D. When fluid pressure is supplied to the lower end of the lower piston, fluid pressure is also supplied to the lower end of the upper piston.

Frames and Fifth Wheels

Upon completion and review of this chapter, you should be able to:

- ❑ Explain section modulus as it relates to truck frames.
- ❑ Define the yield strength of a truck frame.
- ❑ Describe resisting bending moment related to truck frames.
- ❑ Explain applied moment and bending moment in relation to truck frames.
- ❑ Describe the design of a C-channel truck frame.
- ❑ List several necessary precautions when working on truck frames.
- ❑ Describe three different types of frame reinforcements.
- ❑ Explain six different defective frame conditions.
- ❑ Describe the purpose of a fifth wheel.
- ❑ Describe seven types of fifth wheels, and explain the application for each type of fifth wheel.
- ❑ Explain the operation of a sliding fifth wheel.
- ❑ Explain the advantages of a sliding fifth wheel.
- ❑ Describe the operation of a fifth wheel with a yoke and secondary lock during the coupling process.
- ❑ Explain the operation of a fifth wheel with a yoke and secondary lock during the uncoupling process.
- ❑ Describe the proper procedure for coupling a tractor and trailer.
- ❑ Explain the proper procedure for uncoupling a tractor and trailer.

Introduction

The frame is the central component in the truck chassis because the frame supports the engine, transmission, axles, cab, and other chassis components. Because the frame supports these components, proper frame strength and alignment are critical to the operation of these other components. For example, improper frame alignment may cause improper driveline angles and serious vibration problems. A distorted frame may cause improper wheel alignment that adversely affects steering control and tire wear.

The **fifth wheel** couples the tractor to the trailer, or the fifth wheel may be considered as the device that connects the power to the payload. The fifth wheel must provide a positive, secure coupling between the tractor and the trailer under all movement conditions between the tractor and the trailer. This movement includes providing a pivot that allows the tractor to pivot around the trailer kingpin as the tractor and trailer turn corners. The fifth wheel not only couples the tractor to the trailer but also supports the weight on the front of the trailer. The name "fifth wheel" is derived from the trailer weight supporting function of the fifth wheel. Without the fifth wheel semitrailers would not be possible with their increased capacity compared with a single-unit truck.

The fifth wheel and the trailer kingpin keep the tractor from going in one direction and the trailer going in another direction. The fifth wheel is next to the brake system in safety importance.

Frame Terms

The technician must be familiar with terms used regarding frames and frame servicing. These terms are commonly used in relation to truck frames.

Section Modulus

Section modulus is an indication of frame strength made on the basis of the height, width, thickness, and shape of the frame side rails. Frames with deeper webs or wider flanges have increased section modulus. The strength of the steel in the frame is not considered in the section modulus.

Yield Strength

Yield strength is a measure of the steel strength used in the frame. The yield strength is the maximum load measured in pounds per square inch (psi) or kilopascals (kPa) that may be placed on a material and the material will still return to its original shape. Medium duty truck frames are made from 36,000 psi (248,220 kPa) yield strength mild steel or 50,000 psi (344,750 kPa) yield strength high-strength, low-alloy (HSLA) steel. Class 8 trucks have frames made from 110,000 psi (758,450 kPa) yield strength heat-treated steel. Frame yield strength may be identified by a symbol punched into the side rails near the front of the rear spring hangers (Figure 6-1). Various frame materials require different welding and service procedures.

Resisting Bending Moment

The **resisting bending moment (RBM)** is calculated by multiplying the section modulus by the yield strength. Stronger frames have higher RBMs. The RBM is the most accurate indication of a frame's strength.

Applied Moment

The **applied moment** is a measurement of a specific load placed in a certain location on the frame. The applied moment is based on a stationary vehicle.

Bending Moment

The term **bending moment** means that a load applied to the frame will be distributed across a given section of the frame. This load tends to distort the frame at the point of heaviest load concentration. This point of heaviest load concentration and frame distortion varies depending on the type of truck. For example, the point of heaviest load concentration and frame distortion is different on a straight box truck compared with a gravel truck (Figure 6-2).

Area and Safety Factor

The frame area is the total cross section of the frame rail in square inches. The safety factor is the amount of load that can be safely absorbed by the truck frame members. The safety factor is the reaction of the applied moment to the RBM. When the applied moment and the RBM are the same, the frame has a safety factor of 1.

Frame Design

The main components in truck frames are the frame rails and the crossmembers (Figure 6-3). Truck frames usually have straight full channel side rails. These side rails are a **C-channel** design with upper and lower flanges and a vertical web between the flanges (Figure 6-4). **I-beam frames** may

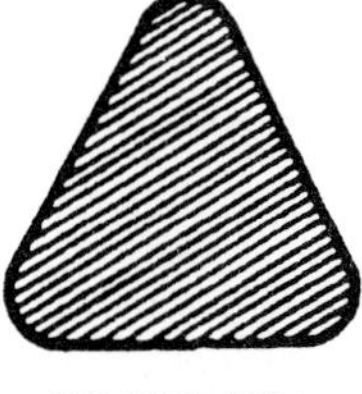

Figure 6-1 Different symbols are punched into the frame side rails to indicate the frame yield strength. (Courtesy of Sterling Truck Corporation.)

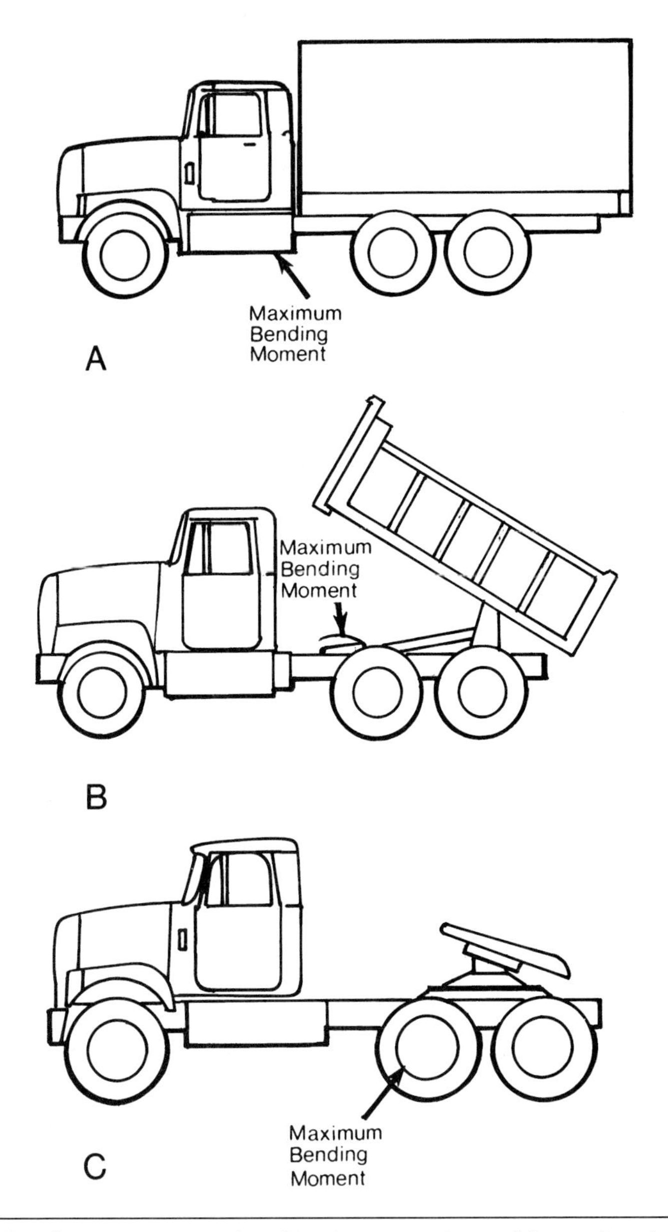

Figure 6-2 Frame bending moment or point of load concentration and frame distortion. (Courtesy of Mack Trucks, Inc.)

be used in some applications such as crane carriers. **Box frames** may be used in some trucks (Figure 6-5).

The **crossmembers** may be C-section, tubular, boxed, or I-beam design. The side rails in the frame carry the load, and the crossmembers stabilize the frame. This type of frame may be referred to as a ladder-type frame. The frame must be made from the proper steel and be designed to support the rated vehicle payload. Frames must also be rigid enough to resist bending and twisting, and provide fatigue resistance from continual flexing. Frames must also be able to resist shocks and vibration.

A ladder-type frame has two full, straight side rails connected by crossmembers.

The metal thickness in the frame side rails varies, depending on the severity of truck operation. For example, the frame rails on an on-road tractor may be $\frac{1}{4}$ in (6.35 mm) thick, and a cement mixer truck chassis may have a frame rail thickness of $\frac{5}{16}$ or $\frac{3}{8}$ in (7.93 or 9.52 mm). The depth of the side rails also varies depending on the severity of operation. An on-road tractor may have frame rails with a depth of 10 $\frac{1}{4}$ in (26 cm), whereas the on/off-road cement mixer truck frame may have a depth of 13 $\frac{1}{2}$ in (34.29 cm).

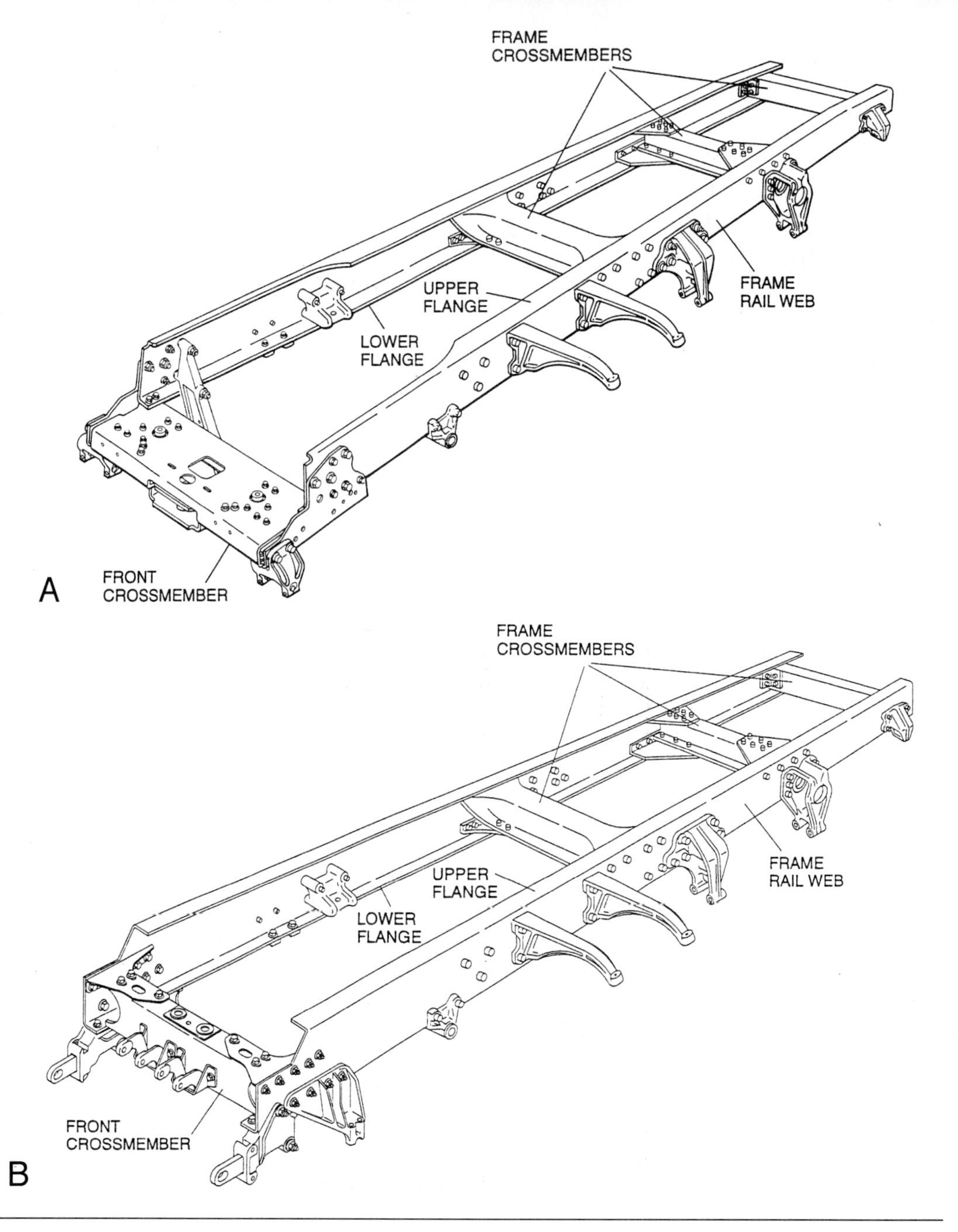

Figure 6-3 Heavy duty truck frames. (This information courtesy of Freightliner Corporation.)

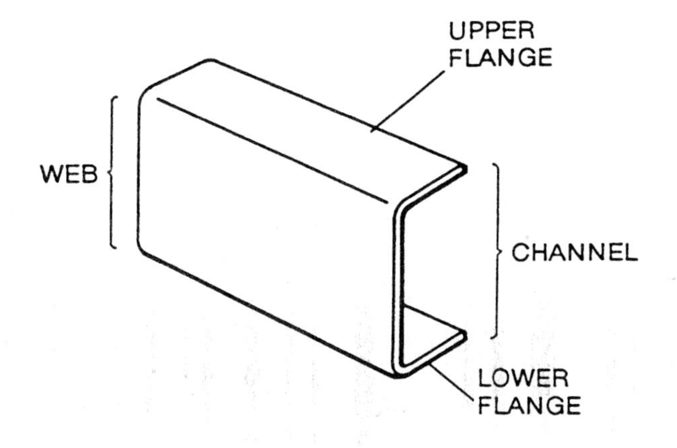

Figure 6-4 C-channel used in a frame rail. (This information courtesy of Freightliner Corporation.)

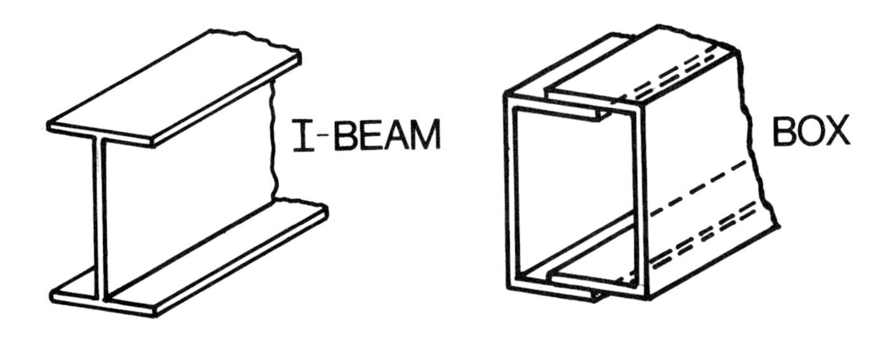

Figure 6-5 I-beam and box type frames.

▲ **WARNING:** Drilling extra holes in a truck frame may weaken the metal in the frame and cause frame cracks or breaks. Use only frame service procedures recommended in the truck manufacturer's service manual. Any deviation from these service procedures may void the truck manufacturer's warranty.

▲ **WARNING:** Heat from arc welding or using a cutting torch on a truck frame may weaken the metal in the frame and cause cracks or breaks in the frame.

During the manufacturing process frame rails are drilled to allow the attachment of crossmembers, engine mounts, spring hangers, and other components. Many frame and truck manufacturers do not recommend drilling extra holes in the frame because this weakens the frame rail. Most truck frames have a decal that warns against welding or using a cutting torch on the side rails (Figure 6-6). The heat from welding or cutting the side rails may weaken these components, causing the frame rail to crack or break. Welding may be allowed for crack repairs on frame rails. Before drilling or welding on a truck frame always refer to the truck manufacturer's recommendations in the appropriate service manual.

Aluminum alloy frames are used in some trucks. These frames are lighter than steel frames, but they are not considered to be as strong. Aluminum frames have greater web and flange thickness compared with steel frames. The type of metal in the frame may be determined by placing a magnet near the frame. The magnet is attracted to a steel frame, but no magnetic attraction exists to an aluminum frame. Aluminum frames are of two common types. One type is a magnesium-silicon-aluminum alloy with a yield strength of 37,000 psi (255,115 kPa). A second common type of aluminum frame is a high-strength copper-aluminum alloy with a yield strength of 60,000 psi (413,700 kPa). Repair procedures on these two types of aluminum frames are different. It is not possible to determine the type of aluminum frame from the appearance. The truck manufacturer's service manual, line-setting ticket, or sales data book must be consulted to determine the frame type.

Figure 6-6 Frame decal warning against heating, welding, or drilling. (Courtesy of Mack Trucks, Inc.)

Frame Reinforcements

CAUTION: Frame reinforcement to support additional truck loads should not be attempted until it is verified that the brake system, suspension system, and steering system can handle the additional load. If additional truck loads cause failure of one of these systems, it may result in an expensive collision or personal injury.

Frame reinforcements are available on most trucks. These reinforcements may be inside or outside the frame rail. Some reinforcements may be the full length of the frame rail or they may be partial reinforcements installed on part of the frame rail in the area of greatest load concentration or bending moment. Frame reinforcements must be attached to the frame rail web. These reinforcements must not be attached to the frame flanges because this is the area of greatest stress.

Frame reinforcement may be a single channel inside the frame rail. This type of reinforcement may be called a two-element frame rail. Other frame reinforcements may have frame rail with a single inside and a single outside channel. This type of frame reinforcement may be referred to as a three element frame rail. Some three-element frame rails have a single L-shaped reinforcement on the outside of the three elements. This type of frame reinforcement is called a four-element frame (Figure 6-7). Some frame reinforcements are a heavy metal plate bolted to the outside of the frame. This type of reinforcement may be called a **fishplate.** Frame reinforcements increase the RBM of the frame, but they add considerable weight to the truck. Frame reinforcements may add 15 to 40 lb (6.8 to 18.14 kg) per foot of reinforcement to the vehicle weight.

Some trucks are available with a deep section reinforcement in the area of greatest load concentration between the axles. Frame reinforcements may also be installed in other areas of heavy load concentration such as under a tractor's fifth wheel, at suspension mounting points such as add-on tag or pusher axles, or where heavy lift gates are installed. Frame reinforcements and crossmembers must be installed with the proper bolts, nuts, and washers specified by the truck or frame manufacturer (Figure 6-8). All frame bolts should be installed in the direction specified by the truck manufacturer. When frame reinforcements are installed, the factory-drilled bolt holes in the frame rails should be used.

Crossmembers are used to connect the frame rails. These crossmembers must provide rigidity and strength, but they must also have enough flexibility to withstand twisting and bending when operating a loaded truck on rough ground. Gussets are used to attach many crossmembers to the frame. These gussets are welded or bolted to the crossmember and then bolted to the frame. The gussets add strength to the frame and help to keep the frame rails from moving rearward or forward.

Frame Defects

Frame sag occurs when the frame or one side rail is bent downward from the original position

Frame sag occurs when the frame or one side rail is bent downward from the original position (Figure 6-9).

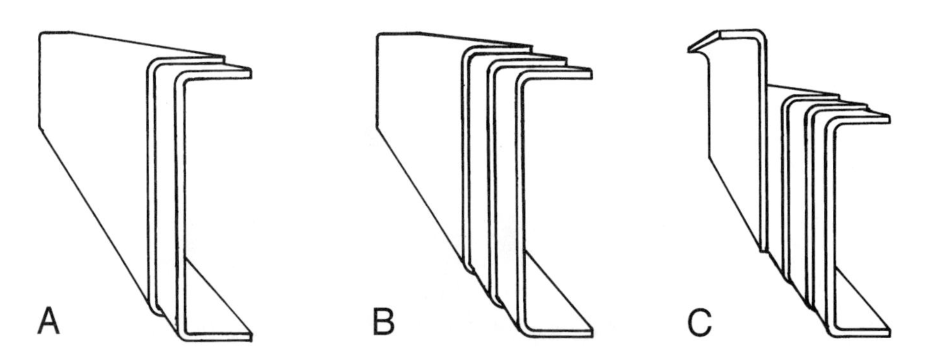

Figure 6-7 Various types of frame reinforcements. (Courtesy of Mack Trucks, Inc.)

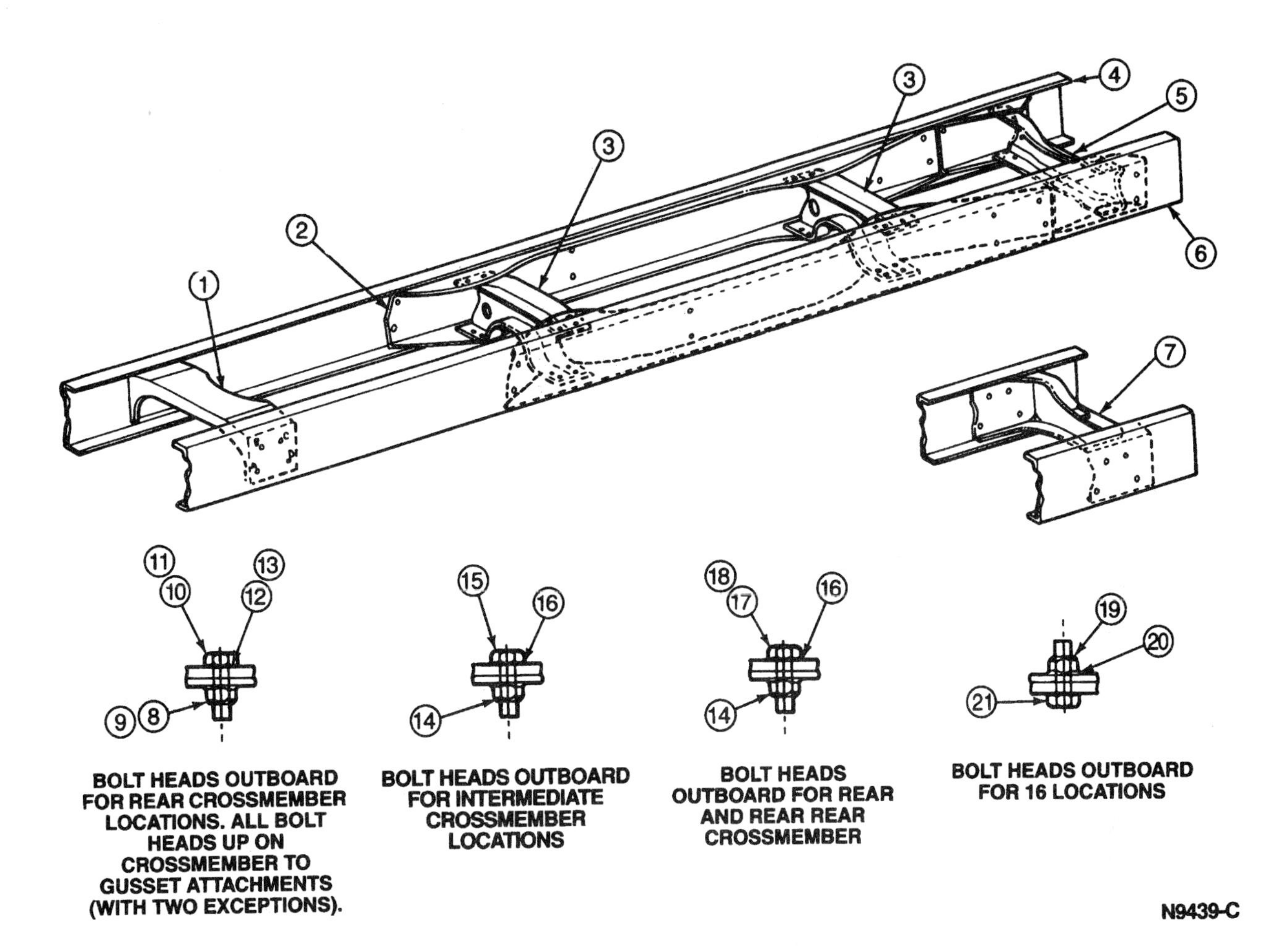

Item	Part Number	Description
1	5028	Intermediate Crossmember
2	5A154	Frame Sidemember Reinforcement
3	5036	No. 5 Crossmember
4	5016	Frame Sidemember (RH)
5	5036	Rear Crossmember
6	5015	Frame Sidemember (LH)
7	3035	Rear Crossmember
8	34991-S2	Nut (Vehicle Dependent)
9	34989-S2	Nut (Vehicle Dependent)
10	58721-S2	Bolt (Vehicle Dependent)

Item	Part Number	Description
11	58677-S2	Bolt (Vehicle Dependent)
12	44881-S2	Washer (Vehicle Dependent)
13	44879-S2	Washer (Vehicle Dependent)
14	34989-S2	Nut
15	58676-S2	Bolt
16	44879-S2	Washer
17	58676-S2	Bolt (Vehicle Dependent)
18	58677-S2	Bolt (Vehicle Dependent)
19	34991-S2	Nut
20	34881-S2	Washer
21	58721-S2	Bolt

Figure 6-8 Frame reinforcements, crossmembers, and proper bolt installation. (Courtesy of Sterling Truck Corporation.)

Buckle is a frame condition that refers to the frame or one side rail that is bent upward from the original position.

A **diamond frame** condition is present when one frame rail is moved rearward or forward in relation to the opposite frame rail (Figure 6-10).

Frame twist occurs when the end of one frame rail is bent upward or downward in relation to the opposite frame rail (Figure 6-11). When frame twist occurs in a tandem axle tractor or truck, the rear part of the frame may appear almost level and the twist may appear to be in the front of the frame. In many cases the twist is being held in by the strong crossmembers in the rear of the frame. This condition is usually present because the front springs are much lighter and weaker compared with the rear springs, and more controlling crossmembers are present from the back of

Buckle is a frame condition that refers to the frame or one side rail that is bent upward from the original position.

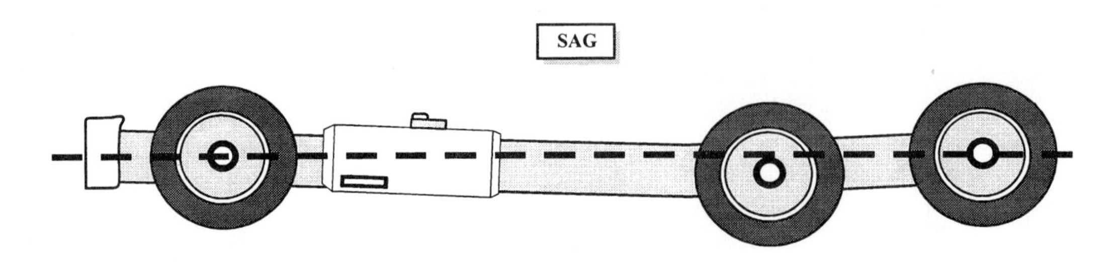

Figure 6-9 Frame sag. (Courtesy of Bee Line Company.)

A diamond frame condition is present when one frame rail is moved rearward or forward in relation to the opposite frame rail.

Frame twist occurs when the end of one frame rail is bent upward or downward in relation to the opposite frame rail.

Sidesway occurs when one or both frame rails are bent inward or outward.

Tracking is the alignment of the truck axles with each other. When the frame and axles are positioned properly to provide correct tracking, all the axles are parallel to each other and each axle is at a 90-degree angle in relation to the centerline of the vehicle.

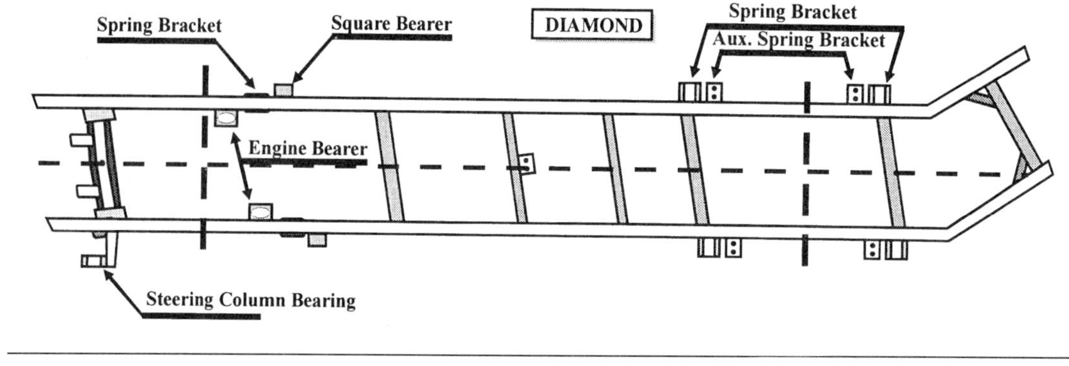

Figure 6-10 Diamond frame condition. (Courtesy of Bee Line Company.)

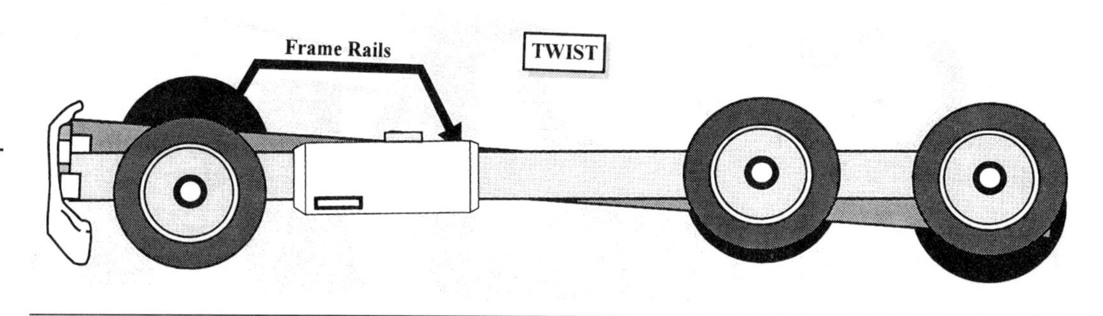

Figure 6-11 Frame twist. (Courtesy of Bee Line Company.)

the cab rearward compared with the number of crossmembers from the back of the cab forward. Therefore the twist may appear to be in the front of the frame when it actually is held in by the stronger crossmembers in the back of the frame.

Sidesway occurs when one or both frame rails are bent inward or outward (Figure 6-12).

Tracking is the alignment of the truck axles with each other. When the frame and axles are positioned properly to provide correct tracking, all the axles are parallel to each other and each axle is at a 90-degree angle in relation to the centerline of the vehicle (Figure 6-13).

Frame Straightening Equipment

See Chapter 6 in the Shop Manual for frame diagnosis and service.

Heavy duty frame straightening equipment is available to correct frame defects. A heavy duty frame press is located between two runways (Figure 6-14). These runways may also be used for heavy duty truck wheel alignment. The runway track width may be easily adjusted for different tractor,

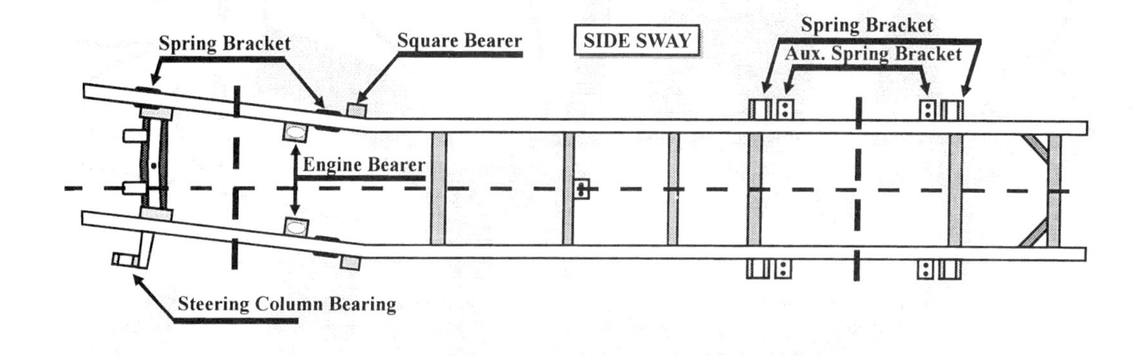

Figure 6-12 Sidesway frame condition. (Courtesy of Bee Line Company.)

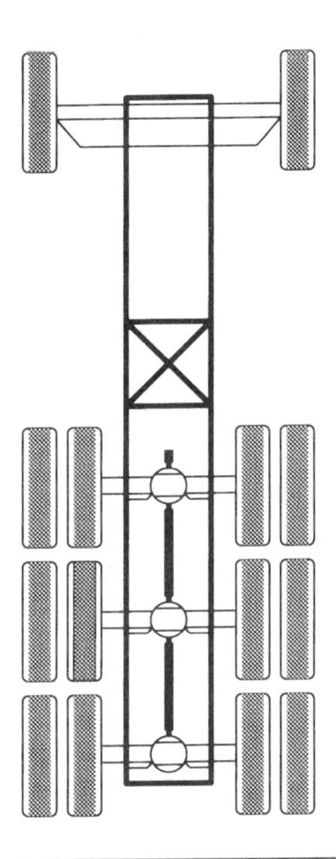

Figure 6-13 When the frame and axles are positioned properly to provide correct tracking, all the axles are parallel to each other and each axle is at a 90-degree angle in relation to the centerline of the vehicle. (Reprinted by permission of Hunter Engineering Company.)

truck, bus, and trailer track widths. After the tractor is driven on the runway, sections of the runway may be removed to install the necessary frame alignment gauges and bending tools. The frame press is raised and lowered by a 20 ton telescoping hydraulic ram at each end of the press to position it at the desired height (Figure 6-15).

Many special tools come with the frame straightening equipment, and these tools are attached to or mounted on the frame press to correct frame defects in trucks, tractors, buses, trailers, and off-road equipment. For example, holding tools are mounted between the frame press and trailer frame on each side of a sagged frame. After these holding tools are properly attached, the frame press must be lowered to remove all slack in the holding tools. Hydraulic rams are positioned in the area of greatest sag, and hydraulic pressure is supplied to these rams to correct the sag (Figure 6-16). Hydraulic pressure is supplied to the rams from a pump driven by an electric motor. The technician uses a fingertip control to operate this pump for precise control. Single or multiple application of hydraulic rams may be selected.

Figure 6-14 Heavy duty frame press and wheel alignment runways. (Courtesy of Bee Line Company.)

Figure 6-15 The frame press may be raised or lowered to position it at the desired height. (Courtesy of Bee Line Company.)

Figure 6-16 Holding tools and hydraulic rams installed at correct trailer frame sag. (Courtesy of Bee Line Company.)

Self-centering frame gauges are installed on the truck frame at various locations from the front to the rear of the frame to measure the frame alignment and to determine the exact frame problems (refer to Figure 6-14). The technician compares the alignment of these gauges to determine frame problems.

See Chapter 6 in the Shop Manual for frame straightening.

Fifth Wheels

Fifth Wheel Purpose

The fifth wheel couples the tractor to the trailer kingpin. When the fifth wheel on the tractor is backed under the front of the trailer, the fifth wheel must support the weight on the front of the trailer. Simultaneously the fifth wheel jaws must clamp securely onto the trailer kingpin. This clamping action must be very strong to maintain the coupling between the tractor and trailer. This clamping action must also be strong enough to withstand the stress caused by movement between the tractor and the trailer. The fifth wheel must also act as a type of suspension that swivels to allow fore-and-aft movement and also side-to-side movement between the tractor and trailer. Various fifth wheel designs and mountings use different methods to provide these swivelling requirements.

Types of Fifth Wheels

Semioscillating Fifth Wheel The **semioscillating fifth wheel** is commonly installed on over-the-road vehicles. This type of fifth wheel oscillates around an axis perpendicular to the tractor centerline (Figure 6-17).

Fully Oscillating Fifth Wheel The **fully oscillating fifth wheel** has a pivot that is perpendicular to the tractor centerline, and a second pivot that is positioned in the same direction as the tractor centerline (Figure 6-18). These two pivots allow front-to-rear oscillation and also side-to-side oscillation. This type of fifth wheel is intended for use with trailers that have a loaded center of gravity at or below the top of the fifth wheel.

Rigid Fifth Wheel A **rigid fifth wheel** is permanently fixed and does not oscillate in any direction (Figure 6-19). A rigid fifth wheel is used with an oscillating kingpin bolster plate on the trailer. This type of kingpin plate allows kingpin oscillation in place of fifth wheel oscillation. A

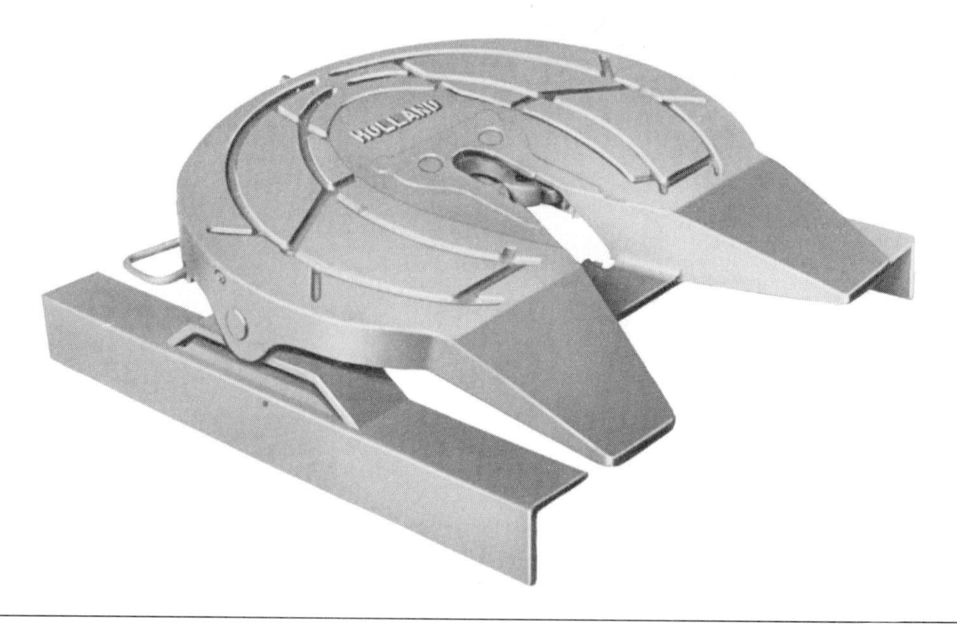

Figure 6-17 Semioscillating fifth wheel. (Courtesy of Holland Hitch Co.)

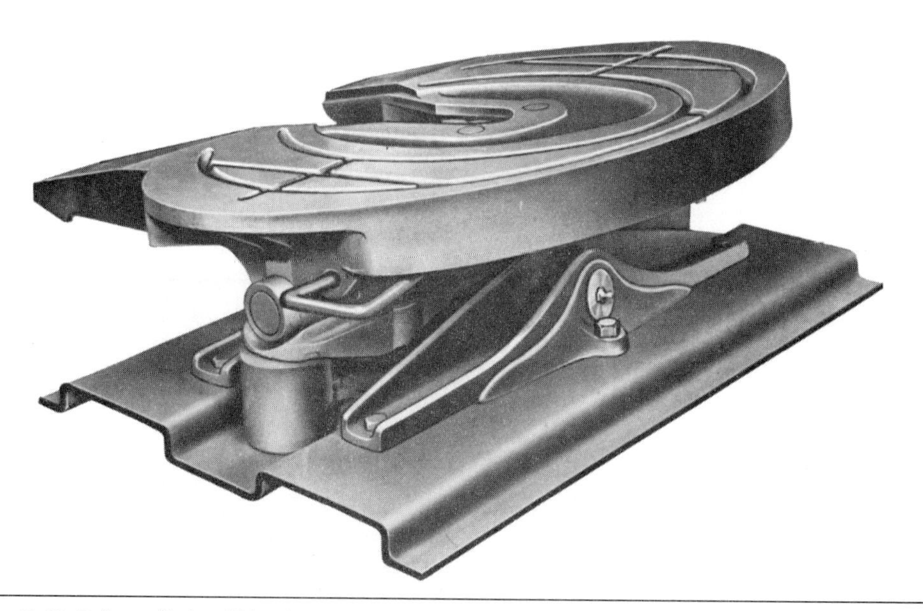

Figure 6-18 Fully oscillating fifth wheel. (Courtesy of Holland Hitch Co.)

rigid fifth wheel and oscillating trailer kingpin bolster plate are used on some applications such as frameless dump trailers (Figure 6-20).

Nontilt Convertible Fifth Wheel A **nontilt convertible fifth wheel** can be changed from a rigid fifth wheel to a semioscillating fifth wheel by removing a steel locking rod at the rear of the fifth wheel (Figure 6-21). This type of fifth wheel is used by trucking companies that have various trailer kingpin mountings.

Stabilized Fifth Wheel On a **stabilized fifth wheel** the top part of the fifth wheel rotates with the trailer. The top plate position is maintained by keys on this plate that lock into the trailer kingpin bolster plate (Figure 6-22). This type of fifth wheel provides four-point support through 90-degree turns between the tractor and trailer. Stops in the fifth wheel limit the turning radius and prevent damage to the tractor cab and trailer during tight turns. The stabilized fifth wheel provides increased stability for loaded trailers with a high center of gravity.

Figure 6-19 Rigid fifth wheel. (Courtesy of Holland Hitch Co.)

Figure 6-20 Frameless dump trailer. (Courtesy of Wabash National.)

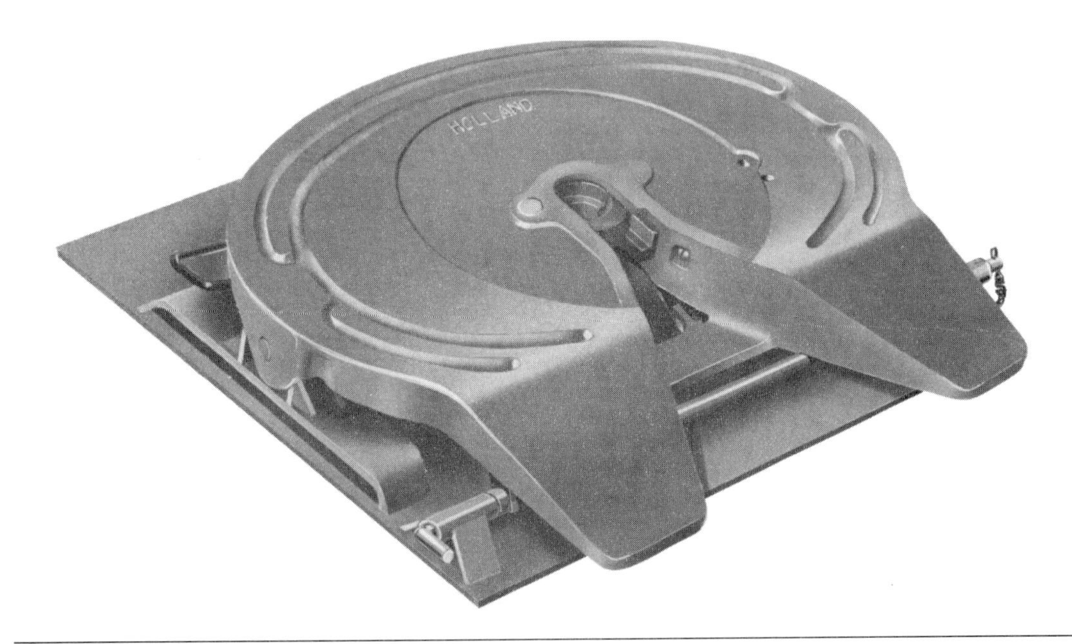

Figure 6-21 Nontilt convertible fifth wheel. (Courtesy of Holland Hitch Co.)

Figure 6-22 Stabilized fifth wheel. (Courtesy of Holland Hitch Co.)

Compensating Fifth Wheel The **compensating fifth wheel** provides both front-to-rear and side-to-side oscillation (Figure 6-23). This type of fifth wheel has shoes on the lower side of the fifth wheel that slide transversely in a track in the mounting plate to provide side-to-side oscillation (Figure 6-24). The compensating fifth wheel has a side-to-side oscillation point that is well above the fifth wheel bearing surface. The compensating fifth wheel also has a transverse shaft and/or trunnions that allow front-to-rear oscillation. This type of fifth wheel is intended for use with trailers with

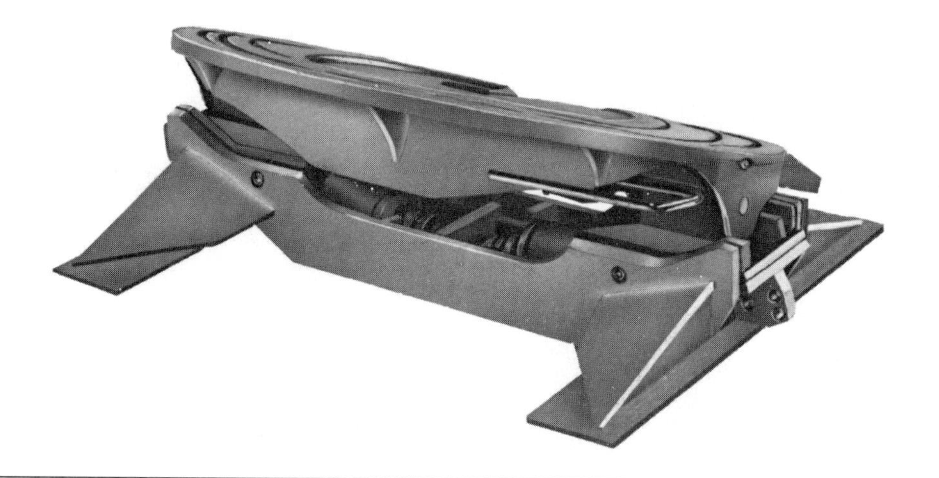

Figure 6-23 Compensating fifth wheel. (Courtesy of Holland Hitch Co.)

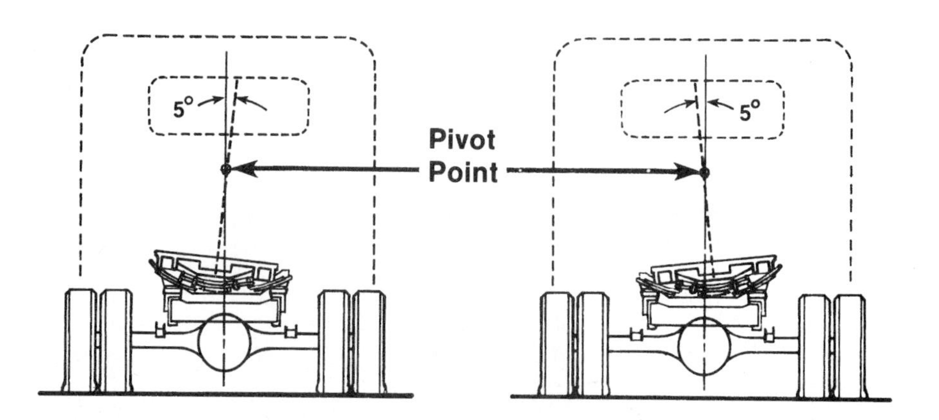

Figure 6-24 Compensating fifth wheel side-to-side oscillation and pivot point. (Courtesy of Heavy Duty Trucking.)

a loaded center of gravity that is a maximum of 39 in (99.06 cm) above the fifth wheel bearing surface. The compensating fifth wheel reduces twisting forces transmitted from the trailer to the tractor.

Elevating Fifth Wheel An **elevating fifth wheel** may be an air-lift type (Figure 6-25) or a hydraulic-lift type (Figure 6-26). An elevating fifth wheel is used to change an over-the-road tractor to a tractor that is used for spotting, switching, and hauling trailers in the yard. Rather than manually raising and lowering the trailer landing gear the driver can operate the elevating fifth wheel control in the cab to raise and lower the fifth wheel when spotting trailers.

Stationary and Sliding Fifth Wheels A **stationary fifth wheel** is permanently attached to the tractor frame and cannot be moved forward or rearward. This type of fifth wheel should be used in a fleet situation where axle loading, kingpin setting, and combination length remain constant.

A **sliding fifth wheel** is designed to move forward or rearward on its mounting plate. This type of fifth wheel is mounted on tracks and locked in position (Figure 6-27). The sliding fifth wheel locking mechanism may be released mechanically with a lever or by air pressure supplied to an air cylinder. The air pressure release is more expensive, but it provides increased convenience, especially in trucking operations where the fifth wheel may have to be adjusted frequently. The sliding fifth wheel provides increased maneuverability. In tight turning situations the fifth wheel may be moved forward for increased maneuverability.

TRADE JARGON: A tandem rear axle may be called bogie.

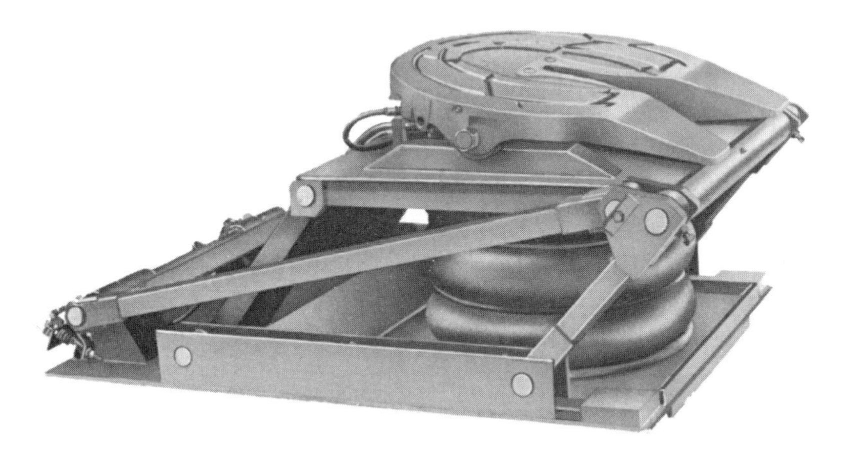

Figure 6-25 Air-lift elevating fifth wheel. (Courtesy of Holland Hitch Co.)

Figure 6-26 Hydraulic-lift elevating fifth wheel. (Courtesy of Holland Hitch Co.)

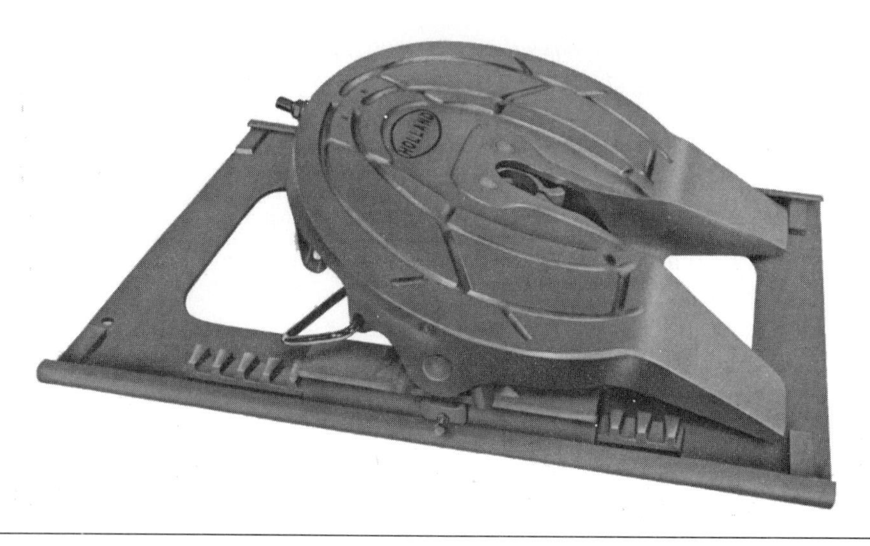

Figure 6-27 Sliding fifth wheel. (Courtesy of Holland Hitch Co.)

Ride quality is greatest when the fifth wheel is positioned near the center of the tandem rear axles. If the axles are not overloaded, the driver may move the sliding fifth wheel rearward to achieve this position and provide improved ride comfort.

When the sliding fifth wheel is moved rearward, trailers with short landing gear may be accommodated. If the slide length on a fifth wheel is not long enough, it may not be possible to move the fifth wheel rearward enough to provide the proper relationship between front and rear axle loading. When the slide length is longer than necessary, the front axle may be overloaded with the fifth wheel moved all the way forward. If the slide length is longer than necessary and the fifth wheel is moved all the way rearward, maneuverability is decreased. The sliding fifth wheel slide length should be selected in relation to the distance from the cab to the forward rear axle (Figure 6-28). Sliding fifth wheels with longer slides have higher cost.

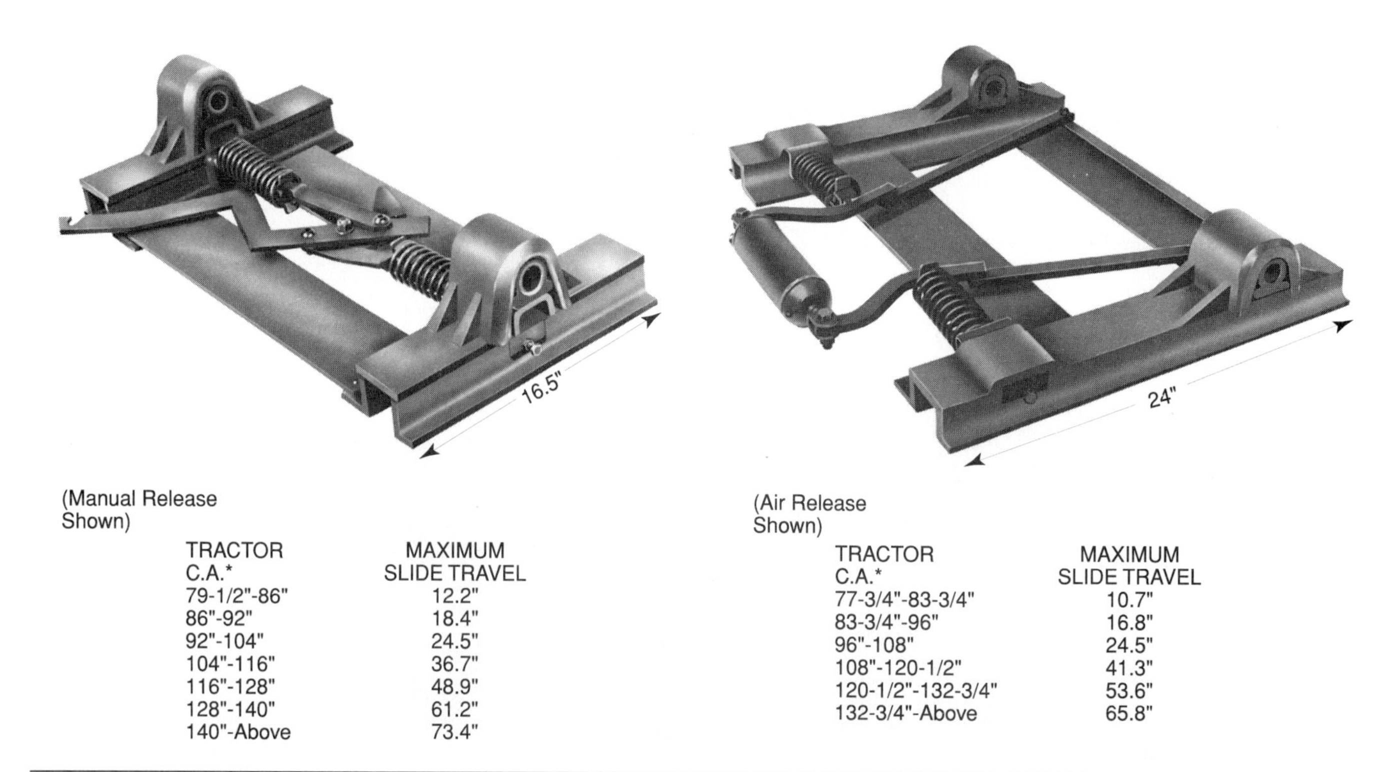

(Manual Release Shown)

TRACTOR C.A.*	MAXIMUM SLIDE TRAVEL
79-1/2"-86"	12.2"
86"-92"	18.4"
92"-104"	24.5"
104"-116"	36.7"
116"-128"	48.9"
128"-140"	61.2"
140"-Above	73.4"

(Air Release Shown)

TRACTOR C.A.*	MAXIMUM SLIDE TRAVEL
77-3/4"-83-3/4"	10.7"
83-3/4"-96"	16.8"
96"-108"	24.5"
108"-120-1/2"	41.3"
120-1/2"-132-3/4"	53.6"
132-3/4"-Above	65.8"

Figure 6-28 Sliding fifth wheel locking mechanisms and slide lengths. (Courtesy of Holland Hitch Co.)

WARNING: Never attempt to slide the fifth wheel when the vehicle is in motion. This action may cause the fifth wheel to break off the stops on the end of the tracks. Under this condition the trailer may move backward and separate from the tractor, or the trailer could move ahead and smash into the back of the cab.

Fifth Wheel Operation

The fifth wheel contains a pair of locking jaws mounted on lock pins. A release handle on the side of the fifth wheel is connected to a cam plate. A yoke assembly is mounted on the front side of the lockjaws. A secondary lock is mounted with the yoke assembly (Figure 6-29).

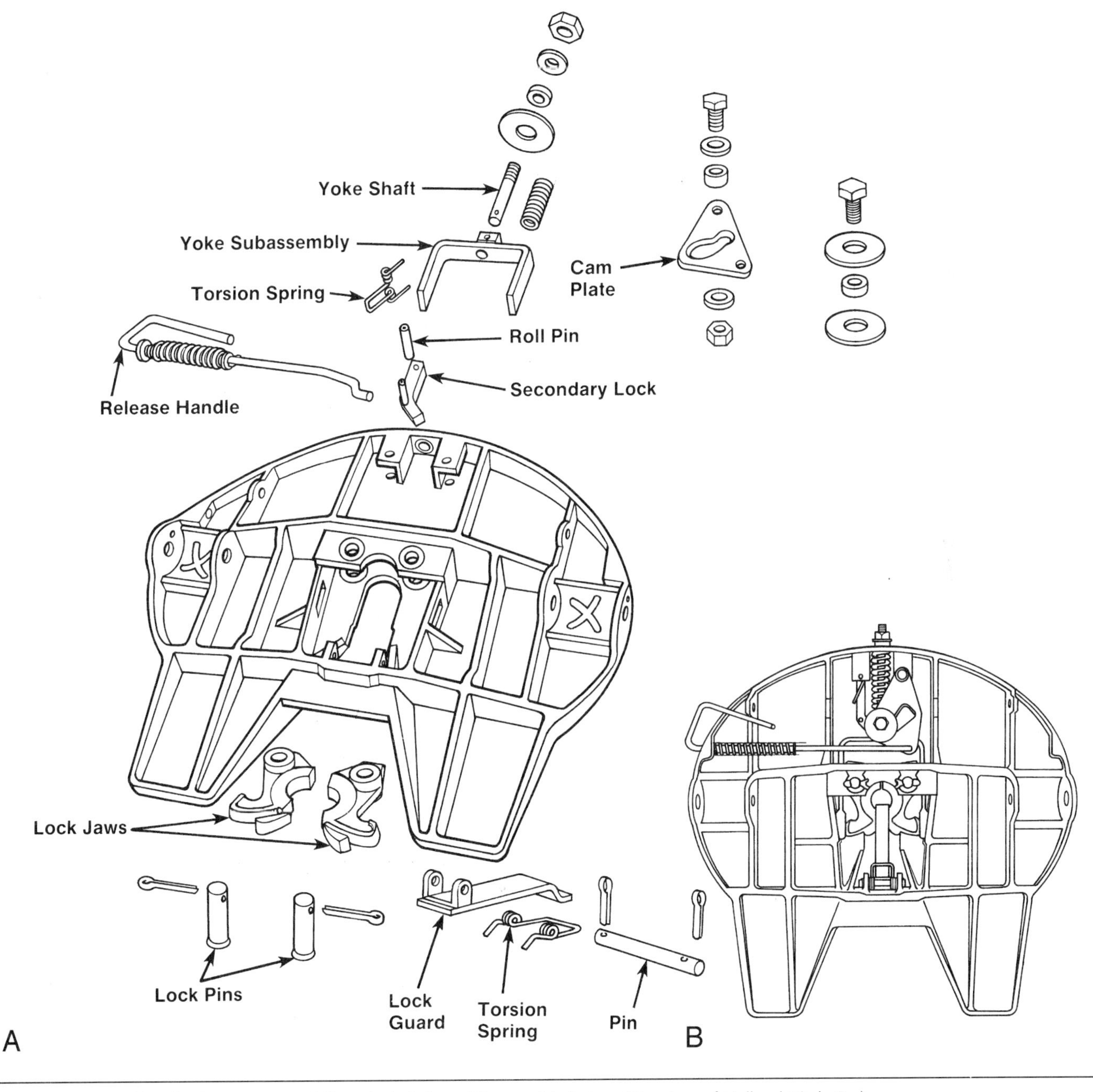

Figure 6-29 Fifth wheel components **(A)** exploded view, **(B)** assembled view. (Courtesy of Holland Hitch Co.)

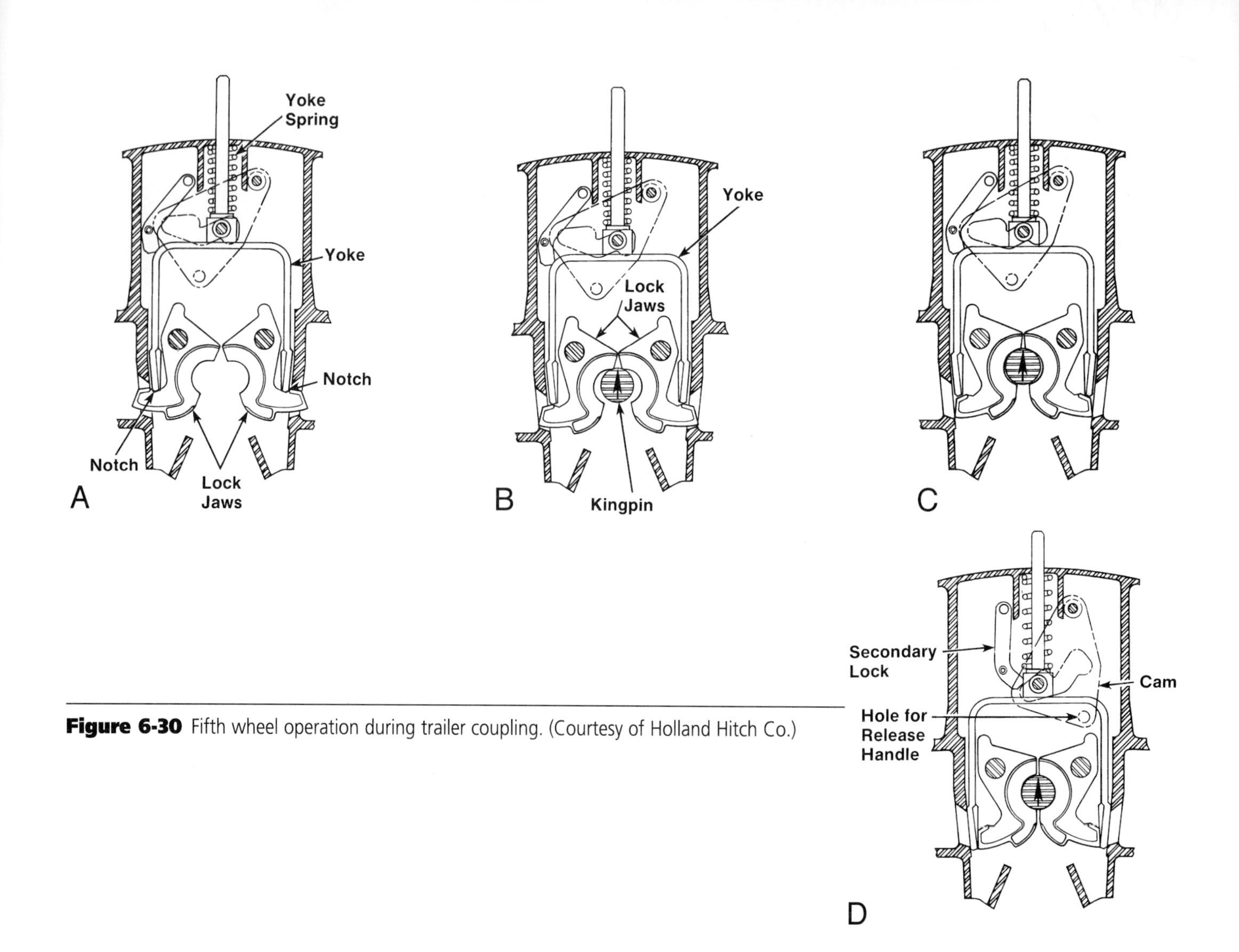

Figure 6-30 Fifth wheel operation during trailer coupling. (Courtesy of Holland Hitch Co.)

When the trailer is uncoupled, the lockjaws are open and the yoke spring is pushing the yoke toward the rear of the fifth wheel. The yoke tips are seated in notches on the lockjaws, and this action prevents the yoke spring from pushing the lockjaws closed (Figure 6-30, *A*).

When trailer coupling begins, the trailer kingpin enters the rear side of the lockjaws and contacts the bore of these jaws. The kingpin pushes the lockjaws until this pushing force moves the yoke tips out of the notches on the front side of the lockjaws. Under this condition the lockjaws begin to close (Figure 6-30, *B*). When the lockjaws close a specific amount, the yoke tips begin to move past the outer end of the lockjaws. In this position the yoke spring pushes the yoke rearward past the end of the lockjaws and helps to move the lockjaws toward the closed position (Figure 6-30, *C*). When the kingpin is fully into the lockjaws, the yoke moves completely past the end of the lockjaws. In this position the yoke tips cover the opening in the fifth wheel occupied by the lockjaws when they are open. In the fully coupled position the secondary lock snaps into place to prevent the yoke from moving forward, and the release handle is pulled in as the cam and yoke move into the closed position (Figure 6-30, *D*).

The driver pulls the release lever to unlock the fifth wheel and uncouple the trailer. As the driver pulls the release lever, the cam moves with the lever until the roller on the yoke is in the notch on the cam (Figure 6-31, *A*). During this cam movement the cam contacts a roll pin on the secondary lock and pushes this lock away from the yoke. This cam and secondary lock movement allows the yoke to move forward so the yoke tips are moved forward away from the outer ends of the lockjaws. The driver moves the tractor ahead slowly, and the kingpin forces the lockjaws outward (Figure 6-31, *B*). The outward lockjaw movement causes the outer end of these jaws to con-

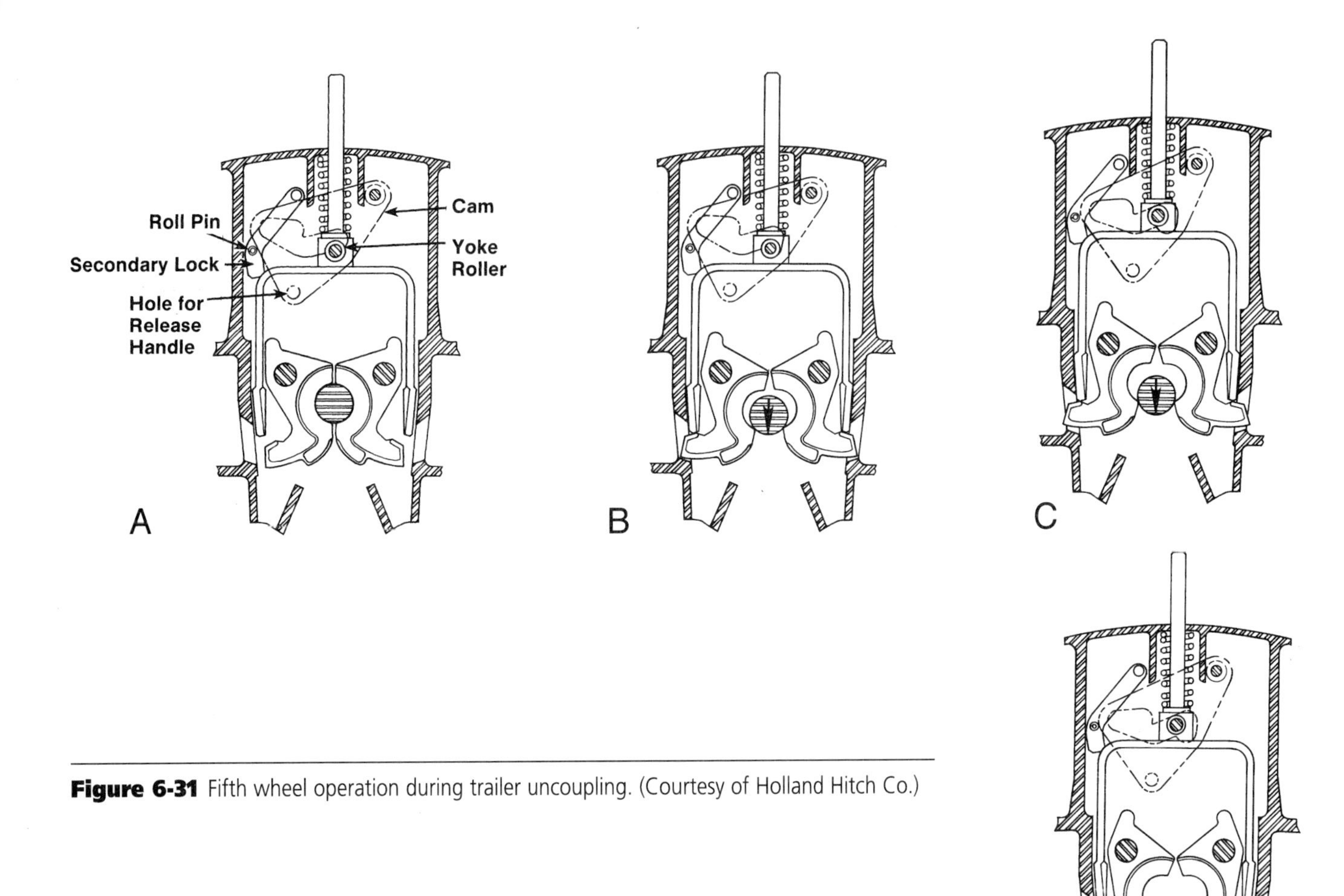

Figure 6-31 Fifth wheel operation during trailer uncoupling. (Courtesy of Holland Hitch Co.)

tact the yoke tips. Because the front side of the lockjaws is slightly tapered, this action pushes the yoke slightly forward against the spring tension (Figure 6-31, *C*). When the kingpin pushes the lockjaws fully outward, the yoke tips drop into the notches on the front side of the lockjaws and this action keeps the lockjaws open (Figure 6-31, *D*). The kingpin is now out of the lockjaws, and the trailer is uncoupled.

Another type of fifth wheel locking mechanism has a longitudinally sliding pair of jaws (Figure 6-32). This type of locking mechanism also has a rubber block and a spring-assisted, gravity-actuated lock. The rubber block is compressed during trailer coupling. When the trailer is coupled and this mechanism is locked, the rubber block exerts a force on the front jaw to provide a pressure fit on the jaws and kingpin. In the locked position the dual lock drops behind the rear jaw to lock the jaws closed (Figure 6-33). In the locked position the safety latch on the outside edge of the fifth wheel is hanging freely, and the operating handle is to the rear of the safety latch (Figure 6-34). The first steps in unlocking the fifth wheel are to rotate the safety latch rearward and move the handle forward to the lock set position (Figure 6-35). The handle movement rotates the cam, and the cam rotation raises the lock arms (Figure 6-36). Once these arms are raised, the kingpin can move out of the jaws. As the tractor is moved ahead, the kingpin moves the front and rear jaws rearward. This jaw movement lifts the lock arms and releases the cam from its detent position on the lock arm (Figure 6-37). This action allows the rear jaw to drop downward, and the kingpin can now move completely out of the jaws as the tractor continues to move ahead (Figure 6-38).

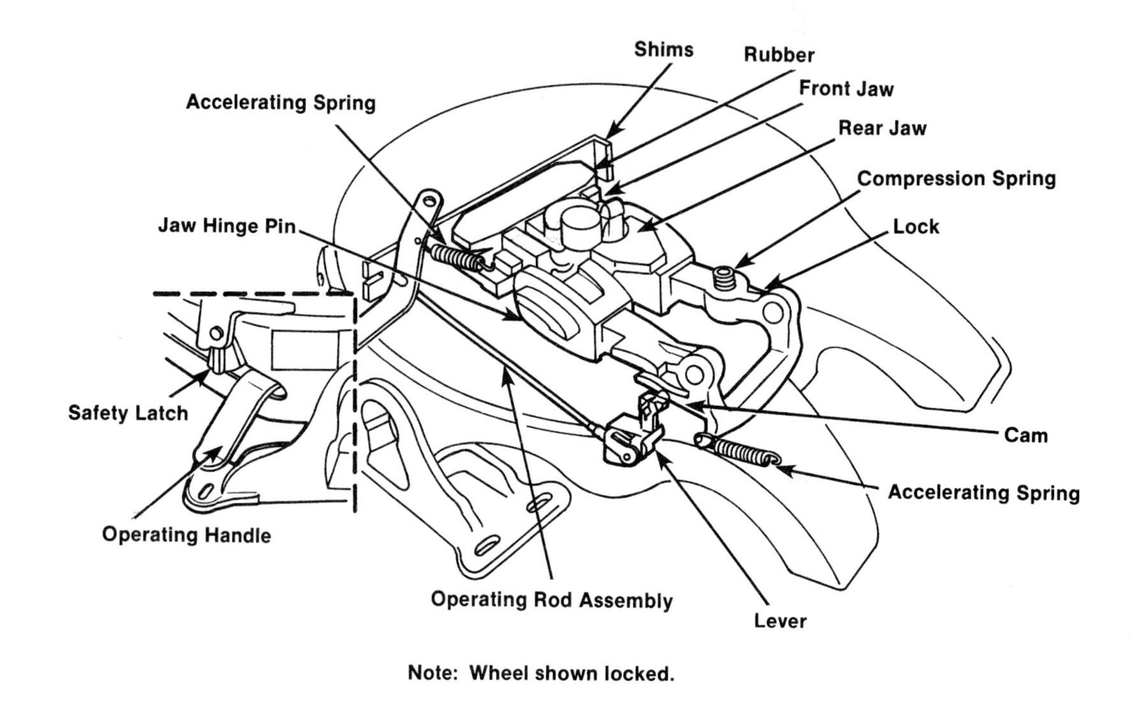

Figure 6-32 Fifth wheel locking mechanism with longitudinally sliding jaws. (Courtesy of American Steel Foundries.)

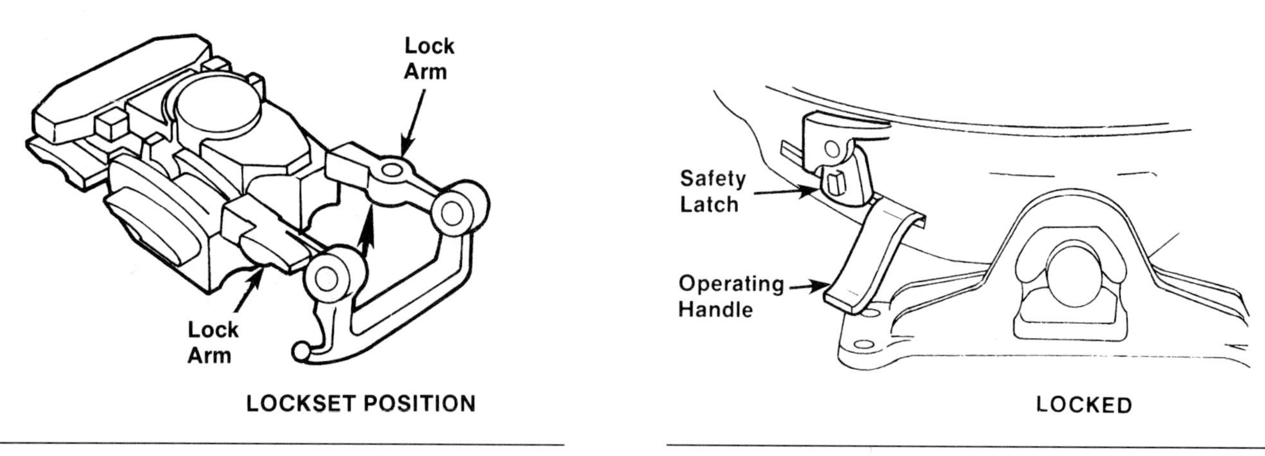

Figure 6-33 Dual locking mechanism in the locked position. (Courtesy of American Steel Foundries.)

Figure 6-34 Safety latch and operating handle in the locked position. (Courtesy of American Steel Foundries.)

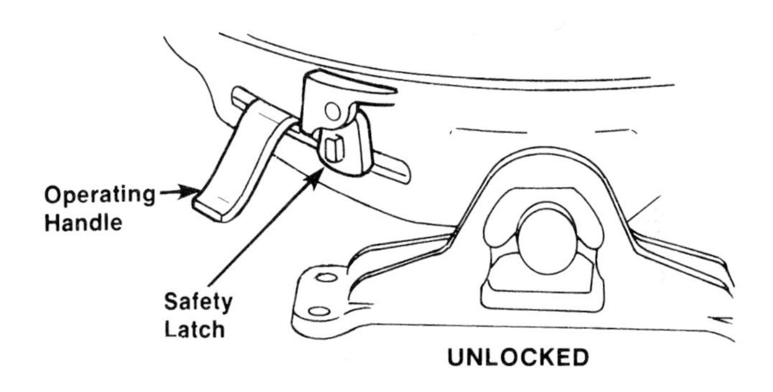

Figure 6-35 Safety latch and operating handle in the unlocked position. (Courtesy of American Steel Foundries.)

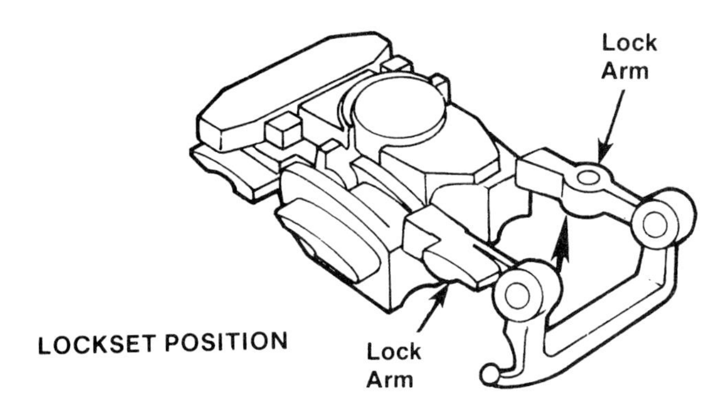

Figure 6-36 Operating handle movement moves the cam and lifts the lock arms to release the rear jaw. (Courtesy of American Steel Foundries.)

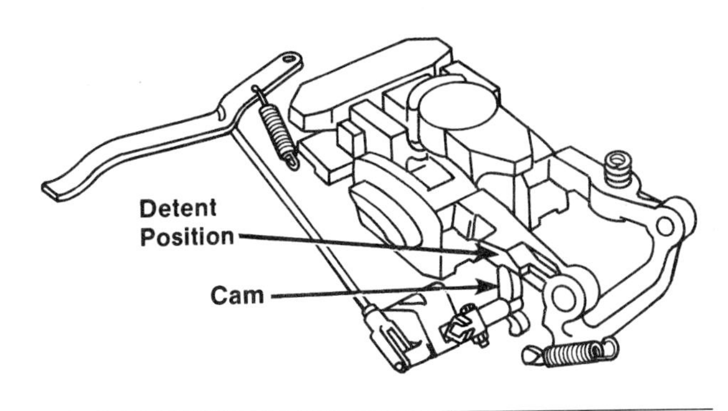

Figure 6-37 When the tractor is moved ahead, the lock arms are lifted and the cam moves out of its detent position on the lock arm. (Courtesy of American Steel Foundries.)

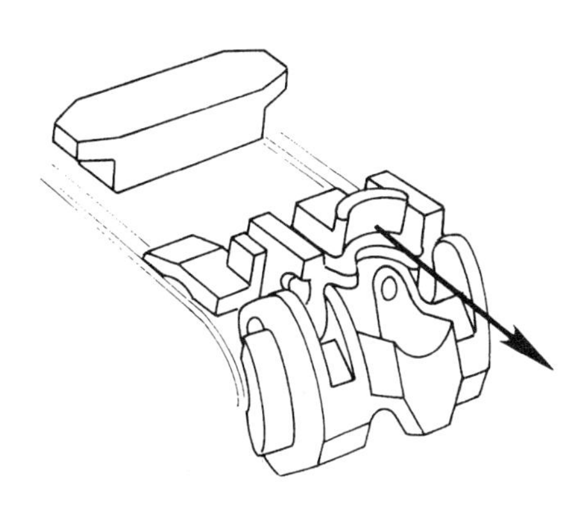

Figure 6-38 Rearward jaw movement allows the rear jaw to drop downward, and the kingpin can now move completely out of the jaws as the tractor continues to move ahead. (Courtesy of American Steel Foundries.)

Figure 6-39 Typical converter dolly. (Courtesy of Dorsey Trailers, Inc.)

Fifth Wheel Coupling

WARNING: Trailer or **converter dolly** coupling and uncoupling procedures vary depending on the type of fifth wheel. Always follow the procedure in the truck manufacturer's or fifth wheel manufacturer's service manual. Improper trailer or converter dolly coupling may cause a trailer separation, resulting in expensive trailer damage or personal injury (Figure 6-39).

Trailer landing gear refers to the manually operated metal legs that may be lowered to support the front of the trailer during and after uncoupling from the tractor.

This coupling procedure is based on the fifth wheel illustrated in Figure 6-32. Before coupling the tractor to the trailer inspect the fifth wheel for damaged, worn, or loose components and mountings. The fifth wheel must be adequately lubricated and tilted down at the rear. Before backing the tractor or converter dolly under the trailer, adjust the trailer landing gear (Figure 6-40) so

Figure 6-40 Trailer landing gear.

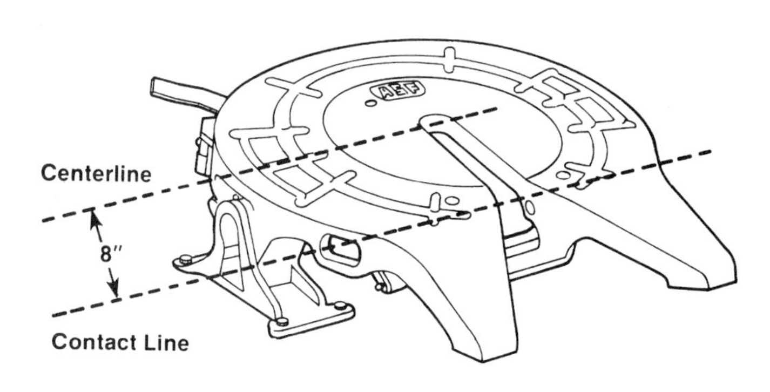

Figure 6-41 Initial contact between the trailer bolster plate and the fifth wheel should occur at approximately 8 in (20.32 cm) to the rear of the center on the fifth wheel bracket. (Courtesy of American Steel Foundries.)

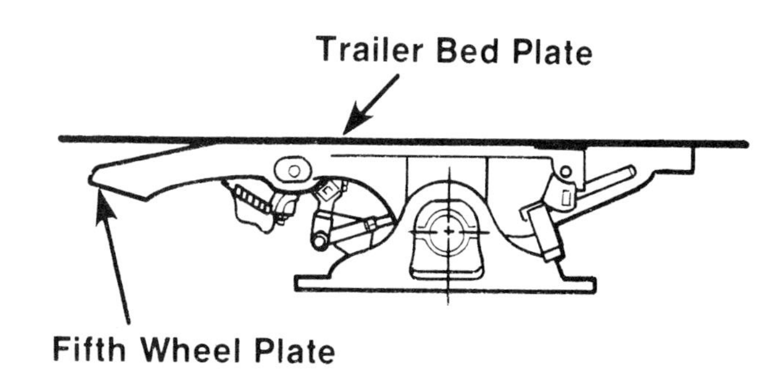

Figure 6-42 Proper trailer bolster (bed) plate position on the fifth wheel. (Courtesy of American Steel Foundries.)

A converter dolly is an axle with a heavy duty hitch and a fifth wheel mounted on the axle. The converter dolly is used to couple the towing trailer to another trailer. The converter dolly axle must have air brakes.

the contact between the **trailer bolster plate** and the fifth wheel will occur at approximately 8 in (20.32 cm) to the rear of the center on the fifth wheel bracket (Figure 6-41). The **fifth wheel jaws** must be open, and the handle must be positioned in the unlocked position ahead of the safety latch. Be sure the trailer brakes are applied.

CAUTION: Before backing up the tractor to couple the trailer, always be sure nobody is behind the tractor. Anyone behind the tractor could be injured.

Slowly back up the tractor or converter dolly while maintaining proper alignment between the fifth wheel and the trailer kingpin. Continue backing up until the trailer bolster plate contacts the fifth wheel. Back up the tractor with just enough power to lift the trailer and level the fifth wheel. Continue to back up until there is full trailer resistance. The fifth wheel locks automatically, but this action requires the fifth wheel to be forced against the kingpin with a slight coupling force. With the trailer brakes applied, place the transmission in the lowest gear and partially apply the clutch momentarily to create a pulling force between the fifth wheel and the kingpin. The fifth wheel jaws should remain securely locked on the kingpin. Connect all the air lines and electrical connectors between the tractor and trailer and apply the tractor and trailer brakes. After the coupling is completed, verify these conditions:

1. The trailer bolster (bed) plate must be supported evenly on top of the fifth wheel (Figure 6-42). No air gap should be between this bolster plate and the fifth wheel surface.

2. The safety latch must swing freely, and the operating handle must be behind the safety latch and positioned to the rear of its operating slot.
3. Crawl under the tractor and use a flashlight to verify that the kingpin is in the proper position in the fifth wheel slot surrounded by the jaws.
4. With only the trailer brakes applied, place the transmission in the lowest gear and partially apply the clutch momentarily to create a pulling force between the fifth wheel and the kingpin. The fifth wheel jaws should remain securely locked on the kingpin.
5. When coupling is completed, apply the tractor and trailer brakes and lift the trailer landing gear to the fully upward position.

During the coupling procedure if the trailer bolster plate and kingpin are too high and do not contact the fifth wheel in the proper position, the kingpin may be too high in the jaws. This condition may be called a high hitch. Under this condition some gap is present between the bolster plate and the fifth wheel. If this condition is present and the trailer wheels hit a severe bump in the road surface, the kingpin may bounce upward out of the fifth wheel jaws. This action results in a trailer separation from the tractor that will likely result in personal injury, and some very expensive trailer damage, plus damage to other vehicles and property. Therefore it is extremely important that the driver always follows the proper fifth wheel coupling procedure and verifies proper coupling by visually inspecting the fifth wheel and coupling after coupling is completed.

A high hitch occurs when the trailer kingpin and bolster plate are positioned too high in the fifth wheel. This condition is usually caused by improper bolster plate position during the coupling process.

WARNING: Improper fifth wheel coupling may cause a separation of the trailer from the tractor, resulting in trailer damage, property damage, vehicle damage, or personal injury.

WARNING: The only accurate way to verify proper fifth wheel coupling is to get under the tractor and visually observe the kingpin position in the fifth wheel jaws. An improperly coupled fifth wheel and kingpin resulting in a high hitch will likely stay coupled when the clutch is applied momentarily with the tractor transmission in gear.

Fifth Wheel Uncoupling

The trailer uncoupling procedure may vary depending on the type of fifth wheel. Always follow the uncoupling procedure in the truck manufacturer's or fifth wheel manufacturer's service manual. This procedure is based on the fifth wheel shown in Figure 6-32.

1. Apply the tractor and trailer parking brakes with the tractor backed against the trailer to relieve the king pin to jaw force.
2. Block the trailer wheels.

CAUTION: Always be sure the trailer brakes are applied and the wheels are blocked on a parked trailer. If this action is not taken, the trailer may roll backward or forward resulting in personal injury or trailer and property damage.

WARNING: If the trailer is loaded and the trailer landing gear legs will be supported on earth or pavement, place lengths of 2 in (5.08 cm) thick wooden planks under these legs to distribute the weight over a larger surface. If both legs sink a considerable distance into the support surface, the front of the loaded trailer may have to be raised with a crane to couple the tractor and trailer. If only one leg sinks into the support surface, the trailer may upset, causing expensive trailer and payload damage.

3. Lower the trailer landing gear until the front of the trailer is raised ½ in (12.7 mm).
4. Disconnect the air lines and electrical connections between the tractor and trailer.

See Chapter 6 in the Shop Manual for fifth wheel selection, inspection, adjustment, and service.

Figure 6-43 Converter dolly with a draw bar. (Courtesy of Dorsey Trailers Inc.)

5. Raise the safety latch, and pull the operating handle fully forward. When this handle is hard to pull, back the tractor slightly to relieve the kingpin to jaw force.
6. Release the tractor parking brakes and drive the tractor forward away from the trailer.

Pintle Hooks and Draw Bars

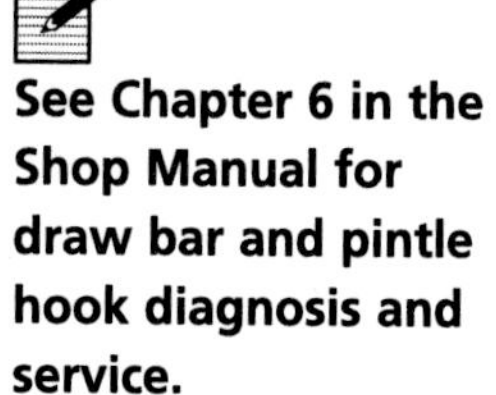

See Chapter 6 in the Shop Manual for draw bar and pintle hook diagnosis and service.

When one trailer is connected behind another trailer, a dolly with a **draw bar** may be used to couple the rearmost trailer to the forward trailer in an A-train configuration. A heavy steel ring on the front of the draw bar is connected to a **pintle hook** that is bolted to the forward trailer. The pintle hook is securely bolted at the centerline of the forward trailer. The steel ring is securely bolted to the draw bar (Figure 6-43). The pintle hook has a spring-loaded latch that prevents the steel ring on the draw bar from disengaging from the pintle hook.

A BIT OF HISTORY

Before 1910 Sternberg and later Sterling trucks had wood frame side members reinforced by a large steel angle section. After 1910 these trucks had channel steel frame side members reinforced with a wood insert. These wood inserts were made from white oak, and the total thickness of the frame side members was 2 in (5.08 cm) (Figure 6-44). All Sterling trucks with a payload capacity more than 1500 lb (680.4 kg) had frame side members with wood inserts. The advantages claimed for this type of frame design were:

1. Improved shock-absorbing qualities
2. Increased rigidity and strength
3. Improved sound-deadening qualities
4. Lower maintenance costs
5. Improved driver comfort

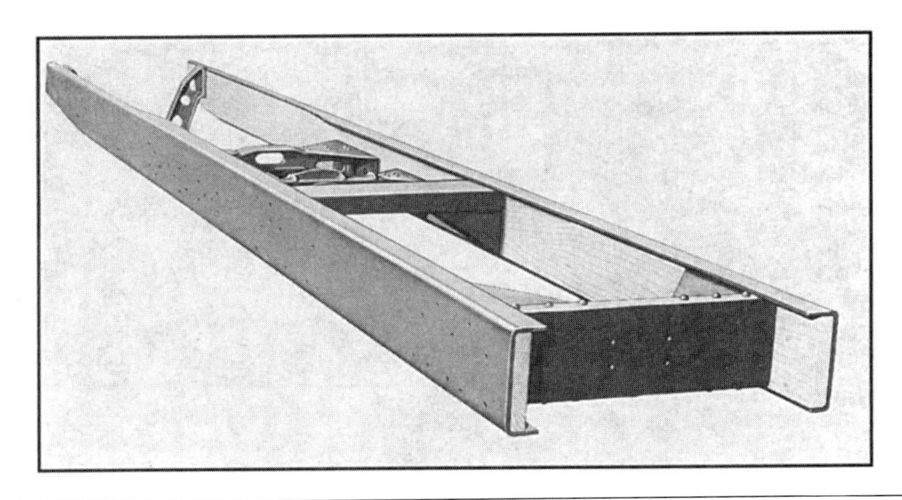

Figure 6-44 Sterling truck frame with wood reinforced side members. (Reprinted with permission from SAE Publication History of Sterling Trucks © 1993 Society of Automotive Engineers, Inc.)

Summary

- Section modulus is an indication of frame strength based on the height, width, thickness, and shape of the frame side rails.
- Yield strength is a measure of the steel strength used in the frame.
- RBM is the most accurate indication of frame strength. RBM is calculated by multiplying the yield strength and the section modulus.
- Frame area is the total cross section of the frame rail in square inches.
- The C-channel is the most common truck frame design, but I-beam and box frame designs are used in some applications.
- Crossmembers are mounted between the frame rails.
- Aluminum alloy frames are used in some trucks. These frames are lighter, but they must be thicker to provide adequate strength.
- Frame reinforcements are installed on the frame in the area of greatest load concentration to increase the RBM.
- Frame sag occurs when the frame or one side rail is bent downward from the original position.
- Buckle is a frame condition that refers to the frame or one side rail that is bent upward from the original position.
- A diamond frame condition is present when one frame rail is moved rearward or forward in relation to the opposite frame rail.
- Frame twist occurs when the end of one frame rail is bent upward or downward in relation to the opposite frame rail.
- Sidesway occurs when one or both frame rails are bent inward or outward.
- Tracking is the alignment of the truck axles with each other.
- A fifth wheel is used to couple the tractor to the trailer kingpin.
- Types of fifth wheels include semioscillating, fully-oscillating, rigid, nontilt convertible, stabilizing, compensating, and elevating.

Terms to Know

Applied moment
Bending moment
Box frames
Buckle
C-channel
Compensating fifth wheel
Converter dolly
Crossmembers
Diamond frame
Draw bar
Elevating fifth wheel
Fifth wheel
Fifth wheel jaws
Fishplate
Frame reinforcements
Frame sag
Frame twist
Fully-oscillating fifth wheel
I-beam frames
Nontilt convertible fifth wheel
Pintle hook
Resisting bending moment (RBM)
Rigid fifth wheel

Section modulus
Semioscillating fifth wheel
Sidesway
Sliding fifth wheel
Stabilized fifth wheel
Stationary fifth wheel
Tracking
Trailer bolster plate
Yield strength

- ❑ A fifth wheel may be stationary or sliding.
- ❑ During the tractor-to-trailer coupling process the trailer height should be adjusted so the trailer bolster plate makes initial contact with the fifth wheel at a point 8 in (20.32 cm) to the rear of the center on the fifth wheel mounting bracket.
- ❑ After the tractor is coupled to the trailer, the driver should get under the tractor and trailer and use a flashlight to visually inspect the position of the kingpin in the fifth wheel jaws.

Review Questions

Short Answer Essays

1. Explain the term *section modulus* in relation to truck frames.
2. Explain yield strength as it relates to truck frames.
3. Explain the most accurate way of calculating frame strength.
4. Describe bending moment in relation to truck frames.
5. Describe the parts of a C-channel.
6. Explain three requirements for a truck frame.
7. Describe three precautions that must be observed when working on truck frames.
8. Explain the purpose of a fifth wheel.
9. Explain the proper procedure for coupling a tractor and trailer.
10. Describe a high hitch condition when coupling a tractor and trailer, and explain how this condition occurs.

Fill-in-the-Blanks

1. A semioscillating fifth wheel oscillates around an axis that is perpendicular to the tractor ____________________.
2. A fully oscillating fifth wheel provides front-to-rear oscillation and ____________________ oscillation.
3. A nontilt convertible fifth wheel can be converted from a rigid fifth wheel to a ____________________ fifth wheel.
4. On a stabilizing fifth wheel the top part of the fifth wheel rotates with the ____________________.
5. An elevating fifth wheel may be operated by ____________________ pressure or ____________________ pressure.
6. The locking mechanism on a sliding fifth wheel may be operated mechanically or by ____________________ pressure.
7. The locking mechanism on a sliding fifth wheel should not be released when the tractor is ____________________.
8. Safety factor on a truck frame is the amount of ____________________ that can be safely absorbed by the truck frame members.
9. The web on a C-channel frame is the area between the ____________________.
10. A fishplate is a heavy steel frame reinforcement bolted to the ____________________ of the truck frame.

ASE Style Review Questions

1. While discussing truck frames:
Technician A says the section modulus is an indication of frame strength based on frame rail height, width, thickness, and shape.
Technician B says the section modulus is an indication of the metal strength in the truck frame.
Who is correct?
A. A only **C.** Both A and B
B. B only **D.** Neither A nor B

2. While discussing truck frames:
Technician A says frame yield strength is indicated by a number stamped on the outside of the frame rails.
Technician B says class 8 truck frames have a 50,000 psi (344,750 kPa) yield strength.
Who is correct?
A. A only **C.** Both A and B
B. B only **D.** Neither A nor B

3. While diagnosing truck frame problems:
A. The resisting bending moment is calculated by multiplying the applied moment by the bending moment.
B. The applied moment is the measurement of a specific load placed on the frame for a certain length of time.
C. The safety factor is the amount of load that can be safely absorbed by the frame members.
D. A diamond-shaped frame occurs when one frame rail is bent inward in relation to the opposite frame rail.

4. When working on truck frames:
A. When installing accessory equipment, an arc welder may be used to weld the equipment brackets to the frame.
B. A deep section frame reinforcement is usually installed near the front of the frame.
C. Gussets may be used when attaching the reinforcement plate to the frame.
D. When installing frame reinforcement plates, the original bolt holes in the frame should be used.

5. When installing fifth wheels:
A. A semioscillating fifth wheel is commonly installed on over-the-road tractors.
B. A fully oscillating fifth wheel is intended for use on trailers with a center of gravity above the top of the fifth wheel.
C. A rigid fifth wheel is used with a rigid bolster plate on the trailer.
D. On a stabilizing fifth wheel the bolster plate on the trailer rotates with the fifth wheel.

6. When discussing fifth wheels:
Technician A says an elevating fifth wheel may be air or hydraulically operated.
Technician B says an elevating fifth wheel may be used to change an over-the-road tractor to a trailer spotting tractor in the yard.
Who is correct?
A. A only **C.** Both A and B
B. B only **D.** Neither A nor B

7. All of these statements about truck frame defects are true *except:*
A. Frame sag occurs when the frame rails are bent downward in relation to the ends of the rails.
B. Frame buckle occurs when one or both frame rails are bent upward in relation to the ends of the rails.
C. Vehicle tracking is not affected when one frame rail is pushed rearward in relation to the opposite frame rail.
D. Frame twist occurs when the end of one frame rail is bent upward or downward in relation to the opposite frame rail.

8. While diagnosing sliding fifth wheels:
A. Ride comfort is greatest with the sliding fifth wheel moved all the way forward.
B. Maneuverability is improved by moving the sliding fifth wheel all the way rearward.
C. When the sliding fifth wheel is moved rearward, the front axle load is decreased.
D. If the slide length is longer than necessary and the fifth wheel is moved all the way rearward, the cab may hit the trailer during a tight turn.

9. When discussing fifth wheel operation:
Technician A says when the trailer is uncoupled the lockjaws should be open.
Technician B says with the fifth wheel in the fully locked position the secondary lock contacts the lockjaws.
Who is correct?
A. A only **C.** Both A and B
B. B only **D.** Neither A nor B

10. When coupling a trailer to the tractor, the trailer height should be set so the trailer bolster plate begins striking the fifth wheel at:
A. The centerline of the fifth wheel viewed from the side.
B. 2 in (5.08 cm) to the rear of the fifth wheel centerline viewed from the side.
C. 6 in (15.24 cm) to the rear of the fifth wheel centerline viewed from the side.
D. 8 in (20.32 cm) to the rear of the fifth wheel centerline viewed from the side.

Suspension Systems

Upon completion and review of this chapter, you should be able to:

- ❑ Describe the design of a short- and long-arm front suspension system.
- ❑ Explain the difference between a compression-loaded and tension-loaded ball joint.
- ❑ Describe front suspension design, including the axle I-beam, steering knuckles, steering arms, and Ackerman arms.
- ❑ Explain the effect of kingpin inclination and kingpin offset on steering effort and returnability.
- ❑ Explain the advantages of a front axle with unitized bearing assemblies and integrated steering knuckles.
- ❑ Describe three different types of springs that may be used in front suspension systems.
- ❑ Describe all the spring mountings on a front suspension system.
- ❑ Explain wheel jounce and rebound.
- ❑ Describe the purposes of shock absorbers.
- ❑ Explain shock absorber operation during wheel jounce and rebound.
- ❑ Describe the difference between conventional and heavy duty shock absorbers.
- ❑ Explain shock absorber ratios.
- ❑ Describe single axle rear suspension system with multileaf springs and auxiliary springs.
- ❑ Describe rear tandem axle suspension systems with multileaf springs and torque rods.
- ❑ Explain rear tandem axle suspension systems with multileaf springs and equalizing beams.
- ❑ Describe rear tandem axle suspension systems with equalizing beams and rubber cushions.
- ❑ Explain rear tandem axle suspension systems with inverted, parabolic leaf springs and rubber springs. Describe the advantages of this type of rear suspension system.

Introduction

In a suspension system **sprung weight** refers to the total vehicle weight supported by the suspension system. This sprung weight includes the vehicle weight and the weight of the cargo. Truck suspension systems must be strong enough to support very heavy sprung weights. **Unsprung weight** refers to the weight that is not supported by the springs. The unsprung weight includes the weight of the suspension system, hubs, wheels, tires, and brake components. Because truck suspension systems have a considerable amount of unsprung weight, this weight must be properly controlled to provide an adequate ride for the driver and cargo and eliminate problems such as wheel hop and vibration. Truck suspension systems must maintain proper lateral and fore-and-aft wheel position to provide accurate wheel alignment and steering control.

Short-and-Long-Arm (SLA) Front Suspension Systems

Upper and Lower Control Arms

A **short-and-long-arm (SLA) front suspension system** has coil springs with upper and lower control arms. Because wheel jounce, or rebound, movement of one front wheel does not directly affect the opposite front wheel, the control arm suspension is an independent system. Some medium duty trucks have SLA front suspension systems. Early SLA suspension systems had equal length upper and lower control arms. On these early suspension systems the bottom of the tire moved in and out with wheel jounce and rebound travel. This action constantly changed the tire tread width and caused tire scuffing and wear problems (Figure 7-1).

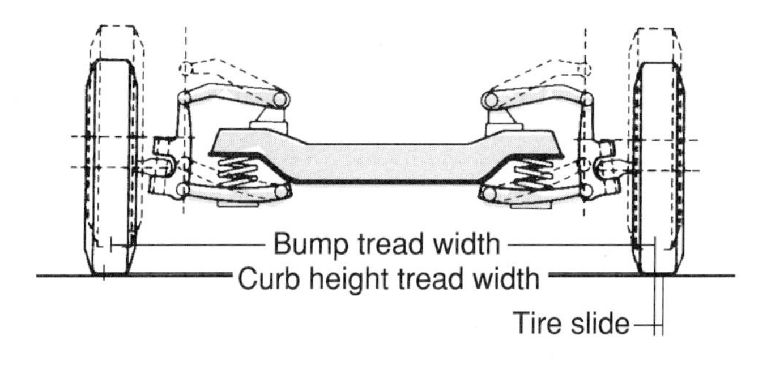

Figure 7-1 Early front suspension system with equal-length upper and lower control arms. (Courtesy of Sealed Power Corporation.)

Figure 7-2 Short-and-long-arm front suspension system. (Courtesy of Sealed Power Corporation.)

In later SLA front suspension systems the upper control arm is shorter than the lower control arm. During wheel jounce and rebound travel in this suspension system the upper control arm moves in a shorter arc compared with the lower control arm. This action moves the top of the tire in and out slightly, but the bottom of the tire remains in a more constant position (Figure 7-2). This SLA front suspension system provided reduced tire tread wear, improved ride quality, and provided better directional stability compared with suspension systems with equal length upper and lower control arms.

The inner end of the lower control arm contains large rubber insulating bushings, and the ball joint is attached to the outer end of the control arm. The lower control arm is bolted to the front crossmember, and the attaching bolts are positioned in the center of the lower control arm bushings (Figure 7-3). The ball joint may be riveted, bolted, pressed, or threaded into the control arm. A spring seat is located in the lower control arm. An upper control arm shaft is bolted to the frame, and rubber insulators are located between this shaft and the control arm.

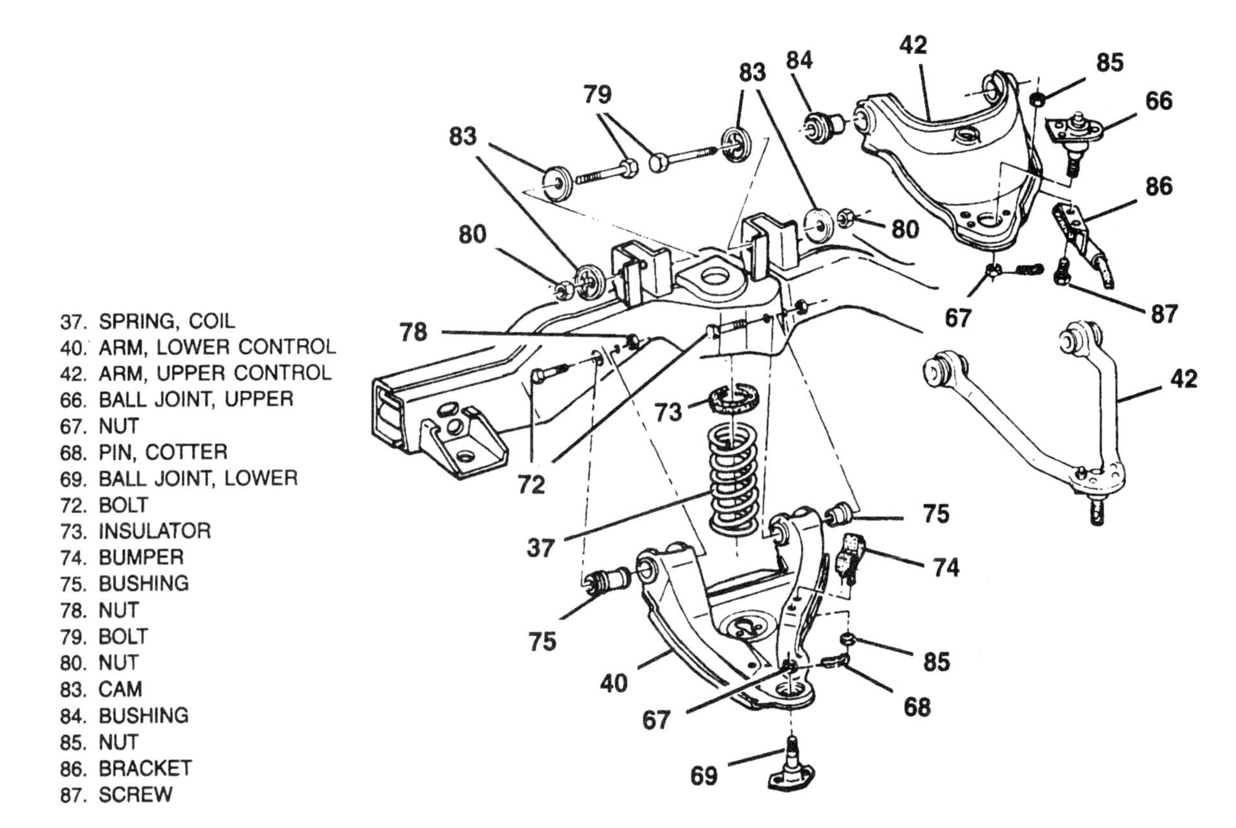

Figure 7-3 Upper and lower control arms and related components, SLA front suspension system. (Courtesy of Chevrolet Motor Division, General Motors Corporation.)

Steering Knuckle

The upper and lower ball joint studs extend through openings in the steering knuckle. Nuts are threaded onto the ball joint studs to retain the ball joints in the knuckle, and the nuts are secured with cotter keys. The wheel hub and bearings are positioned on the steering knuckle extension, and the wheel assembly is bolted to the wheel hub. In some trucks the front wheel bearings are mounted in unitized wheel bearing hubs (Figure 7-4).

When the steering wheel is turned, the steering gear and linkage turn the steering knuckle. During this turning action the steering knuckle pivots on the upper and lower ball joints. The upper and lower control arms must be positioned properly to provide correct tracking and wheelbase between the front and rear wheels. The control arm bushings must be in satisfactory condition to position the control arms properly.

CAUTION: Worn ball joints create a safety hazard. Refer to the ball joint checking procedures in Chapter 7 of the ***Shop Manual.***

The ball joints act as pivot points that allow the front wheels and spindles, or knuckles, to turn between the upper and lower control arms.

Load-Carrying Ball Joint

Ball joints may be grouped into two classifications, load-carrying and nonload-carrying. Ball joints may be manufactured with forged, stamped, cold-formed, or screw-machined housings. Ball joints are designed to precisely fit the opening in the control arm and the steering knuckle. Various ball joints supplied by one manufacturer of truck ball joints are illustrated in Figure 7-5. A load-carrying ball joint supports the vehicle weight. The coil spring is seated on the control arm to which the load-carrying ball joint is attached. For example, when the coil spring is mounted between the lower control arm and the chassis, the lower ball joint is a load-carrying joint.

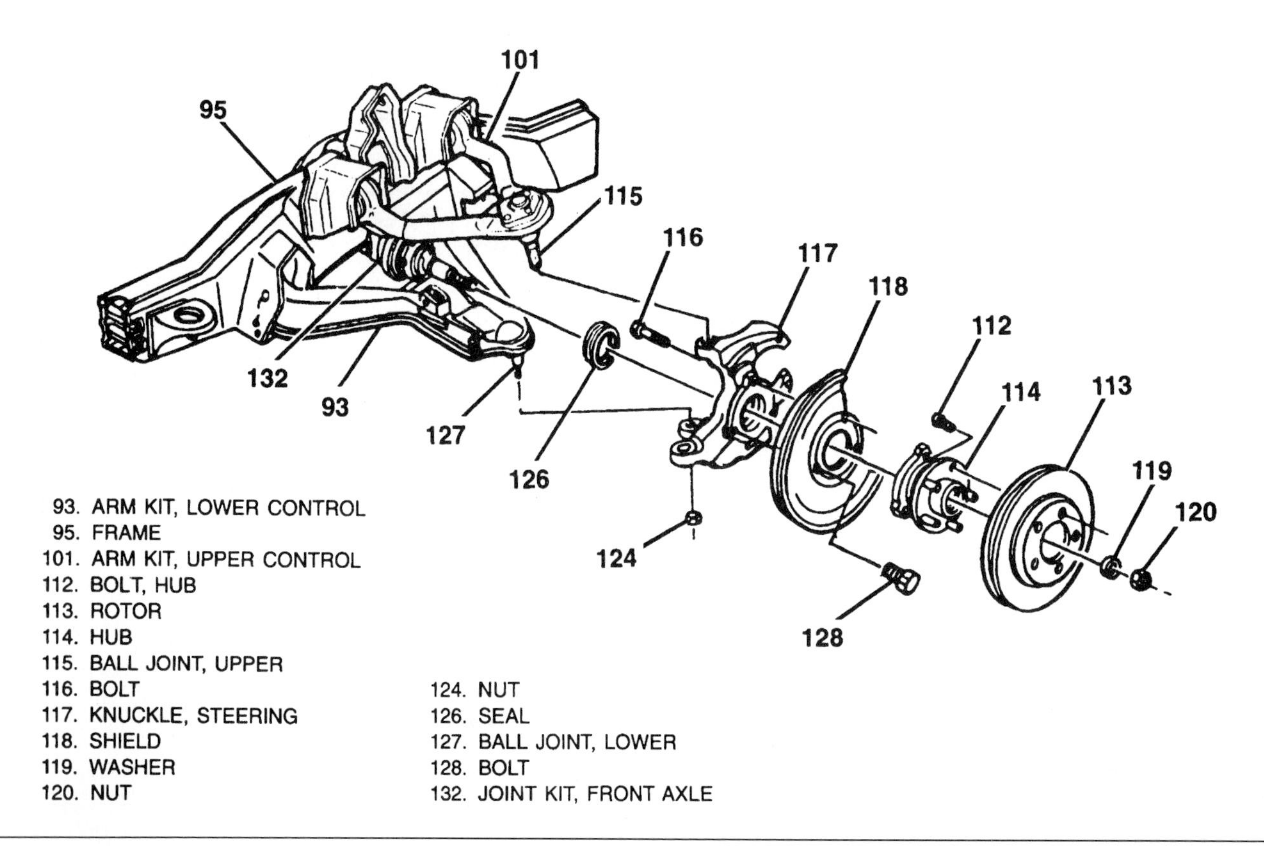

Figure 7-4 Steering knuckle in an SLA front suspension. (Courtesy of Chevrolet Motor Division, General Motors Corporation.)

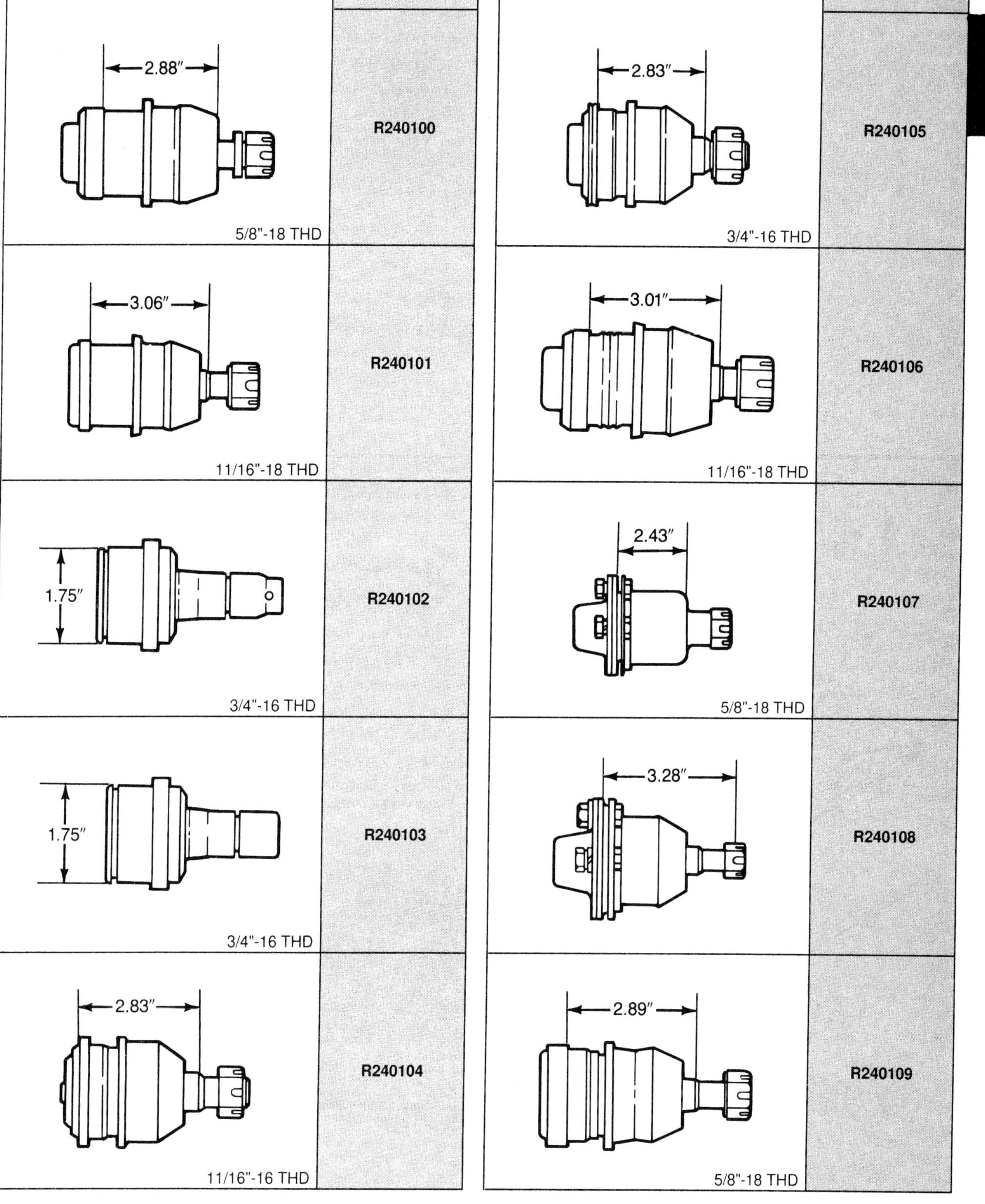

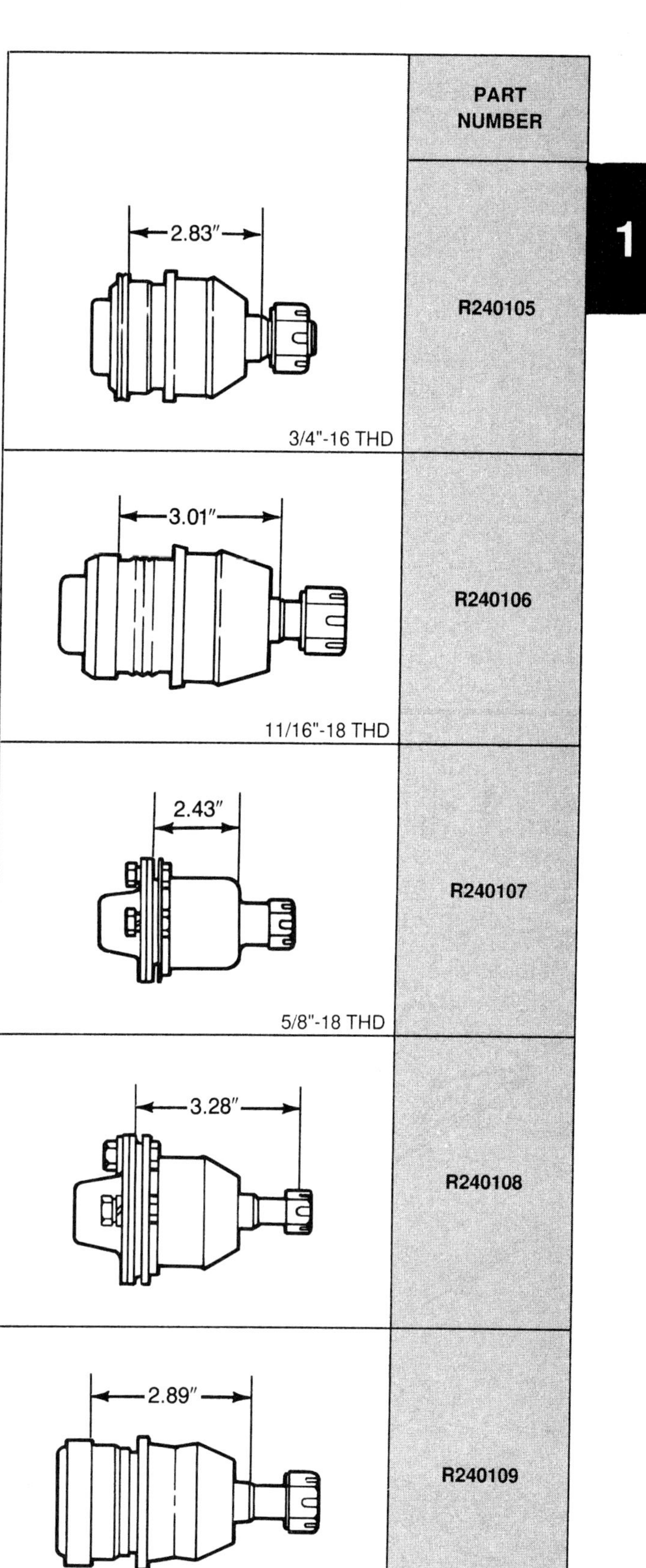

Figure 7-5 Various types of ball joints. (Courtesy of Meritor Automotive, Inc.)

In a load-carrying ball joint the vehicle weight forces the ball stud into contact with the bearing surface in the joint. Load-carrying ball joints may be compression-loaded or tension-loaded. If the control arm is mounted above the lower end of the knuckle and rests on the knuckle, the ball joint is compression-loaded. In this type ball joint the vehicle weight is pushing downward on the control arm, and this weight is supported on the tire and wheel, which are attached to the steering knuckle. Because the ball joint is mounted between the control arm and the steering knuckle, the vehicle weight squeezes the ball joint together (Figure 7-6). In this type of ball joint mounting the ball joint is mounted in the lower control arm and the ball joint stud faces downward.

When the lower control arm is positioned below the steering knuckle, the vehicle weight is pulling the ball joint away from the knuckle. This type of ball joint mounting is referred to as tension-loaded. This type ball joint is mounted in the lower control arm with the ball joint stud facing upward into the knuckle (Figure 7-7).

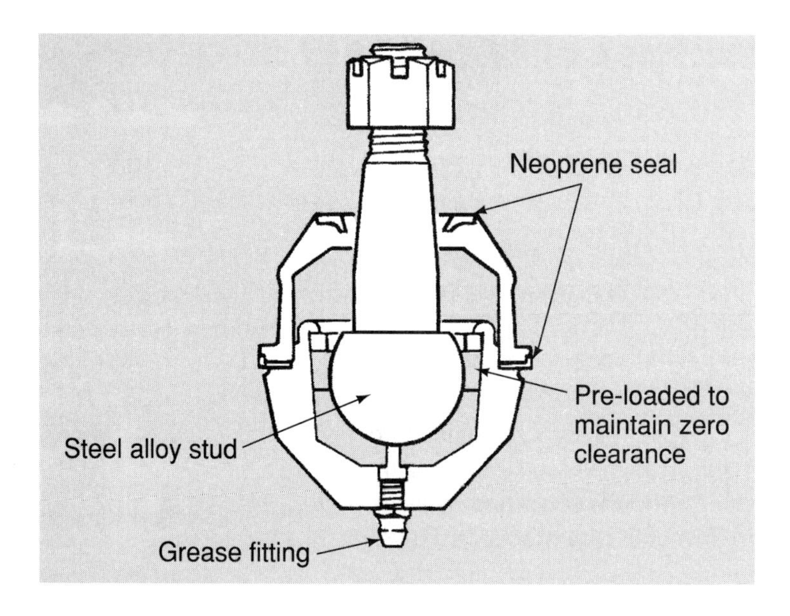

Figure 7-6 Compression-loaded ball joint. (Courtesy of Sealed Power Corporation.)

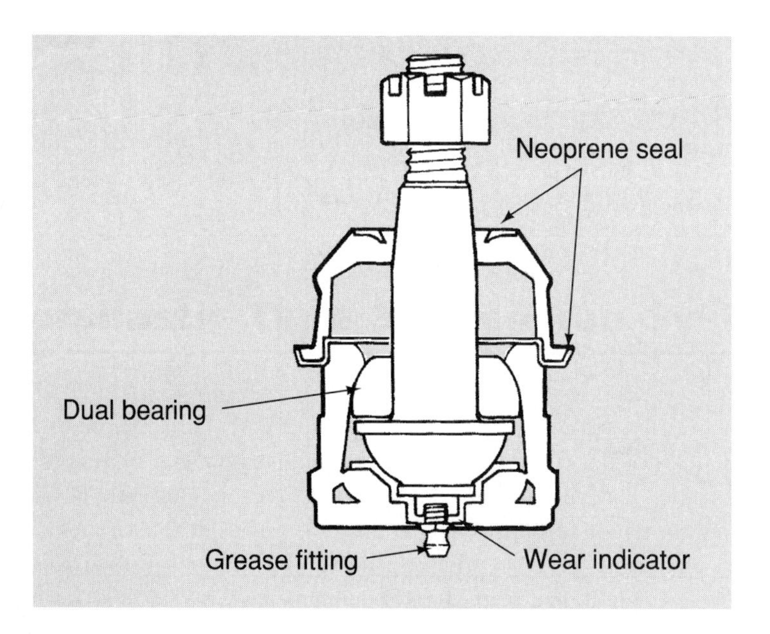

Figure 7-7 Tension-loaded ball joint. (Courtesy of Sealed Power Corporation.)

Because the load-carrying ball joint supports the vehicle weight, this ball joint wears faster compared with a nonload-carrying ball joint. Many load-carrying ball joints have built-in wear indicators. These ball joints have an indicator on the grease nipple surface that recedes into the housing as the joint wears. If the ball joint is in good condition, the grease nipple shoulder extends a specified distance out of the housing. If the grease nipple shoulder is even with or inside the ball joint housing, the ball joint is worn and replacement is necessary.

Nonload-Carrying Ball Joint

A nonload-carrying ball joint may be referred to as a stabilizing ball joint. A nonload-carrying ball joint is designed with a preload that provides damping action (Figure 7-8). This ball joint preload provides improved steering quality and vehicle stability.

Coil Springs

CAUTION: A large amount of energy is stored in a compressed coil spring. Always follow the spring service procedures in the vehicle manufacturer's service manual to avoid personal injury.

CAUTION: Never disconnect a suspension component that quickly releases the tension on a compressed coil spring. This action may cause serious personal injury.

The coil spring is positioned between the lower control arm and the spring seat in the frame. A spring seat is located in the lower control arm and an insulator may be positioned between the top of the coil spring and the spring seat in the frame. The shock absorber is usually mounted in the center of the coil spring, and the lower shock absorber bushing is bolted to the lower control arm. The top of the shock absorber extends through an opening in the frame above the upper spring seat. Washers, grommets, and a nut retain the top of the shock absorber to the frame. Side roll of the front suspension is controlled by a steel stabilizer bar, which is mounted to the lower control arms and the frame with rubber bushings.

Coil springs are actually coiled spring steel bars. When a vehicle wheel strikes a road irregularity, the coil spring compresses to absorb shock and then recoils back to its original installed height. Many coil springs contain a steel alloy that contains different types of steel mixed with other elements such as silicon or chromium. Coil springs may be manufactured by a cold or hot coiling

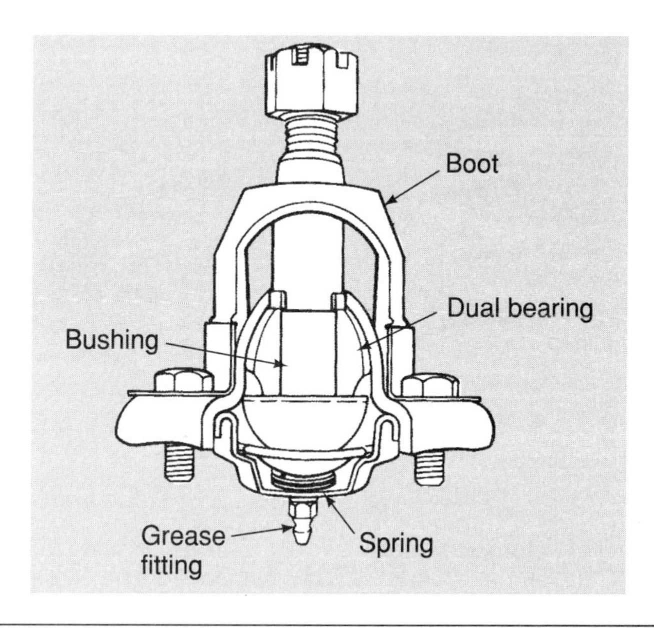

Figure 7-8 Nonload-carrying ball joint. (Courtesy of Sealed Power Corporation.)

process. The hot coiling process includes procedures for tempering and hardening the steel alloy. Coil springs are designed to carry heavy loads, but they must be lightweight. Many coil springs have a vinyl coating that increases corrosion resistance and reduces noise.

Coil spring failures may be caused by the following conditions:

1. Constant overloading
2. Continual jounce and rebound action
3. Metal fatigue
4. A crack or nick on the surface layer or coating

Linear Rate Springs

Coil springs are classified into two general categories, linear rate and variable rate. Linear rate springs have equal spacing between the coils, and these springs have one basic shape with a consistent wire diameter. When the load is increased on a linear rate spring, the spring compresses and the coils twist or deflect. As the load is removed from the spring, the coils unwind or flex back to their original position. The spring rate is the load required to deflect the spring 1 in. Linear coil springs have a constant spring rate regardless of the load. For example, if 200 lb deflect the spring 1 in, 400 lb deflect the spring 1 in. The spring rate on linear springs is usually calculated between 20% and 60% of the total spring deflection. Coil springs do not have much ability to resist lateral movement.

Heavy Duty Springs

Heavy duty springs are designed to carry 3% to 5% greater loads compared with regular duty springs. The wire diameter may be up to 0.100 in greater in a heavy duty spring compared with a regular duty spring. This larger diameter wire increases the load-carrying capacity of the spring. The free diameter of a heavy duty spring is shorter than a regular duty spring for the same application.

Variable-Rate Springs

Variable-rate springs have a variety of wire sizes and shapes. The most common variable-rate springs have a consistent wire diameter with a cylindric shape and unequally spaced coils (Figure 7-9).

The inactive coils at the end of the spring introduce force into the spring when the wheel strikes a road irregularity. When the transitional coils are compressed to their point of maximum

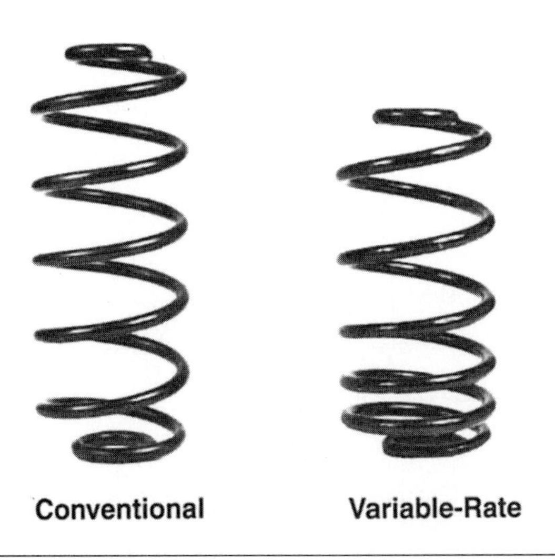

Figure 7-9 Variable-rate springs have consistent wire diameter and unequally spaced coils. (Courtesy of Sealed Power Corporation.)

load-carrying capacity, these coils become inactive. The active coils operate during the complete range of spring loading. When a stationary load is applied to a variable-rate spring, the inactive coils theoretically support the load. If the load is increased, the transitional coils support the load until they reach maximum load-carrying capacity, and the active coils carry the remaining overload. This spring action provides automatic load adjustment while maintaining vehicle height.

Some variable-rate springs have a tapered wire in which the active coils have the larger diameter and the inactive coils have a smaller diameter. Other variable-rate spring designs include truncated cone, double cone, and barrel shape. A variable-rate spring does not have a standard spring rate. This type of spring has an average spring rate based on the load at a predetermined spring deflection. *It is impossible to compare variable spring rates and linear spring rates because of this difference in spring rates.* Variable-rate springs usually have more load-carrying capacity compared with linear rate springs in the same application.

Selecting Replacement Springs

When replacement coil springs are required, the technician must select the correct spring. The original part number is usually on a tag wrapped around one of the coils. However, this tag may have fallen off if thc spring has been in service for very long. Some aftermarket suppliers stamp the part number on the end of the coil spring. If the original part number is available, the replacement springs may be ordered with the same part number. Most vehicle manufacturers recommend that both front springs be replaced at the same time. The replacement springs must have the same type of ends as the springs in the vehicle. Coil spring ends may be square tapered, square untapered, or tangential (Figure 7-10).

Full wire open end springs have the ends cut straight off, and sometimes these ends are flattened, squared, or ground to a D-shape. Taper wire closed spring ends are ground to a taper and wound to ensure squareness. Pigtail spring ends are wound to a smaller diameter. Springs are generally listed for front or rear suspensions.

Regular duty springs are a close replacement for the original spring in the vehicle, and these springs may replace several different original equipment (OE) springs in the same vehicle. Linear rate springs are usually found in regular duty, heavy duty, and sport suspension packages. Heavy duty springs are required when the vehicle is carrying a continuous heavy load.

Variable-rate springs are generally used when automatic load levelling is required under increased loads. Variable-rate springs maintain the correct vehicle height under various loads and provide increased load-carrying capacity compared with heavy duty springs. The technician must select the correct spring to meet the requirements of the vehicle and load conditions.

Stabilizer Bar

TRADE JARGON: A stabilizer bar may be called a sway bar.

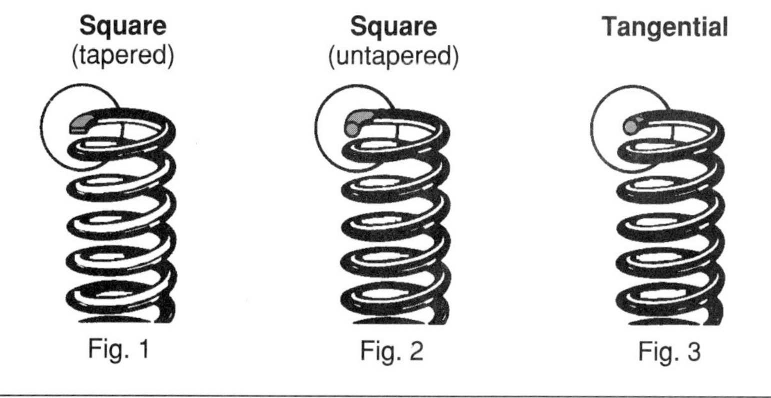

Figure 7-10 Types of coil spring ends. (Courtesy of Sealed Power Corporation.)

The stabilizer bar is attached to the crossmember and interconnects the lower control arms. Rubber insulating bushings are used at all stabilizer bar attachment points (Figure 7-11). When jounce and rebound wheel movements affect one front wheel, the stabilizer bar transmits part of this movement to the opposite lower control arm and wheel, which reduces and stabilizes body roll.

See Chapter 7 in the Shop Manual for short- and long-arm front suspension diagnosis and service.

Strut Rod

On some front suspension systems a strut rod is connected from the lower control arm to the chassis. The strut rod is bolted to the control arm, and a large rubber bushing surrounds the strut rod in the chassis opening. The outer end of the strut rod is threaded, and steel washers are positioned on each side of the strut rod bushing. Two nuts tighten the strut rod into the bushing. The strut rod prevents fore-and-aft movement of the lower control arm.

I-Beam and Leaf Spring Front Suspension Systems

Compared with the I-beam front suspension systems, SLA suspension systems provide improved ride quality, reduced steering effort, and reduced unsprung weight. However, SLA front suspension systems do not provide the weight-carrying capacity required in heavy duty trucks. Therefore heavy duty trucks and many medium duty trucks are equipped with I-beam front suspension systems.

The vehicle weight is transferred through the front springs to the axle, and this weight is applied through the axle and spindles to the wheels and tires. Finally the vehicle weight is supported by the tires on the road surface. The front axle must be strong enough to support the truck weight without bending or distorting.

Most medium and heavy duty trucks have I-beam front axles manufactured from forged steel. This type of front axle provides the necessary rigidity and weight-carrying capacity for medium and

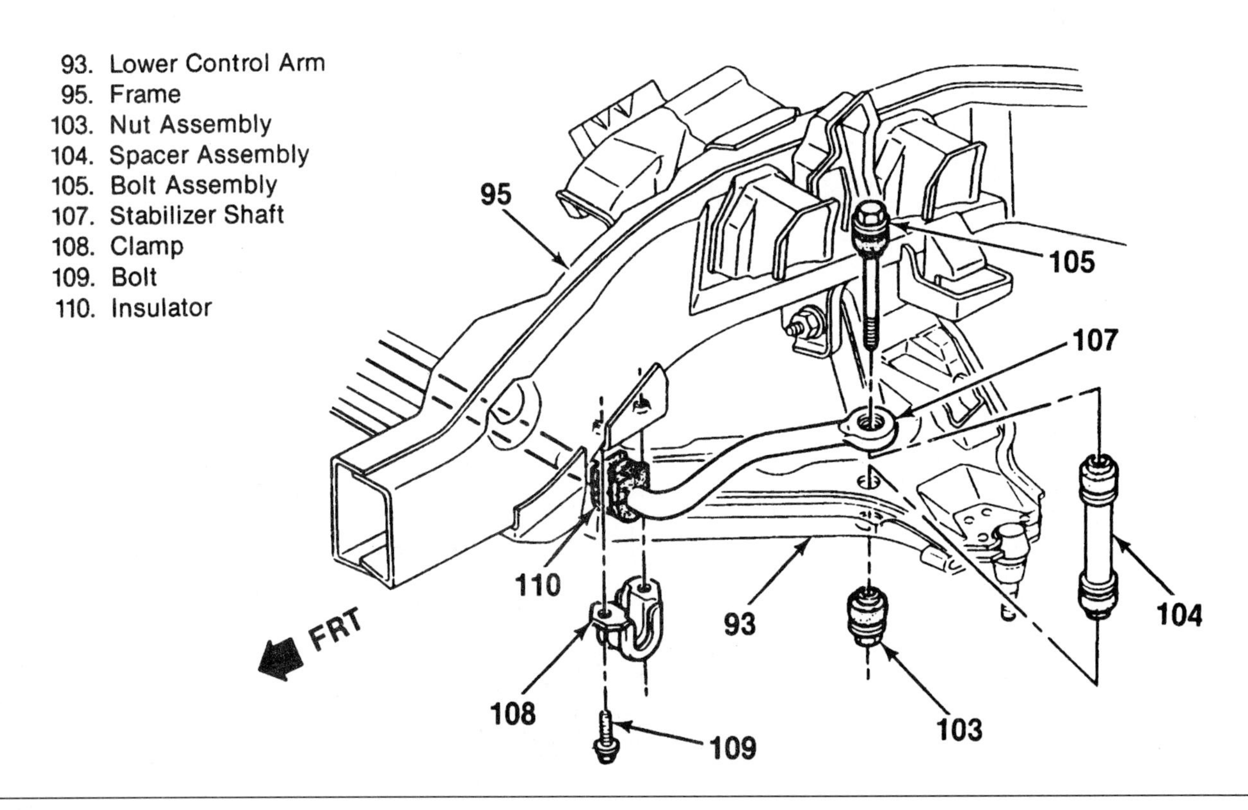

Figure 7-11 Stabilizer bar. (Courtesy of Chevrolet Motor Division, General Motors Corporation.)

heavy duty trucks. The axle I-beam contains machined openings in each end. **Kingpins** are installed in these openings to retain the spindles to the axle. An identification plate is attached to the center of the front side of the axle beam (Figure 7-12). The axle identification plate provides the axle model and specification number, customer number, axle assembly date, and the axle assembly plant and serial number. The numbers and letters in the axle model number indicate front or rear axle, weight capacity, axle series, brake use, specification number, and variation number (Figure 7-13). Some trucks have **tubular steel front axles** rather than **I-beam axles** (Figure 7-14).

A kingpin may be called a knuckle pin.

Steering Knuckles and Related Components

Steering knuckles provide a mounting surface for the front wheel bearings and wheel hubs. Steering knuckles provide a rotating mechanism for the front wheels when the steering wheel is turned to the right or left. The steering knuckle also provides mounting openings for the steering arm and Ackerman arm.

Steering knuckles are attached to the outer ends of the front axle beam with a knuckle (king) pin. The kingpin fits snugly in the axle beam opening, and a draw key fits horizontally through the side of this opening (Figure 7-15). The draw key fits in a slot in the side of the kingpin to prevent vertical or lateral kingpin movement. The kingpin fits through bushings in the steering knuckle. Kingpin bushings may be bronze or nylon. When the front wheels are turned to the right or left, the steering knuckle and bushings turn on the kingpin. A thrust bearing and grease seal are positioned between the lower end of the axle beam opening and the steering knuckle. This bearing reduces friction and steering effort. Knuckle caps and gaskets are bolted to the kingpin openings on the top and bottom of the steering knuckle. On some applications the knuckle cap gaskets are not used. Grease fittings are threaded into the knuckle caps to provide lubrication for the knuckle pins and bushings.

Kingpin inclination (KPI) is the tilt of a line through the center of the kingpin in relation to the vertical centerline of the tire (Figure 7-16). The amount of KPI affects steering effort and

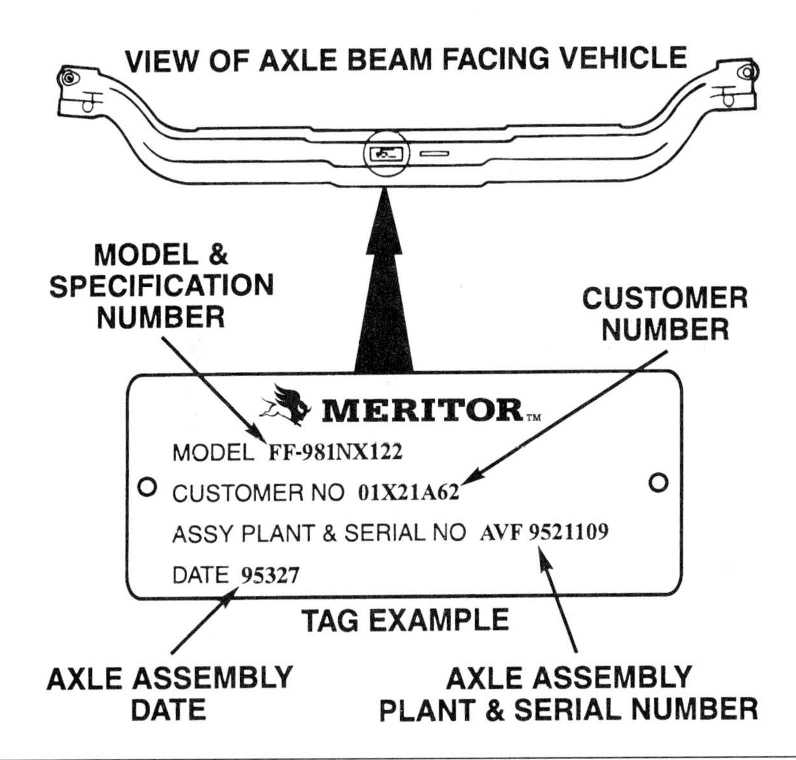

Figure 7-12 Front axle identification plate. (Courtesy of Meritor Automotive, Inc.)

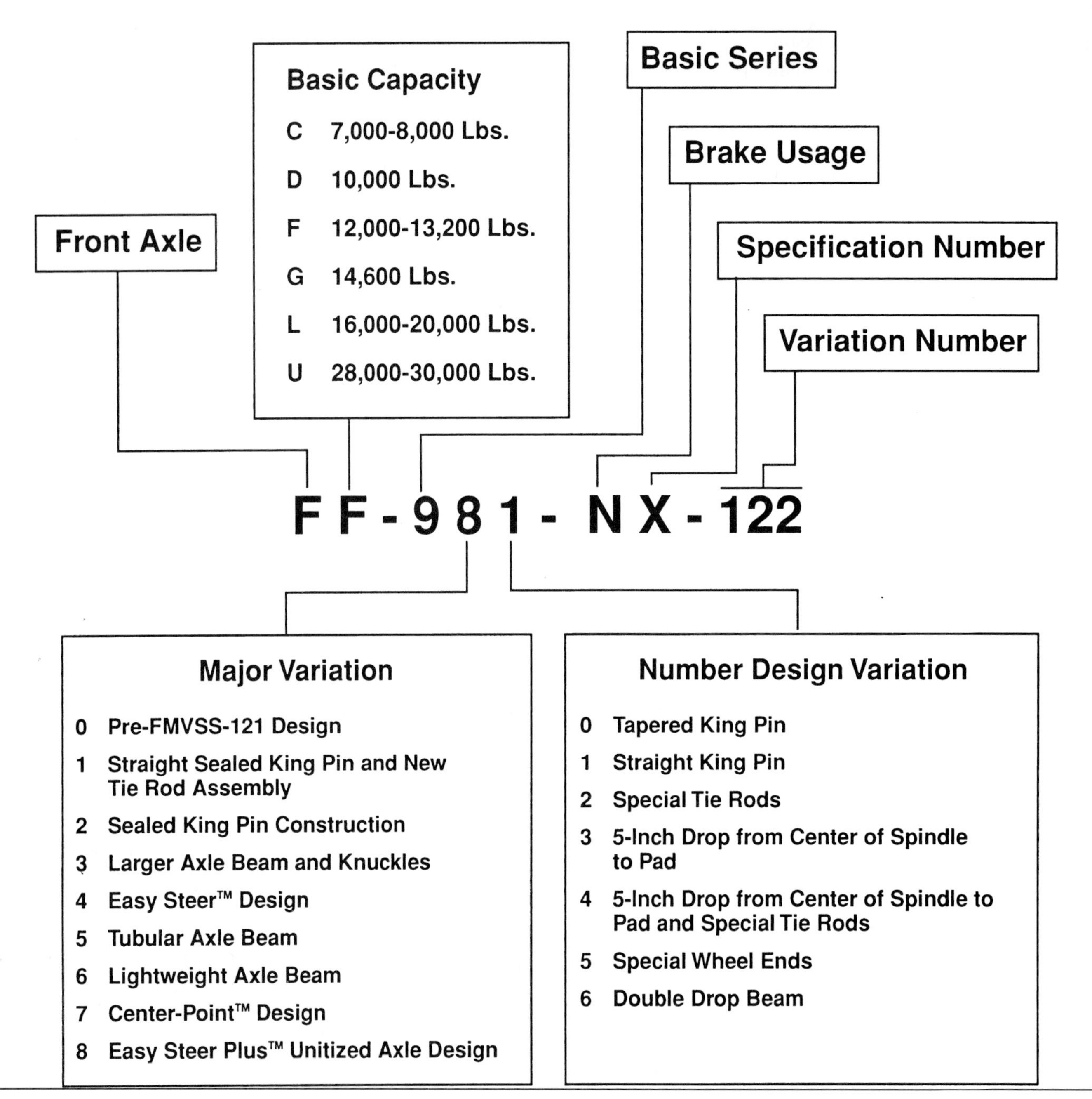

Figure 7-13 Interpretation of front axle model number. (Courtesy of Meritor Automotive, Inc.)

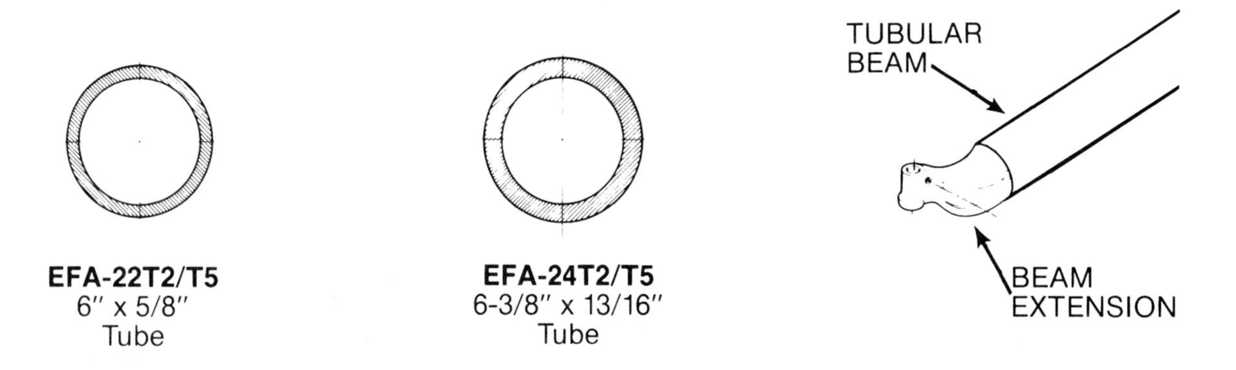

Figure 7-14 Tubular front axle. (Courtesy of Eaton Corporation.)

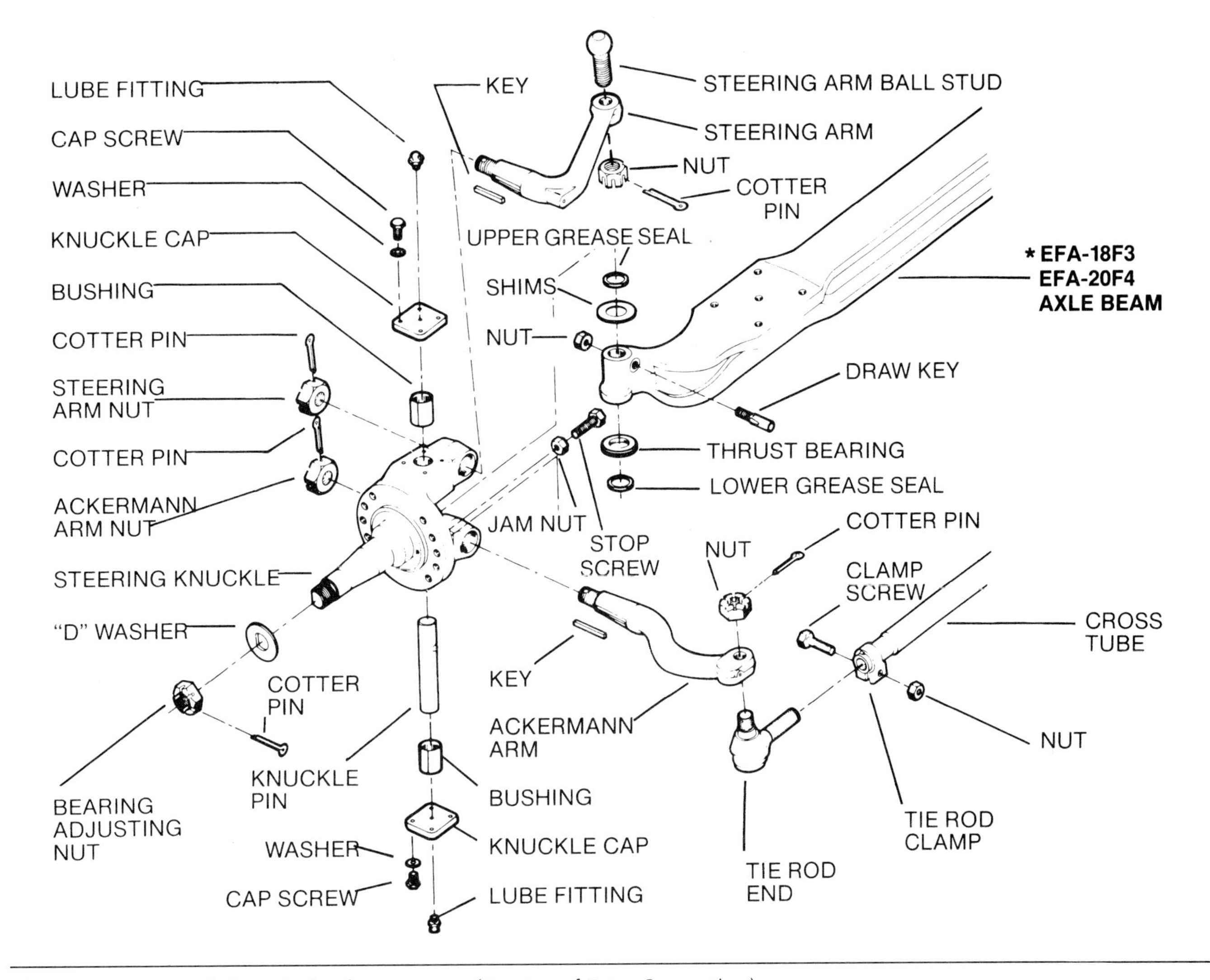

Figure 7-15 Front spindle and related components. (Courtesy of Eaton Corporation.)

steering wheel returnability after a turn. When a front axle is designed with some KPI, the tire and wheel move through an arc as the steering wheel and front wheels are turned. Because the tire cannot move downward into the road surface, the front of the truck is lifted as the steering wheel is turned and the front wheels move through an arc. This action increases steering effort. When the steering wheel is returned to the straight-ahead position, the weight of the truck helps to return the front wheels to the straight-ahead position. Therefore as KPI is increased, steering effort is greater and steering wheel returning force is also increased. Many I-beam front axles have a small degree of KPI.

The distance between the centerline of the tire and the KPI line at the road surface is called **kingpin offset.** Kingpin offset affects static and dynamic steering effort. Static steering effort is the amount of energy required to turn the steering wheel when the truck is stationary. Dynamic steering effort refers to the amount of effort required to turn the steering wheel when the truck is in motion. If an I-beam front axle is designed with some kingpin offset, the front tires tend to roll around the point where the KPI line contacts the road surface when the steering wheel is turned on a stationary vehicle. This action reduces static steering effort. If the I-beam axle is designed with no kingpin offset, the front tires must scrub around the center of the point where the KPI line and tire centerline contact the road surface, and this condition increases static steering effort (Figure 7-17). When there is no kingpin offset, reduced space exists to mount brake components such as brake chambers and slack adjusters.

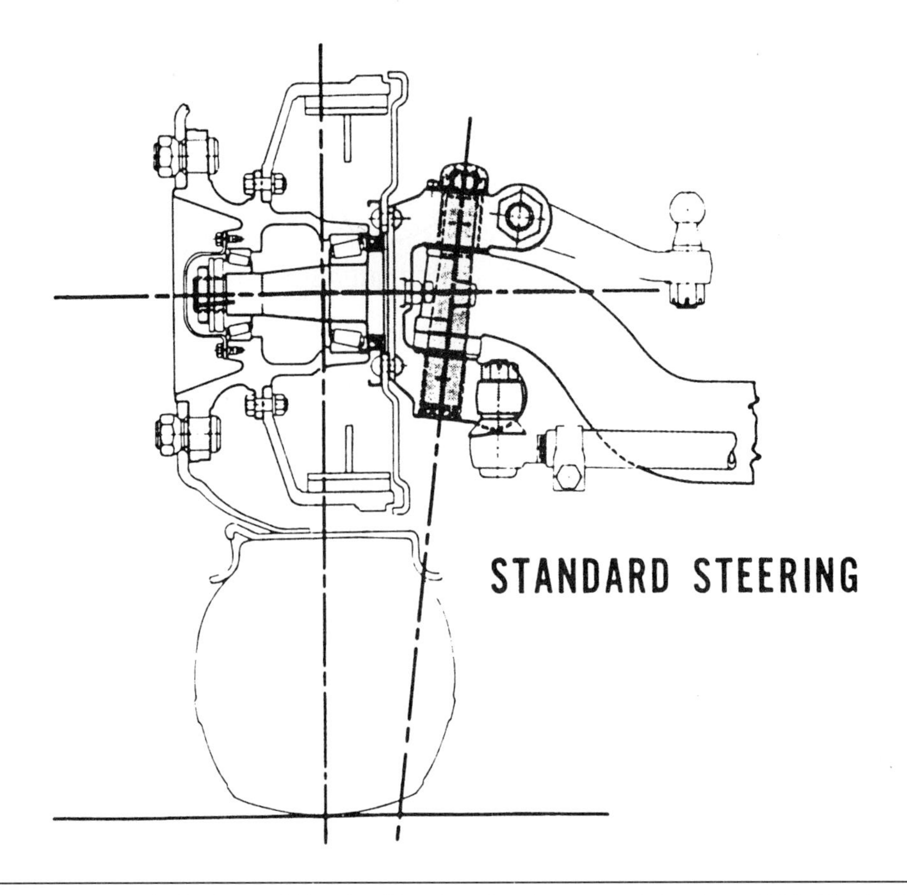

Figure 7-16 Kingpin inclination and kingpin offset. (Reprinted with permission from SAE Publication The Truck Steering System from Hand Wheel to Road Wheel © 1973 Society of Automotive Engineers, Inc.)

Steering effort on a truck in motion is also affected by kingpin offset. When the truck is in motion, the rolling resistance and cornering force of the front tires act like a lever through the kingpin offset. This lever action must be overcome when the steering wheel is initially turned. Therefore an I-beam axle with more kingpin offset requires more initial steering effort when the truck is in motion. Suspension and steering engineers must design the front axle with the proper amount of KPI and kingpin offset to provide the proper balance of static and dynamic steering effort and steering wheel returnability in relation to the weight that is placed on the axle in service.

The **steering arm** is bolted into the upper knuckle opening, and a Woodruff key in this arm fits in a slot in the knuckle opening to prevent steering arm rotation. The steering arm connects the drag link from the steering gear to the steering knuckle. The **Ackerman arm** is bolted into the lower steering knuckle opening, and this arm is also retained with a Woodruff key. The Ackerman arm connects the tie rod to the steering knuckle.

Front Axles with Unitized Hubs and Integrated Knuckles

Some front axles are now available with **unitized hub assemblies** (Figure 7-18). These hub assemblies are a one-piece unit containing the front wheel bearings and seals. Each hub unit contains two tapered roller bearings. Because these hub assemblies are permanently lubricated and sealed, they do not require lubrication or adjustment. The bearings in these hub units are preloaded to eliminate end-play and reduce tire wear. The tie rod arm and spider are integral with the steering knuckle.

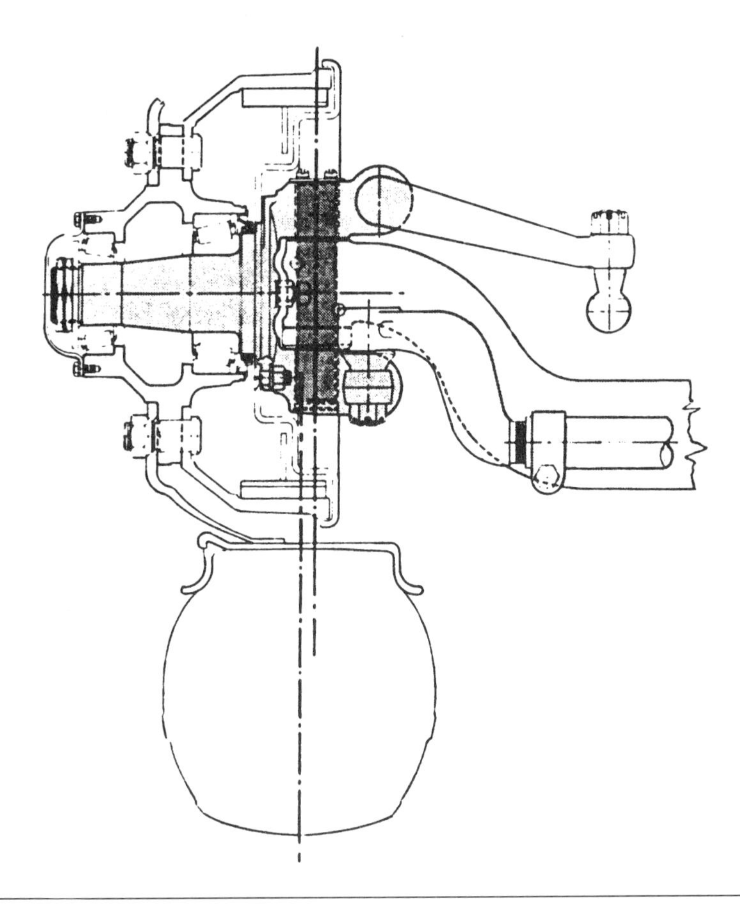

Figure 7-17 I-beam front axle with very little kingpin offset. (Reprinted with permission from SAE Publication The Truck Steering System from Hand Wheel to Road Wheel © 1973 Society of Automotive Engineers, Inc.)

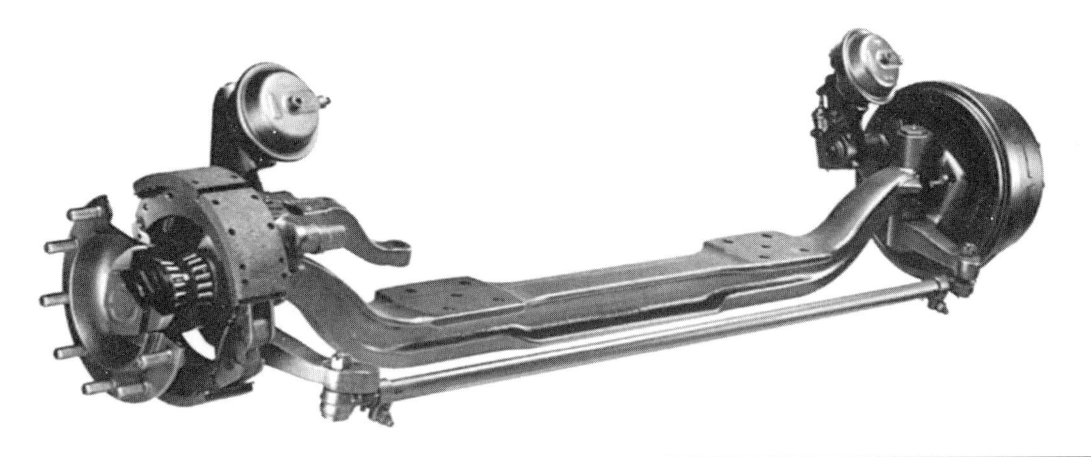

Figure 7-18 Front axle assembly with unitized hubs and integrated steering knuckles. (Courtesy of Meritor Automotive, Inc.)

Front Springs and Mountings

Front springs transfer the vehicle weight to the front axle. When one of the front wheels strikes a road irregularity, the spring transfers road shock to the chassis. The front springs must support the truck weight without permanent distortion or bending. These springs must also reduce road shock transferred from the tires through the axle and springs to the chassis.

One of the most common types of front springs is the **multileaf shackle-type spring** (Figure 7-19). This type of spring may be referred to as a constant rate spring. If 1,000 lb (453.6 kg) causes the spring to deflect 1 in (2.54 cm), 2,000 lb (907.2 kg) deflects the spring 2 in (5.08 cm). In

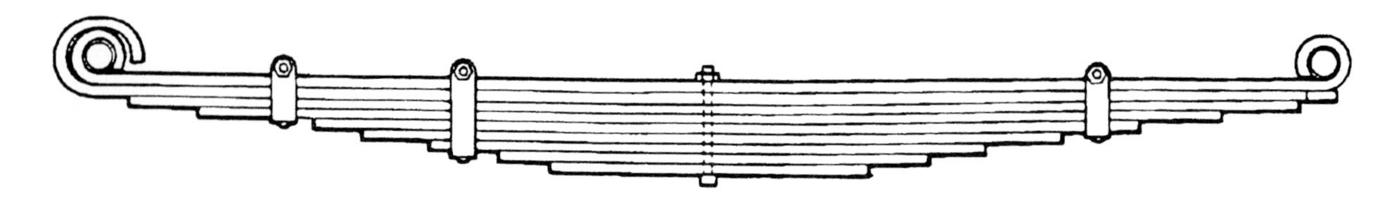

Figure 7-19 Multileaf shackle-type constant rate spring. (Courtesy of Volvo Trucks North America, Inc.)

a multileaf spring the leaves are the same width for the whole length of the spring. In all types of leaf springs a center bolt extends through an opening in the center of the spring leaves. Clamps surround the spring leaves at several locations to prevent the leaves from moving sideways. Other types of front springs include the **parabolic spring** (Figure 7-20) and the **parabolic taperleaf spring** (Figure 7-21). In a parabolic spring the leaf width is usually greater at the center of the spring, and the width decreases toward the outer ends of the spring. The leaves in a taperleaf spring are thicker in the center, and this thickness decreases toward the end of the spring.

An upper plate is placed over the top of the center bolt. Two U-bolts are placed in grooves in the upper plate. These U-bolts straddle the spring, and the lower, threaded end of the U-bolts extends through holes in the axle I-beam (Figure 7-22). Nuts on the ends of the U-bolts are tightened to retain the upper plate and spring to the axle I-beam. The U-bolt nuts must be tightened to the specified torque for proper spring retention.

Caster may be defined as the angle between a line through the center of the kingpin and a line through the vertical center of the wheel and tire viewed from the side.

On some applications a shock absorber mounting plate is positioned between the lower side of the spring and the axle I-beam. Two shock absorbers may be connected from this plate to brackets on the frame (Figure 7-23). On other applications a single shock absorber plate may be mounted between the U-bolt nuts and the lower side of the axle I-beam (Figure 7-24). A **tapered caster shim** may be positioned between the lower side of the spring and the axle I-beam (Figure 7-25). These caster shims are available with different angles to provide a caster adjustment.

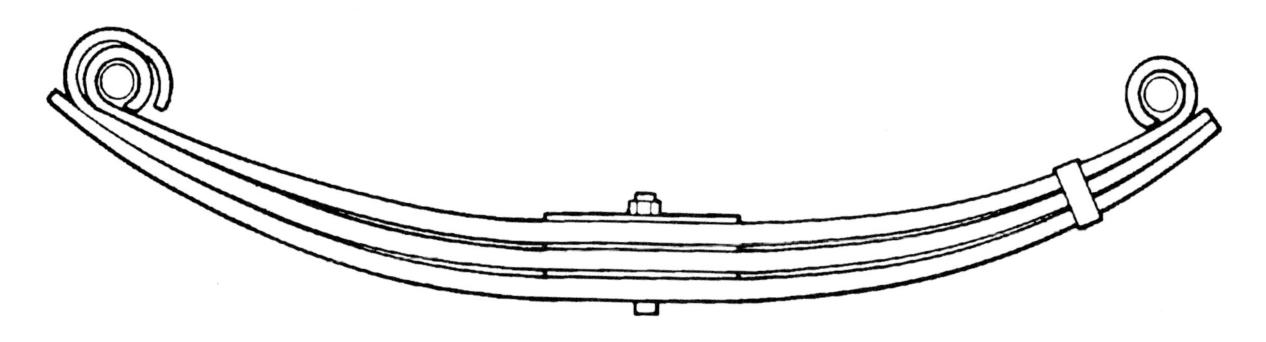

Figure 7-20 Parabolic spring. (Courtesy of Volvo Trucks North America, Inc.)

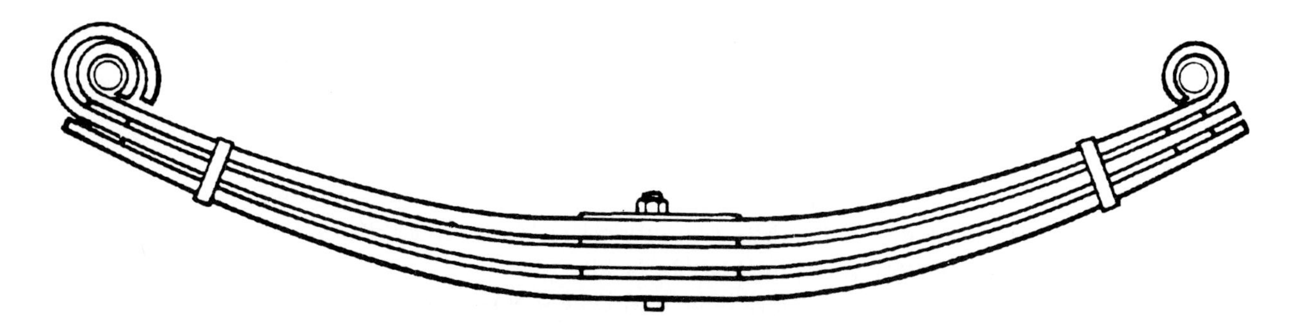

Figure 7-21 Parabolic taper leaf spring. (Courtesy of Volvo Trucks North America, Inc.)

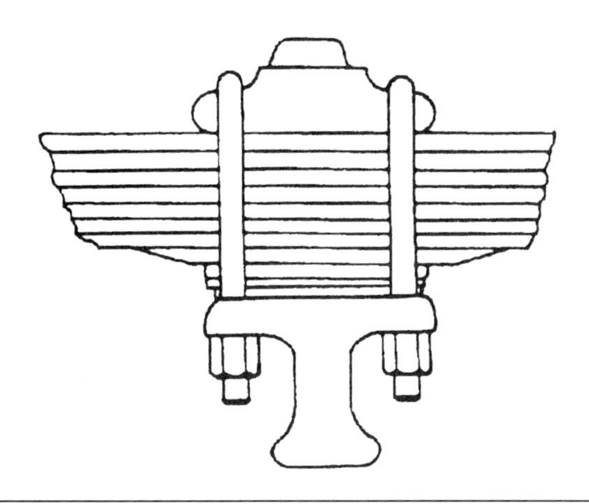

Figure 7-22 Typical upper plate and U-bolt mounting on a front spring. (Courtesy of General Motors Corporation, Service Technology Group.)

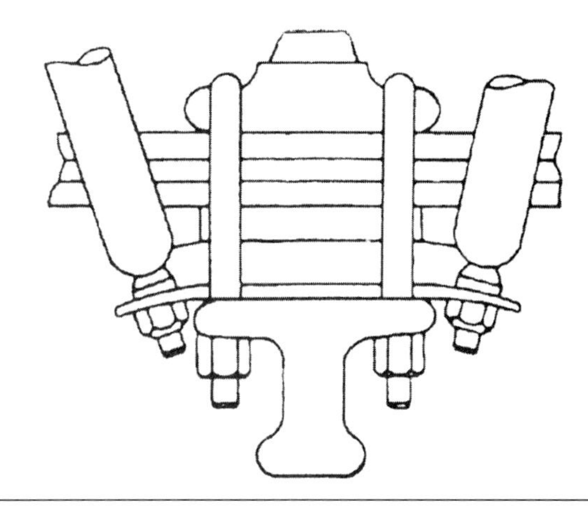

Figure 7-23 Dual shock absorbers with shock absorber plate mounted between the lower side of the spring and the axle I-beam. (Courtesy of Volvo Trucks North America, Inc.)

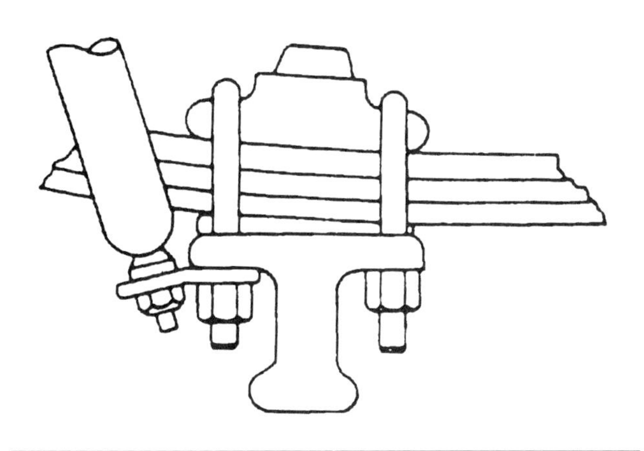

Figure 7-24 Single shock absorber and shock absorber plate mounted between the U-bolt nuts and the axle I-beam. (Courtesy of General Motors Corporation, Service Technology Group.)

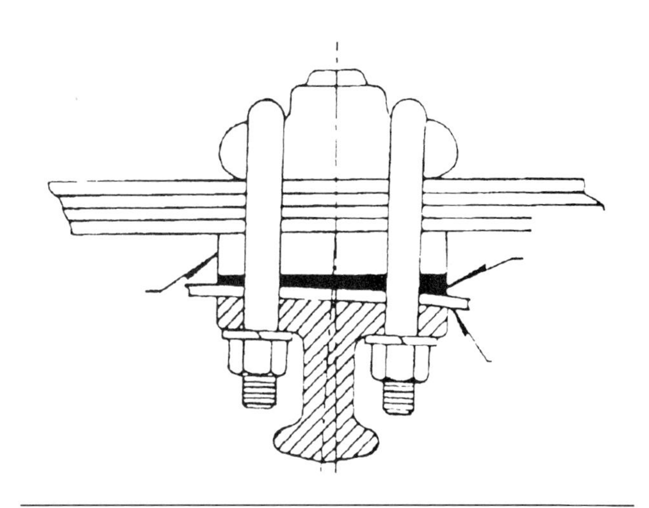

Figure 7-25 Caster shim positioned between the lower side of the spring and the axle I-beam. (Courtesy of Volvo Trucks North America, Inc.)

Front Spring Brackets and Shackles

The front spring brackets, shackles, and related components position the front suspension properly so the springs are parallel with the frame. The front axle must be positioned properly to maintain steering quality and normal tire tread life.

The spring is attached to the frame by a stationary bracket at the front of the spring and a **swinging shackle** at the rear of the spring (Figure 7-26). The stationary bracket is bolted to the frame. On the end of the spring that fits in the stationary bracket, the two top leaves are wrapped to form an eye. A replaceable bushing is pressed into this spring eye. A steel pin extends through the bracket and spring bushing to retain the spring in the bracket. Thrust washers are positioned between the bushing and the bracket (Figure 7-27). Two bolts retain the pin in the bracket. The pin is drilled and a grease fitting in one end of the pin provides lubrication for the pin and bushing.

The swinging shackle permits lengthening and shortening of the spring as the front axle moves vertically while driving the truck over road irregularities. The upper shackle bracket is bolted to the frame. A bushing is mounted in an opening in this bracket, and a steel pin extends through this bushing and the shackle plates mounted on each side of the bracket. Two pinch bolts retain the shackle plates on the steel pin (Figure 7-28). Thrust washers are positioned between the shackle bracket and the shackle plates.

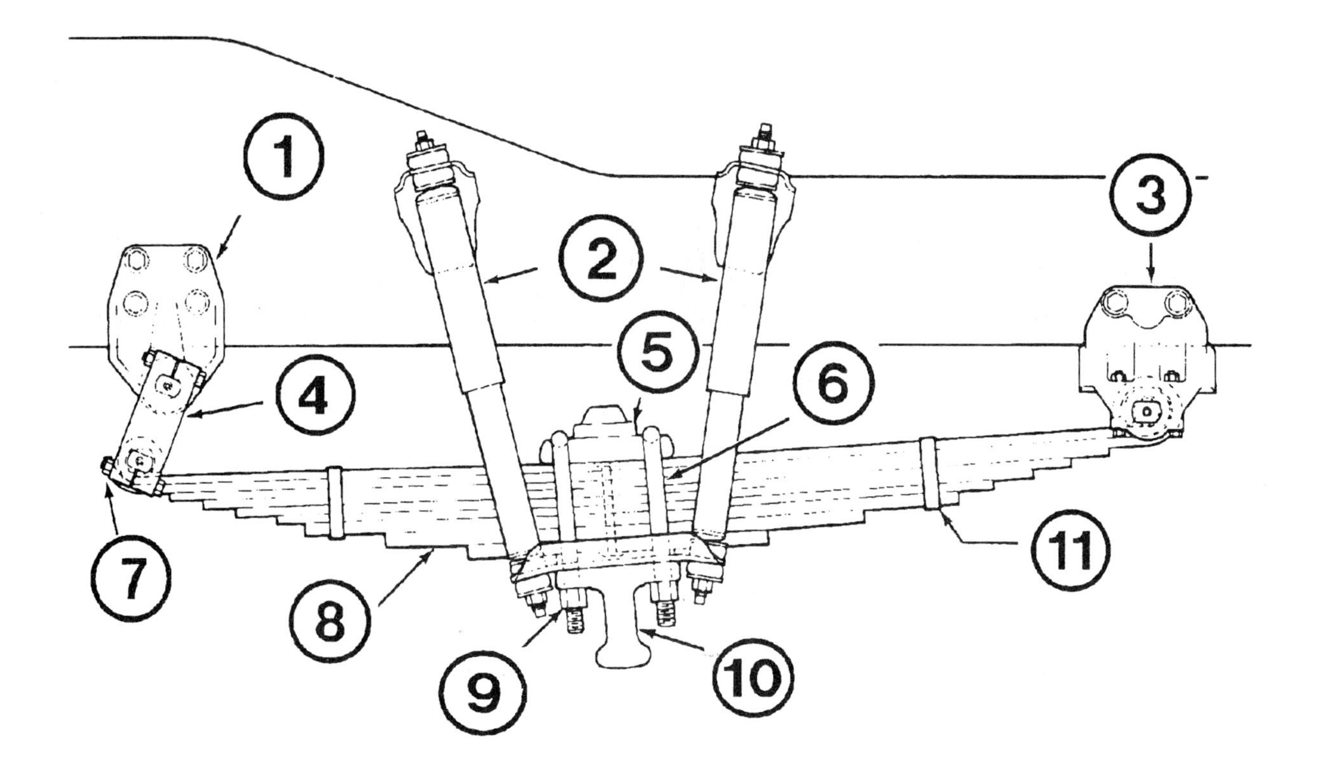

1 - Rear hanger bracket 2 - Shocks 3 - Front spring bracket 4 - Swinging shackle 5 - Upper plate 6 - U-bolt 7 - Pinch bolt 8 - Spring assembly 9 - U-bolt nut 10 - Axle I beam 11 - Spring clip

Figure 7-26 Front spring mounting brackets and shackle. (Courtesy of Volvo Trucks North America, Inc.)

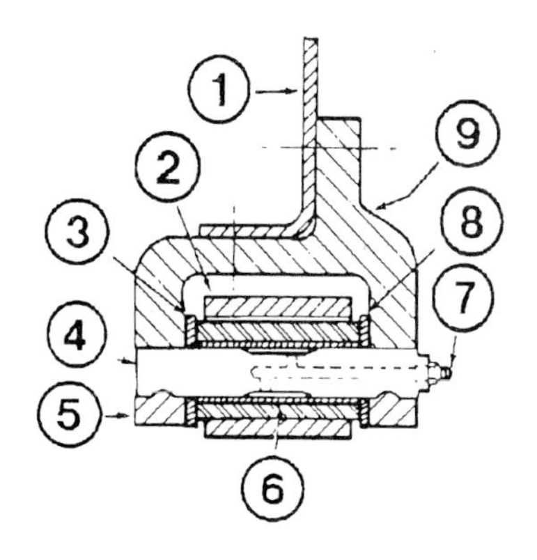

1 - Frame 2 - Spring 3 - Thrust washer 4 - Spring pin 5 - Cap 6 - Bushing 7 - Grease fitting 8 - Thrust washer 9 - Spring bracket

Figure 7-27 Stationary spring bracket, bushing, pin, and related components. (Courtesy of Volvo Trucks North America, Inc.)

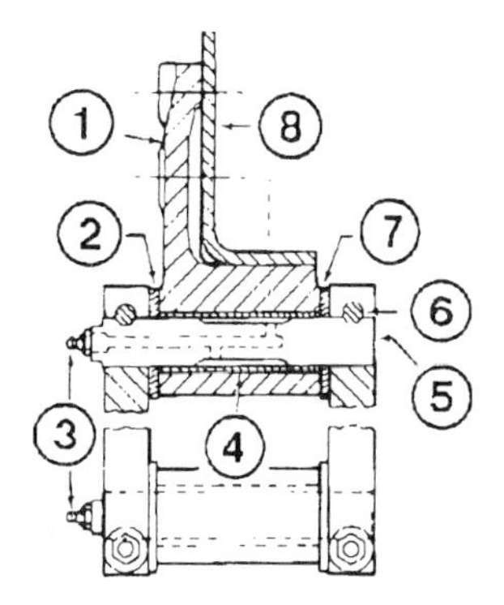

1 - Spring bracket 2 - Thrust washer 3 - Grease fittings 4 - Bushing 5 - Spring pin 6 - Pinch bolt 7 - Thrust washer 8 - Frame

Figure 7-28 Rear spring shackle and related components on a front spring. (Courtesy of Volvo Trucks North America, Inc.)

A bushing is pressed into the eye on the rear of the spring, and a steel pin extends through this bushing and the lower openings in the shackle plates. Pinch bolts retain the steel pin in the lower openings in the shackle plates. The upper and lower steel pins are drilled and threaded so a grease fitting may be installed in one end of these pins. These grease fittings provide lubrication for the pins and bushings.

Fiber Composite Leaf Springs

Fiber composite leaf springs have been used in the automotive industry for many years. This type of spring is now used in some truck suspension systems. Fiber composite spring leaves are made from layers of fiberglass that is laminated and bonded together with polyester resins. These spring leaves may be manufactured by filament winding, a wrapping process in which the fiberglass and resins are bundled together and wrapped. Compression molding is another fiber composite spring leaf manufacturing process in which the fiberglass and resins are squeezed together under pressure.

Fiber composite springs are much lighter than steel springs. Therefore fiber composite springs reduce the vehicle weight and the unsprung suspension weight. Lighter springs reduce the spring effort and amount of shock absorber control that is required to maintain wheel contact with the road. Fiber composite springs provide improved ride quality, better handling, and a faster responding suspension. Fiber composite springs may be used on steer, drive, and trailer axles. A fiber composite spring for a drive axle is shown in Figure 7-29. Fiber composite springs also have these advantages:

1. Because fiber composite spring leaves do not resonate and transmit sound like a steel spring leaf, fiber composite springs provide a quieter ride.
2. Fiber composite spring leaves do not sag with continual loading stress, whereas steel spring leaves do sag. Sagged springs affect ride quality, wheel alignment, and steering control.
3. Fiber composite springs are considerably lighter than steel leaf springs.
4. Because very little danger of a fiber composite spring suddenly breaking exists, these springs provide improved safety.

See Chapter 7 in the Shop Manual for I-beam front suspension diagnosis and service.

Gas-Filled and Conventional Shock Absorbers

CAUTION: New **gas-filled shock absorbers** are wired in the compressed position for shipping purposes. Exercise caution when removing the shipping wire because the shock extends when this strap is cut. After the upper attaching bolt is installed on the shock absorber, the wire may be cut to allow the unit to extend.

WARNING: Never drill a hole in a gas-filled strut or shock absorber and do not throw these units in a fire. Either of these procedures could cause an explosion.

CAUTION: If excessive heat or flame is applied to gas-filled shock absorbers, they may explode and cause serious injury.

On a truck front suspension each side of the suspension has one or two **shock absorbers** connected from plates on the axle to brackets on the frame. Shock absorber purposes may be summarized as follows:

1. Controls spring action to improve ride quality.
2. Controls body sway.
3. Reduces the tendency of a tire tread to lift off the road, which improves tire life and directional stability.

Figure 7-29 Fiber composite spring. (Reprinted with permission from SAE Publication Advantages of Structural Composition in Class 8 Truck Suspensions © 1996 Society of Automotive Engineers, Inc.)

Because shock absorbers help to control spring action during wheel jounce and rebound, they contribute to vehicle safety and driver comfort. If the shock absorbers are worn out, excessive road shock is transferred to the chassis, particularly on rough road surfaces. This excessive road shock results in reduced steering control and driver comfort. Shock absorbers are extremely important to provide longer tire life, vehicle handling and steering quality, and ride quality.

Shock Absorber Design

In many shock absorbers the lower half of a shock absorber is a twin tube steel unit filled with hydraulic oil and nitrogen gas (Figure 7-30). Gas-filled units are identified with a warning label. If a gas-filled shock absorber is removed and compressed to its shortest length, it should re-extend when it is released. Failure to re-extend indicates that shock absorber replacement is necessary. In some shock absorbers, the nitrogen gas is omitted. A relief valve is located in the bottom of the unit, and a circular lower mounting is attached to the lower tube. This mounting contains a rubber isolating bushing or grommets. A piston and rod assembly is connected to the upper half of the shock absorber. This upper portion of the shock absorber has a dust shield that surrounds the lower twin tube unit. The piston is a precision fit in the inner cylinder of the lower unit. A piston rod guide and seal are located in the top of the lower unit. A circular upper mounting with a rubber bushing is attached to the top of the shock absorber.

Shock Absorber Operation

Jounce travel refers to upward wheel movement when the wheel strikes a hump on the road surface. Upward force is supplied to the spring during jounce travel.

Rebound travel refers to downward wheel movement, and the spring is exerting a downward force on the axle during this rebound travel.

When a vehicle wheel strikes a bump, the wheel and suspension move upward in relation to the chassis. Upward wheel movement is referred to as **jounce travel.** This jounce action exerts upward force on the spring. Under this condition, the spring stores energy and springs back downward with all the energy absorbed when it is deflected upward. This downward spring and wheel action is called **rebound travel.** If this spring action is not controlled, the wheel would strike the road with a strong downward force, and the wheel jounce would occur again. Therefore some device must be installed to control the spring action, or the wheel would bounce up and down several times after it hit a bump, causing driver discomfort, directional instability, and suspension component wear.

Shock absorbers are installed on suspension systems to help control spring action. When a wheel strikes a bump and jounce travel occurs, the shock absorber lower tube unit is forced upward. This action forces the piston downward in the lower tube unit. Because oil cannot leak past the piston, the oil in the lower unit is forced through the piston valves to the upper oil chamber. These valves provide precise oil flow control and control the upward action of the wheel and suspension, which is referred to as a **shock absorber compression stroke** (Figure 7-31).

When the spring expands downward in rebound travel, the lower shock absorber unit is also forced downward. When this occurs, the piston moves upward in the lower tube unit, and hydraulic oil is forced through the piston valves from the upper oil chamber to the lower oil chamber. Because the valves restrict oil flow with precise control, the downward suspension and wheel movement is controlled. The **shock absorber extension stroke** occurs during downward wheel movement.

When the shock absorber piston moves, oil is forced through the piston. Because the piston valves and orifices resist the flow of oil, friction and heat are created. The resistance of the oil moving through the piston must be calibrated as closely as possible to the spring's deflection rate or strength. Wheels and suspension systems deflect at many different speeds, depending on the type and size of bump and the vehicle speed. The resistance of a shock absorber piston increases with the square of its speed. For example, if the wheel deflection speed increases four times, the piston resistance is sixteen times as great. Therefore if a wheel strikes a large bump at high speed, the wheel deflection and rebound could be effectively locked by the shock absorber. Shock absorber engineers prevent this action by precisely designing shock absorber valves and orifices to provide enough friction to provide the necessary spring control. These piston valves and orifices must not create excessive friction, which slows the wheel from returning to its original position.

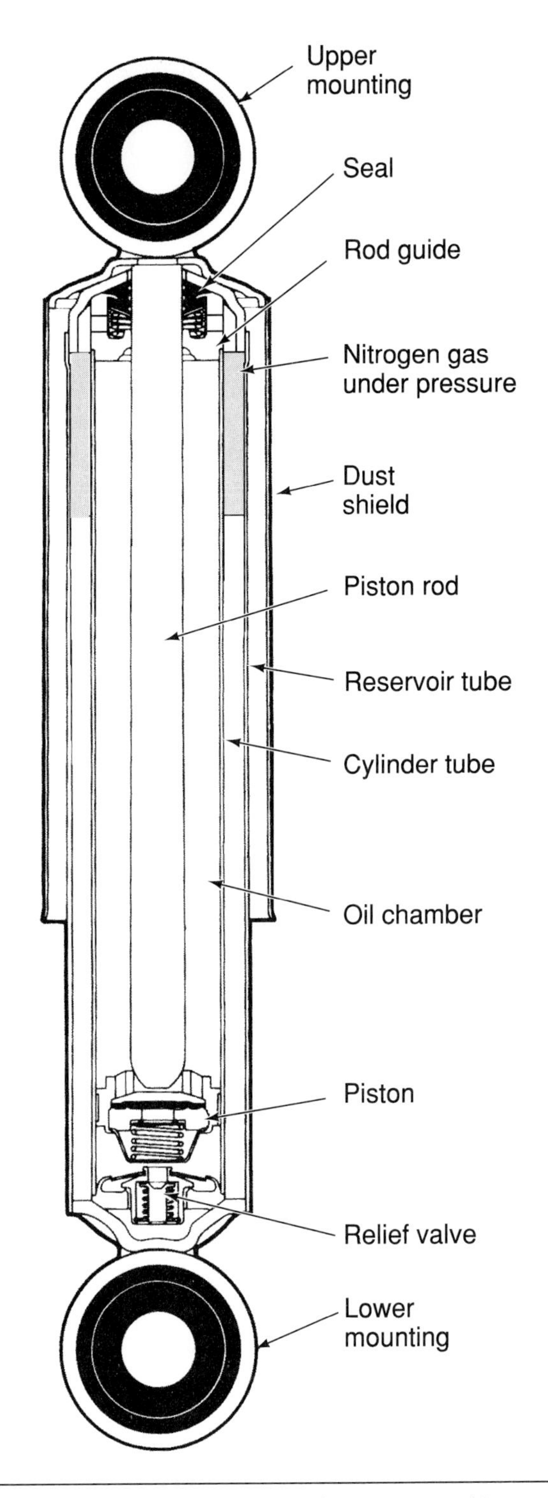

Figure 7-30 Shock absorber filled with hydraulic oil and nitrogen gas. (Courtesy of Delco-Remy, Division of GMC.)

Shock absorber pistons have many different types of valves and orifices. In some pistons small orifices control the oil flow during slow wheel and suspension movements. Stacked steel valves control the oil flow during medium speed wheel and suspension movements. During maximum wheel and suspension movements, larger orifices between the piston valves provide oil flow control. On other shock absorber pistons, the stacked steel valves alone provide oil flow control. Regardless of the piston orifice and valve design, the shock absorber must be precisely matched to absorb the spring's energy.

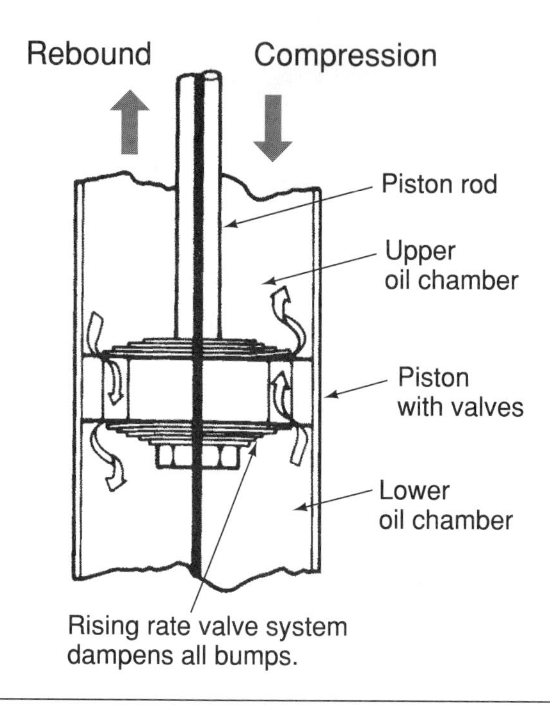

Figure 7-31 Shock absorber action. (Courtesy of Delco-Remy, Division of GMC.)

During fast upward wheel movement on the compression stroke, excessive pressure in the lower oil chamber forces the base valve open and thus allows oil to flow through this valve to the reservoir. The nitrogen gas provides a compensating space for the oil that is displaced into the reservoir on the compression stroke and when the oil is heated. Because the gas exerts pressure on the oil, cavitation, or foaming, of the oil is eliminated. When oil bubbles are eliminated in this way, the shock absorber provides continuous damping for wheel deflections as small as 0.078 in (2.0 mm). A rebound rubber is located on top of the piston. If the wheel drops downward into a hole, the shock absorber may become fully extended. Under this condition, the rebound rubber provides a cushioning action.

Heavy Duty Shock Absorber Design

Some **heavy duty shock absorbers** have a dividing piston in the lower oil chamber. The area below this position is pressurized with nitrogen gas to 360 pounds per square inch (psi) (2,482 kilopascals [kPa]). Hydraulic oil is contained in the oil chamber above the dividing piston. The other main features of the heavy duty shock absorber are as follows:

1. High-quality seal for longer life.
2. Single tube design to prevent excessive heat buildup.
3. Rising rate valve provides precise spring control under all conditions.

The operation of the heavy duty shock absorber is similar to the conventional type (Figure 7-32).

Shock Absorber Ratios

Many truck shock absorbers are a double acting-type that control spring action during jounce and rebound wheel movements. The piston and valves in many shock absorbers are designed to provide more extension control than compression control. An average shock absorber may have 70% of the total control on the extension cycle, and thus 30% of the total control is on the compression cycle. Shock absorbers usually have this type of design because they must control the heavier sprung body weight on the extension cycle. The lighter unsprung axle, wheel, and tire weight is

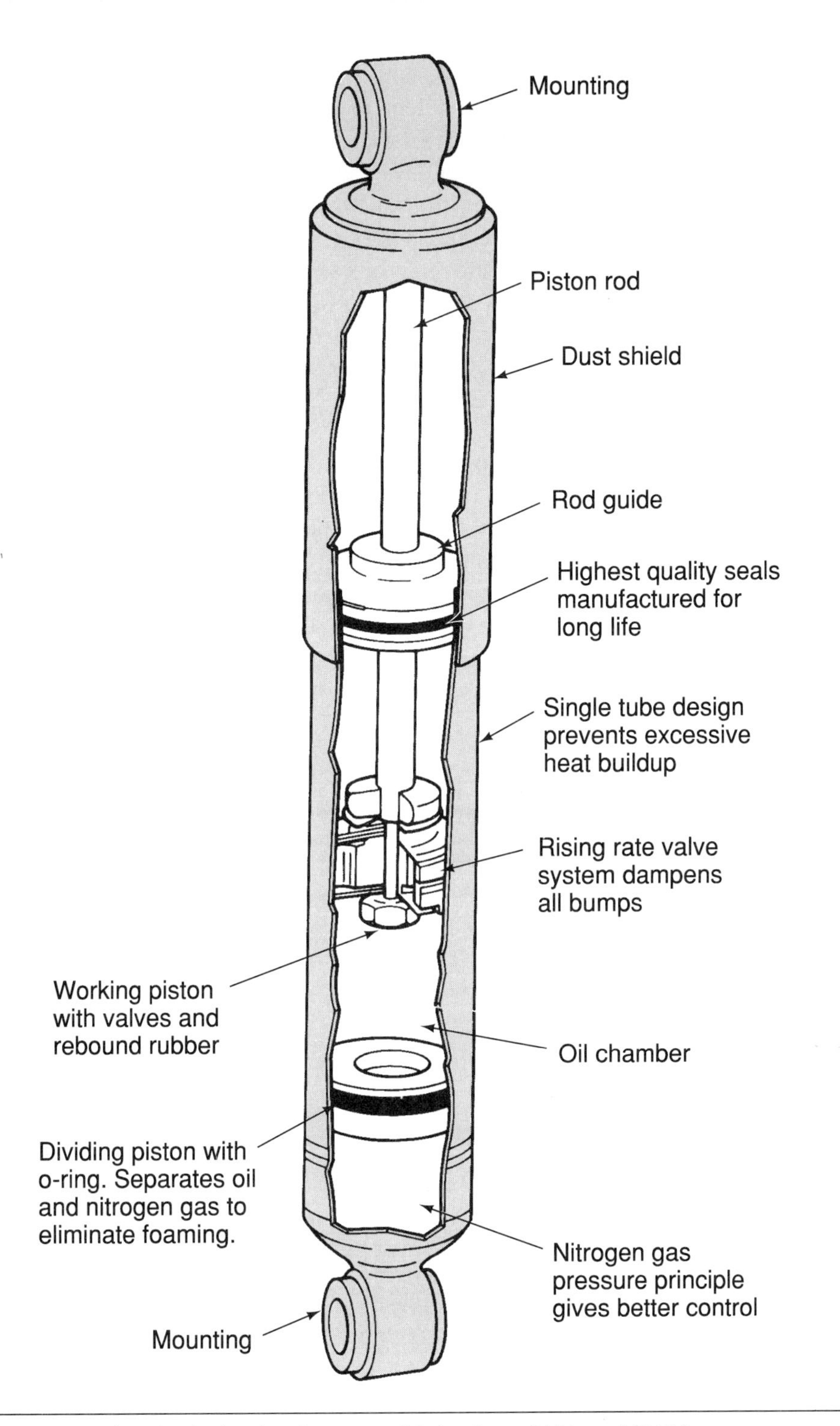

Figure 7-32 Heavy duty shock absorber. (Courtesy of Delco-Remy, Division of GMC.)

See Chapter 7 in the Shop Manual for shock absorber diagnosis and service.

controlled by the shock absorber on the compression cycle. A shock absorber with this type of design is referred to as a 70/30 type. **Shock absorber ratios** vary from 50/50 to 80/20.

Single Axle Leaf Spring Rear Suspension Systems

Single axle suspension systems are used on medium duty trucks. These suspension systems are lighter and have less load-carrying capacity compared with tandem axle suspension systems on heavy duty trucks. Single axle rear suspension systems may have taperleaf, parabolic springs or multileaf springs. Taperleaf springs are usually mounted with a front eye and bushing and a rear shackle. Multileaf springs are often mounted with a front eye and bushing and a rear slider. Some multileaf springs are mounted with a cam bracket at the front and rear of the spring, and a torque leaf near the bottom of the spring is connected to a steel pin in the front cam bracket (Figure 7-33). This torque leaf prevents fore-and-aft axle movement. This design allows the end of the spring to slide on the cam bracket as the truck is loaded and unloaded. This type of spring may be referred to as a **vari-rate spring.** When the vehicle load is increased, the ends of the spring slide through the cam bracket, and this shortens the effective length of the spring to provide a variable deflection rate.

A vari-rate spring changes its deflection rate as the vehicle load is increased.

TRADE JARGON: An auxiliary spring may be called a helper spring.

Many single axle springs have an auxiliary spring mounted on top of the conventional spring. The **auxiliary spring** may be multileaf or parabolic, single leaf (Figure 7-34). Because the ends of the auxiliary spring slide on cam brackets as the vehicle load increases, this spring is a vari-rate spring. The conventional and auxiliary springs are retained on the rear axle with two U-bolts. These U-bolts extend through an anchor plate mounted under the rear axle. Rubber insulating bushings in the spring eye and shackle eliminate the need for lubrication. A stabilizing shaft is connected by two links from a plate mounted between the axle and the spring to the frame. A shock absorber is connected from each anchor plate to a chassis bracket. The shock absorbers are optional on suspension systems with multileaf springs.

Tandem Rear Axle Suspension System with Leaf Springs and Torque Rods

Some rear tandem axle suspension systems have four semielliptical leaf springs retained on the rear axles with U-bolts. On both sides of the tractor the forward end of the forward spring and the rear end of the rear spring ride in brackets attached to the frame rails. These aluminum brackets contain steel wear shoes. Between the front and rear springs on each side of the tractor, the springs

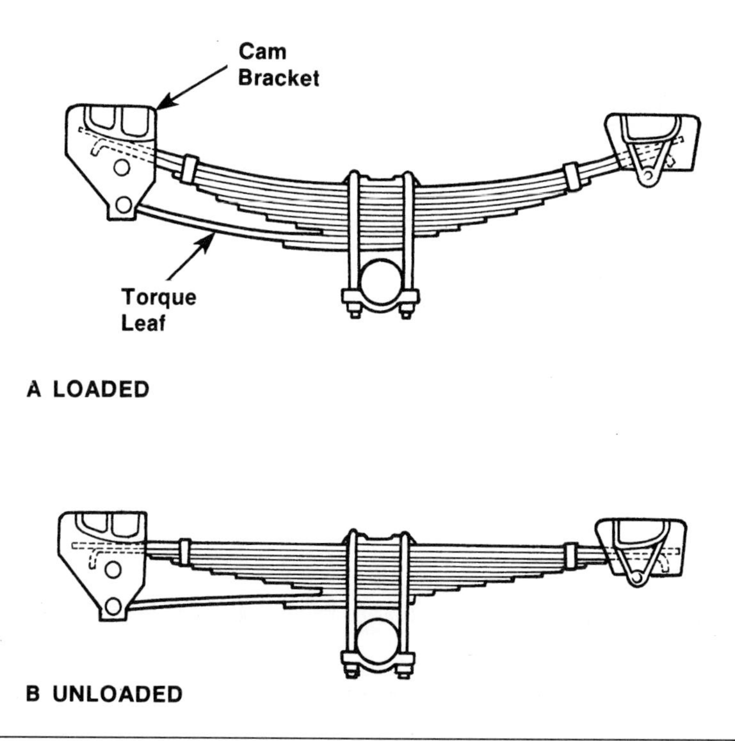

Figure 7-33 Vari-rate spring action. (Courtesy of Navistar International Transportation Corp.)

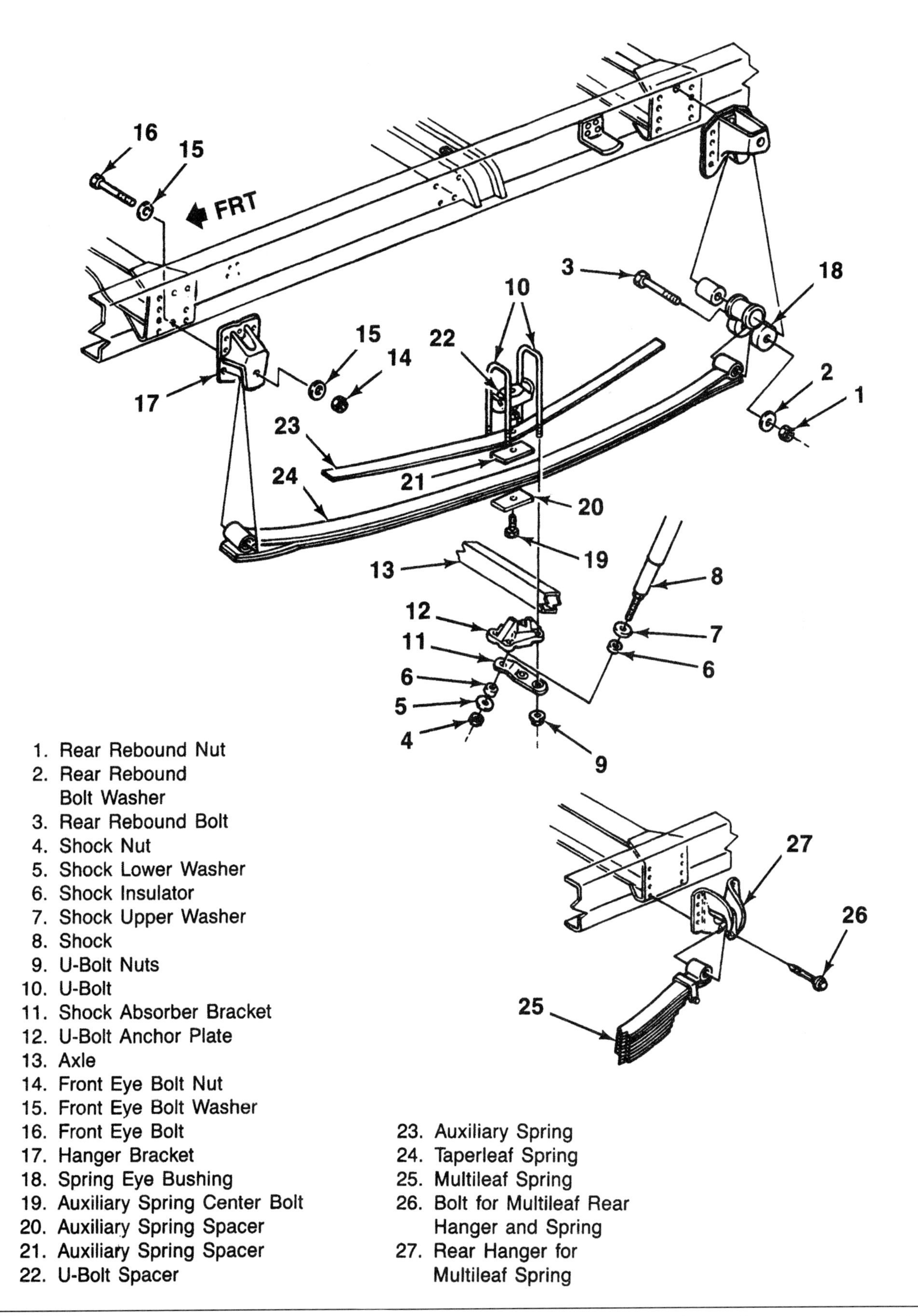

Figure 7-34 Single axle rear spring assembly with parabolic, single leaf auxiliary spring. (Courtesy of General Motors Corporation, Service Technology Group.)

ride on an **equalizer** that pivots on a sleeve in the equalizer bracket (Figure 7-35). When the top of the equalizer contacts the equalizer bracket, equalizer movement is stopped.

A heavy **torque rod** is connected from each forward axle seat between the U-bolt anchor plate and the spring to the front spring brackets. These torque rods prevent fore-and-aft movement of the forward axle. Two torque rods are also connected from each rearmost axle seat to the equalizer bracket to prevent fore-and-aft movement of the rearmost axle. The torque rods on both axles minimize axle windup, resulting from driveline torque, to provide axle stability. The torque rods also provide axle stability when braking forces are applied to the axles. Some rear axles have a third torque rod connected from the center of the axle to a bracket on the frame crossmember. Shims may be installed between the ends of the two outer torque rods and the spring brackets on each axle to provide rear axle alignment.

See Chapter 7 in the Shop Manual for rear suspension with leaf springs and torque rods diagnosis and service.

Some tandem axle tractor suspensions with leaf springs and torque rods such as the Reyco model 102 or 102W have adjustable, cast torque rods (Figure 7-36). Large rubber bushings in the torque rods reduce the transmission of road shock from the suspension to the frame.

Tandem Rear Axle Suspension System with Leaf Springs and Equalizing Beams

In a rear tandem axle suspension system with **equalizing beams,** the leaf springs are mounted on saddles above the equalizer beams (Figure 7-37). A cross tube is mounted between the two equal-

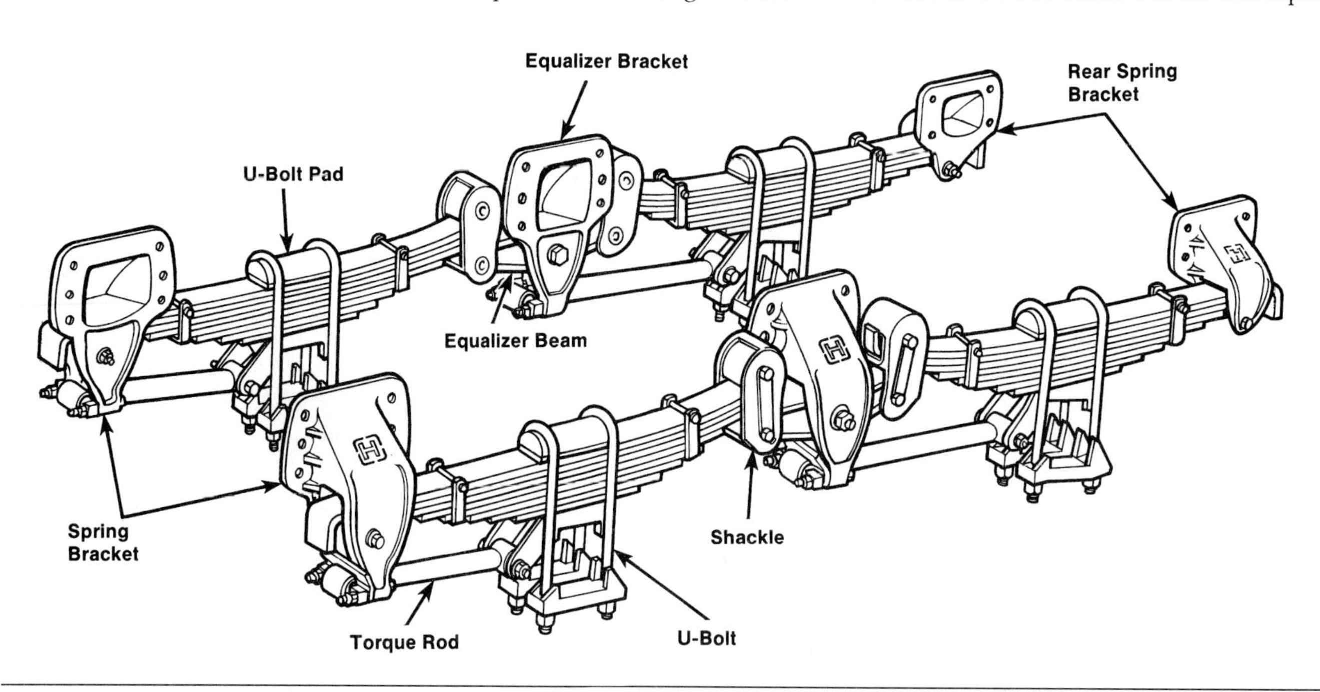

Figure 7-35 Rear tandem axle suspension system with torque rods. (Courtesy of Hendrickson International.)

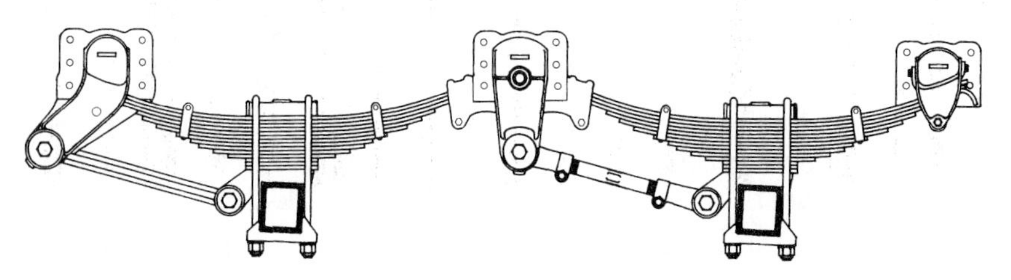

Figure 7-36 Tractor tandem rear axle suspension system with leaf springs and cast, adjustable torque rods. (Courtesy of Reyco Industries Inc.)

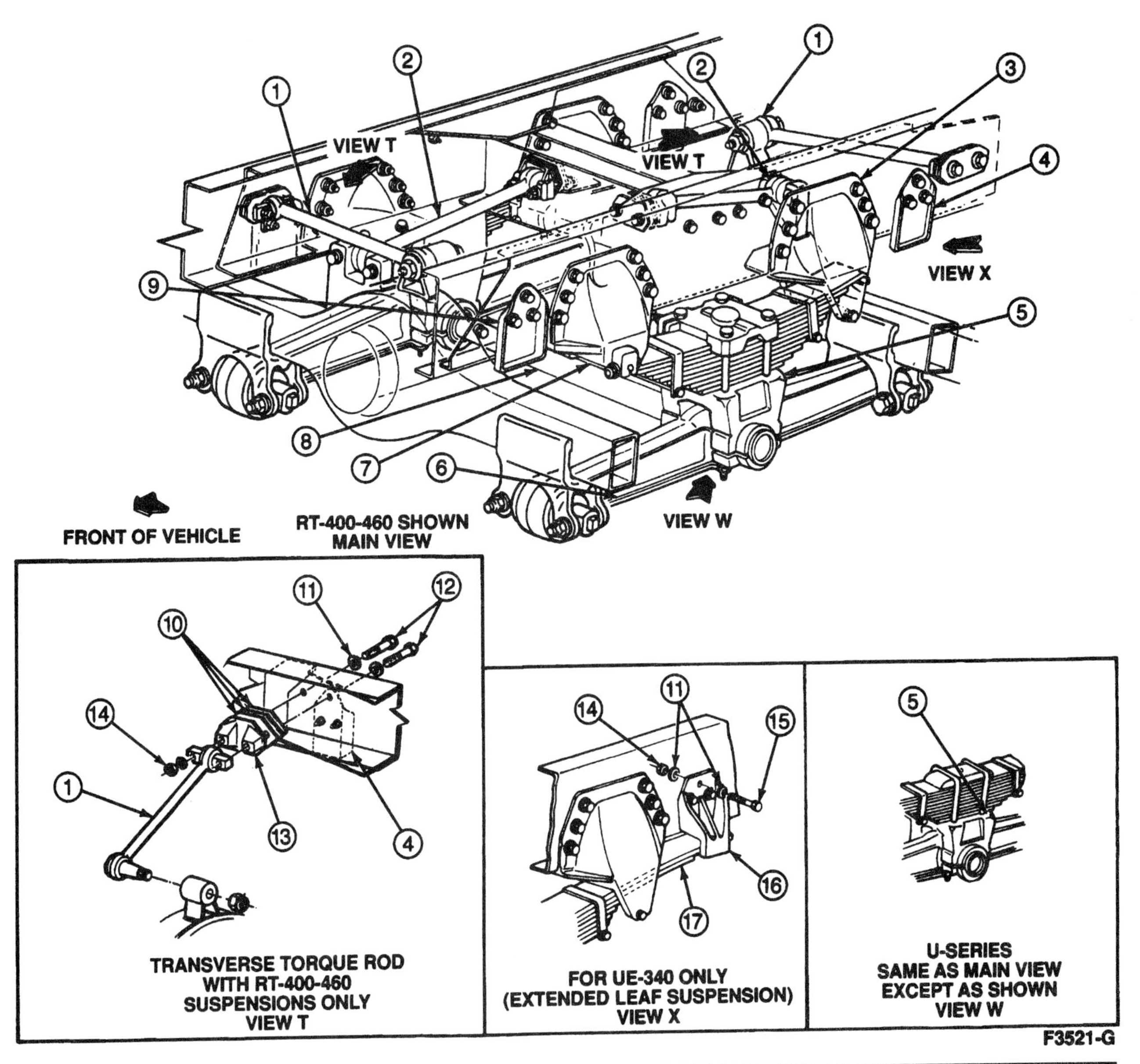

Item	Part Number	Description
1	4962	Rod, Transverse Torque
2	4962	Rod, Longitudinal Torque
3	5894	Bracket, Rear Spring, Rear
4	4730	Axle Bumper, Rear
5	5888/5890	Spring and Saddle Assembly
6	5889	Equalizing Beam
7	5785	Bracket, Rear Spring, Front
8	5884	Crosstube
9	4730	Axle Bumper, Front

Item	Part Number	Description
10	4A002	Spacer
11	44882-S2	Flatwasher
12	384848-S2	Bolt
13	5883	Bracket, Torque Rod
14	34992-S2	Nut
15	58743-S2	Bolt. All Bolt Heads Inboard RH Side of Vehicle.
16	5894	Bracket, Extended Rear Spring
17	5888/5890	Extended Rear Spring

Figure 7-37 Rear tandem axle suspension system with equalizing beams and leaf springs. (Courtesy of Sterling Truck Corporation.)

See Chapter 7 in the Shop Manual for rear suspension with leaf springs and equalizing beams diagnosis and service.

izing beams on each axle. Each equalizing beam is pivoted on a cross tube bushing in the saddle directly below the center of the spring (Figure 7-38). Bushings in the outer ends of the equalizing beams are attached to brackets on the rear axle housings. A transverse torque rod may be connected from the top of the differential housing to a frame bracket. A bolt extends through a bushing in the eye on the front of the spring and the front spring bracket. The rear end of the spring slides on a support in the rear spring bracket.

The equalizing beams act as levers to distribute the load equally between the two rear axles and reduce the effect of road shock caused by road irregularities. The cross tube between the equalizing beams provides correct alignment of the tandem axles. The equalizing beams also stabilize the rear axles when driveline torque or braking forces tend to rotate the axles forward or backward.

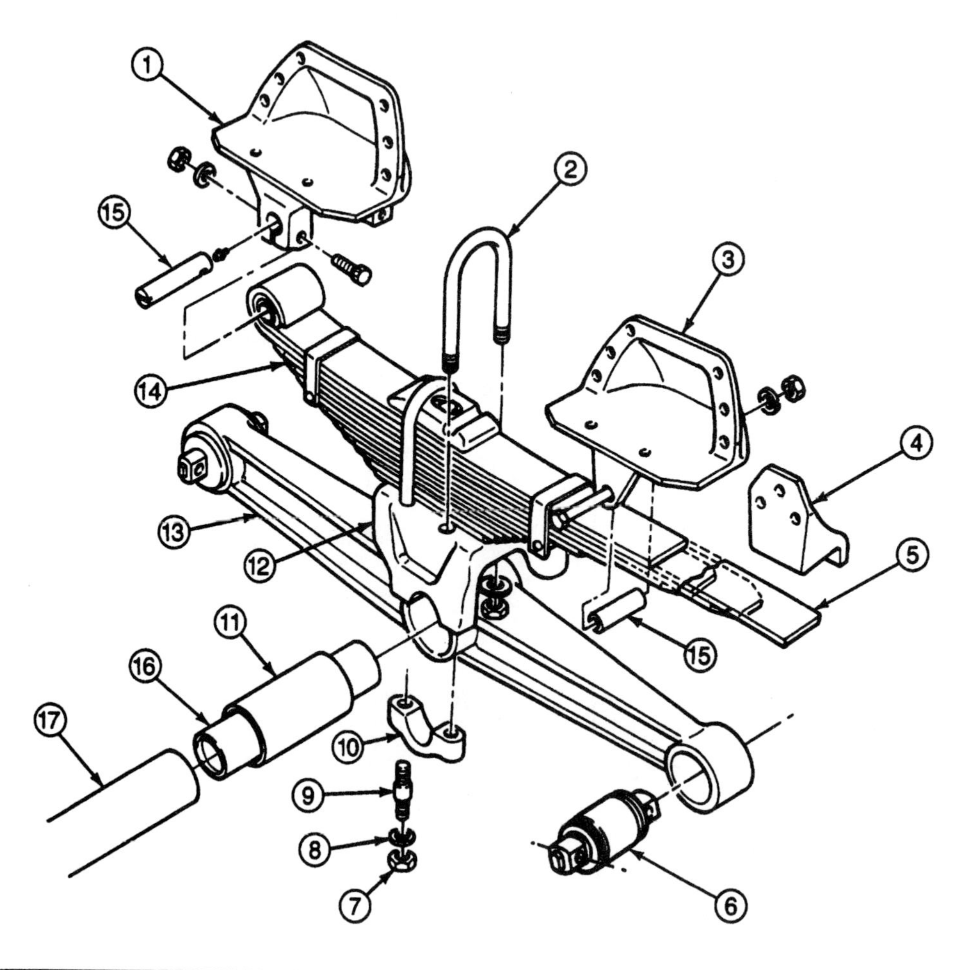

Item	Part Number	Description
1	5785	Bracket, Rear Spring, Front
2	—	U-Bolt (Part of 5888/5890)
3	5894	Bracket, Rear Spring, Rear
4	5894	Bracket, Extended Rear Spring
5	5888/5890	Extended Rear Spring
6	5873	Bar Pin Bushing, Equalizer Beam
7	353705	Nut
8	354639	Washer

Item	Part Number	Description
9	377817	Stud
10	5875	Saddle Cap
11	5872	Rubber Center Bushing
12	5888/5890	Saddle
13	5889	Equalizing Beam
14	5888	Spring
15	5780	Spring Eye Pin
16	5A553	Sleeve, Center Beam
17	5884	Crosstube

Figure 7-38 Equalizing beam and spring mounting. (Courtesy of Sterling Truck Corporation.)

Tandem Rear Axle Suspension System with Rubber Cushions and Equalizing Beams

Some tandem rear axle suspension systems have **rubber cushions** mounted on top of the saddles above the equalizing beams. These rubber cushions are used in place of the leaf springs. Frame hangers attached to the top of the rubber cushions allow these cushions to be bolted to the frame. The rear suspension with rubber cushions weighs considerably less than a rear suspension with leaf springs. On some applications longitudinal and transverse torque rods are connected from the top of the differential housing to the frame (Figure 7-39). Other applications have six rubber cushions and longitudinal torque rods (Figure 7-40). Various types of rear tandem axle suspension systems and their intended application and weight capacity are provided in Figure 7-41 and Figure 7-42.

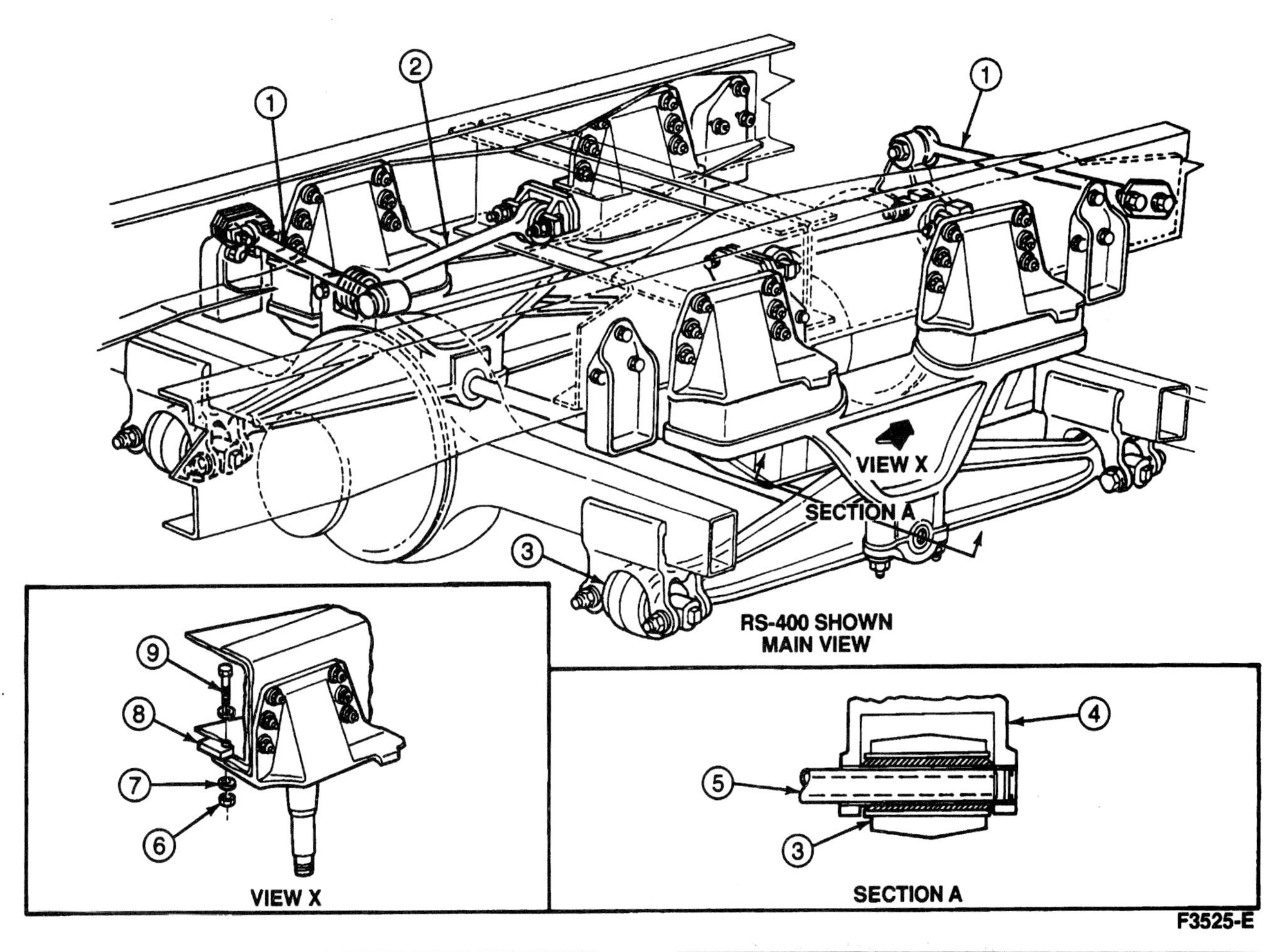

Item	Part Number	Description
1	4962	Rod, Transverse Torque
2	4962	Rod, Longitudinal
3	5889	Equalizing Beam
4	5888/5890	Saddle

Item	Part Number	Description
5	5884	Crosstube
6	34992-S2	Nut
7	44882-S2	Flatwasher
8	5753	Spacer
9	389525	Bolt

Figure 7-39 Rear tandem axle suspension system with four rubber cushions plus longitudinal and transverse torque rods. (Courtesy of Sterling Truck Corporation.)

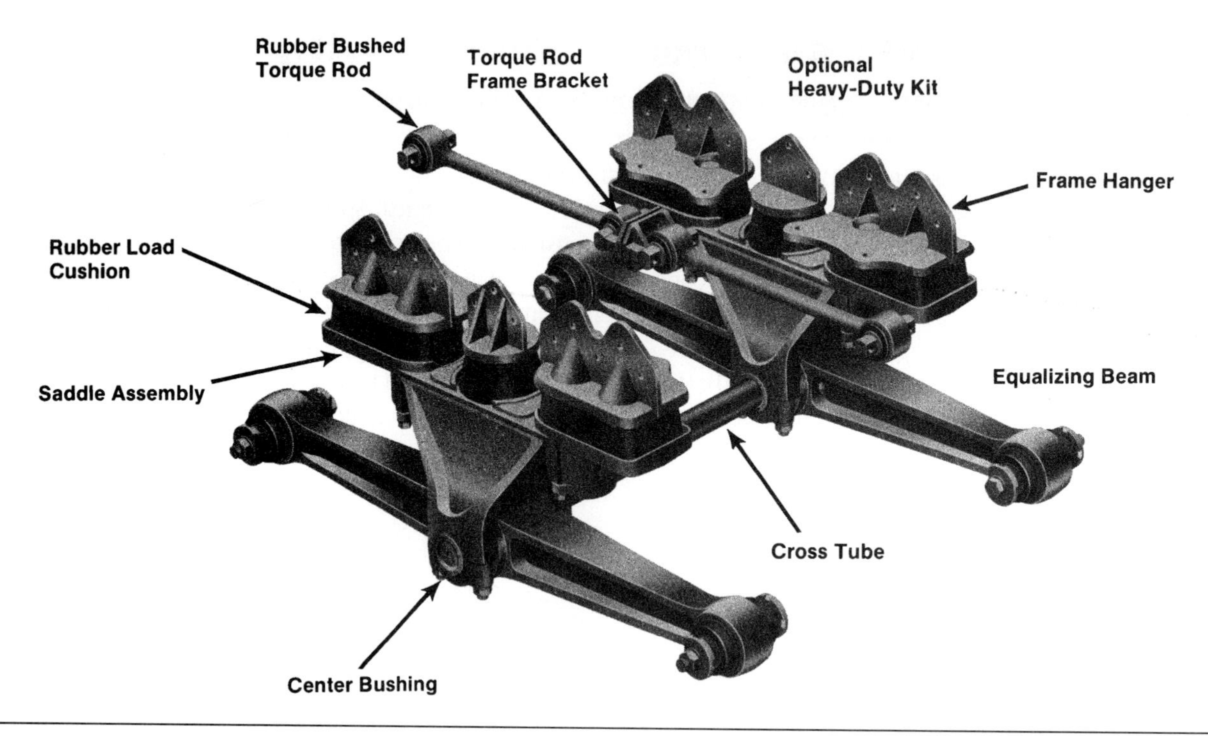

Figure 7-40 Rear tandem axle suspension system with six rubber cushions and longitudinal torque rods. (Courtesy of Hendrickson International.)

Tandem Rear Axle Suspension System with Inverted Springs

The rear tandem axle suspension system with **inverted springs** may be called a T-ride suspension system. This type of suspension system provides maximum articulation for improved traction and excellent ride characteristics for on-highway operation. This suspension system is lighter weight compared with other leaf spring suspension systems.

The inverted, parabolic, tapered spring leaves are attached by means of U-bolts to a **spring cradle** that is bolted to the outside of the tractor frame. Rubber springs are bolted to the outer ends of each spring, and the lower side of these springs rests on saddles attached to the axle housings (Figure 7-43). The rubber springs absorb shock and vibration.

This suspension system has eight torque rods. Two torque rods are connected from the top of each differential housing to the tractor frame in a V-shaped pattern. The V-shaped patterns formed by these torque rods are opposite to each other on the front and rear differentials in the tandem axle. These upper torque rods with V-shaped patterns distribute laterally transmitted forces, along with starting and braking torque forces, to the tractor frame. Four lower torque rods are connected from the rear axles to the lower end of the cradles. The lower torque rods also transmit starting and braking forces from the rear axles to the tractor frame.

Each spring cradle is mounted on a saddle bracket. A trunnion tube is positioned in the center of the saddle bracket, and a roll pin through these components prevents the trunnion tube from rotating in the saddle bracket (Figure 7-44). A nylon bushing is located in the center of the spring saddle, and this bushing is mounted on the trunnion tube. This bushing allows the spring saddle to rotate on the trunnion tube. Thrust washers on both sides of the spring saddle provide the correct saddle end-play. A snap ring retains the saddle on the trunnion tube, and two grease seals prevent lubricant leaks between the spring cradle and the saddle bracket. These seals also keep contaminants out of the cradle bushing area. The spring cradle is lubricated through a grease fitting in the lower side of the cradle.

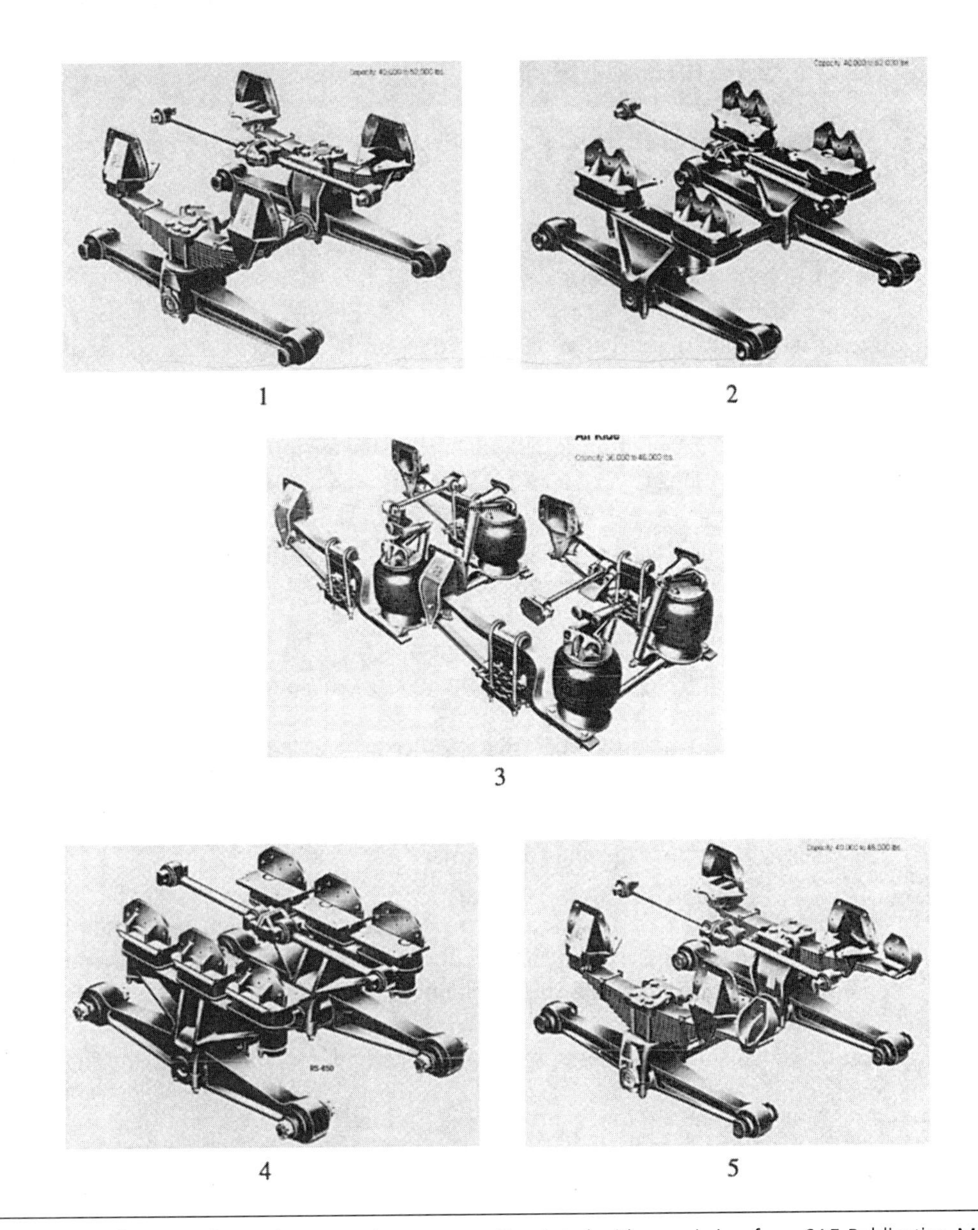

Figure 7-41 Various types of rear tandem axle suspension systems. (Reprinted with permission from SAE Publication Motor Truck Engineering Handbook © 1994 Society of Automotive Engineers, Inc.)

Model	Type	Capacity Ratings (lb)	Use
1) RT Series	Spring Suspensions Alum. Beam Opt. For 34,000-38,000 lb units	40,000-52,000	On/Off-highway
2) RS	Rubber Load Cushions Alum. Beam Opt. For 34,000-38,000 lb units	40,000-52,000	On/Off-highway Transit mix, Refuse, Logging
3) HA	Air Springs	36,000-46,000	Highway use, Tractor-trailer
4) HD RS-650	Solid Mount	52,000-240,000	Logging, Heavy hauling, Quarry
5) RTE-RTU	Extended Leaf Alum. Beam Opt. For 34,000-38,000 lb units	40,000-52,000	Tankers, bulk

Figure 7-42 Rear tandem axle suspension system applications and weight capacities. (Reprinted with permission from SAE Publication Motor Truck Engineering Handbook © 1994 Society of Automotive Engineers, Inc.)

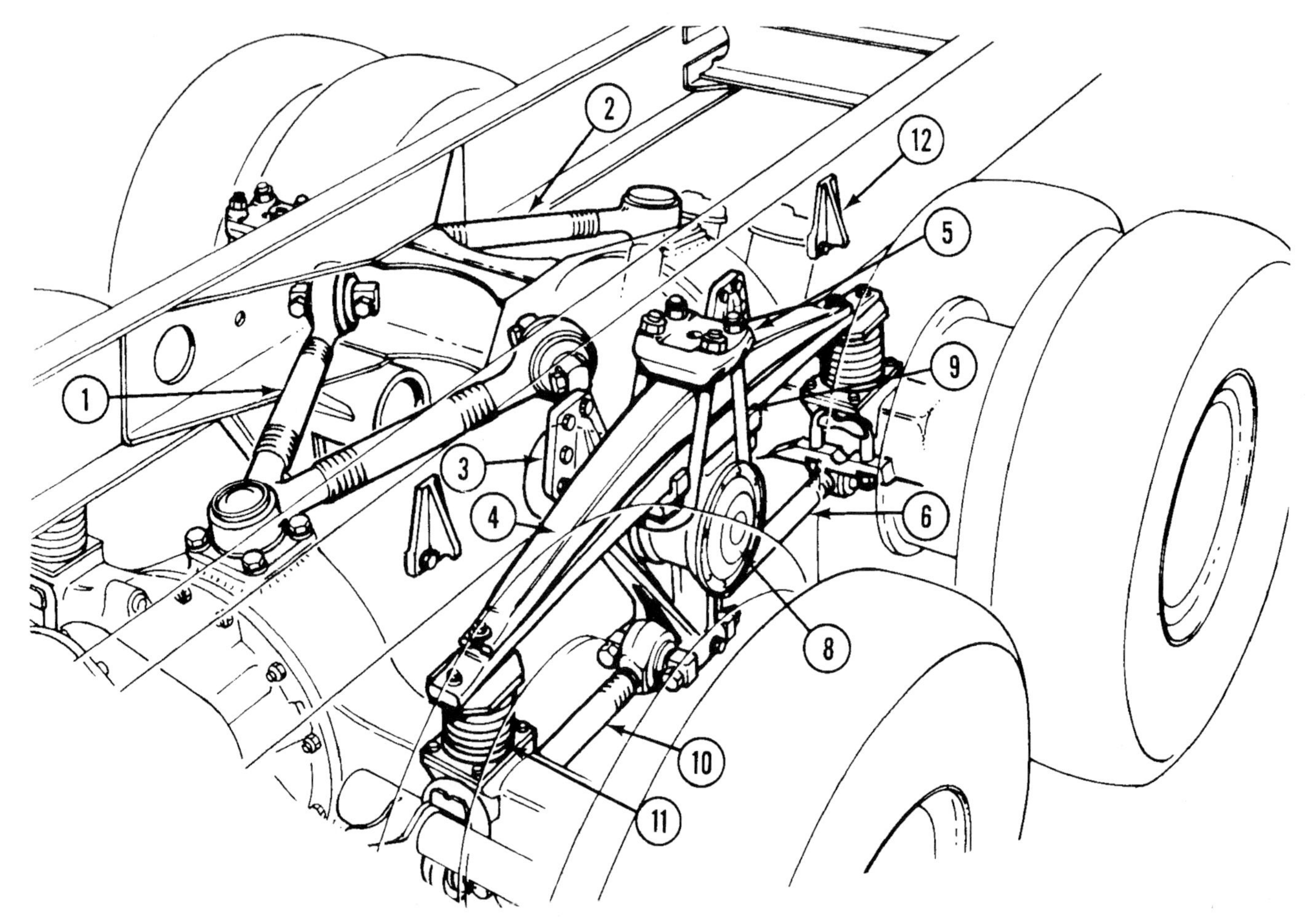

1. Upper V-Torque Rod (front), 2. Upper V-Torque Rod (rear), 3. Saddle Bracket, 4. Spring Assembly, 5. Spring Plate, 6. Lower Torque Rod (rear), 7. Torque Rod Bracket (not shown), 8. Cradle, 9. U-Bolt, 10. Lower V-Torque Rod (front), 11. Rubber Spring, 12. Axle Stop

Figure 7-43 Rear tandem axle suspension system with inverted springs. (Courtesy of Volvo Trucks North America, Inc.)

See Chapter 7 in the Shop Manual for rear suspension with inverted springs diagnosis and service.

Some leaf spring suspension systems have inverted springs with an upward hump where the springs are mounted above the axle. This type of spring may be called a camelback spring. This type of suspension system may be lightweight or heavy weight, depending on the application. A lightweight suspension with camelback springs is illustrated in Figure 7-45.

Tandem Rear Axle Torsion Bar Suspension System

Some Kenworth trucks are equipped with a rear tandem axle torsion bar suspension system (Figure 7-46). A longitudinal torsion bar on each side of the rear suspension supports vehicle weight. The torsion bar suspension is lighter than a tandem axle leaf spring suspension system. The vehicle weight is supported by the twisting of the torsion bars. Each torsion bar is attached to the frame with four brackets containing bushings and sleeves. Seals in these brackets keep contaminants out of the bushings. The ends on the torsion bars are locked in position so these bars cannot rotate.

On the forward rear axle shackles are connected from the axle to the torsion bars on the outboard side of the frame. On the rearmost axle similar shackles are connected from the axle to the torsion bars on the inboard side of the frame. The shackles are locked to the torsion bars with pin locks. Because the outer ends of the torsion bars are locked in place, upward or downward axle and shackle movement twists the torsion bar, and this twisting torsion bar action supports the ve-

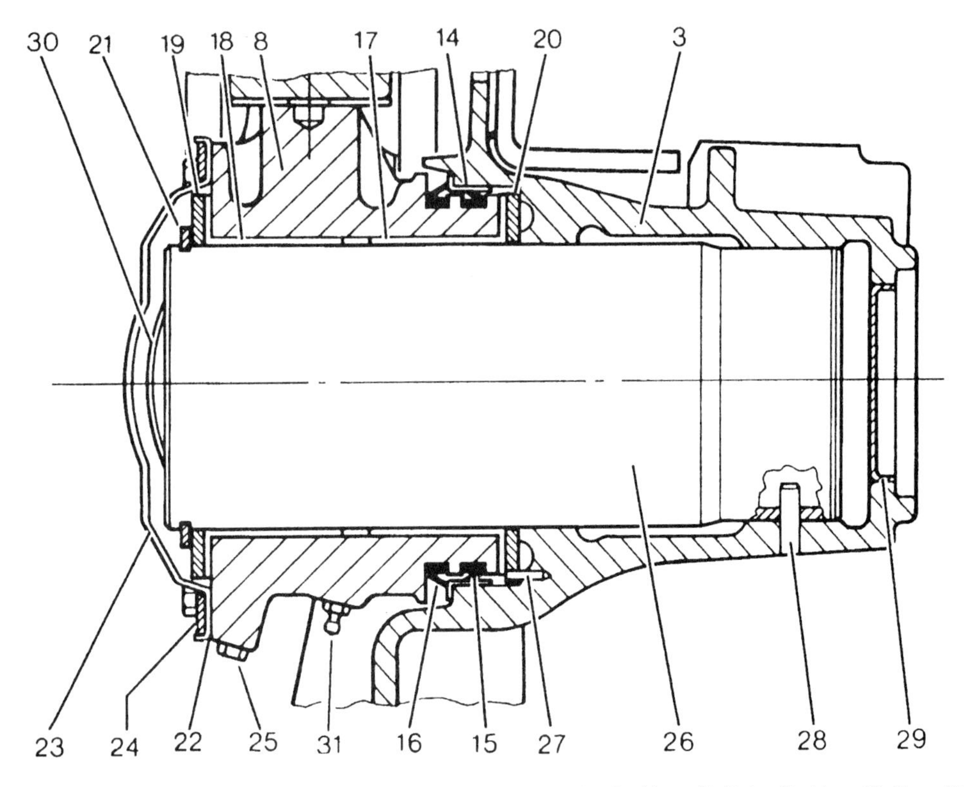

3. Saddle Bracket, 8. Cradle, 14. Wear Ring, 15. Grease Seal, 16. Grease Seal, 17. Nylon Bushing, 18. Nylon Bushing, 19. Outer Thrust Washer, 20. Inner Thrust Washer, 22. Gasket, 23. Hub Cap, 24. Retaining Ring, 25. Relief Valve, 26. Trunnion Tube, 27. Roll Pin, 28. Roll Pin, 29. Plug, 30. Plug, 31. Grease Zerk

Figure 7-44 Spring cradle and saddle bracket assembly. (Courtesy of Volvo Trucks North America, Inc.)

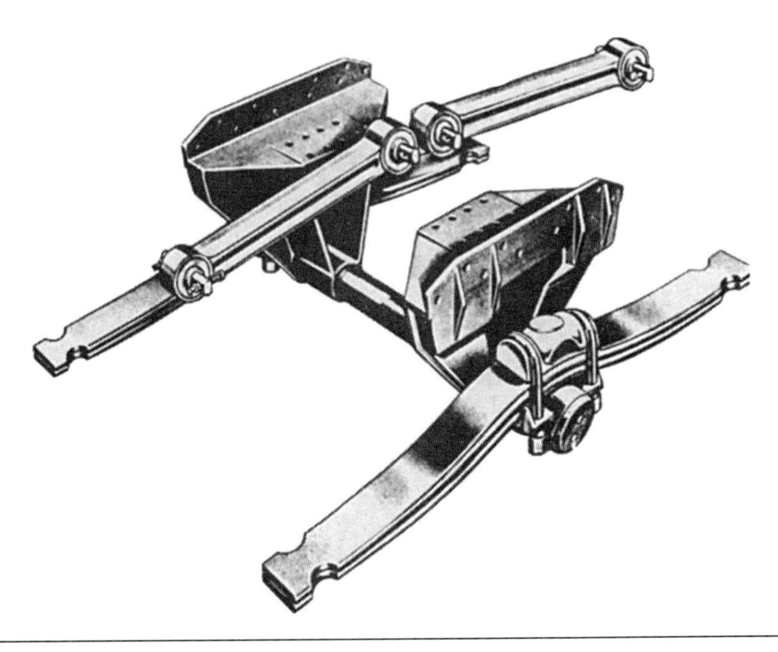

Figure 7-45 Lightweight suspension system with camelback springs. (Reprinted with permission from SAE Publication Motor Truck Engineering Handbook © 1994 Society of Automotive Engineers, Inc.)

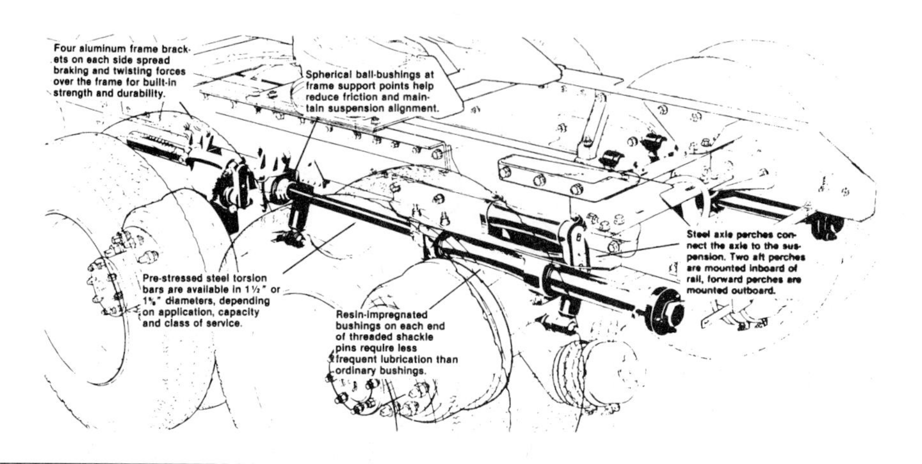

Figure 7-46 Tandem rear axle torsion bar suspension system. (Reprinted with permission from SAE Publication Motor Truck Engineering Handbook © 1994 Society of Automotive Engineers, Inc.)

hicle weight. On the front end of each torsion bar the hex-shaped end of the torsion bar is mounted in a hex opening in an adjustable coupling that is bolted to the front frame bracket. When the adjustable coupling mounting bolts are loosened, this coupling may be rotated to adjust torsion bar tension and suspension height.

CAUTION: Never loosen the adjustable coupling mounting bolts or any other suspension components until the vehicle frame has been lifted with a floor jack and supported on safety stands on both sides so all the vehicle weight is removed from the suspension system. Failure to observe this precaution may result in personal injury.

Trailer Suspension Systems

Leaf Spring Trailer Suspension Systems

Some tandem trailer suspension systems have inverted leaf springs. A trunnion tube is mounted between the springs on each side of the trailer. Each end of the trunnion tube is connected to the leaf spring with U-bolts, and a trunnion seat and cap. These U-bolts are installed over the leaf spring and through the trunnion seat and cap (Figure 7-47 and Figure 7-48). A spring saddle is positioned between the upper side of the spring and the U-bolts. Four trunnion clamps connect the trunnion tube to the trailer frame. The outer ends of the spring are attached to the axles with U-bolts, end caps, axle seats, spring end pads, and a rubber pad. Square or round axles may be used in this type of suspension. In some leaf spring suspensions the leaf spring is mounted above the trunnion tube, and the axles are positioned below the leaf springs (Figure 7-49). Other leaf spring trailer suspensions have the leaf springs mounted below the trunnion tube and the axles positioned above the ends of the springs (Figure 7-50).

Some tandem axle trailer suspension systems such as the Reyco model 21B have four leaf springs with adjustable, cast torque rods (Figure 7-51). These suspension systems are available with composite or steel leaf springs and various spring capacities from 18,000 lb to 26,000 lb (7,257 to 11,793 kg).

Spring Beam Trailer Suspension Systems

Some trailer suspension systems have equalizing beams mounted on each end of the trunnion tube. U-bolts extend over the trunnion tube cap on each side of the equalizing beam. Rubber bushings are positioned between the trunnion tube and the tube cap and equalizing beam. The U-bolts

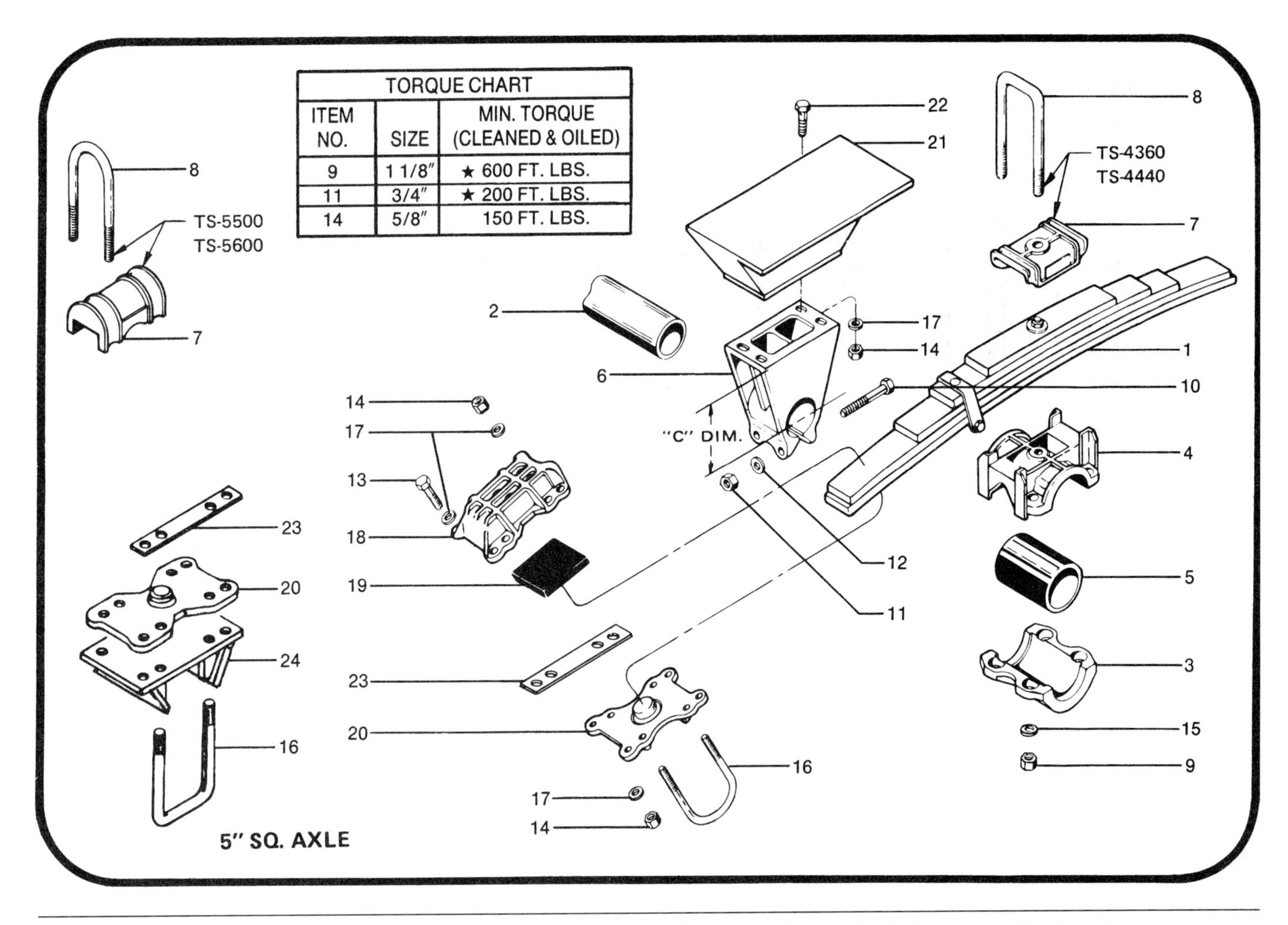

TORQUE CHART		
ITEM NO.	SIZE	MIN. TORQUE (CLEANED & OILED)
9	1 1/8"	★ 600 FT. LBS.
11	3/4"	★ 200 FT. LBS.
14	5/8"	150 FT. LBS.

Figure 7-47 Trailer leaf spring tandem suspension system components. (Courtesy of Neway Anchorlock International, Inc.)

extend through the spring saddle on the underside of the spring. Trunnion clamps attach the trunnion tube to the trailer frame. The two-leaf spring is mounted under the equalizing beam (Figure 7-52). Spring clips are bolted near the outer ends of the equalizing beams, and a bolt extends through these spring clips under the leaf spring. A spacer is mounted between the lower ends of the spring clips. The outer ends of the spring are attached to the axles with U-bolts, spring end caps, spring pads, and a rubber pad. The outer ends of the equalizing beams fit into a recess in the spring end caps (Figure 7-53 and Figure 7-54).

Tri-Axle Trailer Suspension Systems

In some tri-axle trailer suspension systems the leaf springs are attached to the center axle with U-bolts, a spring seat, and spring clamp (Figure 7-55). The outer ends or the leaf springs are mounted in the equalizing beams. The ends of the springs are bent downward, and this portion of each spring is positioned over a bolt and spacer in the end of the equalizing beam. Radius rods are connected from clamps on the center axle to the rear trunnion tube. Front and rear trunnion tubes are mounted in the equalizing beams with trunnion caps and bolts. Trunnion clamps attach the front and rear trunnion tubes to the trailer chassis. The forward and rearmost axles are mounted in the outer ends of the equalizing beams with axle caps, adapters, rubber wrappers, rubber pads, and bolts (Figure 7-56 and Figure 7-57).

See Chapter 7 in the Shop Manual for general suspension system and ride diagnosis.

Item No.	TS-4360 Part No.	TS-4440 Part No.	TS-5500 Part No.	TS-5600 Part No.	Description	Qty.
1	915 57 271	915 57 272	915 57 273	915 57 277	Spring Assembly	2
2	910 38 003				Trunnion Tube 4″ O.D. × 1/2″ Wall × 48″	1
		910 38 004	910 38 004	910 38 004	Trunnion Tube 4″ O.D. × 1/2″ Wall × 48″ H.T.	1
3	910 10 047	910 10 047	910 01 047	910 01 047	Trunnion Rubber Cap Casting	2
4	910 01 051	910 01 051	910 01 051	910 01 051	Trunnion Spring Seat Casting	2
5	910 08 022	910 08 022	910 08 022	910 08 022	Trunnion Rubber Bushing	2
6	910 01 054	910 01 054	910 01 054	910 01 054	Trunnion Clamp Casting 2 1/2″ High-1	2
	910 01 055	910 01 055	910 01 055	910 01 055	Trunnion Clamp Casting 4 1/2″ High-2	2
	910 01 056	910 01 056	910 01 056	910 10 056	Trunnion Clamp Casting 6 1/2″ High-3	2
	910 01 058	910 01 058	910 01 058	910 01 058	Trunnion Clamp Casting 8 1/2″ High-4	2
7	910 10 021	910 10 021	910 01 022	910 01 022	Spring Saddle Casting	2
8	900 41 690	900 41 694	900 41 474	900 41 478	U-Bolt 1 1/8-7	4
9	934 00 506	934 00 506	934 00 506	934 00 506	Lock Nut 1 1/8-7	8
*10	930 03 621	930 03 621	930 03 621	930 03 621	Cap Screw 3/4-10 × 4 1/2″	4
11	934 00 494	934 00 494	934 00 494	934 00 494	Lock Nut 3/4-10	4
12	936 00 156	936 00 156	936 00 156	936 00 156	Washer 3/4″	4
13	930 03 353	930 03 353	930 03 353	930 03 353	Cap Screw 5/8-11 × 2 1/4″, 5″ & 5 3/4″ Rd. Axles	16
	930 03 365	930 03 365	930 03 365	930 03 365	Cap Screw 5/8-11 × 2 3/4″, 5″ Sq. Axle	16
14	934 00 490	934 00 490	934 00 490	934 00 490	Lock Nut 5/8-11	40
15	936 00 174	936 00 174	936 00 174	936 00 174	Washer 1 1/8″	8
16	900 41 067	900 41 067	900 41 067	900 41 067	U-Bolt 5/8-11, 5″ Rd. Axle	8
	900 42 064	900 42 064	900 42 064	900 42 064	U-Bolt 5/8-11, 5 3/4″ Rd. Axle	8
	900 42 544	900 42 544	900 42 544	900 42 544	U-Bolt 5/8-11, 5″ Sq. Axle	8
17	939 00 025	939 00 025	939 00 025	930 00 025	Washer H.T. 5/8″	56
18	910 10 043	910 10 043	910 10 043	910 10 043	Spring End Cap	4
*19	910 28 152	910 28 152	910 28 152	910 28 152	Spring End Rubber Pad	4
20	910 01 044	910 01 044	910 01 044	910 01 044	Spring End Pad 5″ Rd. Axle	4
	910 01 047	910 01 047	910 01 047	910 01 047	Spring End Pad 5 3/4″ Rd. Axle	4
	910 01 049	910 01 049	910 01 049	910 01 049	Spring End Pad 5″ Sq. Axle	4
21	910 18 008	910 18 008	910 18 008	910 18 008	Frame Bracket 9 1/4″ High Std. (Optional Specify)	2
22	930 03 345	930 03 345	930 03 345	930 03 345	Cap Screw 5/8-11 × 2″	8
23	910 31 570	910 31 570	910 31 570	910 31 570	Spacer	16
24	915 01 063	915 01 063	915 01 063	915 01 063	Axle Seat 5″ Sq. Axle	4
25	900 44 236	900 44 236	900 44 236	900 44 236	Torque Decal (not shown)	1

Figure 7-48 Trailer leaf spring tandem suspension system parts list. (Courtesy of Neway Anchorlok International, Inc.)

Figure 7-49 Trailer leaf spring tandem suspension with the axles mounted below the springs. (Courtesy of Neway Anchorlok International, Inc.)

Figure 7-50 Trailer leaf spring tandem suspension with the axles mounted above the springs. (Courtesy of Neway Anchorlok International, Inc.)

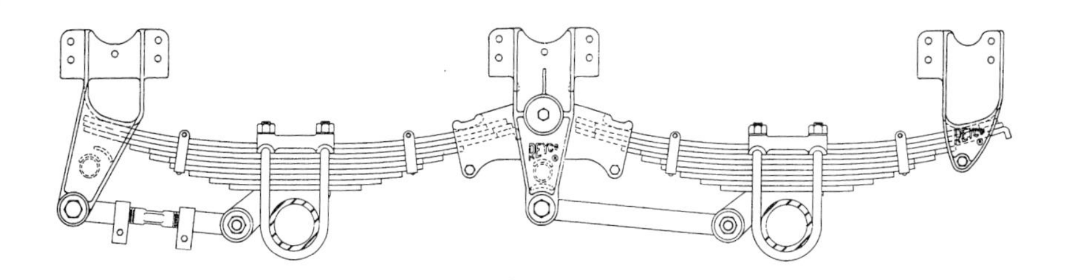

Figure 7-51 Tandem axle leaf spring trailer suspension with cast, adjustable torque rods. (Courtesy of Reyco Industries Inc.)

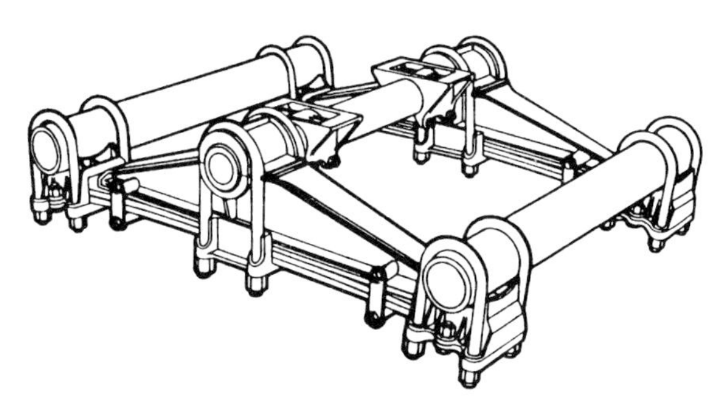

Figure 7-52 Trailer tandem spring beam suspension. (Courtesy of Neway Anchorlok International, Inc.)

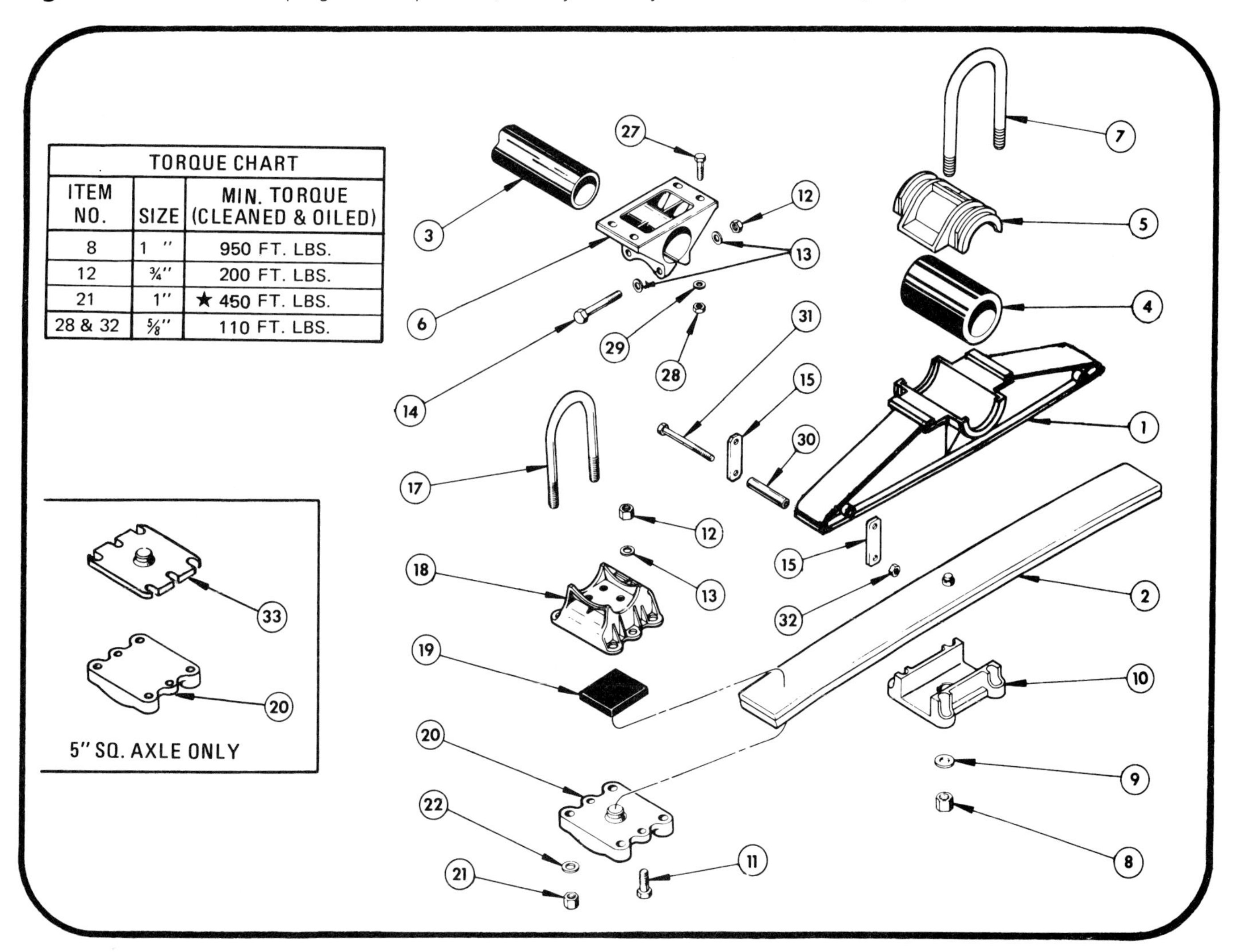

TORQUE CHART

ITEM NO.	SIZE	MIN. TORQUE (CLEANED & OILED)
8	1 "	950 FT. LBS.
12	3/4"	200 FT. LBS.
21	1"	★ 450 FT. LBS.
28 & 32	5/8"	110 FT. LBS.

Figure 7-53 Trailer tandem spring beam suspension components. (Courtesy of Neway Anchorlok International, Inc.)

Item No.	Description	Part No.	Qty.
*1	Equalizing Beam	915 15 108	2
2	Spring—2 Leaf (Models SB-360-B & SB-500-B)	915 57 100	2
	Spring—2 Leaf (Model SB-600-B)	915 57 136	2
3	Trunnion Tube—40 O.D. 3 1/2" Wall × 48" Long (Model SB-360-B)	910 38 003	1
	Trunnion Tube—40 O.D. 3 1/2" Wall × 48" Long, Alloy H.T. (Models SB-500-B & SB-600-B)	910 38 004	1
4	Trunnion Rubber Bushing	910 08 088	2
5	Trunnion Rubber Cap Casting	910 10 089	2
6	Trunnion Clamp Casting	910 01 054	2
7	1 1/8"—7 Spring U-Bolt	*900 41 478	4
8	1 1/8"—7 Lock Nut	934 00 506	8
9	1 1/8" Washer	936 00 174	8
10	Spring Saddle Casting	910 01 103	2
11	3/4"—1" × 2 1/4" Cap Screw	930 03 567	8
	3/4"—1" × 2 1/2" Cap Screw (5" Sq. Axle)	930 03 573	8
12	3/4"—1" Lock Nut	934 00 492	12
13	3/4" Washer	936 00 156	16
*14	3/4"—1" 3 4 1/2" Cap Screw	930 03 621	4
15	Spring Clip	*910 23 056	8
17	10—8 U-Bolt 5" Rd. Axle (Model SB-360-B)	900 41 323	8
	10—8 U-Bolt 5" Rd. Axle (Models SB-500-B & SB-600-B)	900 41 324	8
	10—8 U-Bolt 5 3/4" & 6" Rd. Axle (Model SB-600-B)	900 41 350	8
	10—8 U-Bolt (5" Sq. Axle)	900 41 971	8
18	Spring End Cap Casting—5" Rd. Axle	910 10 041	4
	Spring End Cap Casting—5 3/4" & 6" Rd. Axle (Model SB-600-B)	910 10 130	4
	Spring End Cap Casting—5" Sq. Axle	910 01 188	4
19	Spring End Rubber Pad—(Models SB-360-B & SB-500-B)	910 28 089	4
	Spring End Rubber Pad—(Model SB-600-B)	910 28 013	4
20	Spring End Pad Casting—5" Rd. Axle	910 01 038	4
	Spring End Pad Casting—5 3/4" & 6" Rd. Axle (Model SB-600-B)	910 01 050	4
	Spring End Pad Casting—5" Sq. Axle	910 01 182	4
21	10—8 Lock Nut	*934 00 500	16
22	10 Washer	936 00 168	16
27	5/8"—11 3 20 Cap Screw	930 03 345	8
28	5/8"—11 Lock Nut	934 00 488	8
29	5/8" Washer	*939 00 025	*16
30	Spring Clip Spacer	910 36 008	4
31	5/8"—11 3 6 3/4" Cap Screw	930 03 459	8
32	5/8"—11 Hex Nut	934 00 144	8
33	Adjustable Spring End Pad-50 SQ. AXLE	910 01 181	4

NOTE: When ordering parts, be sure to specify item No., Part No., Serial No. and Model No.

Figure 7-54 Trailer tandem spring beam suspension parts list. (Courtesy of Neway Anchorlok International, Inc.)

Coach Torsilastic-Type Suspension Systems

Some coaches have front and rear torsilastic-type suspension systems. In these suspension systems a torsilastic unit is mounted at the end of each axle. On a front suspension system the outer housing on the torsilastic unit is attached to the front axle. The torsion bar in the center of the unit is connected through an adjustable linkage to a plate that is bolted to the chassis (Figure 7-58). Upward wheel movement rotates the outer housing on the torsilastic unit, and this action twists the torsion bar. The torsilastic unit supports the vehicle weight. On a rear drive axle suspension system the outer housing on the torsilastic unit is attached to the rear axle housing, and both ends of the torsion bar are attached through arms to the chassis (Figure 7-59). Ride height adjusting nuts are mounted on the lower side of the torsion bar arms.

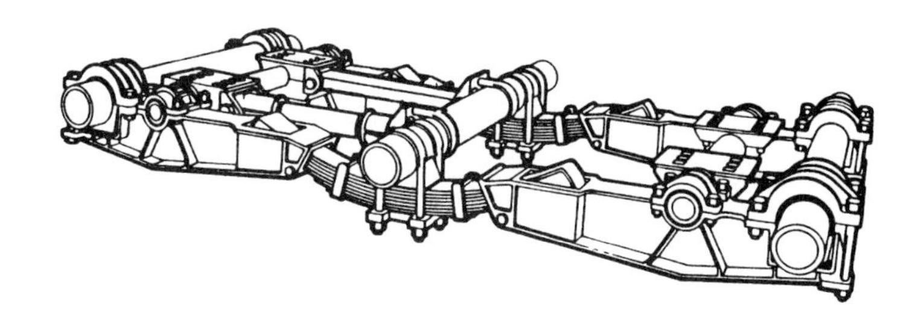

Figure 7-55 Tri-axle trailer suspension with equalizing beams and leaf springs. (Courtesy of Neway Anchorlock International, Inc.)

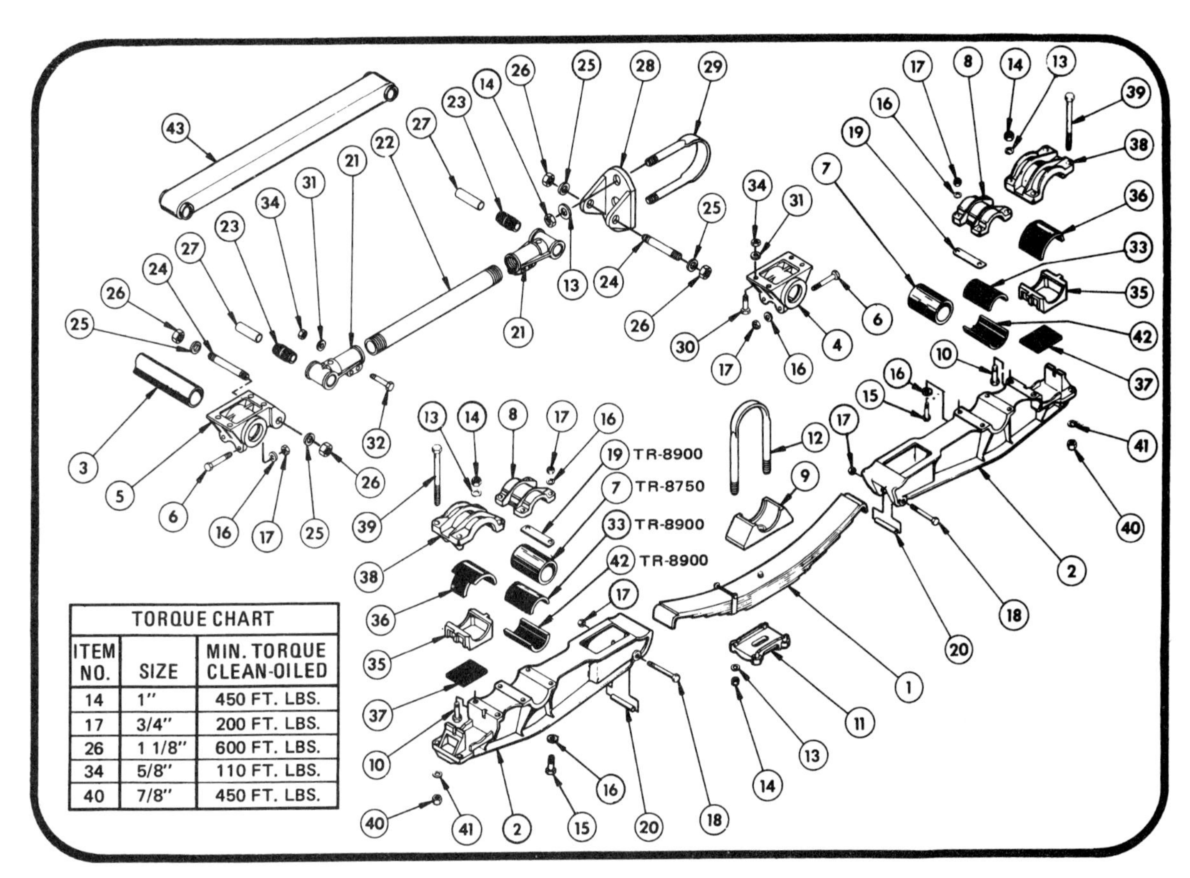

TORQUE CHART		
ITEM NO.	SIZE	MIN. TORQUE CLEAN-OILED
14	1"	450 FT. LBS.
17	3/4"	200 FT. LBS.
26	1 1/8"	600 FT. LBS.
34	5/8"	110 FT. LBS.
40	7/8"	450 FT. LBS.

ITEM NO.	DESCRIPTION	PART NO.	QTY.	MODELS
30	5/8"—11 X 2" Cap Screw	930 03 345	16	ALL
31	5/8" Flat Washer	939 00 025	20	ALL
32	5/8"—11 X 3 1/4" Cap Screw	930 03 375	4	ALL
33	Trunnion Bushing—Upper (Rubber)	910 08 005	4	TR-8900
34	5/8"—11 Crown Lock Nut	934 00 488	20	ALL
35	Axle Adapter for 5" Round Axles	910 01 089	4	ALL
	Axle Adapter for 5 3/4" Round Axles	910 01 120	4	ALL
36	Rubber Wrapper for 5" Round Axles	910 28 051	4	ALL
	Rubber Wrapper for 5 3/4" Round Axles	910 28 053	4	ALL
37	Rubber Pad Lower for All Axles	910 28 089	4	ALL
38	Axle Cap for 5" Round Axles	910 10 060	4	ALL
	Axle Cap for 5 3/4" Round Axles	910 10 059	4	ALL
39	7/8"—9 X 8 3/4" Cap Screw	930 03 941	8	ALL
40	7/8"—9 Lock Nut	934 00 498	8	ALL
41	7/8" Flat Washer	936 00 162	8	ALL
42	Trunnion Bushing—Lower (Fiber)	910 08 004	4	TR-8900
43	Radius Rod — Fixed (Includes 2 of item 23)	915 44 147	1	ALL

Figure 7-56 Components in tri-axle trailer suspension with equalizing beams and leaf springs. (Courtesy of Neway Anchorlok International, Inc.)

Item No.	Description	Part No.	Qty.	Models
1	Spring Assembly	915 57 172	2	ALL
2	Equalizing Beam for 5" Round Axles	910 15 015	4	ALL
	Equalizing Beam for 5 3/4" Round Axles	915 15 011	4	ALL
3	Trunnion Tube × 1/2" O.D. × 1/2" W. × 44 1/4" Long H.T.	910 38 290	2	ALL
4	Trunnion Clamp Rear	915 01 011	2	ALL
5	Trunnion Clamp-Front (Left Hand)	915 18 040	1	ALL
	Trunnion Clamp-Front (Right Hand)	915 18 041	1	ALL
6	3/4"—1" × 5" Cap Screw	930 03 633	8	ALL
7	Trunnion Rubber Bushing	900 08 011	4	TR-8750
8	Trunnion Rubber Cap	910 10 157	4	ALL
9	Axle Spring Seat for 5" Round Axles	910 01 075	2	ALL
	Axle Spring Seat for 5 3/4" Round Axles	910 01 073	2	ALL
10	1"—8 × 3 1/2" Cap Screw for 5" Round Axles	930 04 239	8	ALL
	1"—8 × 4" Cap Screw for 5" Round Axles	930 04 251	8	ALL
11	Spring Clamp Plate for 5" Round Axles	910 10 108	2	ALL
	Spring Clamp Plate for 5 3/4" Round Axles	910 10 093	2	ALL
12	1"—8 U-Bolt for 5" Round Axles	900 41 823	4	ALL
	1"—8 U-Bolt for 5 3/4" Round Axles	900 41 857	4	ALL
13	1" Flat Washer	936 00 168	20	ALL
14	1"—8 Lock Nut	934 00 500	20	ALL
15	3/4"—1" × 2 3/4" Cap Screw	930 03 579	16	ALL
16	3/4" Flat Washer	936 00 156	32	ALL
17	3/4"—1" Lock Nut	934 00 492	28	ALL
18	3/4"—1" 3 6" Cap Screw	930 03 657	4	ALL
19	Trunnion Cap Spacer	910 36 001	8	TR-8900
20	Pipe Spacer	910 36 078	4	ALL
*21	Radius Rod End R.H. (Includes items 23)	915 44 138	1	ALL
	Radius Rod End L.H. (Includes items 23)	915 44 137	1	ALL
22	Radius Rod Screw	910 38 475	1	ALL
23	Rubber Bushing	900 08 002	4	ALL
24	1 1/8"—7 × 9 1/4" Rod Bolt	932 00 945	4	ALL
25	1 1/8" Flat Washer	936 00 174	8	ALL
26	1 1/8"—7 Lock Nut	934 00 504	4	ALL
27	1 1/8" Delrin Liner	900 38 083	4	ALL
28	Radius Rod Axle Bracket	910 18 039	2	ALL
29	1"—8 U-Bolt for 5" Round Axles	900 41 807	2	ALL
	1"—8 U-Bolt for 5 3/4" Round Axles	900 41 841	2	ALL
30	5/8"—11 × 2" Cap Screw	930 03 345	16	ALL
31	5/8" Flat Washer	939 00 025	20	ALL
32	5/8"—11 × 3 1/4" Cap Screw	930 03 375	4	ALL
33	Trunnion Bushing-Upper (Rubber)	910 08 005	4	TR-8900
34	5/8"—11 Crown Lock Nut	934 00 488	20	ALL
35	Axle Adapter for 5" Round Axles	910 01 089	4	ALL
	Axle Adapter for 5 3/4" Round Axles	910 01 120	4	ALL
36	Rubber Wrapper for 5" Round Axles	910 28 051	4	ALL
	Rubber Wrapper for 5 3/4" Round Axles	910 28 053	4	ALL
37	Rubber Pad Lower for All Axles	910 28 089	4	ALL
38	Axle Cap for 5" Round Axles	910 10 060	4	ALL
	Axle Cap for 5 3/4" Round Axles	910 10 059	4	ALL
39	7/8"—9 × 8 3/4" Cap Screw	930 03 941	8	ALL
40	7/8"—9 Lock Nut	934 00 498	8	ALL
41	7/8" Flat Washer	936 00 162	8	ALL
42	Trunnion Bushing—Lower (Fiber)	910 08 004	4	TR-8900
43	Radius Rod—Fixed (Includes 2 of item 23)	915 44 147	1	ALL

Figure 7-57 Parts list for tri-axle trailer suspension with equalizing beams and leaf springs. (Courtesy of Neway Anchorlok International, Inc.)

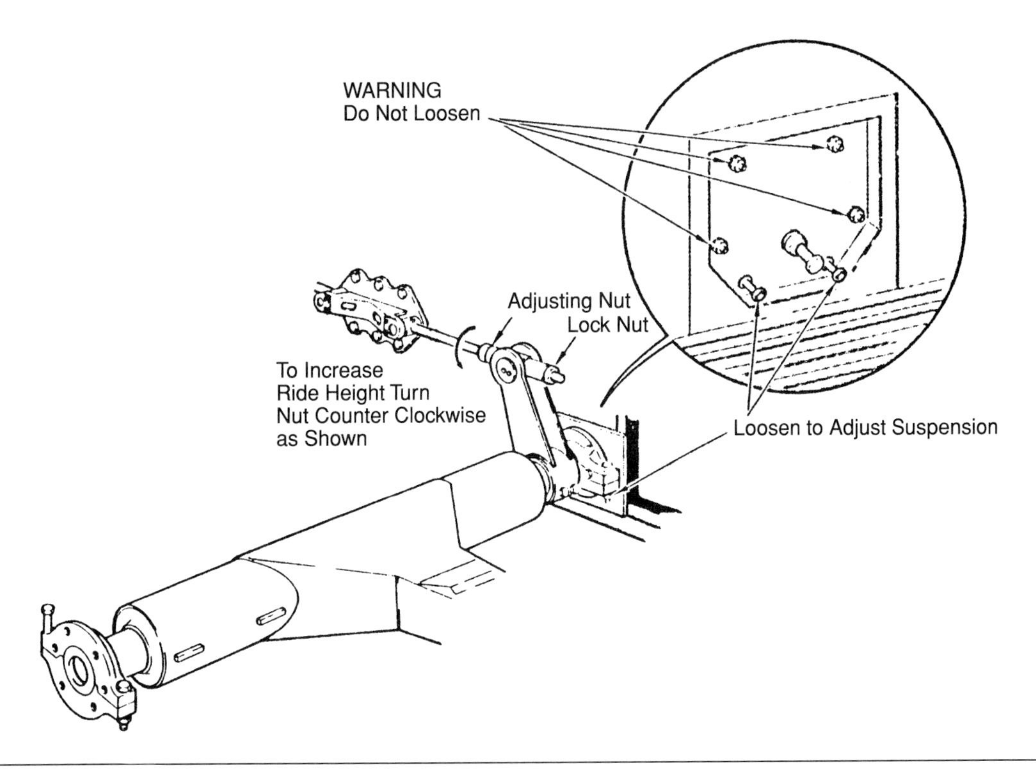

Figure 7-58 Coach torsilastic-type front suspension system (Reprinted with Permission from SAE Publication History of Steering Trucks © 1993 Society of Automotive Engineers, Inc.)

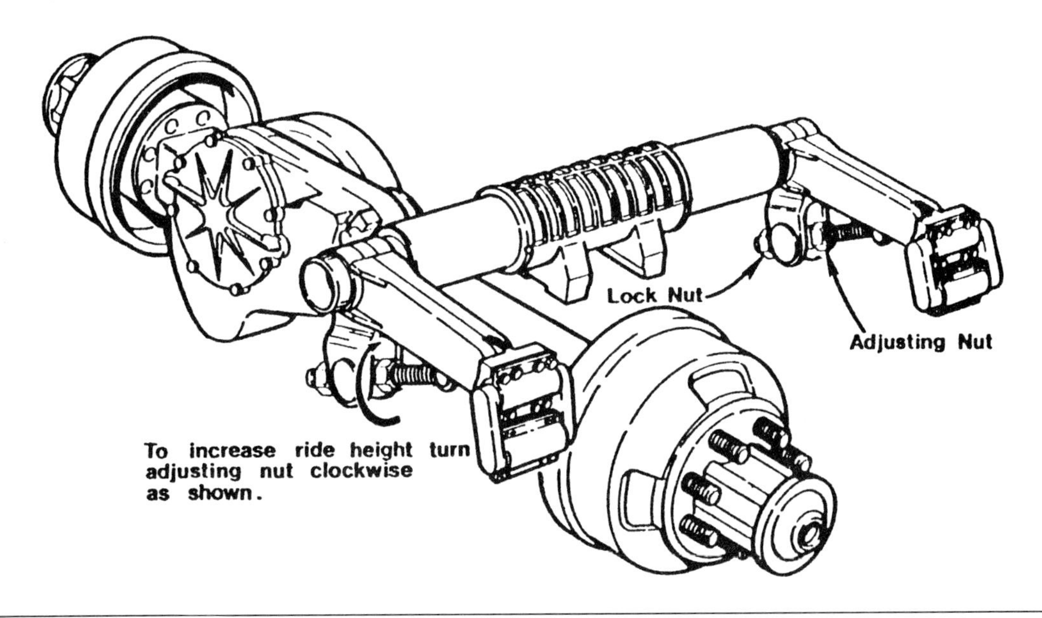

Figure 7-59 Coach torsilastic-type rear suspension system. (Courtesy of Motor Coach Industries.)

A BIT OF HISTORY

Leaf spring suspension systems have been used since the early days of the trucking industry. In Figure 7-60 the front leaf springs, rear leaf springs, auxiliary springs, and torque rods can be plainly seen on the model E series Sterling truck with a concrete mixer body manufactured from 1924 to 1927.

Figure 7-60 E series Sterling truck manufactured from 1924 to 1927. (Reprinted with permission from SAE History of Sterling Trucks SP941 © 1993 Society of Automotive Engineers.)

Summary

- ❑ A front axle I-beam has an identification plate attached near the center on the front of the axle. This plate provides the axle model and specification numbers, customer number, axle assembly date, assembly plant, and serial number.
- ❑ Knuckle pins or kingpins retain the steering knuckles to the axle I-beam on a front axle.
- ❑ On a front suspension system the hub and bearings are mounted on the steering knuckles, and the steering arms and Ackerman arms are attached to these knuckles.
- ❑ Some front suspension systems are available with unitized front wheel bearing assemblies.
- ❑ Front springs may be multileaf, parabolic, or taperleaf.
- ❑ Truck springs may be constant rate or vari-rate.
- ❑ Many front spring eyes are attached to a stationary bracket at the front of the spring, and the rear of the spring is attached to the truck frame through a swinging shackle.
- ❑ Front suspension systems may have one or two shock absorbers connected from each side of the axle to the chassis.
- ❑ Shock absorbers help control spring action, reduce body sway, and improve tire life, directional stability and driver comfort.
- ❑ Shock absorber ratio is the relationship between the shock absorber control on the extension cycle and the control on the compression cycle.
- ❑ Some single axle rear suspension systems have auxiliary springs mounted on top of the conventional springs.
- ❑ Some rear tandem axle suspension systems have four multileaf springs and four torque rods. Between the front and rear springs on each side of the tractor the springs ride on a equalizer that pivots on a sleeve in the equalizer bracket.

Terms to Know

Ackerman arm
Auxiliary spring
Equalizer
Equalizing beams
Gas-filled shock absorbers
Heavy duty shock absorbers
I-beam axles
Inverted springs
Jounce travel
Kingpins
Kingpin inclination
Kingpin offset
Multileaf shackle-type spring
Parabolic spring
Parabolic taperleaf spring
Rebound travel
Rubber cushions
Shock absorbers
Shock absorber compression stroke

- ❑ Some rear tandem axle suspension systems have equalizing beams on each side of the suspension. Bushings in the equalizing beams are attached to brackets on the rear axle housings, and a cross tube is mounted between the two equalizing beams. Multileaf springs are mounted on saddles above the equalizing beams.
- ❑ Some rear tandem axle suspension systems have equalizing beams and four or six rubber cushions mounted between the equalizing beam saddles and the tractor frame in place of multileaf springs.
- ❑ Other rear tandem axle suspension systems have inverted parabolic, tapered leaf springs mounted on cradles that pivot in a saddle bracket. Rubber springs are mounted between the ends of the springs and saddles on the rear axle housings. Four upper torque rods are mounted between the rear axle housings and the tractor frame, and four lower torque rods are positioned between the rear axle housings and the saddle brackets.

Terms to Know (Continued)

Shock absorber extension stroke
Shock absorber ratios
Short-and-long-arm (SLA) front suspension system
Spring cradle
Sprung weight
Steering arm
Steering knuckles
Swinging shackle
Tapered caster shim
Torque rod
Tubular steel front axles
Unitized hub assemblies
Unsprung weight
Vari-rate spring

Review Questions

Short Answer Essays

1. Explain how the steering knuckles are attached to the ends of the I-beam axle in a front suspension system.
2. Explain the advantages of unitized hubs and integrated knuckles on an I-beam front suspension system.
3. Describe the purpose of a swinging shackle on a front leaf spring.
4. Describe the purposes of shock absorbers.
5. Explain the purpose of the nitrogen gas charge in a shock absorber.
6. Describe shock absorber action during wheel jounce and rebound.
7. Explain the difference between conventional and heavy duty shock absorbers.
8. Explain shock absorber ratios.
9. Describe the design of a single axle suspension system with conventional and auxiliary leaf springs.
10. Describe the design of a rear tandem axle suspension system with equalizing beams.

Fill-in-the-Blanks

1. On an I-beam front axle the identification plate is located at the ____________________ on the ____________________ side of the axle.
2. A thrust bearing is positioned between the ____________________ side of the axle I-beam and the ____________________.
3. On some steering arms a ____________________ positions the steering arm properly in the ____________________.
4. A tapered caster shim may be positioned between the ____________________ side of the spring and the ____________________.
5. A swinging shackle permits ____________________ and ____________________ of the spring as the front axle moves vertically.
6. Shock absorbers usually have more control on the ____________________ stroke compared with the ____________________ stroke.

7. In a rear tandem axle suspension system with four leaf springs and four torque rods, the torque rods provide ______________________ when accelerating and braking.

8. In a rear tandem axle suspension system with equalizing beams and leaf springs the load is distributed equally between the two rear axles by the ______________________.

9. In a rear tandem axle suspension system with equalizing beams and rubber cushions, the rubber cushions are mounted between the ______________________ and the tractor ______________________.

10. In a rear tandem axle suspension system with inverted leaf springs, U-bolts attach each spring to a ______________________ and ______________________ are attached to the outer ends of the springs.

ASE Style Review Questions

1. While discussing I-beam front axles:
   Technician A says the weight capacity of the front axle is stamped on the axle identification plate.
   Technician B says the customer number is stamped on the axle identification plate.
   Who is correct?
   A. A only **C.** Both A and B
   B. B only **D.** Neither A nor B

2. On an I-beam front suspension system:
   A. The kingpins are held in the I-beam axle with a draw key.
   B. The steering arm in the upper knuckle opening is connected to the tie rod.
   C. The Ackerman arm in the lower knuckle opening is connected to the drag link.
   D. The kingpins contact the machined steel openings in the knuckles.

3. While discussing unitized front wheel hubs:
   Technician A says these hubs require periodic lubrication.
   Technician B says these hubs require an end-play adjustment.
   Who is correct?
   A. A only **C.** Both A and B
   B. B only **D.** Neither A nor B

4. While discussing fiber composite leaf springs:
   Technician A says a spring with fiber composite leaves requires more shock absorber control than a spring with steel leaves.
   Technician B says a spring with fiber composite leaves sags from loading stresses after a period of time.
   Who is correct?
   A. A only **C.** Both A and B
   B. B only **D.** Neither A nor B

5. While considering shock absorber design and operation:
   A. The nitrogen gas charge in shock absorber eliminates oil foaming.
   B. During wheel jounce the wheel and suspension are moving downward.
   C. Heavy duty shock absorbers have a double tube design for more strength.
   D. During the shock absorber compression stroke the upper part of the shock absorber is moving downward.

6. All these statements about shock absorbers are true *except:*
   A. Shock absorbers control spring action and body sway.
   B. Shock absorbers help to maintain tire tread contact on the road surface.
   C. Shock absorbers provide more control on the compression cycle than the extension cycle.
   D. Shock absorbers improve vehicle handling and steering control.

7. While discussing a single axle rear suspension system with a multileaf spring that has a front eye and bushing and a rear slider plus an auxiliary spring:
   Technician A says the auxiliary spring is a constant rate spring with an effective length that remains constant in relation to vehicle load.
   Technician B says as the truck load is increased the effective length of the conventional multileaf spring becomes shorter.
   Who is correct?
   A. A only **C.** Both A and B
   B. B only **D.** Neither A nor B

8. On a rear tandem axle suspension system with equalizing beams and leaf springs:
   A. The leaf springs are mounted on the lower side of the equalizing beams.
   B. The cross tube is connected between the leaf springs.
   C. The outer ends of the equalizing beams are connected to the tractor frame.
   D. The rear end of the leaf spring slides on a support in the rear spring bracket.

9. While discussing a rear tandem axle suspension system with equalizing beams and rubber cushions:
   Technician A says this type of suspension system is lighter than the same type of suspension with leaf springs.
   Technician B says this type of suspension system may have four or six rubber cushions connected between the saddles and the frame.
   Who is correct?
   A. A only
   B. B only
   C. Both A and B
   D. Neither A nor B

10. On a rear tandem axle suspension system with inverted leaf springs:
   A. The spring cradle can rotate on the saddle bracket.
   B. The upper torque rods are mounted parallel to the leaf springs.
   C. The rubber springs are mounted under the center of the leaf springs.
   D. The lower torque rods are connected from the differential housings to the tractor frame.

Air Suspension Systems

Upon completion and review of this chapter, you should be able to:

- ❑ Explain the advantages of an air spring suspension system.
- ❑ Describe the design of a tractor rear axle air suspension system.
- ❑ Explain the purpose of a pressure protection valve in an air suspension system.
- ❑ Describe the operation of a height control valve.
- ❑ Explain the operation of the control valve and relay valves in an air suspension system.
- ❑ Describe the safety features designed into an air suspension system.
- ❑ Describe the advantages of the Volvo optimized air suspension system.
- ❑ Explain the design and operation of an electronically controlled air suspension system.
- ❑ Describe the operation of a lift axle with air suspension and coil spring lift.
- ❑ Explain the operation of a lift axle with air suspension and air lift.
- ❑ Describe the design and operation of a trailer air suspension system.
- ❑ Explain the operation of a slider-type trailer air suspension system with an air pin puller.
- ❑ Describe the operation of an external dock lock mechanism.
- ❑ Explain the operation of a trailer axle with an air suspension system and an air-operated lift mechanism.
- ❑ Describe the design and advantages of a cab air suspension system.
- ❑ Explain the design and advantages of an air suspended seat and related air control valves.

Introduction

Air suspension systems may be used on single or tandem rear axles and trailer suspensions. The air springs in an **air suspension system** take the place of the leaf springs in a conventional suspension system. An air suspension system eliminates the interleaf friction encountered in a leaf-spring suspension system. Compared with a leaf-spring suspension system, an air suspension system minimizes road shock transferred from the suspension to the truck frame, driver, and cargo. An air suspension adjusts automatically to different load conditions, and provides a softer suspension system with light loads and a firmer suspension system with heavier loads. An air suspension system provides a constant frame height under various load conditions. Compared with a leaf-spring suspension system, an air suspension system provides a considerable unsprung weight saving. The main components in an air suspension system are the air springs, support beams, height control valve, air control system, torque rods, and insulators or wear pads. The advantages of an air suspension system may be summarized as follows:

1. The improved ride quality provided by an air suspension system reduces driver fatigue.
2. An air suspension system provides improved tire-to-road contact, and this improves braking capability.
3. An air suspension system provides low frequency and viscous damping of the suspension, and this action reduces wheel hop while braking.
4. The improved roll resistance of an air suspension system provides better vehicle stability.
5. An air suspension system provides pneumatic equalizing between the forward and rearmost rear axles for improved weight distribution between these axles. This helps to reduce overloading of one axle and one set of tires.
6. An air suspension reduces suspension noise and allows improved monitoring of the truck by the driver.

7. Because an air suspension system provides a constant frame height, headlight and mirror aim is constant.
8. An air suspension provides increased tire life.

Tractor Rear Axle Air Suspension System Design

Main Support Beams and Mountings

The **main support beams** support the axles and **air springs.** On some rear axle air suspension systems the main support beam is mounted on top of the rear axle housing. The main support beam is retained on the rear axle housing with U-bolts. A top pad is mounted between the upper side of the main support beam and the U-bolts. The U-bolts extend through a bottom cap positioned under the rear axle housing. A spring seat and a delrin liner are mounted between the main support beam and the top of the rear axle housing. The spring seat is mounted on the rear axle housing, and the delrin liner is positioned between the spring seat and the main support beam (Figure 8-1).

A main support beam may be called a rigid beam.

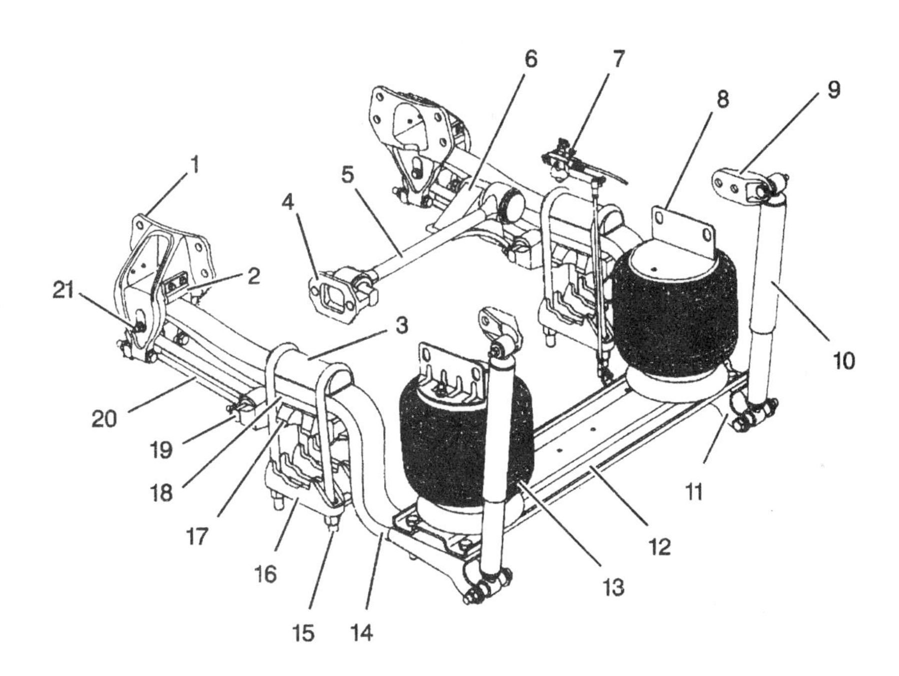

1. SPRING HANGER
2. INSULATOR
3. TOP PAD
4. TRANSVERSE ROD FRAME BRACKET
5. TRANSVERSE ROD
6. TRANSVERSE ROD AXLE BRACKET
7. HEIGHT LEVELING VALVE
8. AIR SPRING FRAME BRACKET
9. SHOCK ABSORBER UPPER MOUNT
10. SHOCK ABSORBER
11. SHOCK ABSORBER LOWER MOUNT
12. CROSS CHANNEL
13. AIR SPRING
14. MAIN SUPPORT BEAM
15. U – BOLT
16. BOTTOM CAP
17. SPRING SEAT
18. DELRIN LINER
19. TORQUE ROD BAR PIN
20. TORQUE ROD
21. REBOUND ROLLER BOLT

Figure 8-1 Rear single axle air suspension system. (Courtesy of General Motors Corporation, Service Technology Group.)

The front of the main support beam is positioned between the rebound roller and the wiper in the **front bracket** (Figure 8-2). A cross channel is bolted to each main support beam near the rear of these beams. This channel prevents lateral beam movement.

Air Springs, Shock Absorbers, and Torque Rods

The lower ends of the air springs are bolted to the **cross channel,** and a frame bracket is attached to the upper side of the air spring. This frame bracket is bolted to the truck frame. The weight supplied to the tractor frame is transferred through the air spring to the main support beams and the rear axles. Road shock is transmitted from rear axles and main support beams through the air spring to the tractor frame. The air springs support the weight supplied to the tractor frame and effectively cushion road shock applied to the axles and main support beams.

Torque rods are connected from each front bracket to the rear axle housing. These torque rods provide rear axle stability when accelerating and braking. Similar torque rods are used on leaf-spring rear suspension systems explained in the previous chapter. A **transverse rod** is connected from the rear axle housing to the frame to prevent lateral rear axle movement. Shock absorbers are connected from the rear of each main support beam to a bracket attached to the truck frame. Shock absorbers are explained in the previous chapter. In a tandem rear axle air suspension system there is a pair of main support beams and air springs for each axle, but the design of these components is similar in single or tandem axle systems (Figure 8-3).

Tractor Rear Axle Air Suspension Systems with Rigid Beams

On some rear air suspension systems **rigid beams** are mounted under the rear axle (Figure 8-4). These rigid beams provide very stable rear axle position and improved load equalization. Each rigid beam is attached to the rear axle housing by a bracket that is welded to the rear axle housing. A bar pin bushing is mounted in an opening in the rigid beam and this bar pin bushing is attached to the axle-to-beam bracket. The bar pin bushing controls rear axle movement during acceleration and braking and reduces stress to the rear axle housing. On many rear axle air suspension systems, U-bolts are used to attach the rigid beams to the drive axle housing.

A super heavy duty transverse beam is connected to the ends of each pair of rigid beams to provide improved roll stability and resistance to lean and to reduce rear axle stress. The mounting plates on the top side of the air springs are integral with the air springs. These mounting plates are

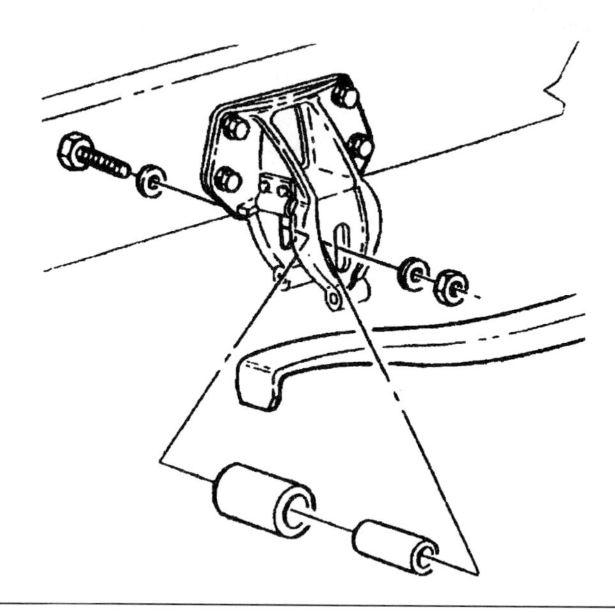

Figure 8-2 Main support beam mounting in the front bracket. (Courtesy of General Motors Corporation, Service Technology Group.)

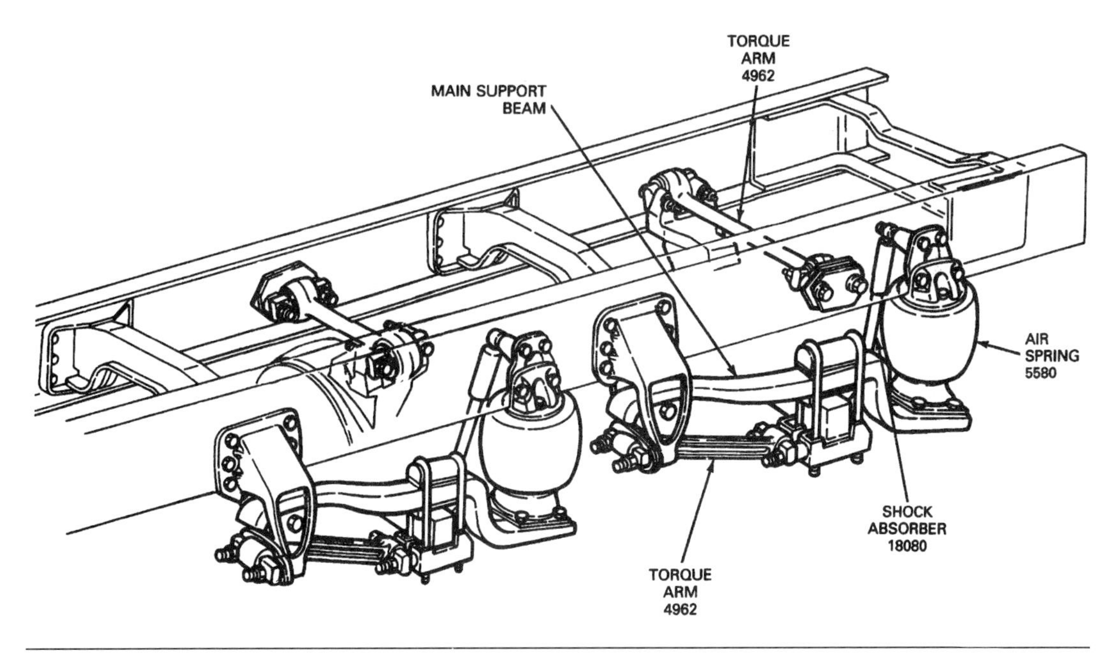

Figure 8-3 Tandem rear axle air suspension system. (Courtesy of Sterling Truck Corporation.)

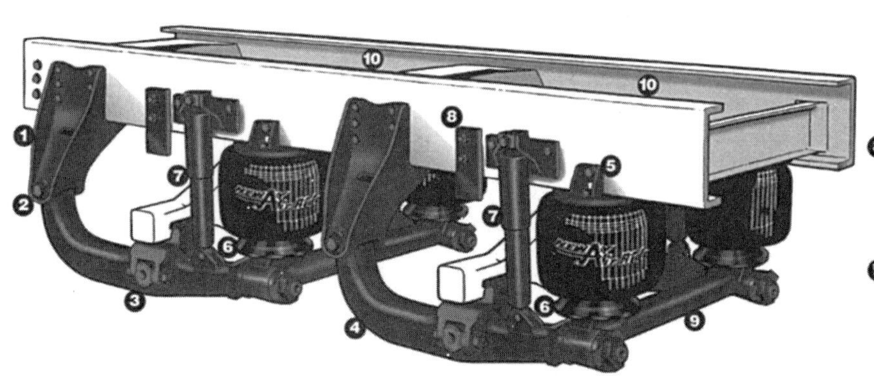

Figure 8-4 Tandem rear axle air suspension system with rigid beams. (Courtesy of Neway Anchorlok International, Inc.)

bolted to the tractor frame, and mounting bolts extend through a reinforcing gusset on the inside of the frame. A composite piston is integral with the lower end of the air spring. This piston is attached to the **transverse beam.** Axle stops attached to the tractor frame prevent excessive upward rear axle movement or downward frame movement.

A high strength cast bracket attaches the front of each rigid beam to the tractor frame (Figure 8-5). The rigid beam is retained in the bracket with a pivot bushing and a bolt. Universal frame brackets are available that provide an adjustment on the beam-to-bracket bolt for rear axle alignment (Figure 8-6).

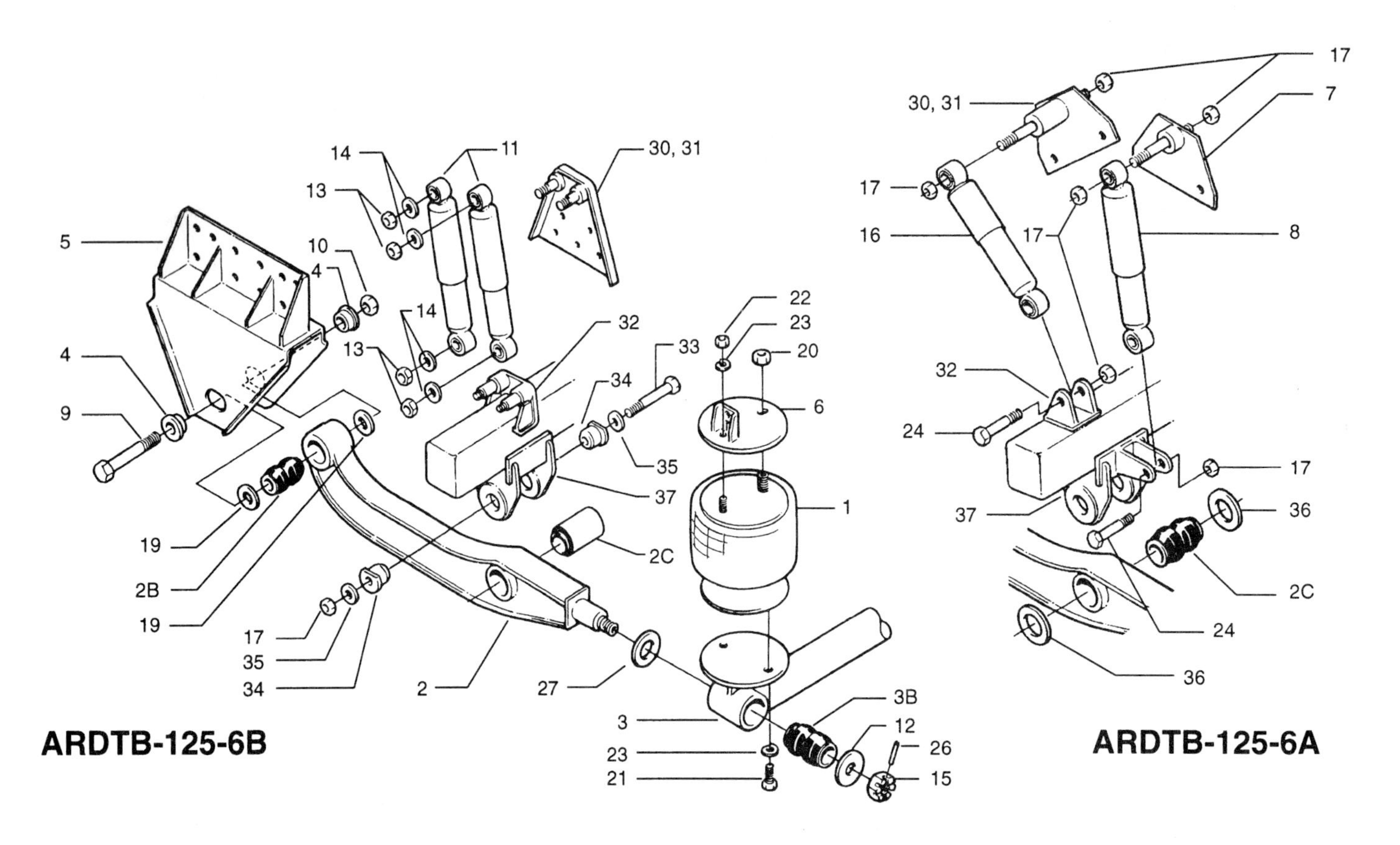

Item No.	Part No.	Description	Qty ARDTB-125-6A	Qty ARDTB-125-6B
*1	905 57 098	Air Spring	2	2
**2	905 16 293	Equalizing Beam Assembly	2	2
2B	900 08 151	Bushing (Front End)	2	2
2C	M00 08 000	Bushing (Beam Center)	2	–
	900 08 183	Bushing (Beam Center)	–	2
3	905 44 876	Transverse Beam Assembly	1	1
3B	900 08 135	Transverse Beam Bushing	2	2
4	900 08 181	Alignment Bushing	4	4
5	905 20 156	Frame Bracket L.H. Adj.	1	1
	905 20 157	Frame Bracket R.H. Fixed	1	1
6	905 31 136	Air Spring Mounting Plate	2	2
7	905 20 258	Shock Absorber Mount Bracket	2	–
8	900 44 448	Shock Absorber - Rear	2	–
9	932 01 046	Rod Bolt 1 1/8" - 7	2	2
10	934 00 506	Lock Nut 1 1/8" - 7	2	2
11	900 45 031	Shock Absorber	–	4
12	900 36 118	Spacer Washer	2	2
13	934 00 553	Lock Nut 3/4" -10	–	8
14	936 00 157	Flat Washer 3/4"	–	8
15	934 00 603	Hex Slotted Nut 2 1/4" - 12	2	2
16	900 44 868	Shock Absorber - Front	2	–
17	934 00 492	Lock Nut 3/4" - 10	14	2
19	936 00 502	Spacer Washer	4	4
20	934 00 417	Lock Nut (light, thin) 3/4" - 16	2	2
21	930 02 893	Cap Screw 1/2" - 13 x 1"	4	4
22	934 00 136	Hex Nut 1/2" - 13	2	2
23	936 00 072	Lock Washer 1/2"	6	6
24	930 03 597	Cap Screw 3/4" - 10 3 1/2"	4	–
26	938 00 235	Roll Pin 3/8" x 3 1/2" Lg.	2	2
27	936 00 498	Washer	2	2
30	905 20 226	Upper Shock Mount. Brkt. L.H.	1	–
	905 20 404	Upper Shock Mount. Brkt. L.H.	–	1
31	905 20 227	Upper Shock Mount. Brkt. R.H.	1	–
	905 20 405	Upper Shock Mount. Brkt. R.H.	–	1
32	905 20 166	Lower Shock Mount. Bracket	2	–
	905 20 406	Lower Shock Mount. Bracket	–	2
33	930 03 693	Cap Screw 3/4" - 10 x 7 1/2"	2	2
34	M00 01 000C	Adapter Bushing	4	4
35	936 00 156	Plain Washer 3/4"	4	4
36	936 00 514	Bearing Washer	4	–
37	905 01 210	Beam Hanger Bracket L.H.	1	–
	905 01 211	Beam Hanger Bracket R.H.	1	–
	900 01 162	Beam Hanger Bracket	–	2

* For Air Spring Identification refer to page 6.

** For Equalizing Beam Identification refer to page 6.

Figure 8-5 Rear axle air suspension components. (Courtesy of Neway Anchorlok International, Inc.)

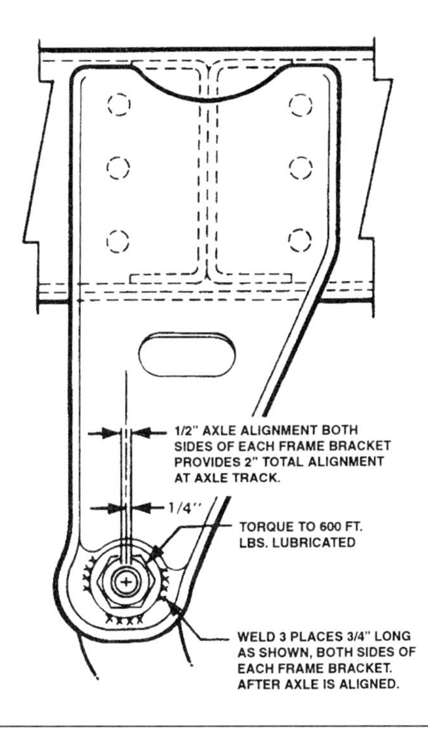

Figure 8-6 Bracket with rear axle adjustment. (Courtesy of Neway Anchorlok International, Inc.)

Super heavy duty shock absorbers are connected from rigid beam brackets to frame brackets. Torque rods and transverse rods may be added by the original tractor manufacturer.

Some rear tandem axle air suspension systems have shock absorbers connected from the lower axle attaching plates to the chassis (Figure 8-7). Some tractors are equipped with air springs on the front I-beam axle (Figure 8-8). On these systems shock absorbers are connected from the rear of the spring beam to the chassis.

Air System for Air Suspension Systems

Pressure Protection Valve The **pressure protection valve** protects the air supply in the air brake system if a leak occurs in the air suspension system. If the air pressure drops below the setting of the pressure protection valve, this valve closes and protects the brake system pressure. Air pressure for the air suspension system is supplied from one of the reservoirs in the air brake system. The pressure protection valve is threaded into the air brake reservoir, and a line is connected from this valve to the air suspension system (Figure 8-9). The pressure protection valve remains

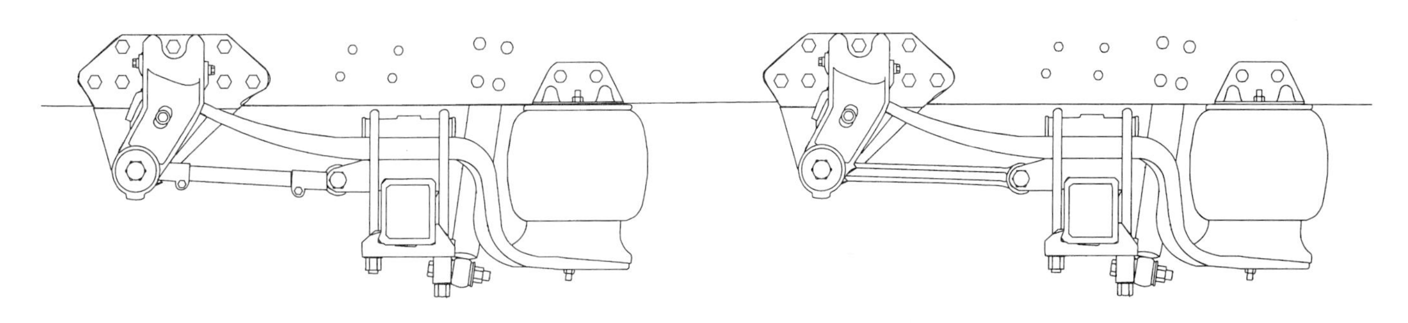

Figure 8-7 Tandem rear axle air suspension with shock absorbers. (Courtesy of Reyco Industries Inc.)

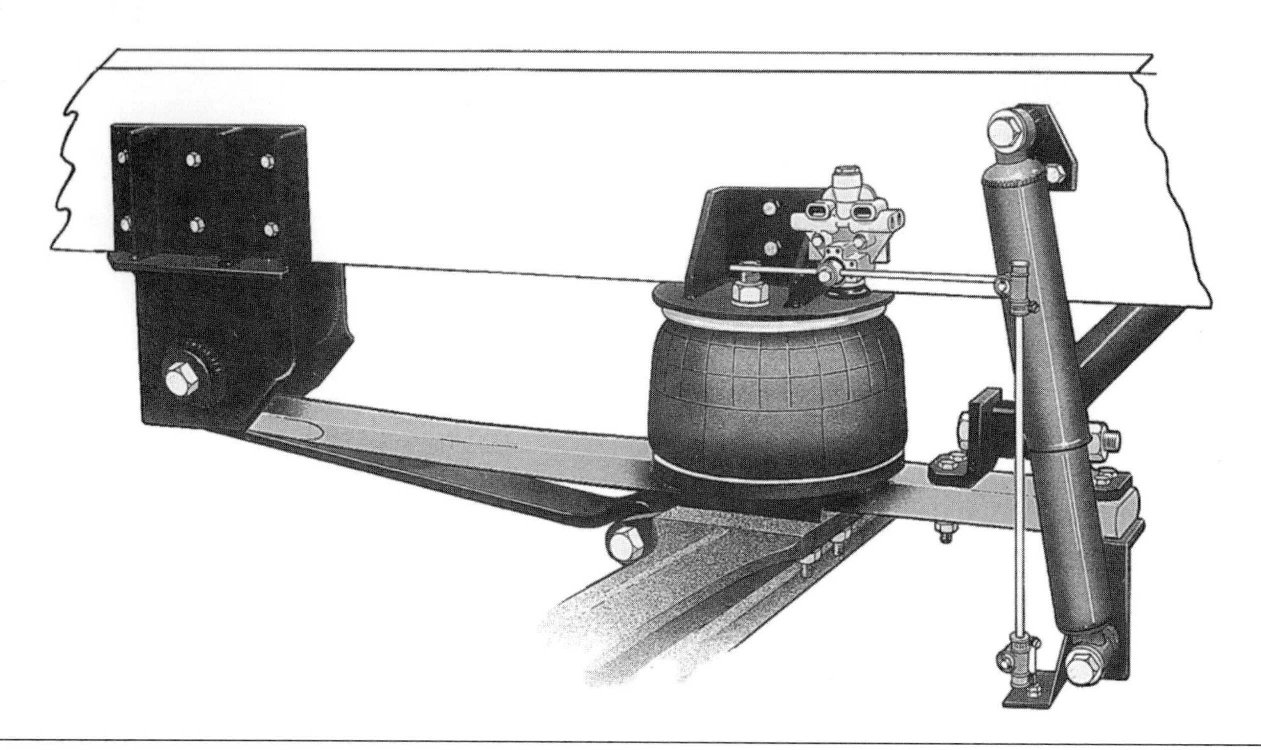

Figure 8-8 I-beam front axle with air suspension. (Courtesy of Reyco Industries Inc.)

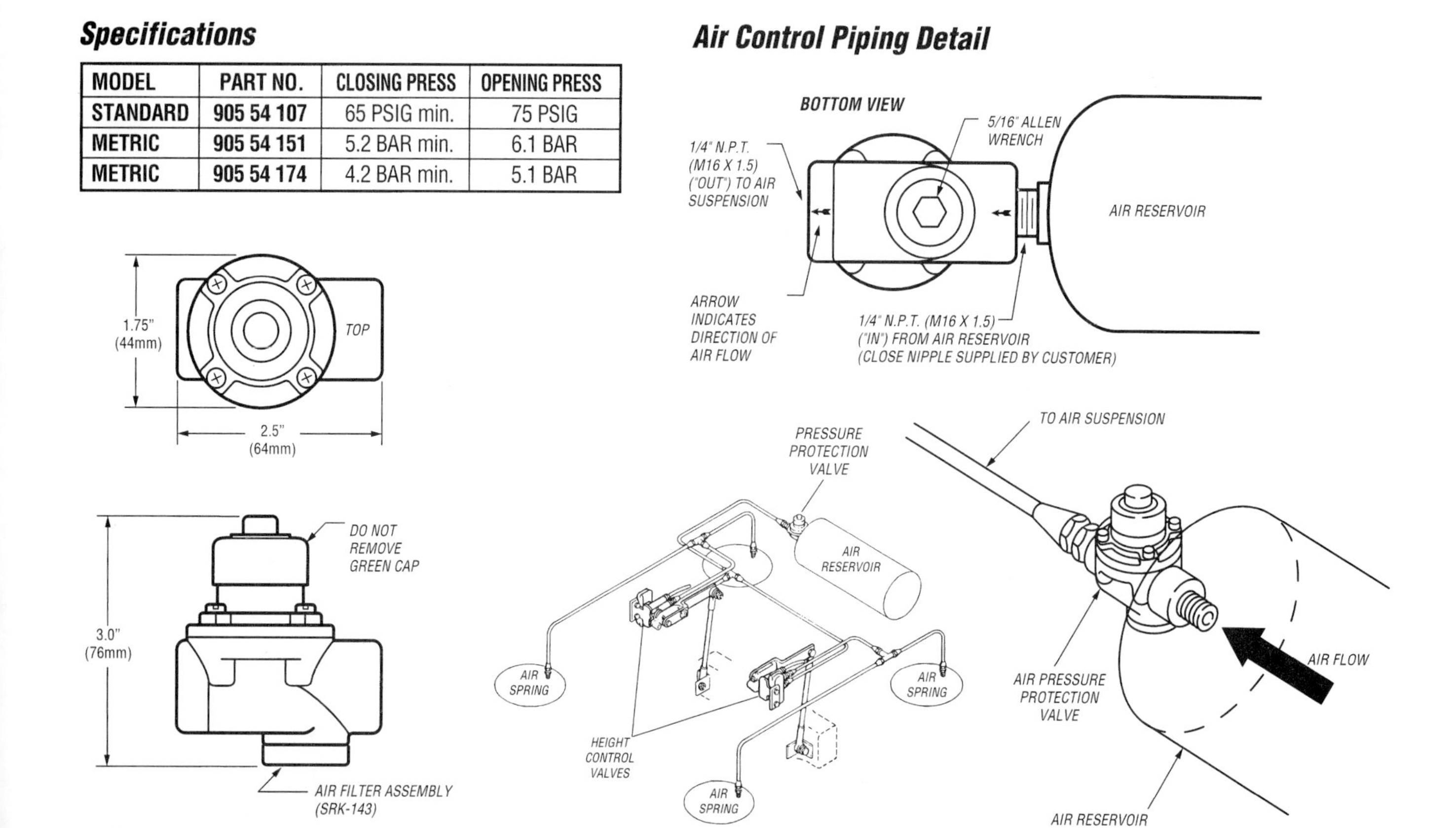

Specifications

MODEL	PART NO.	CLOSING PRESS	OPENING PRESS
STANDARD	905 54 107	65 PSIG min.	75 PSIG
METRIC	905 54 151	5.2 BAR min.	6.1 BAR
METRIC	905 54 174	4.2 BAR min.	5.1 BAR

Figure 8-9 Pressure protection valve mounting in the air brake system reservoir. (Courtesy of Neway Anchorlok International, Inc.)

closed at less than 70 psi (482.65 kPa). When the air brake system pressure increases to more than 70 psi (482.65 kPa), the pressure protection valve opens and supplies air pressure to the air suspension system. If a leak occurs in the air suspension system, and the air pressure decreases to less than 70 psi (482.65 kPa), the pressure protection valve closes to protect the air brake system from any further loss of air.

The pressure protection valve must be installed only in one direction. An arrow on the valve casting indicates the proper direction of air flow through this valve. A filter in the pressure protection valve must be cleaned periodically (Figure 8-10).

Height Control Valve The **height control valve** supplies the proper amount of air pressure to the air springs to maintain the normal ride height regardless of the load on the suspension system. On many tandem rear axle suspension systems a height control valve is mounted on each side of the suspension system. Some rear axle suspension systems may have only one height control valve. The height control valves are bolted to the tractor frame, and a linkage is connected from each of these valve arms to the air suspension system (Figure 8-11). Air lines are connected from the pressure protection valve to the height control valves. Air lines are also connected from the height control valves to the air springs. An optional dump valve may be mounted in or on the height control valve.

When additional weight is added to the tractor frame, the frame is forced downward. This action moves the height control valve linkage and arm upward. Under this condition the height control valve supplies more air pressure to the air springs to bring the tractor frame back up to the normal ride height. If weight is removed from the tractor frame, the frame moves upward. When this action occurs, the linkage and height control valve arm move downward. Under this condition the height control valve exhausts air from the air springs until the frame returns to the normal ride height. When the rear axle strikes a bump or hole in the road surface, the height control valves pro-

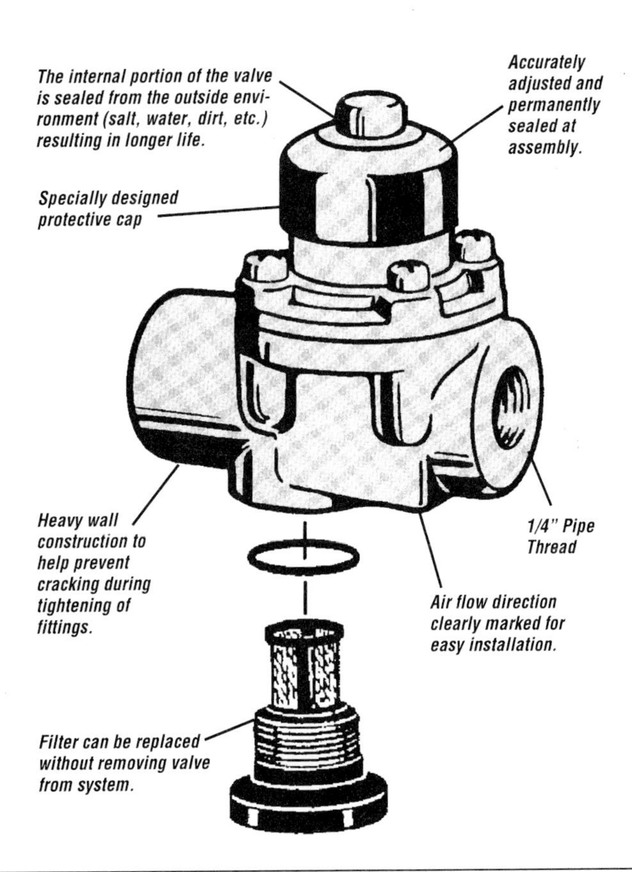

Figure 8-10 Pressure protection valve features. (Courtesy of Neway Anchorlok International, Inc.)

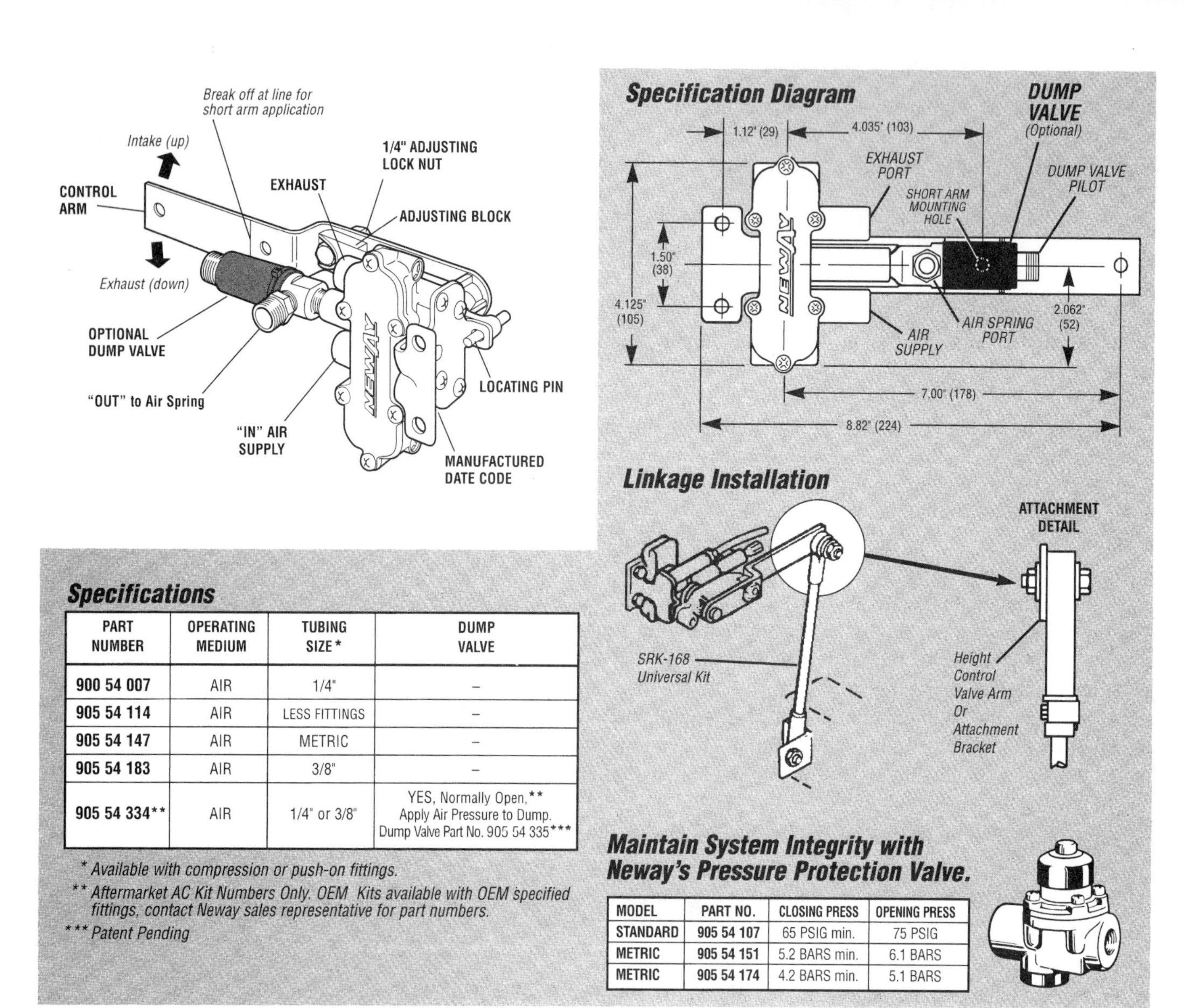

Specifications

PART NUMBER	OPERATING MEDIUM	TUBING SIZE*	DUMP VALVE
900 54 007	AIR	1/4"	–
905 54 114	AIR	LESS FITTINGS	–
905 54 147	AIR	METRIC	–
905 54 183	AIR	3/8"	–
905 54 334**	AIR	1/4" or 3/8"	YES, Normally Open,** Apply Air Pressure to Dump. Dump Valve Part No. 905 54 335***

* Available with compression or push-on fittings.
** Aftermarket AC Kit Numbers Only. OEM Kits available with OEM specified fittings, contact Neway sales representative for part numbers.
*** Patent Pending

Maintain System Integrity with Neway's Pressure Protection Valve.

MODEL	PART NO.	CLOSING PRESS	OPENING PRESS
STANDARD	905 54 107	65 PSIG min.	75 PSIG
METRIC	905 54 151	5.2 BARS min.	6.1 BARS
METRIC	905 54 174	4.2 BARS min.	5.1 BARS

Figure 8-11 Height control valve. (Courtesy of Neway Anchorlok International, Inc.)

vide a delay in air pressure change in the air spring. This action prevents undesirable chassis oscillations. The safety features designed into an air suspension system are the following:

1. If an air leak occurs in an air suspension system, the pressure protection valve closes to protect the pressure in the air brake system to maintain braking capabilities.
2. If an air spring ruptures, solid rubber bumpers inside the air spring support the vehicle load without loss of vehicle control.
3. If a control linkage fails, the height control valve arm remains in a neutral position to maintain the same air supply in the air springs.

Control Valve and Relay Valve The **control valve** and **relay valves** allow air spring deflation when uncoupling the trailer from the tractor. On some air suspension systems a manually operated control valve and two relay valves are used to deflate the air springs. The control valve is usually mounted within reach of the driver, and the relay valves are mounted on the tractor frame near the air springs. Air pressure is supplied from the air brake system to the supply port on the control valve (Figure 8-12). The delivery port on the control valve is connected to the control ports on the relay valves (Figure 8-13). The supply ports on the relay valves are connected to the supply line from the height control valves. The delivery port on the relay valve is connected to the air springs (Figure 8-14). The relay valve also contains an exhaust port that releases air pressure to the atmosphere.

TRADE JARGON: A relay valve may be called a dump relay valve or a pilot valve.

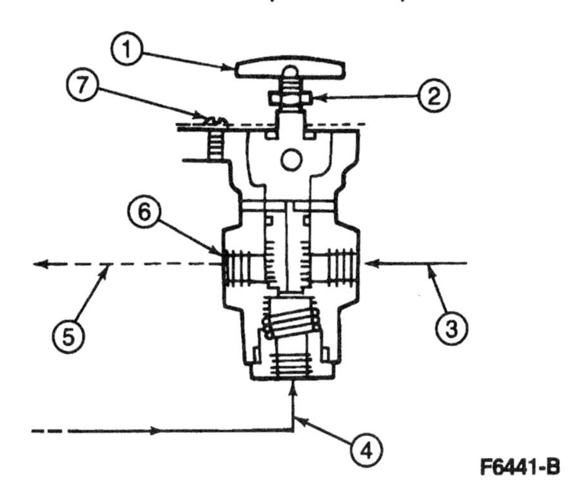

Item	Part Number	Description
1	—	Push Button (Part of 5K862)
2	—	Nut (Part of 5K862)
3	—	Pipe Plug (Unused Second Delivery Port) (Part of 5K862)
4	—	Supply Air (Part of 5K862)
5	—	Delivery to Control Port on Relay Valve (Part of 5K862)
6	—	Delivery Port (Part of 5K862)
7	—	Screw (Part of 5K862)

Figure 8-12 Control valve for air spring deflation. (Courtesy of Sterling Truck Corporation.)

F6443-C

Item	Part Number	Description
1	2B280	Air Source Standard System
2	5K862	Supply
3	5A743	Dump Relay Valves
4	5A687	Height Control Valves
5	—	Pressure Protection Valve (Part of 5A687)
6	5580	Air Springs
7	2B422	Signal
8	5K862	Control Valve

Figure 8-13 Air line connections to control valve and relay valves. (Courtesy of Sterling Truck Corporation.)

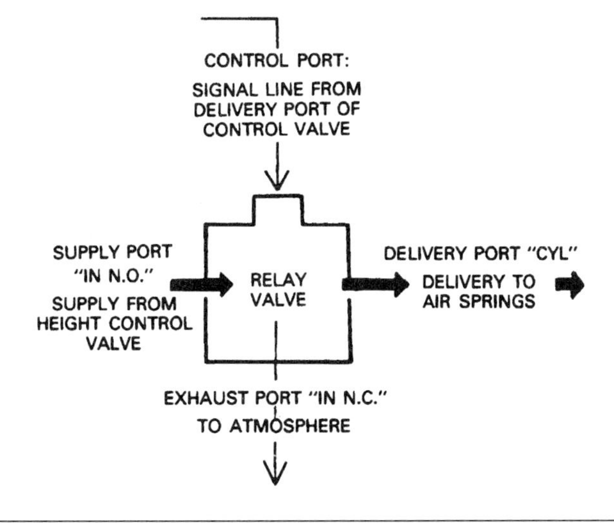

Figure 8-14 Relay valve ports. (Courtesy of Sterling Truck Corporation.)

When the control valve is released by the driver, this valve closes the supply air to the relay valve control port. Under this condition the control valve exhausts air pressure in the line to the relay valve control port through the exhaust port to the atmosphere. With the control valve in the released position, air pressure is supplied through the relay valves to the air springs.

Gladhand connectors are quick-disconnect connectors in the air lines between the tractor and trailer.

Before uncoupling a trailer the trailer parking brakes must be applied, the trailer landing gear should be lowered to the desired height, and the gladhand connectors in the air lines between the tractor and trailer must be disconnected. Under these conditions the air suspension control valve may be depressed. This action supplies air pressure through this valve to the relay valve control ports. When the relay valves receive this pressure signal, these valves exhaust air pressure from the air springs to the atmosphere. The air spring deflation prevents a sudden rise in the truck frame height when the trailer is uncoupled. This sudden rise in truck frame height could damage the shock absorbers.

CAUTION: Be sure no one is under or near the tractor frame when deflating the air springs. The frame may drop suddenly, resulting in personal injury.

WARNING: Never operate the tractor until the air suspension control valve is released and the air suspension has returned to normal ride height. Operating the tractor with the air springs deflated or partially deflated causes rough riding, erratic vehicle handling, and air suspension component damage.

See Chapter 8 in the Shop Manual for tractor rear axle air suspension system maintenance, diagnosis, and service.

When the control valve handle is released, the air passage is closed through this valve to the relay valve control ports. This action allows air pressure to flow through the relay valves to the air springs and return the rear suspension to the normal ride height. When coupling a trailer to the tractor, leave the control valve in the released or off position.

In some air suspension systems a **dump valve** is combined with each height control valve (Figure 8-15). The dump valve operation is basically the same as the relay valve operation. The dump valve may be operated by air pressure from a manually operated valve. Some dump valves are operated by an electrical switch.

Volvo Optimized Air Suspension (VOAS) System

The Volvo optimized air suspension (VOAS) system is a modified version of the previous air suspension on rear axles of Volvo GM heavy duty trucks (Figure 8-16). A number of modifications in the VOAS system provide improved ride, reduced unsprung weight, increased durability, and better alignment ability. The VOAS system is available on all Volvo GM on-highway tractors. On 6 × 4

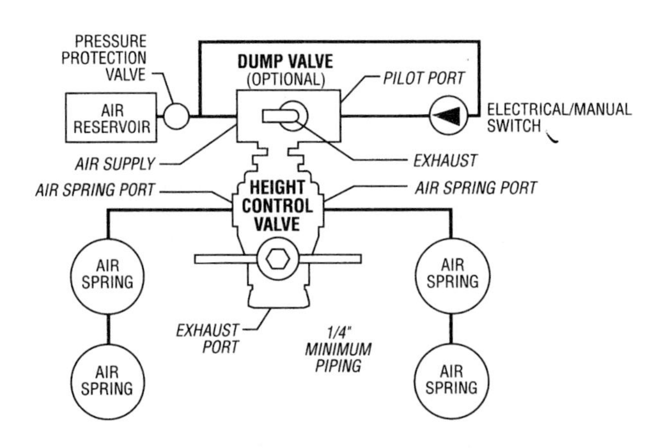

Figure 8-15 Height control valve with integral dump valve. (Courtesy of Neway Anchorlock International, Inc.)

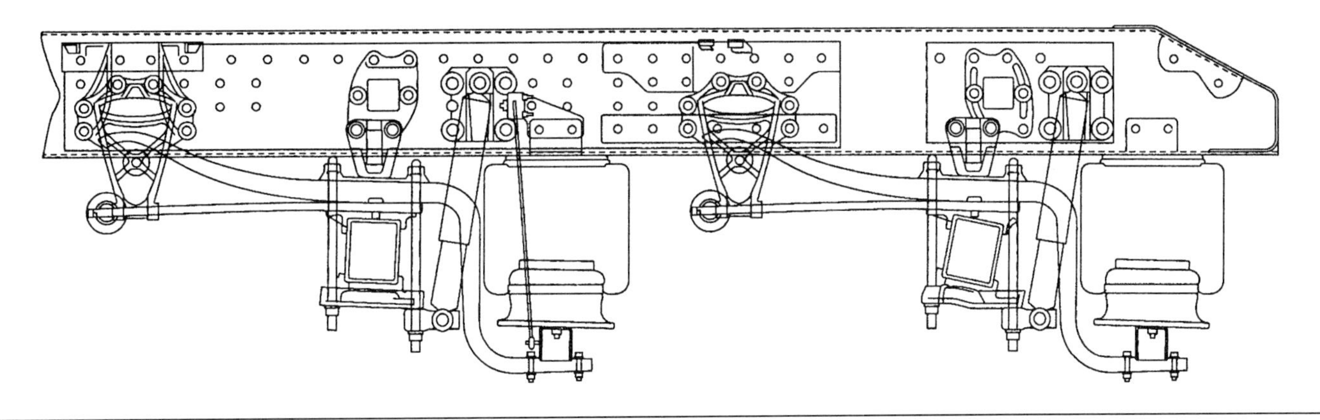

Figure 8-16 Volvo optimized air suspension. (Courtesy of Volvo Trucks North America, Inc.)

tractors the VOAS system will accommodate both Eaton and Rockwell 40,000 lb axles. The VOAS system on 4 × 2 vehicles accommodates both Eaton and Rockwell 23,000 lb axles. The operation of the VOAS system is similar to the operation of other air suspension systems explained previously in this chapter.

The spring brackets in the VOAS system are manufactured from ductile iron, and these brackets have closed sections and cutouts to reduce the weight of each bracket by 4 lb (2 kg) (Figure 8-17).

As in previous models, a polyethylene wear pad provides a very smooth surface for the end of the spring to contact (Figure 8-18). This type of wear pad eliminates noise caused by contact between the spring and metal wear pad in other suspension systems. The wear pad in the VOAS 50% is lighter compared with the wear pad in previous suspension systems. A single fastener retains the wear pad to the spring bracket legs.

The Z-spring is redesigned with a larger clamping surface at the axle seat and improved alignment with other suspension components (Figure 8-19). The Z-spring is also redesigned to provide more clearance between this spring and the lower shock absorber mounting bracket.

The rear of the radius spring is mounted between the axle seat and the Z-spring. The front end of each radius spring contains a bushing that is bolted to openings in the lower end of the spring bracket. In the VOAS system each radius spring has an improved bushing for increased durability (Figure 8-20). Each radius spring has a locating pin for accurate positioning of the radius spring, Z-spring, and axle seat. This design provides improved axle alignment.

The crossbeam in the VOAS system is manufactured from thinner, high-strength steel with cutouts (Figure 8-21). Each crossbeam is 12 lb (5 kg) lighter than previous models.

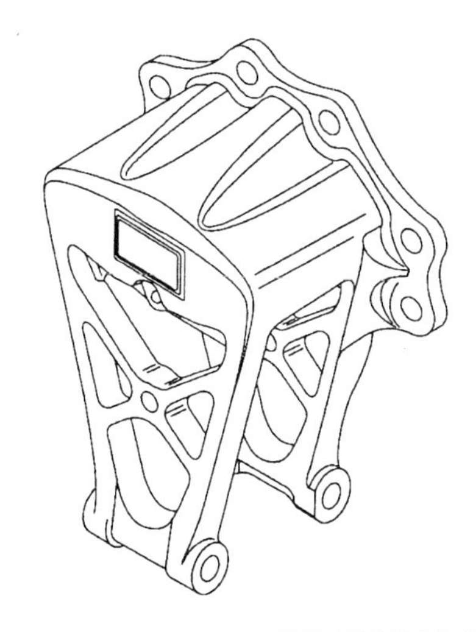

Figure 8-17 Spring bracket, Volvo optimized air suspension. (Courtesy of Volvo Trucks North America, Inc.)

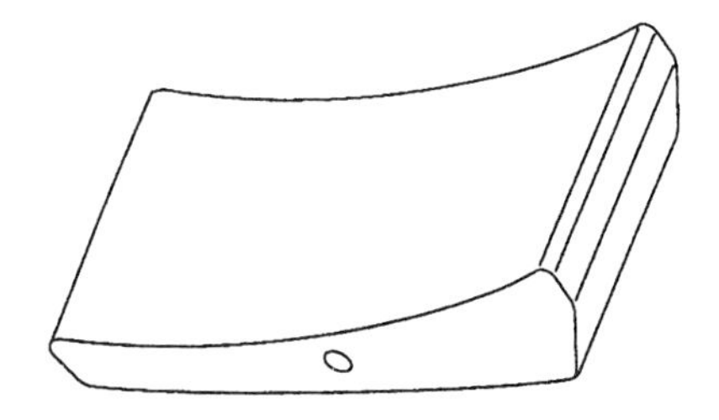

Figure 8-18 Spring bracket wear pad, Volvo optimized air suspension. (Courtesy of Volvo Trucks North America, Inc.)

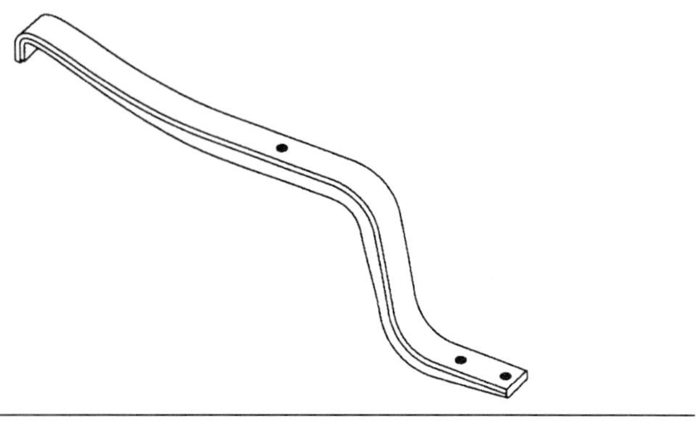

Figure 8-19 Z-spring, Volvo optimized air suspension. (Courtesy of Volvo Trucks North America, Inc.)

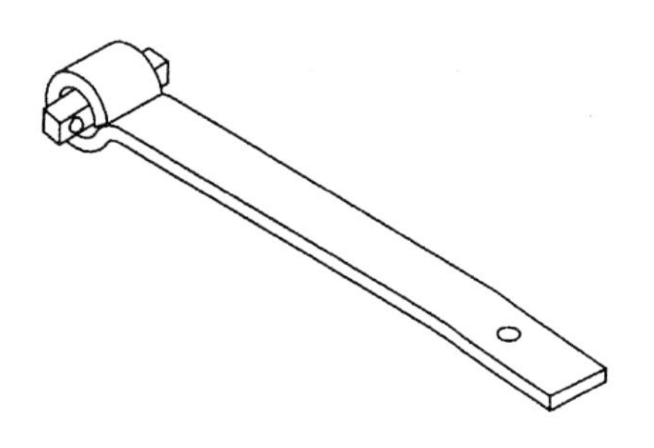

Figure 8-20 Radius spring, Volvo optimized air suspension. (Courtesy of Volvo Trucks North America, Inc.)

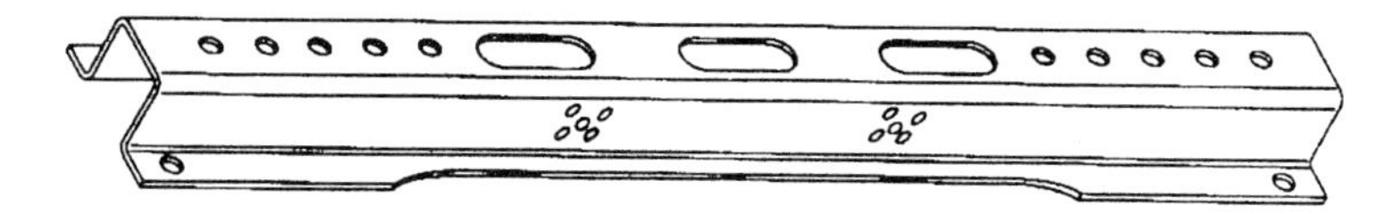

Figure 8-21 Crossbeam, Volvo optimized air suspension. (Courtesy of Volvo Trucks North America, Inc.)

WARNING: Do not interchange air springs in the VOAS system with air springs in previous air suspension systems. This action may cause air suspension system failure.

WARNING: Do not interchange the link rods in the VOAS system with link rods in previous systems. Air suspension system failure may occur if this action is taken.

See Chapter 8 in the Shop Manual for Volvo optimized air suspension diagnosis, adjustment, and service.

The axle seat is redesigned so extra spacers are not required. The air springs in the VOAS system have a composite piston and a rolling lobe, sleeve-type design. The VOAS system has a fixed-length link rod. Each rod is connected from the crossbeam to the height control valve. Because the length of the link rod cannot be adjusted, various openings in the crossbeam for the lower link rod stud provide ride height adjustment. The height control valve in the VOAS system has improved strength and an integral dump valve. This design eliminates the need for a separate dump valve. The VOAS system may have one or two height control valves. Improved torque rod brackets attached to the axle reduce the suspension system weight by 10 to 15 lb (4 to 7 kg) per axle.

Electronically Controlled Air Suspension Systems

Electronically controlled air suspension systems are not widely used as standard equipment on tractors at present. This type of suspension system is in the experimental and developmental stage. However, with the ever-expanding use of electronics in the trucking industry, electronically controlled air suspension systems may be standard equipment on new tractors in the near future. Because electronically controlled air suspension systems are not standard equipment at present, our discussion is brief.

An **electronic control unit (ECU)** in the suspension system receives voltage input signals from various sensors including two rear axle and one front axle height sensor, air pressure sensor, accelerator position sensor, and steering angle sensor. The steering angle sensor is mounted on the steering column shaft and senses the amount and speed of steering wheel rotation. Other ECU inputs may include the brake light switch, ignition switch, vehicle speed, and engine speed. Solenoids mounted in a solenoid valve block are operated electronically by the ECU (Figure 8-22). Air pressure is supplied from the air brake system to these solenoids. If the left rear axle height sensor indicates the left rear frame height is less than specified, the ECU energizes the appropriate sole-

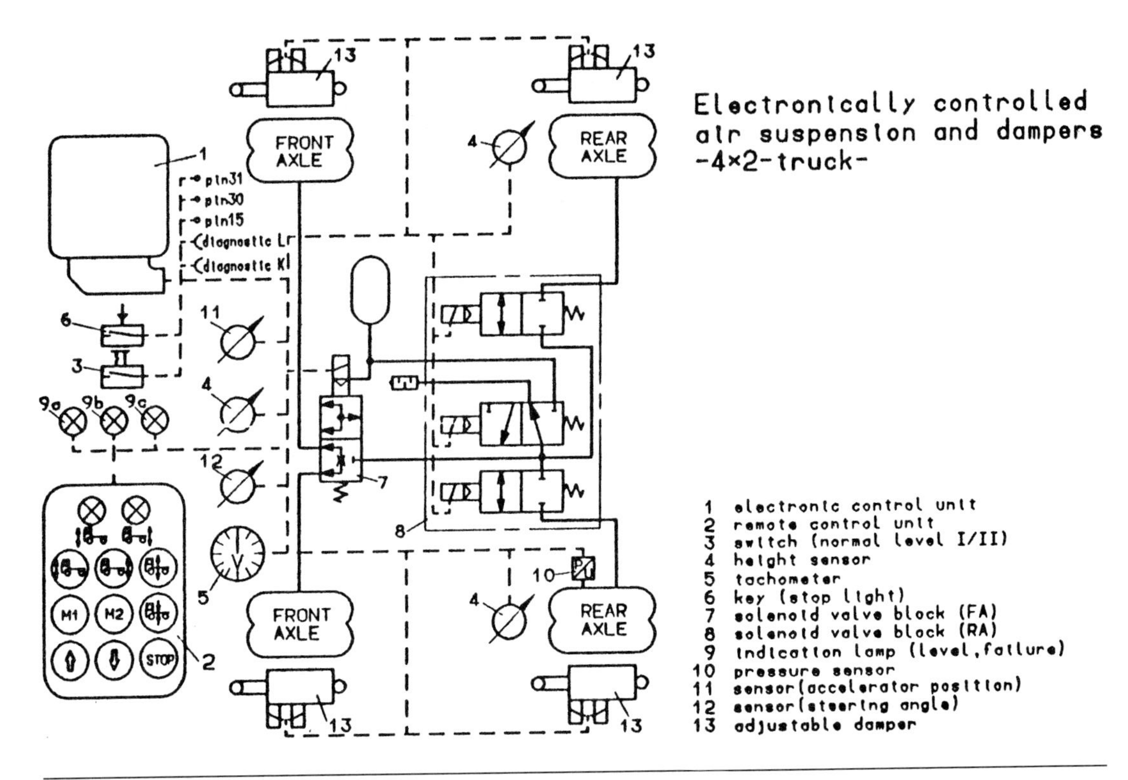

Figure 8-22 Electronically controlled air suspension system. (Reprinted with permission from SAE Paper Electronically Controlled Air Suspension [ECAS] for Commercial Vehicles © 1991 Society of Automotive Engineers, Inc.)

noid. This action opens the solenoid, and supplies air pressure to the left rear air spring to restore the frame height to specifications. When the left rear height sensor signal to the ECU indicates the left rear frame height is within specifications, the ECU de-energizes the solenoid. Under this condition the solenoid closes and maintains the air pressure in the solenoid.

When a height sensor signal indicates to the ECU that the frame height is above specifications, the ECU energizes the appropriate solenoid so it is in the vent mode. Under this condition air pressure is released from the air spring to restore the frame height to specifications.

The ECU can switch the air suspension very quickly to a firm or hard mode by rotating valves in the shock absorbers (dampers). During the soft shock absorber mode the shock absorber valves are positioned so they offer less restriction to the flow of oil. If the ECU enters the firm mode and rotates the shock absorber valves, these valves move to a position that provides more restriction to oil movement. This firm mode may be entered by the ECU and shock absorbers during fast avoidance maneuvers, fast cornering, hard braking, rapid traction changes, continual road irregularities, cross winds, and high static load. The capability of the electronically controlled suspension system to switch quickly to the firm mode provides improved vehicle stability and safety.

A remote control unit allows the driver to manually control the suspension height (Figure 8-23). This feature may be used when coupling the tractor to a trailer or when backing a truck up to a loading dock.

Lift Axles with Air Suspension and Coil Spring Lift

A nondrive **lift axle** may be used to support additional weight when the vehicle is heavily loaded. When the axle is not required to carry the load, the axle may be lifted so the tires are off the road surface. Lifting the axle when it is not required reduces tire wear and scuffing and provides easier maneuverability. Some nondrive axles have an air spring suspension system with a coil spring lift

A lift axle may be called a pusher axle if it is mounted ahead of the drive axle or axles.

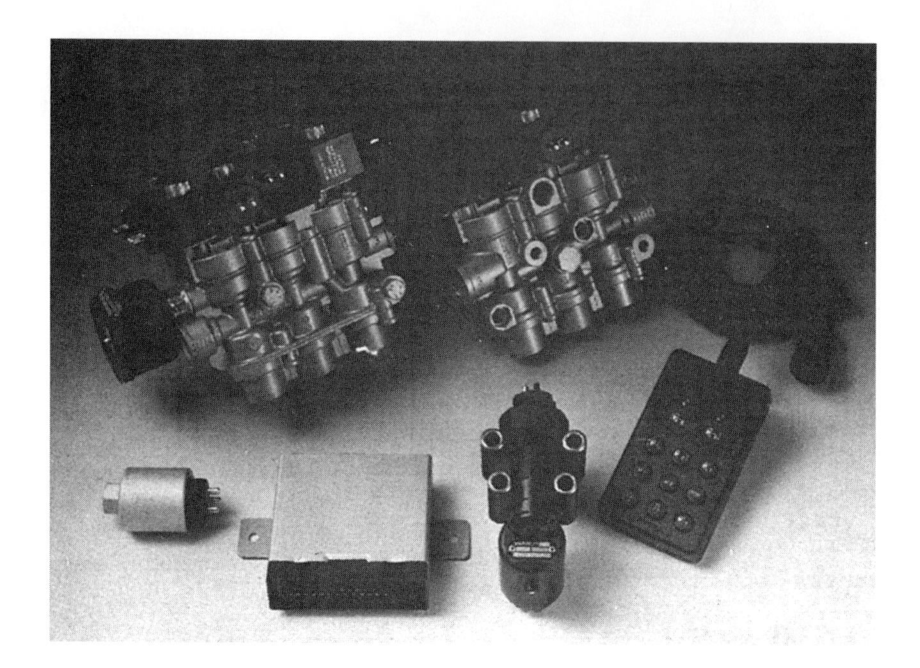

Figure 8-23 Electronically controlled air suspension system components. (Reprinted with permission from SAE Paper Electronically Controlled Air Suspension [ECAS] for Commercial Vehicles © 1991 Society of Automotive Engineers, Inc.)

A lift axle may be called a tag axle if it is mounted behind the drive axle or axles.

(Figure 8-24). When the axle is lowered, air pressure is supplied to the air springs to maintain the proper ride height.

The driver operates a manual control valve to supply or release air pressure from the air springs. Air pressure is supplied from an air brake system reservoir and the pressure protection valve to the manual control valve (Figure 8-25). When the driver places the manual control valve in the inflate position, air pressure is supplied through the manual control valve and the quick release valve to the air springs. The maximum pressure adjustment on the manual control valve must be adjusted to provide the specified pressure to the air springs. On some applications this pressure is 65 psi (448.17 kPa) for a 16,000 lb (7,257.6 kPa) load on the axle. Always refer to the suspension or truck manufacturer's specifications. An air pressure gauge in the air line between the manual control valve and the quick release valve indicates the air pressure supplied to the air springs.

WARNING: Do not operate a loaded vehicle without the specified air pressure in the air springs on the lift axle. This action overloads the drive axle and causes component damage and tire overloading plus excessive tire wear.

When the driver moves the control to the deflate position, the air pressure in the air springs is exhausted quickly through the quick release valve. Air pressure in the line between the quick release valve and the manual control valve is exhausted through the manual control valve. When the air pressure in the air springs is exhausted, the heavy coil springs lift the axle. In the deflate position the minimum pressure screw in the manual control valve is adjusted to maintain 3 psi (20.68 kPa) in the air springs. Observe the precautions in Figure 8-25 regarding operation of the lift axle in the raised position.

WARNING: Do not operate the vehicle without the minimum air pressure in the air springs. This action may damage system components.

Lift Axles with Air Suspension and Air Lift

Some nondrive lift axles have air suspension and **air lift** (Figure 8-26). The same manual control valve operates the air springs as explained previously. Two relay valves are connected in the air

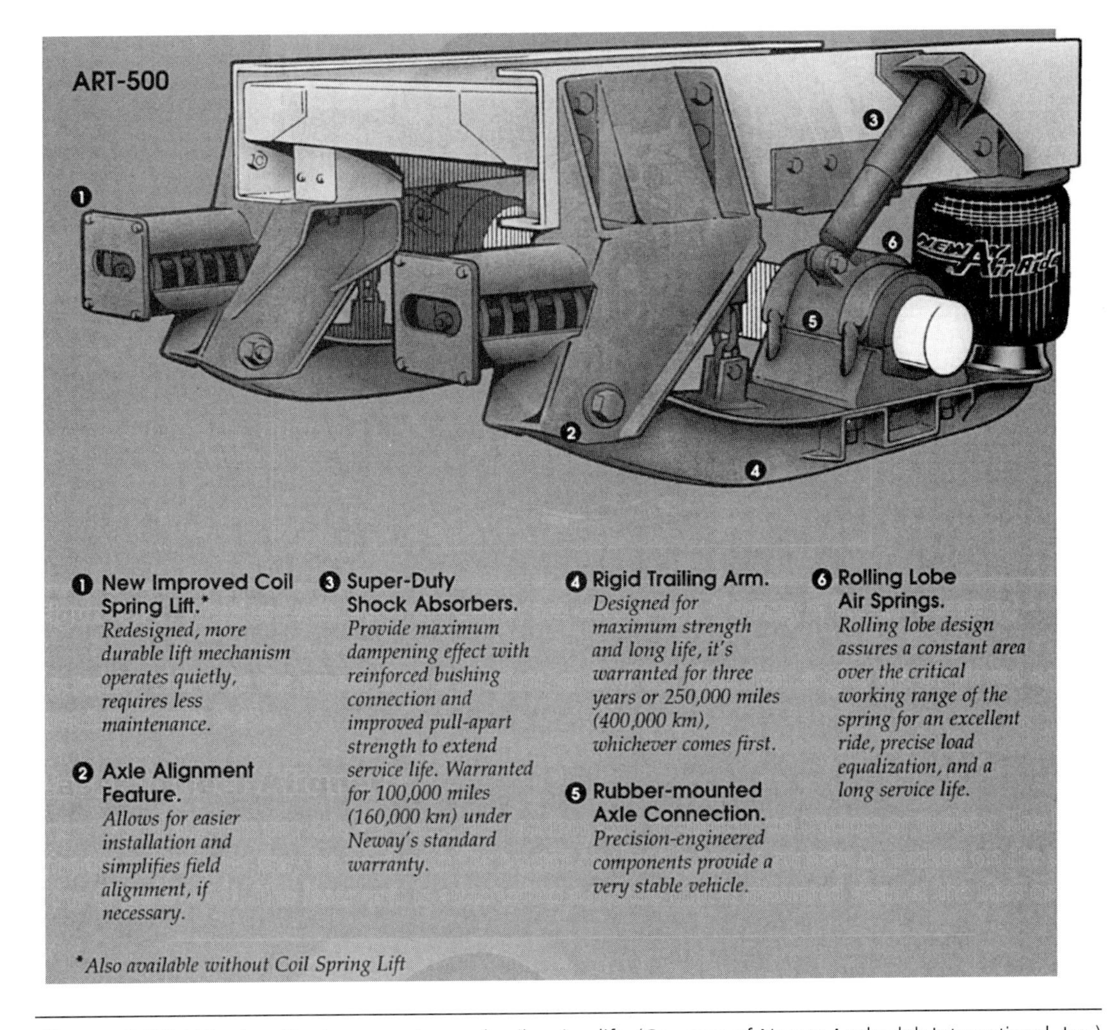

Figure 8-24 Lift axle with air suspension and coil spring lift. (Courtesy of Neway Anchorlok International, Inc.)

system. The air supply to the control ports on these relay valves is controlled by a manual valve or an electric solenoid valve (Figure 8-27). The electric solenoid valve is controlled by a toggle switch mounted within the driver's reach.

When the manual valve or toggle switch is in the off position, no air pressure is supplied through the manual valve or solenoid valve to the relay valves. Under this condition air pressure is supplied through the normally open (NO) port on the relay valve to the air springs to maintain the lift axle in the downward position. When no air pressure is supplied to the control port on the relay valve connected to the lift air springs, this normally closed (NC) relay valve closes off the air supply from the reservoir to the lift air springs.

If the manual valve or toggle switch is in the on position, air pressure is supplied through the manual valve or solenoid valve to the control ports on the relay valves. Under this condition the relay valve connected to the air springs exhausts air pressure to the air springs, and the relay valve connected to the lift springs supplies air pressure to the lift air springs. The air pressure supplied to the lift air springs causes the axle to lift. The same precautions mentioned previously in Figure 8-25 on lift axles with coil spring lift regarding operating the vehicle with the lift axle in the raised position must be observed on lift axles with air lift.

See Chapter 8 in the Shop Manual for lift axle diagnosis, adjustment and service.

WARNING: Many different air suspension systems and various methods of connecting the air lines to the system components exist. When diagnosing, servicing, or installing air suspension systems, always consult the diagrams and specifications supplied by the suspension or truck manufacturer. Failure to consult these diagrams and specifications may cause improper system operation, damage to system components, and personal injury.

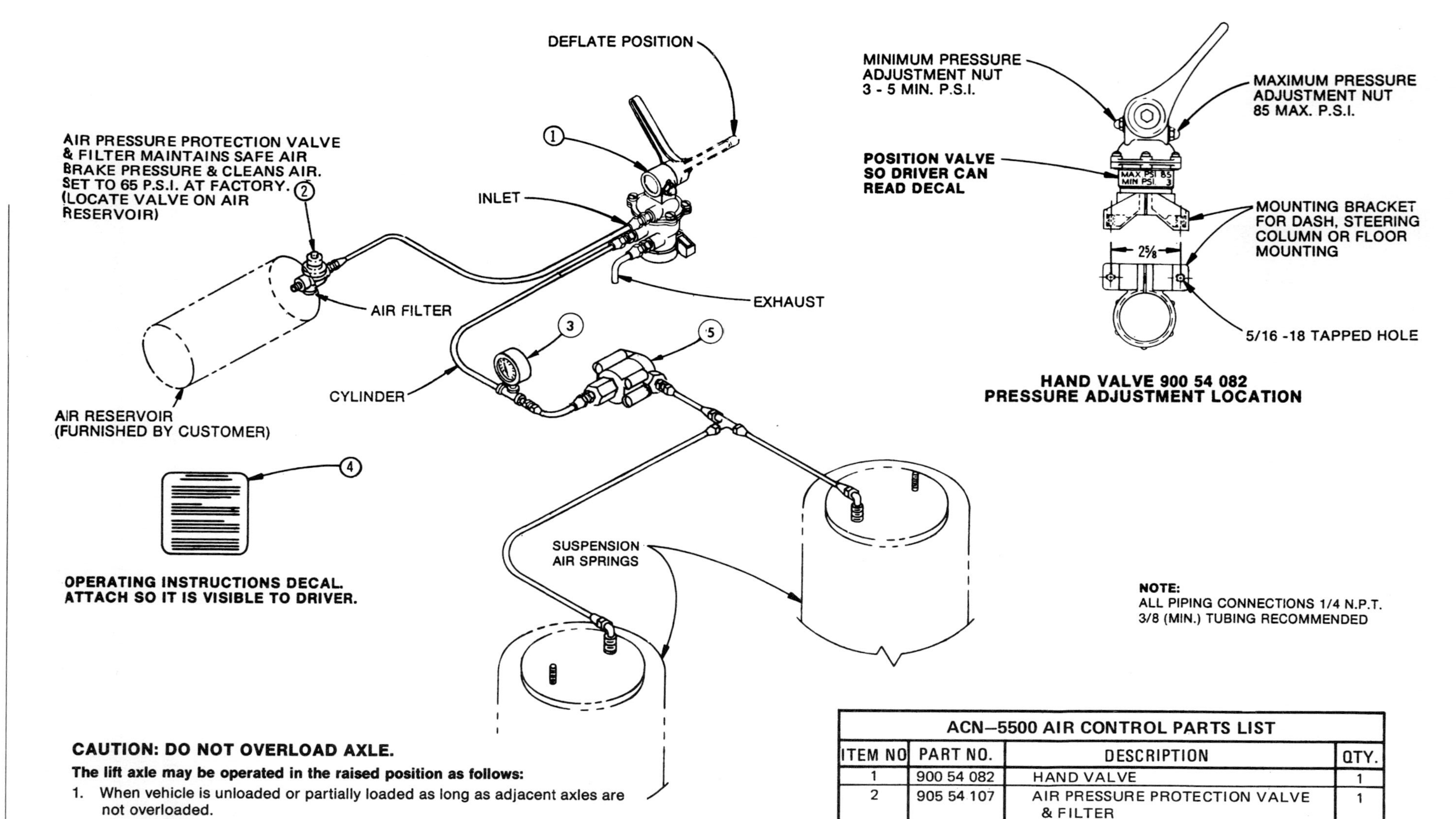

CAUTION: DO NOT OVERLOAD AXLE.

The lift axle may be operated in the raised position as follows:

1. When vehicle is unloaded or partially loaded as long as adjacent axles are not overloaded.
2. When vehicle is loaded and off highway, but at reduced and safe speed.
3. When vehicle is loaded and on highway, but as Government law allows for turning puposes, and then at reduced and safe speed.
4. Whenever it is necessary for reasons of safety to maintain traction or sufficient load on other vehicle driving, steering, braking or cornering axles.

Failure to adhere to the above directions could result in overload fines, premature equipment failure and/or vehicle accident.

ACN–5500 AIR CONTROL PARTS LIST			
ITEM NO	PART NO.	DESCRIPTION	QTY.
1	900 54 082	HAND VALVE	1
2	905 54 107	AIR PRESSURE PROTECTION VALVE & FILTER	1
3	900 54 280	AIR GAUGE (Non-Illuminated)	1
4	900 44 021	DECAL " OPERATING INSTRUCTIONS"	1
5	900 54 096	QUICK RELEASE VALVE	1
6	900 44 755	Decal (Air Pressure vs Load)	1
NOTE:ALL TUBING AND FITTINGS TO BE FURNISHED BY CUSTOMER			

Figure 8-25 Air system for lift axle with coil spring lift. (Courtesy of Neway Anchorlok International, Inc.)

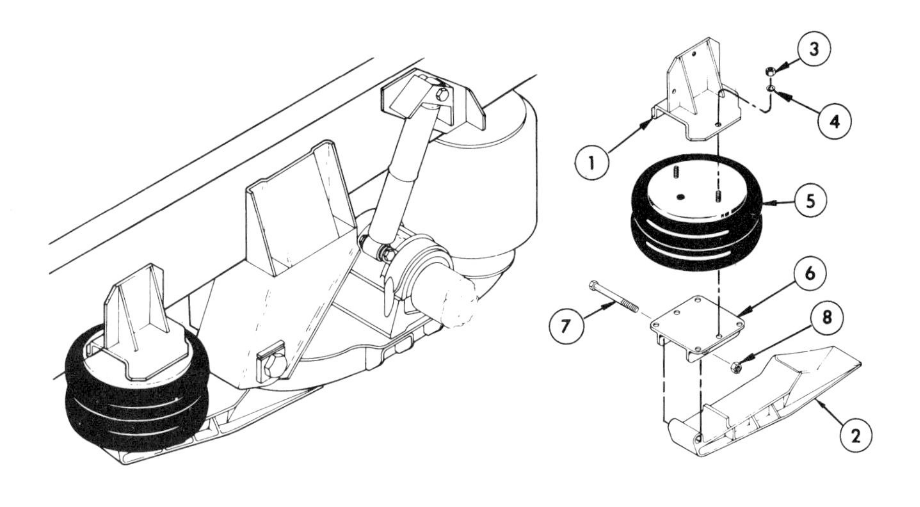

ITEM NO.	DESCRIPTION	PART NO.	QTY.
	MODEL ALT-40 (DRAWING NO. 44000004)		
1	Air Spring Mounting Bracket Assembly	905 18 368	2
2	Lift Beam	900 44 203	2
3	½"–13 Hex. Nut	934 00 136	8
4	½" Lock Washer	936 00 073	8
5	Air Spring Assembly	905 57 014	2
6	Air Spring Mounting Bracket	900 01 012	2
7	¾"–10 x 6" Hex. Head Bolt, Grade 5	930 03 657	2
8	¾"–10 Hex. Lock Nut, Grade B	934 00 492	2

Figure 8-26 Lift axle with air suspension and air lift. (Courtesy of Neway Anchorlok International, Inc.)

Trailer Air Suspension Systems

Trailer air suspension systems maintain a constant trailer ride height regardless of cargo load and provide a smoother ride to help protect the cargo. **Trailer air suspension systems** have a rigid beam on each side of the suspension. The front of this beam is retained in the bracket on a rubber bushing. Some trailer air suspension systems have an eccentric pivot bolt that retains each rigid beam in the bracket. This bolt may be loosened and rotated to provide the necessary suspension alignment (Figure 8-28). Brackets are welded to the trailer axle and these brackets are retained in the rigid beams with two bushings and retaining bolts. These two bushings distribute trailer loads and axle deflections more evenly to provide longer tire life. The air springs are mounted between the rear end of the rigid beams and the frame. Some trailer air suspension systems have integrated axles and brakes with double convoluted air springs (Figure 8-29).

The air supply system for trailer air suspension systems is similar to the tractor air suspension systems explained previously in this chapter. Air pressure is supplied from one of the trailer air brake system reservoirs through the pressure protection valve to the height control valves. The height control valves supply the proper air pressure to the air springs to maintain the specified ride height. Some trailer air suspension systems have dual height control valves, whereas a single height control valve is used on the other systems (Figure 8-30).

Trailer Air Suspension System with Slider and Air Pin Puller

A **slider-type trailer air suspension system** is available. On this type of suspension system the air suspension system may be slid on the trailer frame to provide improved cargo load distribution

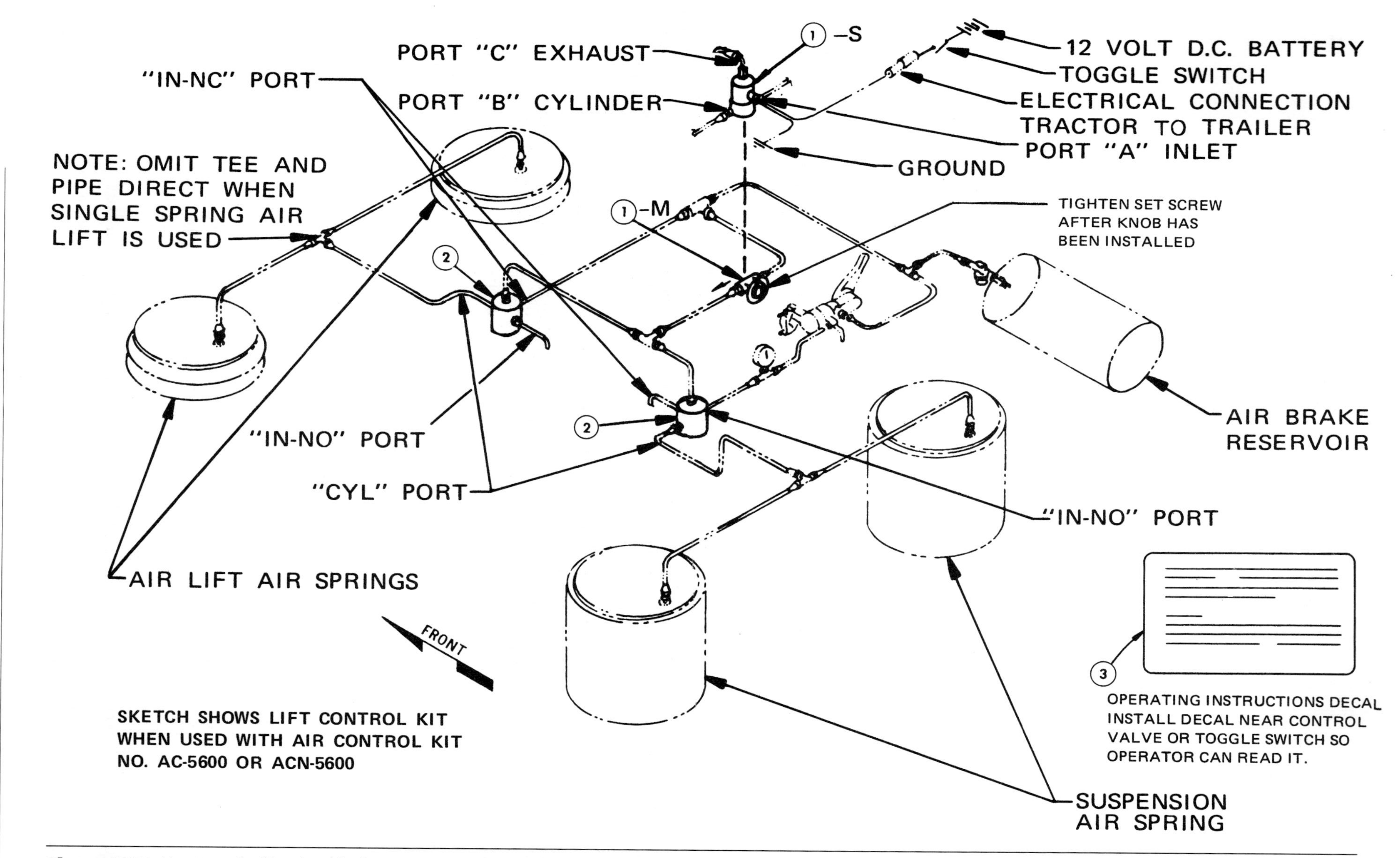

Figure 8-27 Air system for lift axle with air suspension and air lift. (Courtesy of Neway Anchorlok International, Inc.)

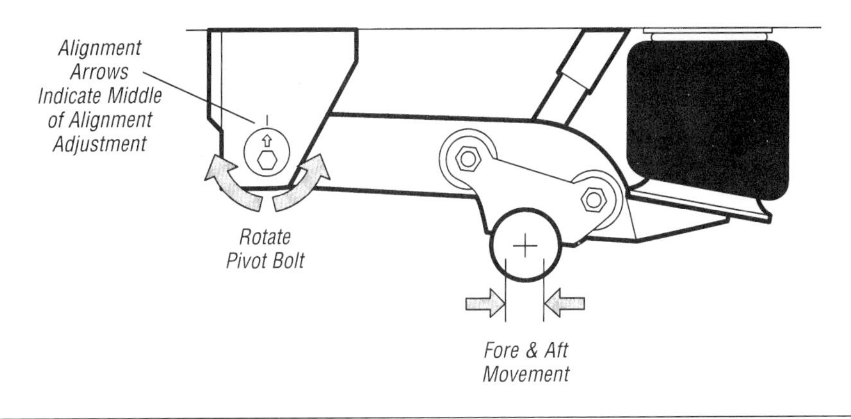

Figure 8-28 Trailer air suspension system. (Courtesy of Neway Anchorlok International, Inc.)

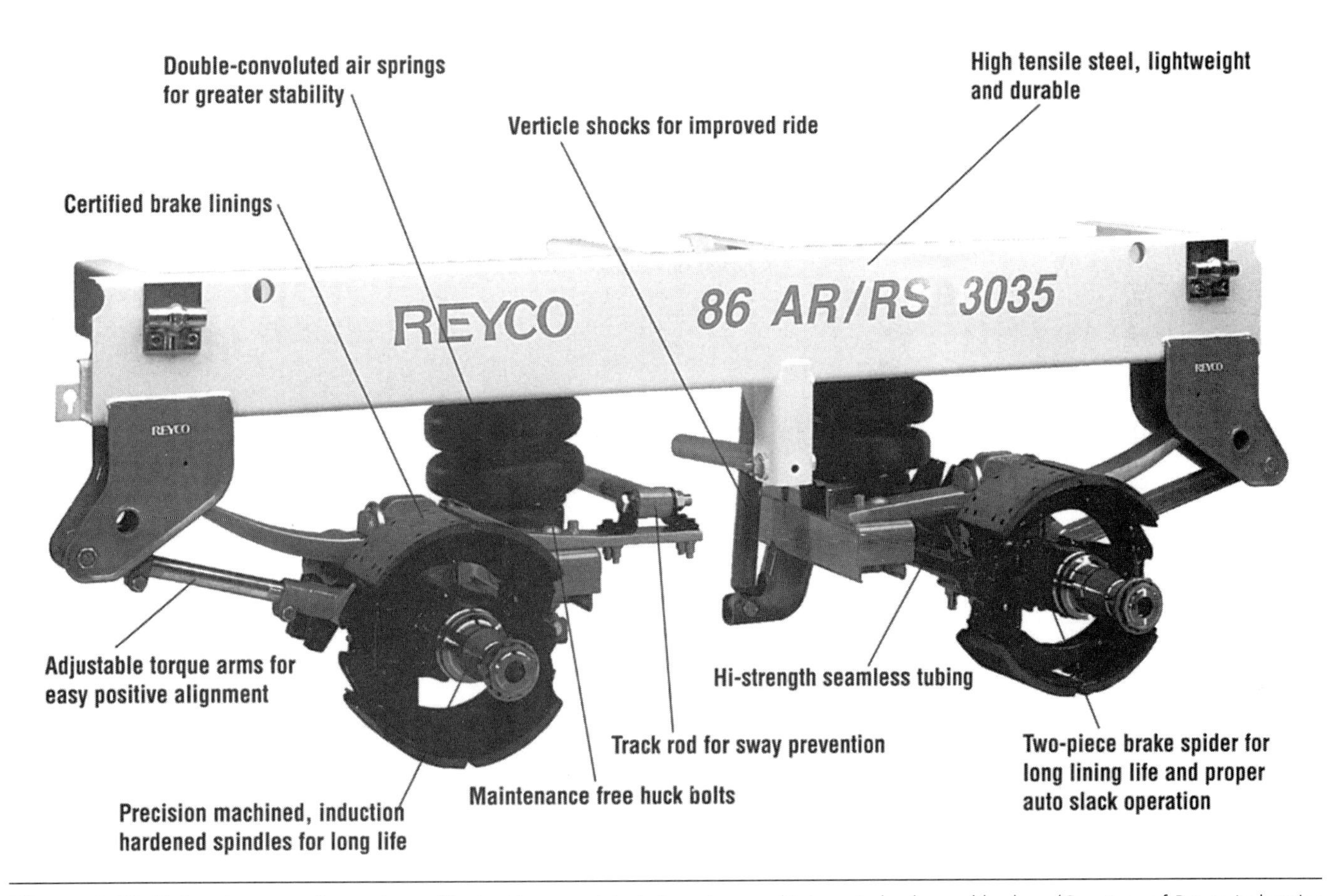

Figure 8-29 Trailer air suspension system with double convoluted air springs and integrated axles and brakes. (Courtesy of Reyco Industries Inc.)

on the suspension system (Figure 8-31). Lock pins are used to lock the suspension frame and the trailer frame together (Figure 8-32). The release mechanism is operated by an air cylinder for easy release operation. A manual override lever provides manual release of the lock pins if air pressure is not available. Once the lock pins are released, the tractor is used to gently pull or push the trailer frame to the desired position.

WARNING: After changing the trailer air suspension position, always be sure the lock pins are completely engaged before driving the tractor and trailer. Failure to observe this precaution may allow the suspension to shift position while driving the vehicle, resulting in serious trailer and suspension damage.

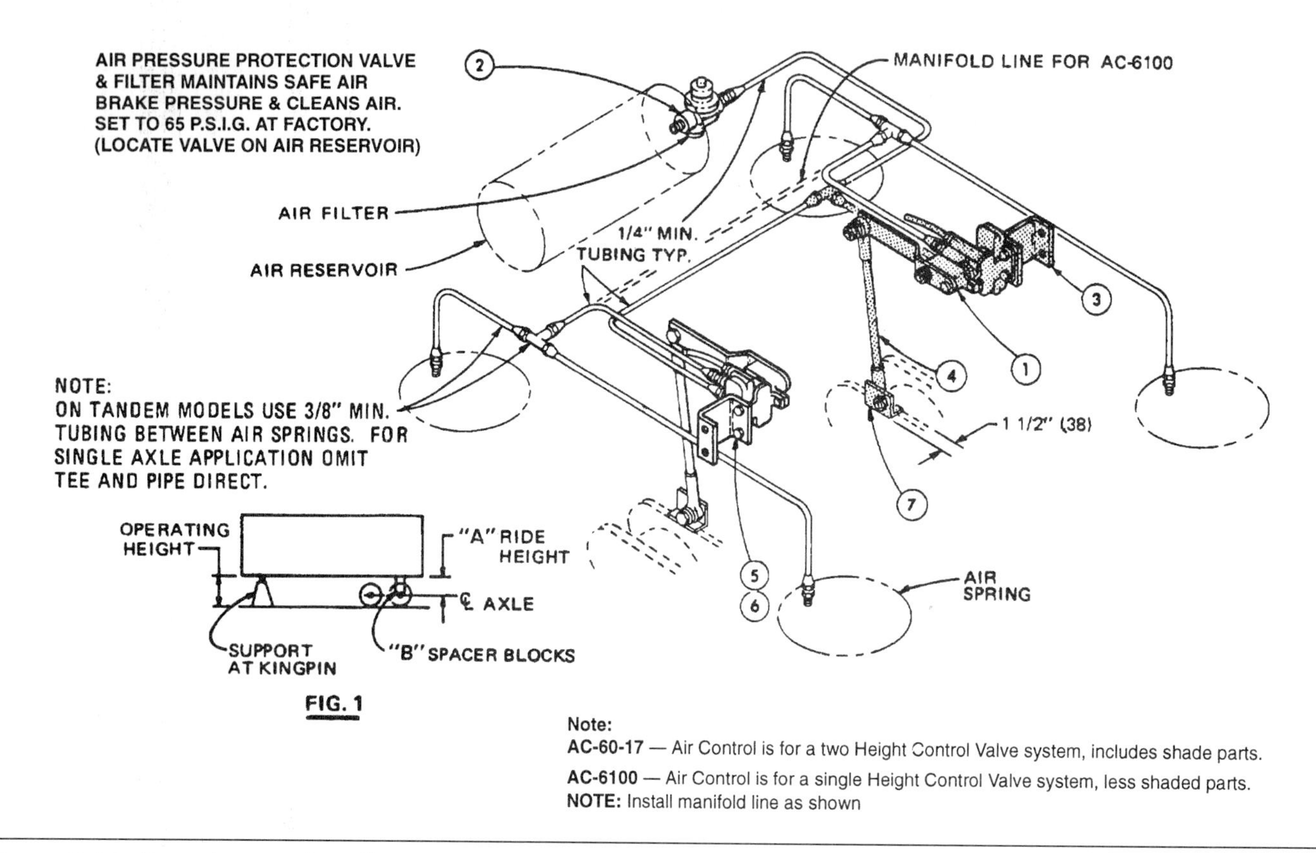

Figure 8-30 Air supply system for trailer air suspension system. (Courtesy of Neway Anchorlok International, Inc.)

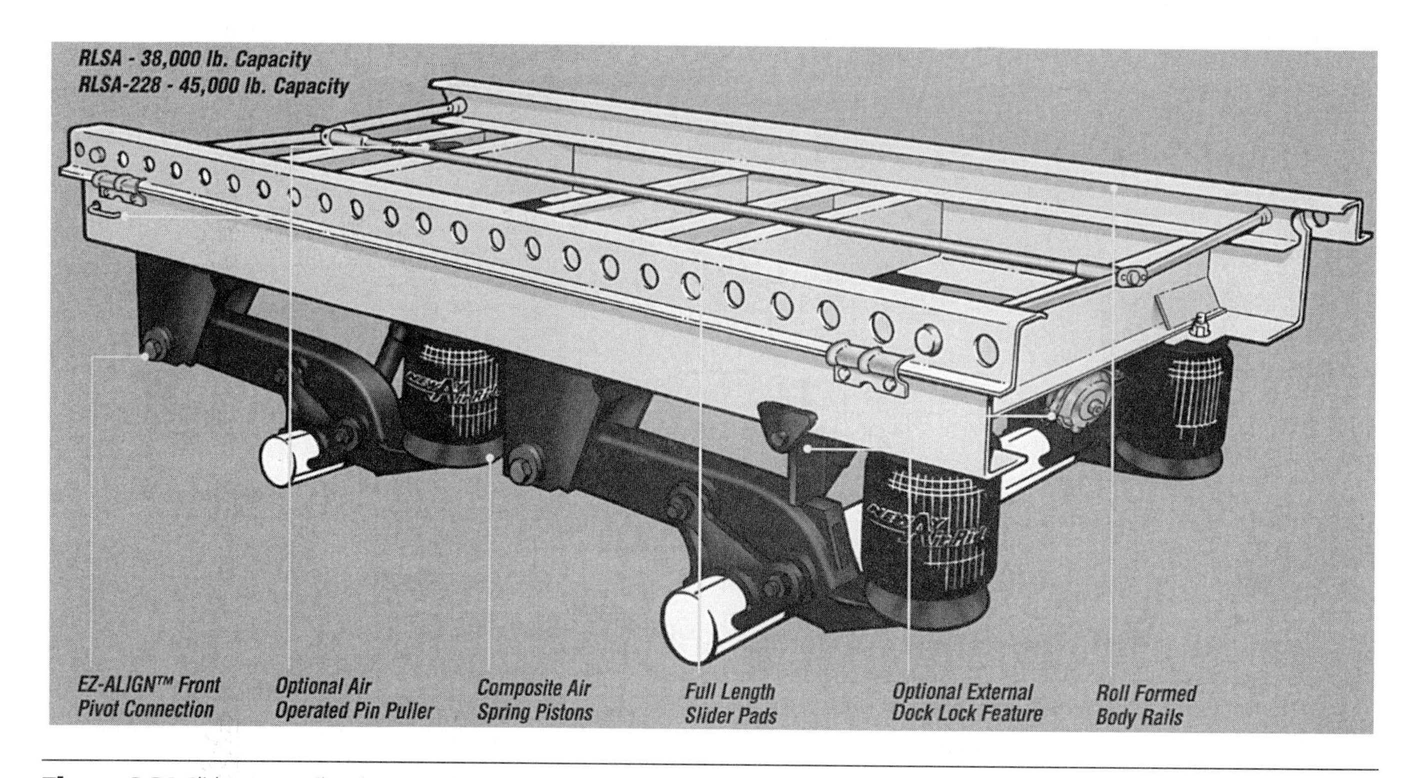

Figure 8-31 Slider-type trailer air suspension system. (Courtesy of Neway Anchorlok International, Inc.)

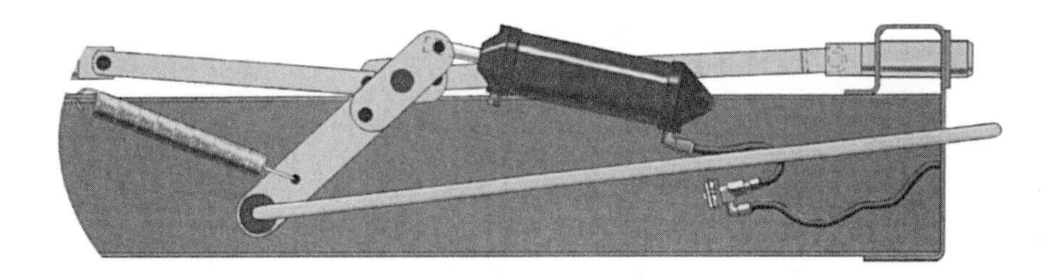

Figure 8-32 Air-operated pin puller for slider-type air suspension system. (Courtesy of Neway Anchorlok International, Inc.)

Trailer Air Suspension System with External Dock Lock Mechanism

An **external dock lock (EDL) system** is available with trailer air suspension systems. During normal trailer operation air pressure from the trailer air brake system is supplied to an EDL chamber. This air pressure forces the chamber diaphragm to extend the pushrod, and this rod movement rotates an EDL shaft and two flip plates so these plates are pulled upward. In the upward position these flip plates do not contact the air suspension system.

When the trailer parking brakes are applied, air is released from the EDL chamber. This action allows a diaphragm spring to retract the chamber pushrod. Pushrod retraction rotates the EDL shaft and rotates the flip plates downward so they are directly above the rigid beams on the suspension system (Figure 8-33). The flip plates maintain the trailer at dock height and prevent the trailer from walking forward. A Reyco model 86AR-RS1015 slider-type trailer air suspension system with double convoluted air springs and dock lock mechanism is illustrated in Figure 8-34.

WARNING: Driving the tractor and trailer with the external dock lock mechanism in the downward position will cause severe component damage and rough riding.

Trailer Air Suspension System with Air-Operated Lift Axle

Some trailer axles are available with a center bag lift (CBL). The lift air spring is mounted on the center of a frame crossmember. The lower side of the lift air spring is attached to a pivoted steel beam. A steel cable surrounds a pin in the outer end of this beam, and the lower end of this cable surrounds the rear axle (Figure 8-35). When the lift air spring control is off, air pressure is supplied

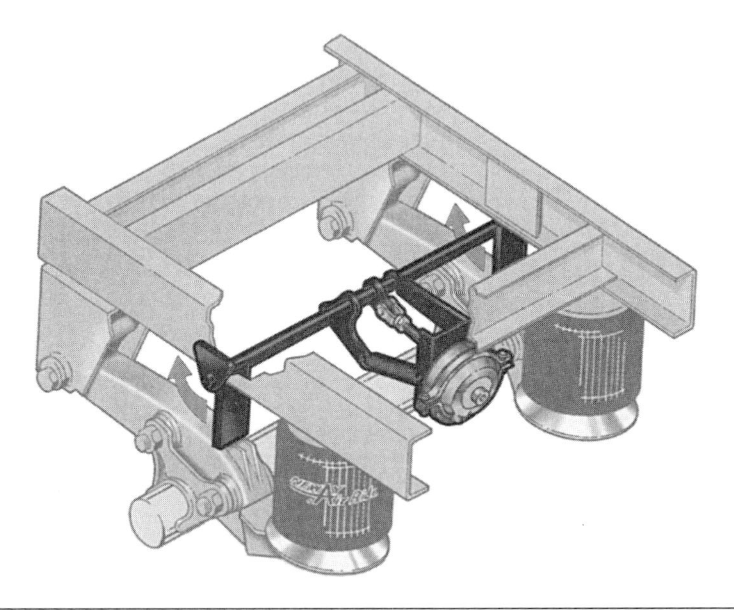

Figure 8-33 External dock lock system. (Courtesy of Neway Anchorlok International, Inc.)

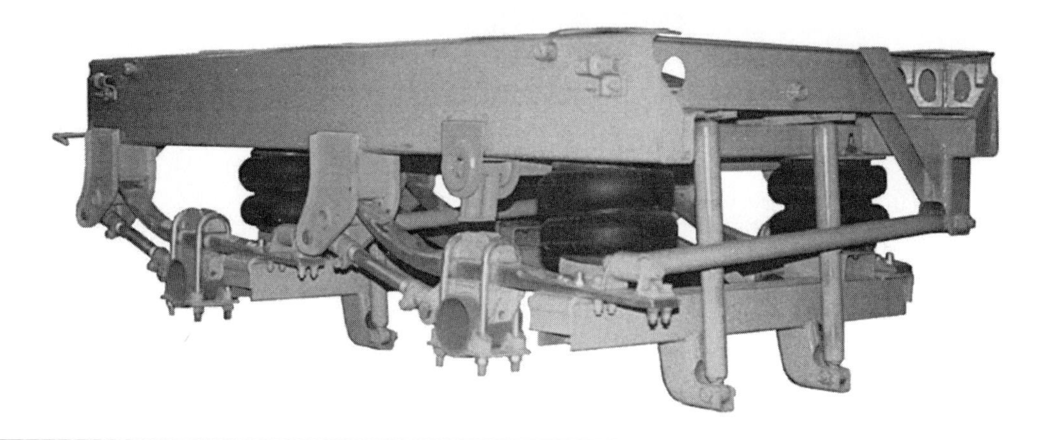

Figure 8-34 Slider-type trailer air suspension with double convoluted air springs and dock lock mechanism. (Courtesy of Reyco Industries Inc.)

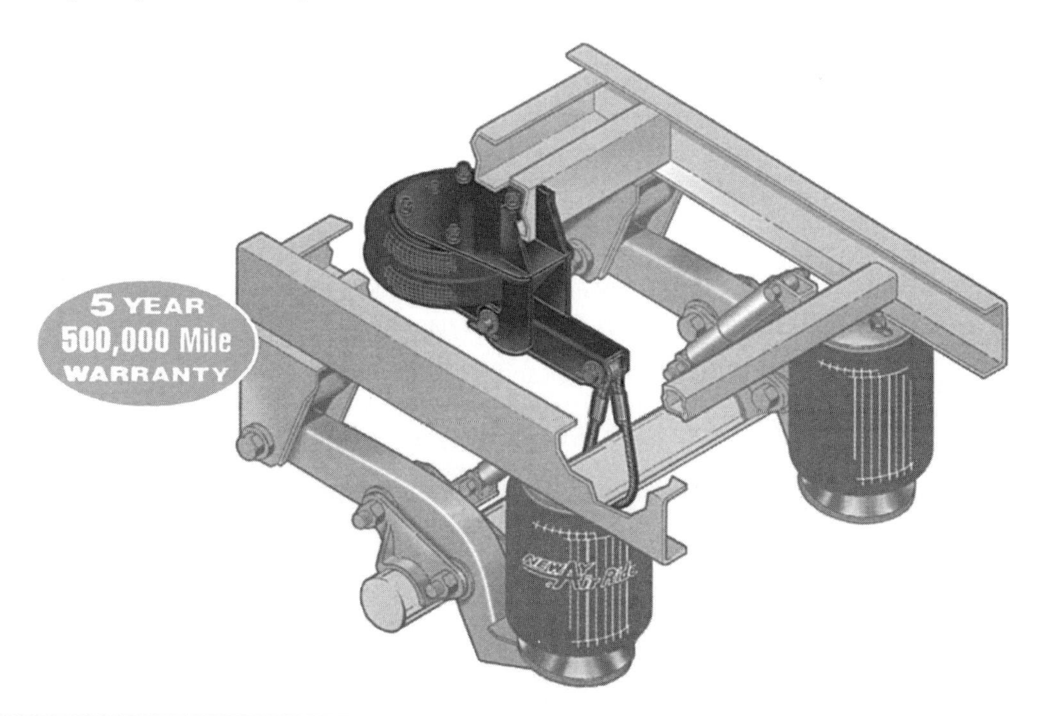

Figure 8-35 Trailer air suspension system with air-operated lift axle. (Courtesy of Neway Anchorlok International, Inc.)

See Chapter 8 in the Shop Manual for trailer air suspension system maintenance, diagnosis, and service.

to the air springs, and air is exhausted from the air lift spring. Under this condition the air pressure in the air springs holds the lift axle in the downward position so the axle is supporting some of the trailer weight. If the lift air spring control is placed in the on position, air pressure is exhausted from the air springs, and air pressure is supplied to the lift air spring. The air pressure in the lift air spring forces the outer end of the steel beam upward to lift the axle.

Cab Air Suspension Systems

Cab air suspension systems provide increased driver comfort and reduced driver fatigue compared with rubber cab mounts. Cabs with rubber mounts may be retrofitted with a cab air suspension system. Many cab air suspension systems contain two air springs, two shock absorbers, and a leveling valve (Figure 8-36). The leveling valve maintains the proper air pressure in the air springs to provide the correct cab height. Some cab air suspension systems have a single air spring and

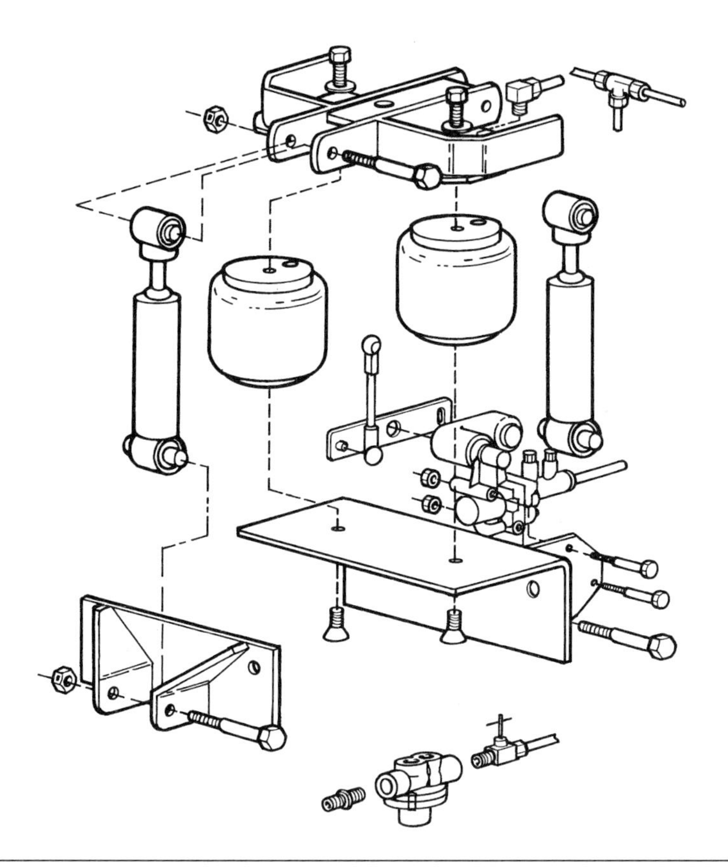

Figure 8-36 Cab air suspension system.

shock absorber. Other cab air suspension systems have a transverse rod to reduce lateral cab movement (Figure 8-37). Many variations exist in cab air suspension system mounting brackets, depending on the design of the cab.

Air pressure is supplied from one of the air brake system reservoirs through a pressure protection valve to the cab air suspension system. If an air leak occurs in the cab air suspension system and the air pressure drops below a specified value, the pressure protection valve closes to protect the air supply in the air brake system.

See Chapter 8 in the Shop Manual for cab air suspension system diagnosis and service.

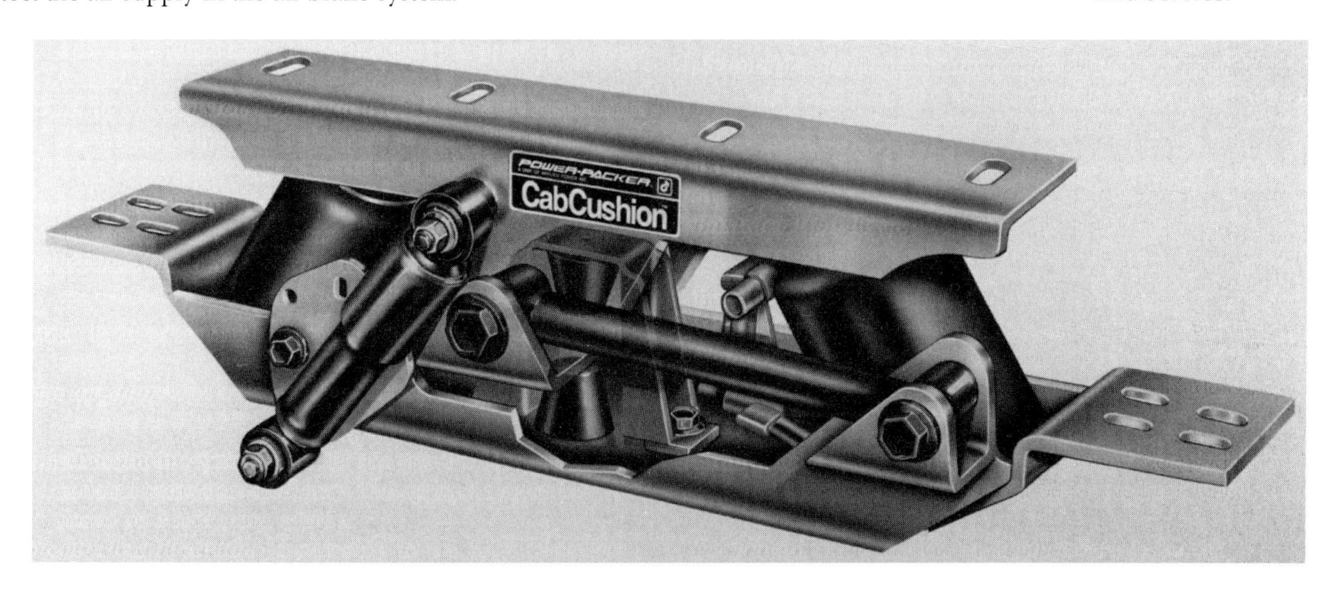

Figure 8-37 Cab air suspension system with a transverse rod, two air springs, one shock absorber, and a leveling valve. (Courtesy of APW Engineered Solutions [Power-Packer].)

Air Suspended Seat Systems

An **air suspended seat** reduces road shock transferred through the chassis, cab, and seat to the driver. Therefore an air suspended seat also increases driver comfort and reduces driver fatigue. An air suspended seat is supported on an air cushion, and a shock absorber also helps to cushion the seat movement (Figure 8-38).

Air pressure is supplied from one of the air brake system reservoirs through a pressure protection valve and a **pressure reducing valve** to the air suspended seat (Figure 8-39). The pressure protection valve serves the same purpose in cab air suspension systems and air suspended seat systems. The pressure reducing valve reduces the air pressure to the pressure specified by the air suspended seat manufacturer.

A BIT OF HISTORY

A wide variety of air spring designs have been used in various applications (Figure 8-40) during the evolution of air suspension systems. Types of air springs include double and single convolution bellows, rolling lobe, laterally restrained rolling lobe, bellobe, reversible diaphragm, and fully supported sleeve. One interesting type of air spring is the hydropneumatic type. The **hydropneumatic spring** is filled with a fluid, and a hose is connected from each spring to an accumulator. The accumulator contains a flexible diaphragm with a permanent gas charge under the diaphragm. When the tire strikes a hump in the road surface and the axle moves upward, fluid is forced from the spring into the area above the accumulator diaphragm. This action moves the diaphragm downward and compresses the gas charge to provide a cushioning action. Hydropneumatic springs may be used as stand alone units or in conjunction with constant rate leaf springs (Figure 8-41). Hydropneumatic suspension systems may be found on some heavy hauling highway vehicles.

Figure 8-38 Air suspended driver's seat. (Courtesy of Bostrom Seating, Inc.)

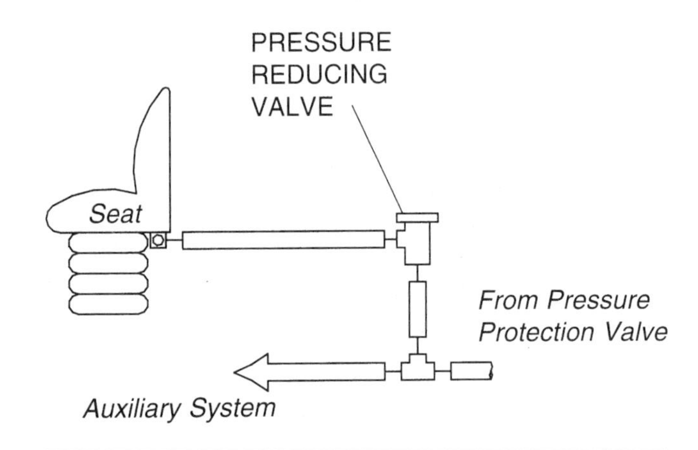

Figure 8-39 Pressure reducing valve in the air line to the air suspended seat. (Courtesy of Allied Signal Truck Brake Systems Co.)

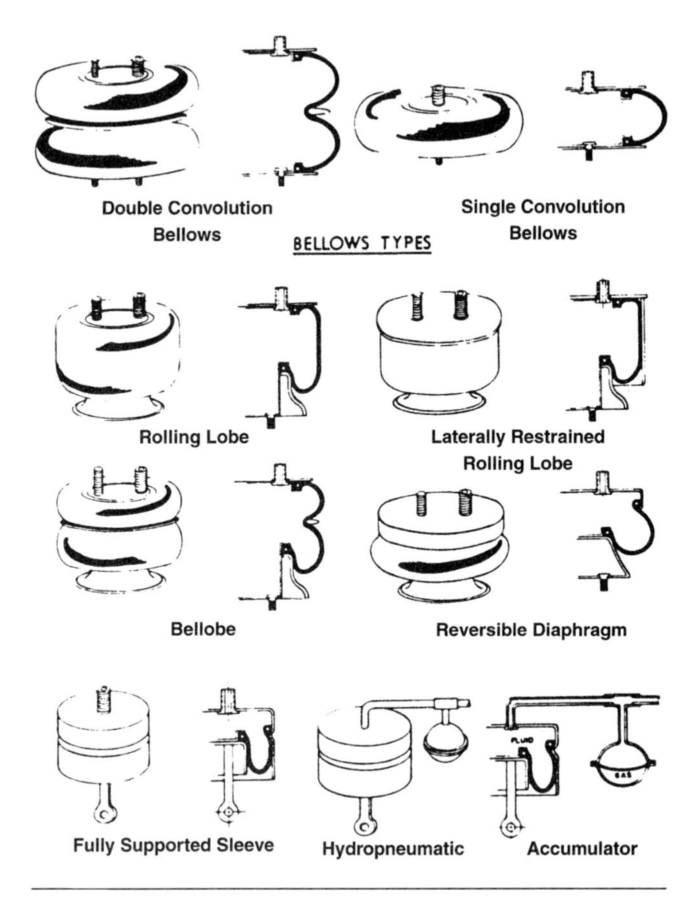

Figure 8-40 Various types of air springs. (Reprinted with permission from SAE Publications Principles and Applications of Pneumatic Springs © 1973 Society of Automotive Engineers, Inc.)

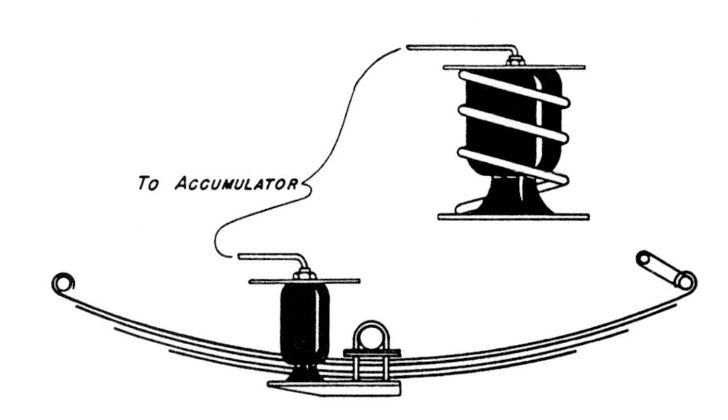

Figure 8-41 Combined hydropneumatic and leaf spring suspension. (Reprinted with permission from SAE Publications Principles and Applications of Pneumatic Springs © 1973 Society of Automotive Engineers, Inc.)

Coach Air Suspension Systems

Many coaches have air suspension systems because these systems provide improved ride quality and passenger comfort. Coach air suspension systems are similar to the air suspension systems described previously in this chapter. In a front axle air suspension system, each air spring support is bolted to the upper side of the front I-beam axle. A lower radius rod is positioned directly below the air spring support on each side of the axle, and the retaining bolts extend through the air spring support into the lower radius rod retainers (Figure 8-42). The rear end of the lower radius rod is attached to the chassis. An upper radius rod is connected from each air spring support to the chassis. The upper and lower radius rods maintain the proper longitudinal axle position. The upper radius rods also reduce vehicle roll while turning. A single transverse radius rod helps to maintain the proper longitudinal and lateral front axle position. Some front air suspension systems have two convoluted air springs on each end of the front axle. These dual air springs are mounted on a common support. This suspension system also has upper and lower radius rods (Figure 8-43).

Some rear air suspension systems have two convoluted air springs mounted on a common support that is bolted to the rear drive axle housing. Rear air suspension systems may also have a trailing axle supported by an air spring on each end of the axle (Figure 8-44). Some trailing axles have an air suspension with an air-operated lift mechanism (Figure 8-45). During normal operation with the trailing suspension lowered, the air spring is inflated and the lift actuator spring is is deflated. The trailing axle lift system may be operated by electric solenoid valves or manual valves. When the driver operates the manual valve or electric solenoid valves to lift the trailing axle, the

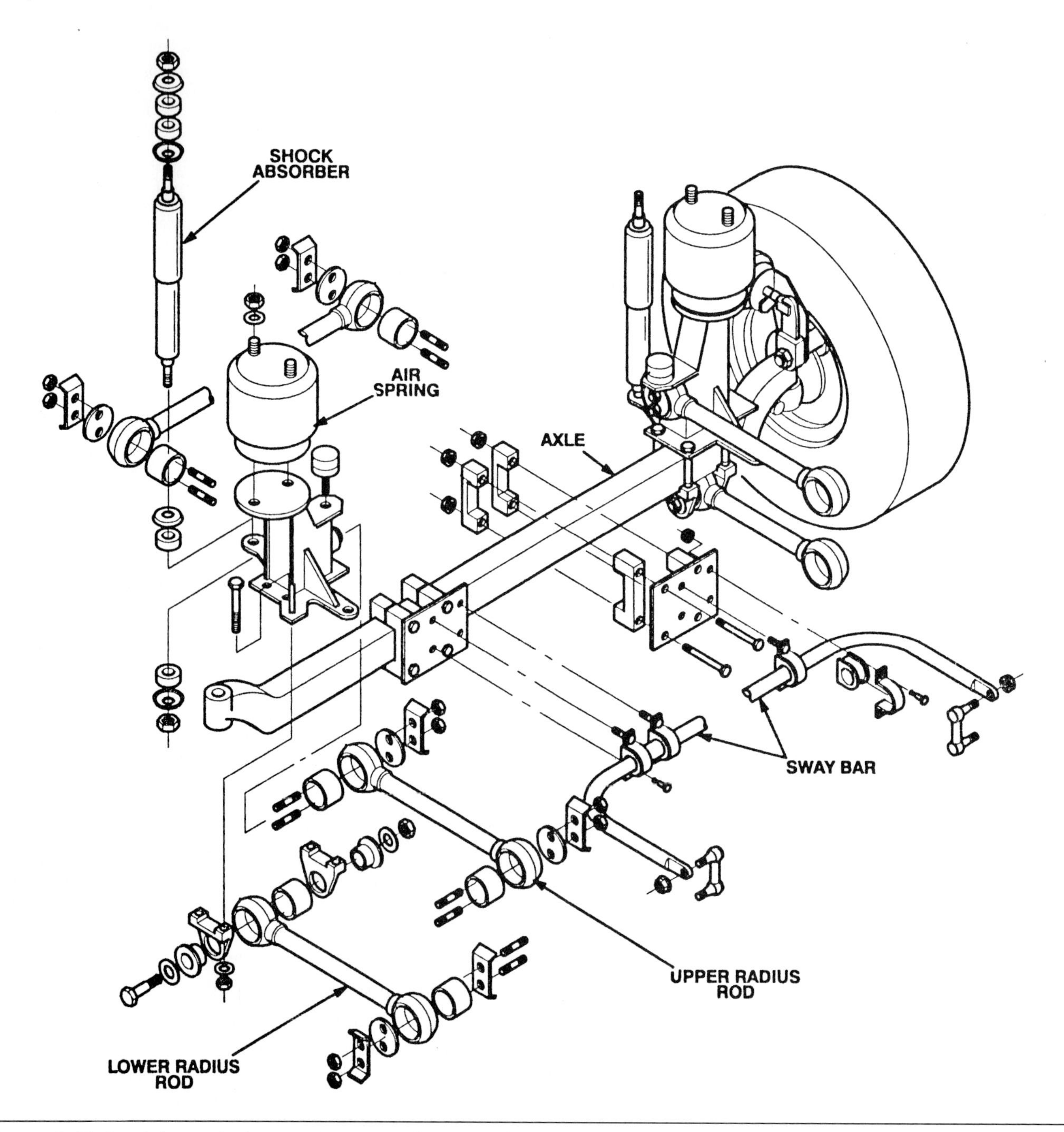

Figure 8-42 Coach front axle air suspension system. (Courtesy of Motor Coach Industries.)

air spring on the trailing axle is slowly deflated, and the lift actuator spring is inflated to lift the axle. If the trailing axle is lowered, the lift actuator spring is deflated and the air spring is inflated. When the trailing axle is lowered, a locking mechanism provides proper axle position. If the trailing axle is not properly locked in position, warning lights are illuminated in the instrument panel.

WARNING: The trailing axle should only be lifted at the speeds, loads, and driving conditions recommended by the coach manufacturer. Lifting the trailing axle while operating at speeds, loads, or driving conditions that are not recommended by the vehicle manufacturer may cause suspension damage, excessive tire wear, and unsafe vehicle operation.

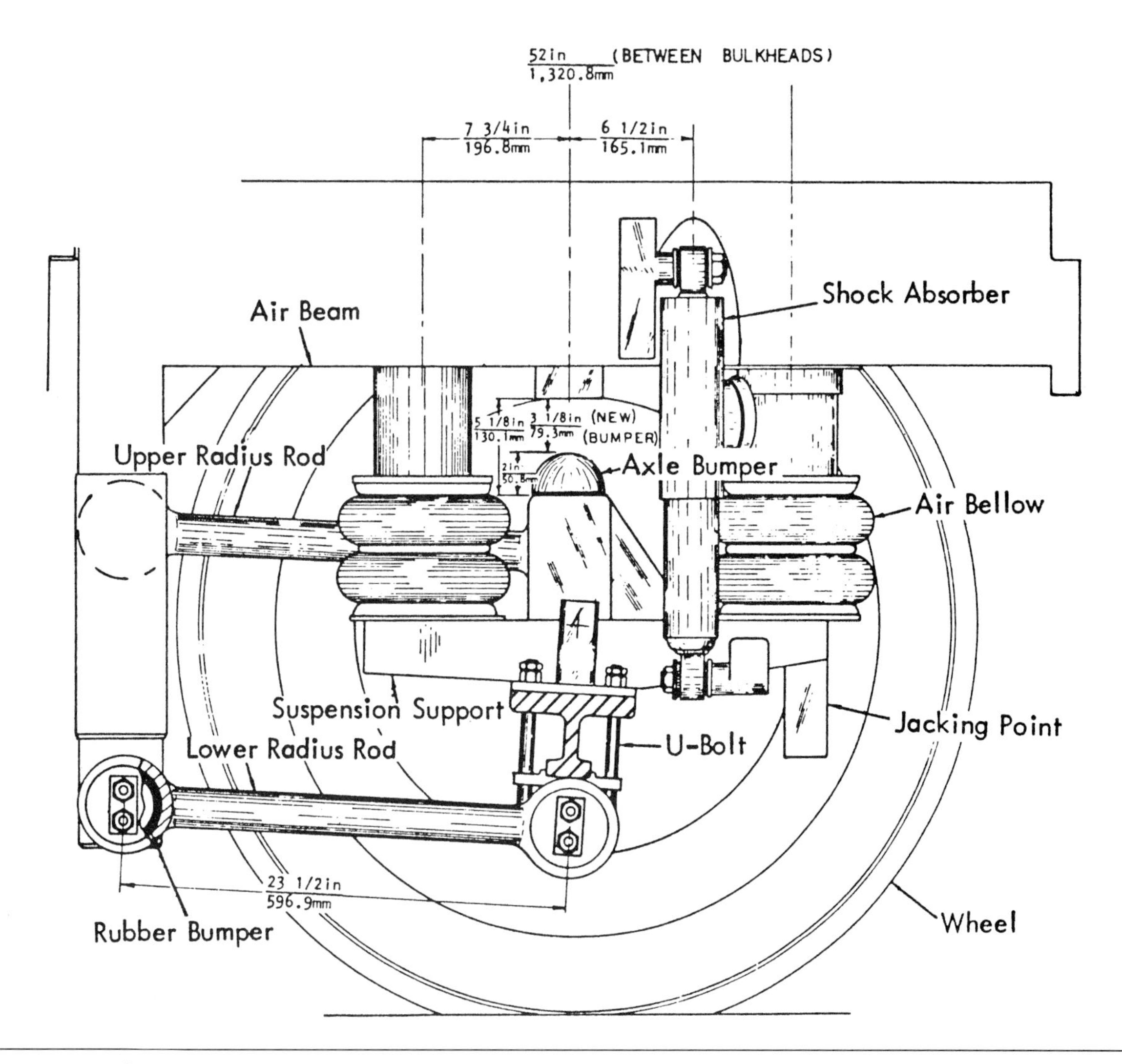

Figure 8-43 Coach front axle air suspension system with dual convoluted air springs at each end of the axle. (Courtesy of Motor Coach Industries.)

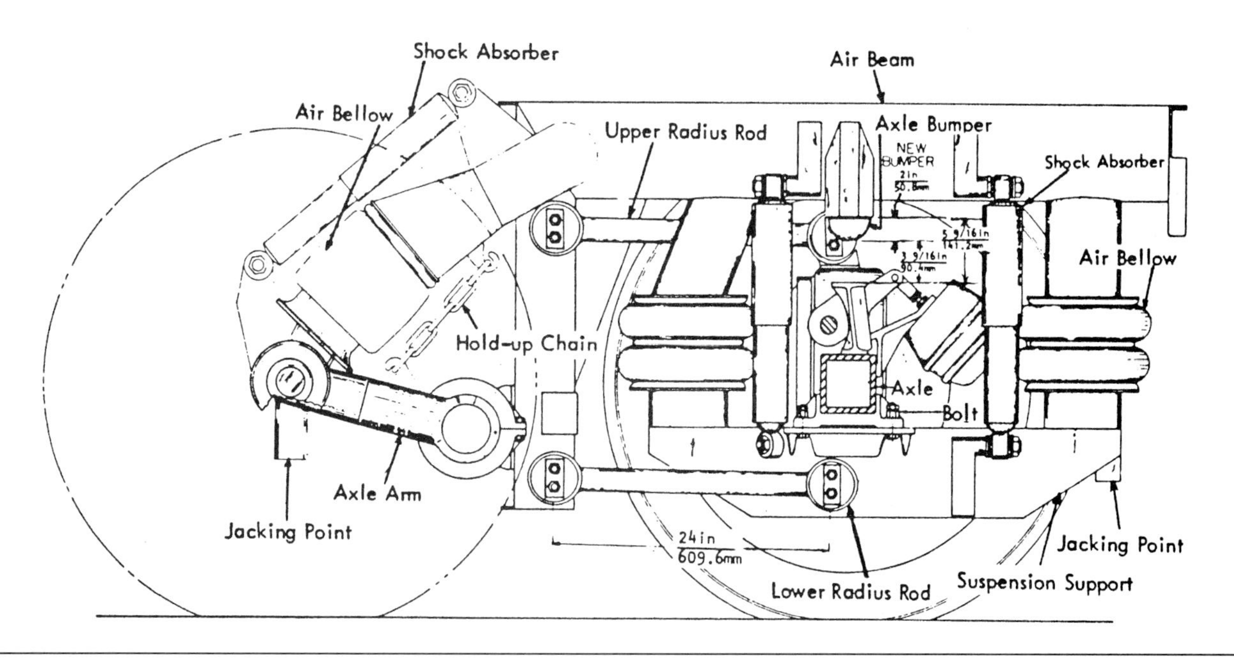

Figure 8-44 Coach rear air suspension with trailing axle. (Courtesy of Motor Coach Industries.)

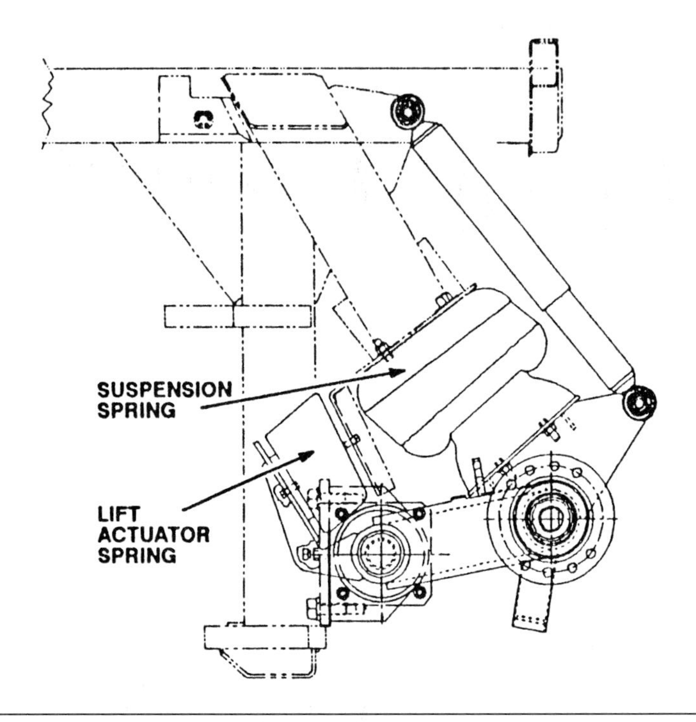

Figure 8-45 Coach trailing, lift axle. (Courtesy of Motor Coach Industries.)

WARNING: Never operate the vehicle with the trailing axle warning lights illuminated. This action may cause suspension damage, excessive tire wear, and unsafe driving conditions.

Summary

- An air suspension system minimizes road shock transferred from the suspension system to the truck frame, driver, and cargo.
- An air suspension system provides a constant frame height under various load conditions.
- In a rear axle air suspension system the air springs are mounted between the frame and the outer end of the main support beams, and these beams are attached to the rear axle. The front of the main support beam is mounted in a frame bracket.
- In a rear axle air suspension system a transverse beam is connected between the rear ends of the main support beams, and heavy duty shock absorbers are connected from the main support beams to frame brackets.
- The pressure protection valve closes and protects the air supply in the air brake system if a leak occurs in the air suspension system.
- A height control valve controls the air supply to the air springs to maintain a constant ride height. Air suspension systems may have single or dual height control valves.
- The control valve and relay valves provide a means of deflating the air springs when uncoupling a trailer from the tractor.
- Some lift axles have an air suspension system with a coil spring lift. In these systems the air pressure in the air springs is determined by the maximum pressure setting on the manual control valve.
- Other lift axles have an air suspension system and an air lift mechanism.
- Some trailer air suspension systems have an eccentric for axle alignment on the bolt that retains the front of the main support beam to the frame bracket.

Terms to Know

Air lift
Air springs
Air suspended seat
Air suspension system
Cab air suspension systems
Control valve
Cross channel
Dump valve
Electronic control unit (ECU)
Electronically controlled air suspension systems
External dock lock (EDL) system
Front bracket
Height control valve
Hydropneumatic spring
Lift axle

- ❑ Slider-type trailer air suspension systems can be moved on the trailer frame.
- ❑ Some trailer air suspension systems have an external dock lock mechanism to maintain the trailer position at loading docks.
- ❑ In many cab air suspension systems the cab is supported by two air springs and two shock absorbers in place of rubber mounts.
- ❑ An air suspended seat is supported by an air cushion for improved driver comfort.
- ❑ Air pressure is supplied to an air suspended seat through a pressure protection valve and a pressure reduction valve.

Terms to Know—cont'd

Main support beams

Pressure protection valve

Pressure reducing valve

Relay valves

Rigid beams

Slider-type trailer air suspension system

Torque rods

Trailer air suspension systems

Transverse beam

Transverse rod

Review Questions

Short Answer Essays

1. Explain the advantages of a tractor rear axle air suspension system.
2. Describe the design of a tractor rear tandem axle air suspension system.
3. Explain the purpose of a pressure protection valve in an air suspension system.
4. Describe the operation of a height control valve in an air suspension system.
5. Explain the operation of the control valve and relay valves during air spring deflation.
6. Explain the main purpose for air spring deflation.
7. Describe two methods of operating a dump valve in an air suspension system.
8. Explain the advantages of a lift axle.
9. Explain the operation of a lift axle with a coil spring lift.
10. Describe the operation of a lift axle with an air-lift mechanism.

Fill-in-the-Blanks

1. In an air suspension system the air springs are mounted near the back of the ____________________.
2. The front of the main support beam is retained in a ____________________.
3. The transverse beam between the rigid main support beams provides improved ____________________ and resistance to ____________________.
4. Air pressure is supplied to the air suspension system from one of the air brake system ____________________.
5. The height control valves are mounted on the tractor ____________________.
6. Depressing the control valve for an air suspension deflation system supplies air pressure to the ____________________ ports on the ____________________.
7. The maximum air pressure supplied to the air springs on a lift axle with a coil spring lift is determined by the maximum pressure setting on the ____________________.
8. In a lift axle with air suspension and an air lift mechanism, air is supplied to the air lift spring to ____________________ the axle.
9. A cab air suspension system usually contains two air springs, two ____________________ and a ____________________ valve.
10. Air pressure is usually supplied to an air suspended seat through a ____________________ valve and a ____________________valve.

ASE Style Review Questions

1. While discussing a tractor rear axle air suspension system:
   Technician A says the weight of the tractor is transmitted through the air spring to the main support beams and the rear axles.
   Technician B says the amount of air pressure in the air springs is controlled by a pressure regulating valve.
   Who is correct?
   A. A only **C.** Both A and B
   B. B only **D.** Neither A nor B

2. In a tractor rear axle suspension system:
   A. The height control valve is bolted to the air spring bracket.
   B. The linkage from the height control valve is attached to the main support beam.
   C. If a leak occurs in an air spring, the air pressure is lost in the air brake system.
   D. If the weight on the tractor frame is reduced, the height control valves supply more air pressure to the air springs.

3. While discussing the control valve and relay valves for air spring deflation:
   Technician A says the control valve is operated by air pressure from the pressure protection valve.
   Technician B says when air pressure is supplied to the control ports in the relay valves, these valves supply air pressure to the air springs.
   Who is correct?
   A. A only **C.** Both A and B
   B. B only **D.** Neither A nor B

4. All of these statements about a tractor rear axle air spring suspension system are true *except:*
   A. A dump valve may be integral with the height control valve.
   B. The dump valve may be electrically operated.
   C. The dump valve performs the same function as a relay valve.
   D. The tractor may be operated safely with the air springs deflated.

5. In a lift axle air suspension system with a coil spring lift:
   A. The maximum air pressure in the air springs is determined by the pressure setting on the pressure protection valve.
   B. When the axle is lifted, the air springs have 0 psi pressure.
   C. A quick release valve may be connected in the air lines to the air springs.
   D. The minimum air pressure in the air springs is determined by the quick release valve setting.

6. While discussing a lift axle with an air suspension system and an air lift mechanism:
   Technician A says an electronically operated solenoid valve may be used to control air pressure to the control port on the relay valve connected to the air springs.
   Technician B says when air pressure is supplied to the control port on the relay valve connected to the air springs, the axle is lifted.
   Who is correct?
   A. A only **C.** Both A and B
   B. B only **D.** Neither A nor B

7. On a lift axle with an air suspension system and air lift:
   A. The inlet air passage is normally open on relay valve connected to the lift air spring.
   B. Air pressure to the control port on the relay valve connected to the lift air spring may be supplied through a manual valve.
   C. When air pressure is supplied to the lift air spring, maximum air pressure is also supplied to the air springs.
   D. When air pressure is shut off to the lift air spring, the axle is lifted.

8. When discussing trailer air suspension systems:
   Technician A says the rigid beam to frame bracket bolt may contain an eccentric for axle alignment.
   Technician B says air pressure for the air springs is supplied from one of the trailer air brake reservoirs through a pressure protection valve.
   Who is correct?
   A. A only **C.** Both A and B
   B. B only **D.** Neither A nor B

9. A cab air suspension system:
   A. Has three air springs.
   B. Has a leveling valve.
   C. Is used with rubber cab mounts at the rear of the cab.
   D. Does not have shock absorbers.

10. While discussing an air suspended seat:
    Technician A says if the seat air cushion has a leak, the pressure reduction valve closes to protect the air brake system from pressure loss.
    Technician B says the pressure protection valve protects the air seat cushion if the air brake system has excessive pressure.
    Who is correct?
    A. A only **C.** Both A and B
    B. B only **D.** Neither A nor B

Wheel Alignment

Upon completion and review of this chapter, you should be able to:

- ❑ Describe the variables that affect wheel alignment.
- ❑ Define wheel alignment.
- ❑ Describe the safety hazards created by incorrect wheel alignment, or worn suspension and steering components.
- ❑ Define front wheel camber.
- ❑ Describe front tire tread wear caused by incorrect camber.
- ❑ Define front wheel caster.
- ❑ Describe the effects of positive and negative caster on directional control and steering effort.
- ❑ Describe positive and negative caster as they relate to ride quality.
- ❑ Describe how higher or lower than specified front or rear suspension height affects front suspension caster.
- ❑ Explain the kingpin inclination (KPI) angle and the included angle.
- ❑ Describe how KPI helps return the front wheels to the straight ahead position.
- ❑ Describe set back on front suspension systems.
- ❑ Describe toe-in and toe-out on front suspension systems.
- ❑ Describe tire tread wear caused by excessive toe-in.
- ❑ Explain how the front suspension system is designed to provide toe-out on turns.

Introduction

Truck and tractor engineers design suspension and steering systems that provide satisfactory vehicle control with acceptable driver effort and road feel. The vehicle should have a tendency to go straight ahead without being steered. This tendency is referred to as **directional stability.** A vehicle must have predictable directional control, which means the steering must provide a feeling that the vehicle will turn in the direction steered. The wheels must be reasonably easy to turn, and tire wear should be minimized. These steering qualities and tire conditions are achieved when front and rear wheel alignment angles are within the vehicle manufacturer's specifications.

Directional stability is the tendency of a vehicle to travel straight ahead without being steered.

The condition of suspension system components, and **wheel alignment,** are extremely important to maintain driving safety and normal tire wear. Worn suspension components, such as tie rod ends, can suddenly fall apart and cause complete loss of steering. This disastrous event may result in expensive property damage and the loss of human life. When alignment angles are incorrect, an uncontrolled vehicle swerve or skid may occur during hard braking, resulting in a serious accident. Severe misalignment reduces normal expected tire life. After suspension components such as tie rod ends and knuckle pins are replaced, wheel alignment is essential. Wheel alignment may be defined as an adjustment and refitting of suspension parts to original specifications, which ensures design performance.

Wheel alignment may be defined as an adjustment and refitting of suspension parts to original specifications that ensure design performance.

Wheel Alignment Theory

Road Variables

Vehicles are subjected to many variables in road surfaces and conditions that affect wheel alignment. These variables must be counteracted by the suspension design and alignment, or steering would be very difficult. Some of the **road variables** that affect wheel alignment and suspension design are the following:

1. Road crown, the curvature of the road surface.
2. Bumps and holes.
3. Natural crosswinds or crosswinds created by other vehicles.
4. Heavy loads or unequal weight distribution.

5. Road surface friction and conditions such as ice, snow, and water.
6. Tire traction and pressure.
7. Side forces while cornering.
8. Relationship of suspension parts as the front wheels turn and move vertically when road bumps and holes are encountered.

A desired plan to reduce tire wear would be to place the front wheels and tires so they are perfectly vertical, and therefore the tires would be flat on the road. However, if the wheels and tires are perfectly vertical, variables may change the wheels and tires from the true vertical position. Therefore tire wear and steering operation would be adversely affected.

Rather than allowing the variables to adversely affect tire wear and steering operation, the suspension and steering are designed with intentional characteristics that provide directional stability, predictable directional control, and minimum tire wear. Wheel alignment angles are designed to provide these desired requirements despite road variables. The rear wheels must also be aligned properly in relation to the front wheels, and tandem axles must be properly aligned in relation to each other.

Camber Fundamentals

Camber Definition

Camber refers to the tilt of a line through the tire and wheel centerline in relation to the true vertical centerline of the tire and wheel.

Positive camber is obtained when the top of the tire and wheel is tilted outward away from the true vertical line of the wheel assembly (Figure 9-1). If the front wheels have a slightly positive camber, when the vehicle load is placed on the front axle the camber angle is zero (Figure 9-2). **Negative camber** occurs when the tire and wheel centerline tilts inward in relation to the wheel assembly true vertical centerline. Improper camber angles on an I-beam front suspension are usually caused by worn components such as knuckle pins and knuckle pin bushings. On this type of suspension system bent components such as front axles or spindles may also cause improper camber angles.

Driving Conditions Affecting Camber

Wheel Jounce and Rebound

Wheel jounce refers to upward wheel movement.

Wheel rebound refers to downward wheel movement.

Wheel jounce may be referred to as vertical wheel movement that occurs when the tire and wheel strike a bump in the road surface. **Wheel rebound** refers to downward wheel movement that occurs when the wheel and tire move downward after wheel jounce. On an I-beam steer axle if one front tire strikes a bump and wheel jounce occurs, the opposite front wheel temporarily moves to a more positive camber position. If a front tire drops into a hole in the road surface on an I-beam steer axle, the opposite front wheel moves to a more negative camber position. This camber change during wheel jounce and rebound is undesirable because it adversely affects tire tread wear. However, the I-beam front axle is necessary to support the load encountered in trucks and tractors.

Tire Tread Wear

The camber angle may be referred to as one of the tire wear alignment angles. When the front wheels have the truck manufacturer's specified camber setting, maximum tire tread life and directional stability are maintained.

If a front wheel has excessive positive camber, the wheel is tilted outward and the vehicle weight is concentrated on the outside edge of the tire. Under this condition the outside edge of the tire has a smaller diameter than the inside tire edge. Therefore the outside tire edge has to complete more revolutions to travel the same distance as the inside tire edge. Because both edges are on the same tire, the outside edge must slip and scuff on the road surface as the wheel and tire revolve (Figure 9-3).

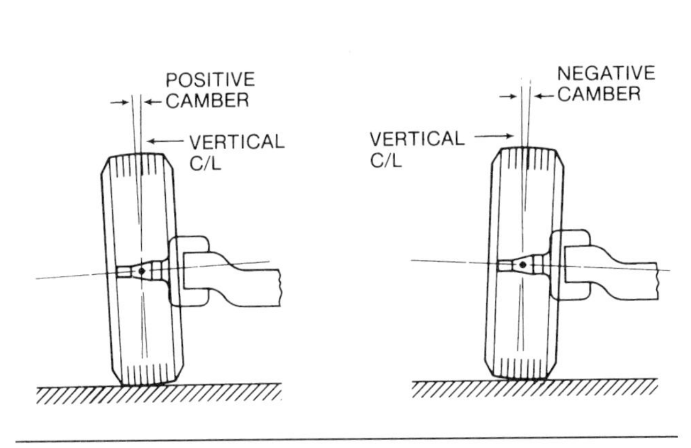

Figure 9-1 Positive and negative camber. (Courtesy of Eaton Corporation.)

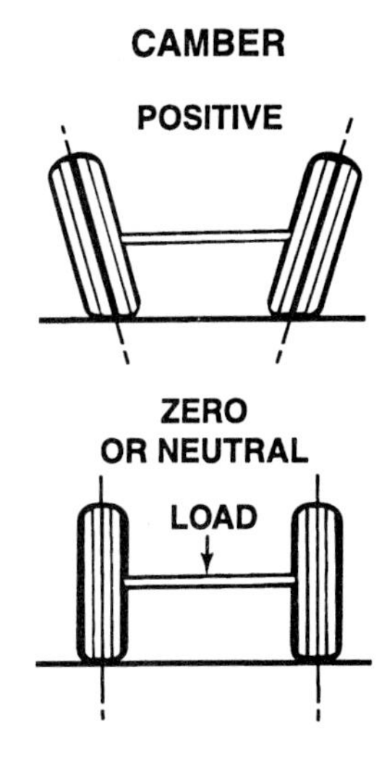

Figure 9-2 The effect of vehicle load on camber. (Courtesy of Meritor Automotive, Inc.)

Excessive negative camber tilts the wheel inward and concentrates the vehicle weight on the inside edge of the tire. This condition causes wear and scuffing on the inside edge of the tire tread. Therefore correct camber adjustment is extremely important to provide normal tire tread life.

Road Crown

A wheel that is tilted tends to steer in the direction it is tilted. For example, a bicycle rider tilts the bicycle in the direction he or she wishes to turn, which makes the turning process easier.

When camber angles are equal on both front wheels, the camber steering forces are equal and the vehicle tends to maintain a straight line position. If the camber on the front wheels is significantly unequal, the vehicle will drift to the side with the greatest degree of positive camber.

Road crown prevents water buildup on the driving surface. When a vehicle is driven on a crowned road it is actually driven on a slight slope, which causes the vehicle steering to pull toward the right.

When the front wheels have the manufacturer's specified camber, vehicle directional stability is maintained.

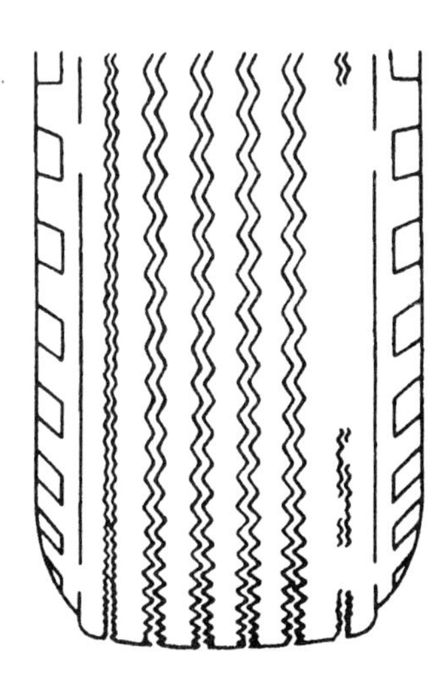

Figure 9-3 Tire tread wear caused by improper camber. (Courtesy of General Motors Corporation, Service Technology Group.)

Caster Fundamentals

Caster Definition

Caster refers to the tilt of a line that intersects the center of the kingpins in relation to a vertical line through the center of the wheel and spindle as viewed from the side.

Caster is the tilt of a vertical line through the center of the knuckle pins in relation to the true vertical centerline of the tire and wheel viewed from the side. **Positive caster** is present when the caster line is tilted backward toward the rear of the vehicle in relation to the vertical centerline of the tire and wheel (Figure 9-4).

Negative caster occurs when the centerline of the knuckle pins is tilted toward the front of the vehicle in relation to the vertical centerline of the tire and wheel.

Effects of Positive Caster

Kingpins may be called knuckle pins.

If a piece of furniture mounted on casters is pushed, the casters turn on their pivots to bring the wheels into line with the pushing force applied to the furniture. Therefore the furniture moves easily in a straight line (Figure 9-5).

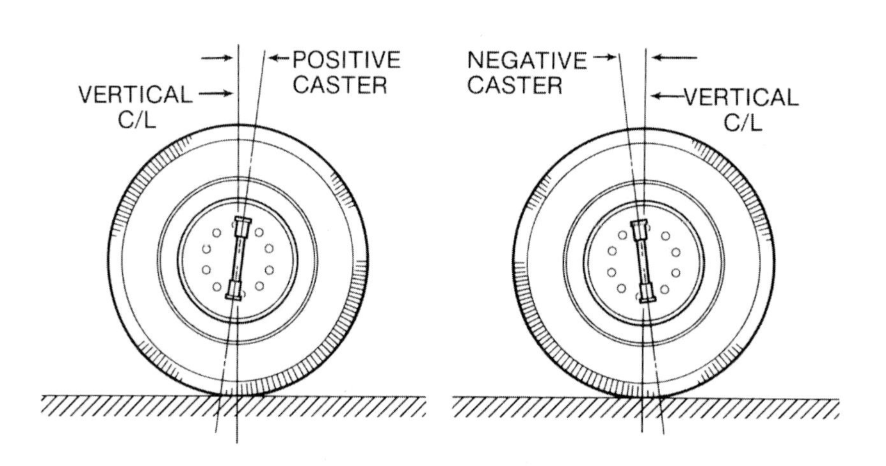

Figure 9-4 Positive and negative caster. (Courtesy of Eaton Corporation.)

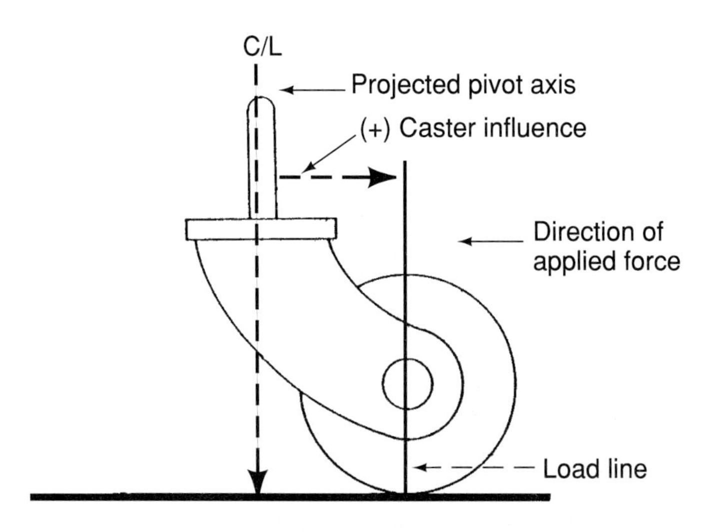

Figure 9-5 Caster action on furniture.

Any force exerted on the pivot causes the wheel to turn until it is lined up with the force on the pivot because the weight on the wheel results in resistance to wheel movement (Figure 9-6).

Most bicycles are designed with positive caster. The weight of the bicycle and rider is projected through the bicycle front forks to the road surface, and the tire pivots on the vertical centerline of the wheel and spindle when the handle bars and front wheel are turned. Notice the caster line through the center of the front forks is tilted rearward in relation to the vertical centerline of the wheel and spindle viewed from the side (Figure 9-7).

Because the pivot point is behind the caster line where the bicycle weight is projected against the road surface, the front wheel tends to return to the straight ahead position after a turn. The wheel also tends to remain in the straight ahead position as the bicycle is driven. Therefore the caster angle on a bicycle front wheel provides the same action as the caster angle on a piece of furniture.

Positive caster projects the vehicle weight ahead of the wheel centerline, whereas negative caster projects the vehicle weight behind the wheel centerline. Because positive caster causes a larger tire contact area behind the caster pivot point, this large contact area tends to follow the pivot point. This action tends to return the wheels to a straight ahead position after a turn and also helps to maintain the straight ahead position. Positive caster increases steering effort because the tendency of the tires to remain in the straight ahead position must be overcome during a turn. The returning force to the straight ahead position is proportional to the amount of positive caster. Positive caster helps to maintain vehicle directional stability.

Excessive positive caster is undesirable because it increases steering effort and creates a very rapid steering wheel return. If the caster angle is 0 degrees, the front spindles will rotate horizontally in relation to the road surface. However, a positive caster angle causes the left front spindle to tilt toward the road surface during a left turn.

This downward spindle movement tends to drive the tire into the road surface. Because this action cannot take place, the left side of the suspension and chassis is lifted. When the driver begins to return the wheel to the centered position, gravity forces the vehicle weight to its lowest position, which helps to return the steering wheel to the straight ahead position. Excessive positive caster increases the left front spindle downward tilt during a left turn, which increases the suspension and chassis lift. Therefore excessive positive caster increases steering effort. The same action occurs at the right front spindle, but this spindle tilts downward during a right turn.

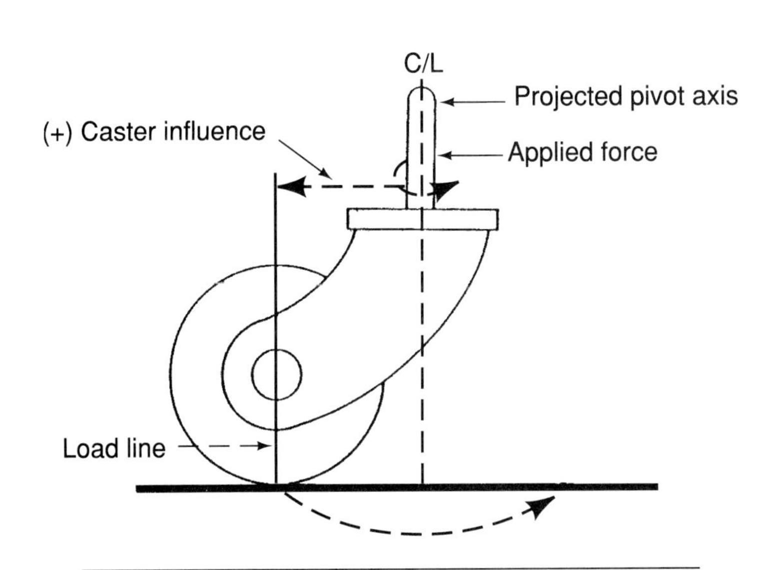

Figure 9-6 Furniture caster wheel aligned with the force on the pivot.

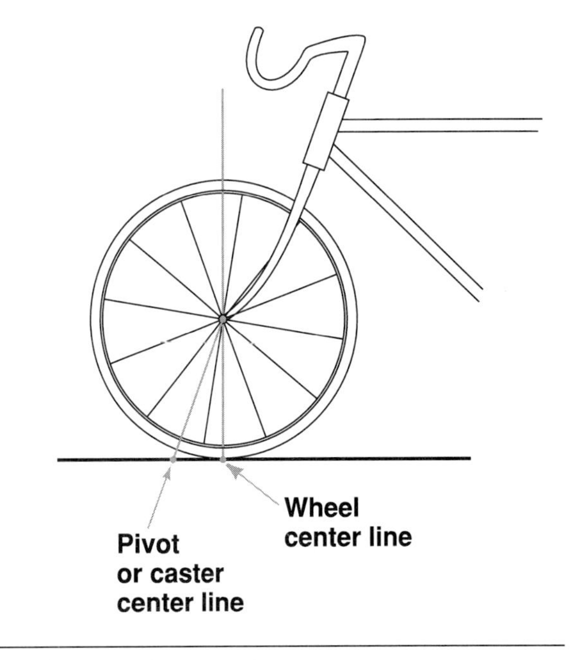

Figure 9-7 Caster line on a bicycle.

Front wheel shimmy refers to rapid side-to-side front wheel movement.

Harsh riding may be caused by excessive positive caster because the caster line is actually aimed at some road irregularities (Figure 9-8). Excessive positive caster may cause front **wheel shimmy** from side to side at low speeds.

When the caster line is aimed directly at the road irregularity, road shock is transmitted through the kingpin to the suspension and chassis. A small degree of positive caster allows the front wheel to roll into a road depression without the caster line being aimed at the hole in the road. Therefore this type of caster line improves ride quality.

If one front wheel has more positive caster than the other front wheel, the steering pulls toward the side that has the least amount of positive caster. The most important facts about positive caster are the following:

1. Positive caster helps the front wheels return to the straight ahead position after a turn.
2. Correct positive caster provides improved directional stability of a vehicle.
3. Excessive positive caster produces harsh riding quality.
4. Excessive positive caster promotes sideways front wheel shimmy.

Effects of Negative Caster

Negative caster moves the centerline of the kingpin behind the vertical centerline of the wheel and spindle at the road surface. If this condition is present, the friction of the tire causes the tire to pivot around the point where the centerline of the kingpin meets the road surface. When this occurs, the wheel is pulled away from the straight ahead position, which decreases directional stability.

Negative caster reduces steering effort. Because excessive positive caster increases road shock transmitted to the suspension and chassis, negative caster reduces this shock and improves ride quality. This improvement occurs because the front wheel rolls into a road depression without the caster line being aimed at the hole in the road.

Effects of Suspension Height on Caster

When the rear springs become sagged or overloaded, the caster on the front wheels becomes more positive (Figure 9-9). This action explains why front wheel shimmy may occur when a trunk is severely overloaded.

If someone raises the rear suspension height above the vehicle manufacturer's specification, the caster on the front wheels becomes less positive (Figure 9-10). Under this condition the front wheel caster may change from positive to negative. This fact explains why a vehicle may have reduced directional stability and control when the rear suspension height is raised. When the front suspension height is raised above the vehicle manufacturer's specification, the front wheel caster becomes less positive. The most important facts about negative caster are these:

1. Negative caster does not help return the front wheels to the straight ahead position after a turn.

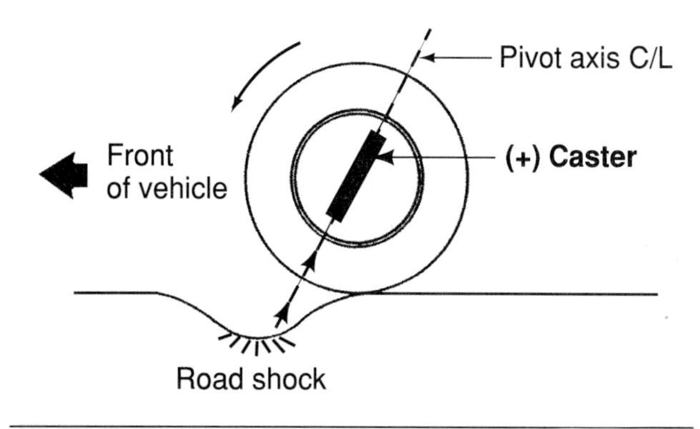

Figure 9-8 Harsh riding quality caused by excessive positive caster.

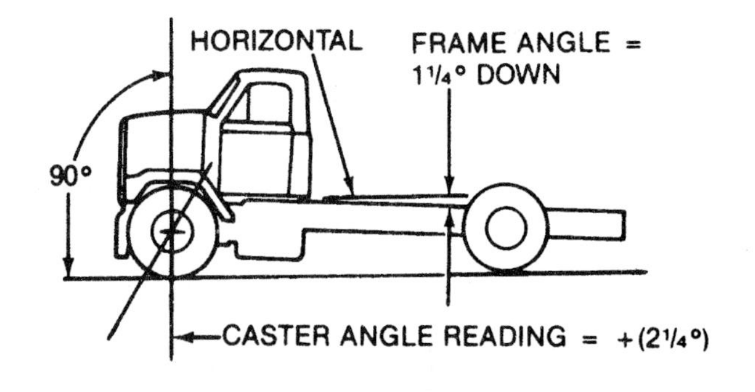

Figure 9-9 Reduced ride height at the rear axle increases positive caster on the front axle. (Courtesy of General Motors Corporation, Service Technology Group.)

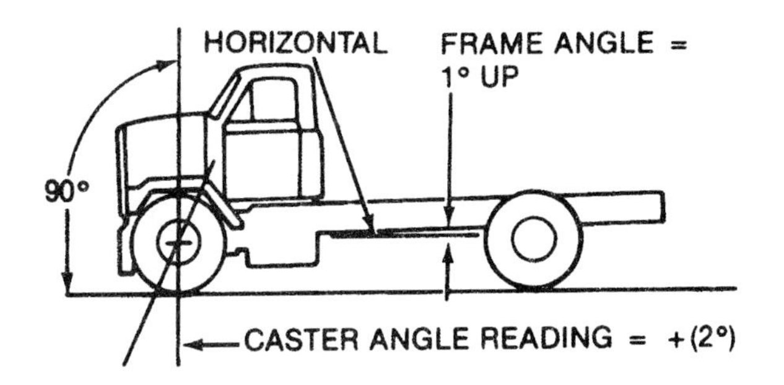

Figure 9-10 Excessive ride height at the rear axle decreases positive caster on the front axle. (Courtesy of General Motors Corporation, Service Technology Group.)

2. Negative caster contributes to directional instability and reduced directional control.
3. Negative caster does not contribute to front wheel shimmy.
4. Negative caster reduces road shock transmitted to the suspension and chassis.

Safety Factors and Caster

Directional Control

As explained previously in this chapter, positive caster provides increased directional stability and control, whereas negative caster reduces directional stability. Therefore front wheel caster must be adjusted to manufacturer's specifications to maintain vehicle directional control and safe handling characteristics.

See Chapter 9 in the Shop Manual for camber and caster diagnosis and adjustment.

Suspension Height

We have already explained how incorrect front or rear suspension height results in changes in front wheel caster. Therefore abnormal suspension heights may contribute to reduced directional control and unsafe steering characteristics.

Steering Terminology

In the truck service industry certain terms are used for specific steering problems. Some of these problems are related to camber and caster, whereas other problems are caused by various suspension or steering defects. Technicians must be familiar with this steering terminology and the cause of these problems.

Torque steer may be defined as the tendency of the steering to pull to one side during hard acceleration.

Bump steer is the tendency of the steering to veer suddenly in one direction when one or both of the front wheels strike a bump.

When **memory steer** occurs, the vehicle does not want to steer straight ahead after a turn because the steering does not return to the straight ahead position.

A front wheel drive vehicle with unequal drive axle lengths produces some torque steer on hard acceleration. On a front wheel drive vehicle, torque steer is aggravated by different tire tread designs on the front tires or uneven wear on the front tires.

Memory steer may be caused by a binding condition in the steering column or in the steering shaft universal joints. A binding upper strut mount may result in memory steer. Negative caster or reduced positive caster also causes memory steer.

Steering pull or drift is the tendency of the steering to gradually pull to the right or left when the vehicle is driven straight ahead on a level road. Steering pull or drift may be caused by improper caster or camber angles or improper rear axle alignment.

When **steering wander** occurs, the vehicle tends to steer in either direction rather than straight ahead on a level road surface. Steering wander may be caused by improper caster adjustment or improper rear axle alignment.

Kingpin Inclination (KPI) Definition

KPI is the angle of a line through the center of the knuckle pins in relation to the true vertical centerline of the wheel and tire viewed from the front of the vehicle.

Kingpin inclination (KPI) may be called steering axis inclination (SAI).

The included angle is the sum of the KPI angle and the positive camber angle.

Improper **kingpin inclination** (KPI) angles on either side of the front suspension may cause hazardous steering conditions while braking or accelerating. Therefore technicians must be familiar with KPI and other related steering geometry.

On medium and heavy duty trucks with I-beam front axles, KPI refers to the inward tilt of a line through the center of the knuckle pins in relation to the true vertical line through the center of the tire and wheel. These two lines are viewed from the front of the vehicle, and the KPI line always tilts inward in relation to the true vertical line (Figure 9-11). Incorrect KPI on an I-beam front suspension is usually caused by bent components such as a bent I-beam axle. The included angle is the sum of the KPI angle and the camber angle.

If the camber angle is positive, this angle is added to the KPI angle to obtain the included angle. A negative camber angle must be subtracted from the KPI angle to calculate the included angle.

KPI Purpose

When the KPI angle is tilted toward the center of the vehicle and the wheels are straight ahead, the height of the spindle is raised closer to the chassis. This action lowers the height of the vehicle because of gravity. When the front wheels are turned, each spindle moves through an arc that tries to force the tire into the ground. Because this reaction cannot take place, the chassis lifts when the wheels are turned. When the steering wheel is released after a turn, thc vehicle weight has a tendency to settle to its lowest point of gravity. Therefore KPI helps return the wheels to the straight ahead position after a turn, and KPI also tends to maintain the wheels in the straight ahead position. However, KPI does increase steering effort because the chassis has to lift slightly on turns.

The vehicle weight is projected through the KPI line to the road surface. Let us assume that a front suspension is designed with a vertical (0 degrees) KPI line, and this line meets the road surface a considerable distance inside the tire vertical centerline. Under this condition, the vehicle weight is projected through the KPI line a considerable distance inside the true vertical tire center-

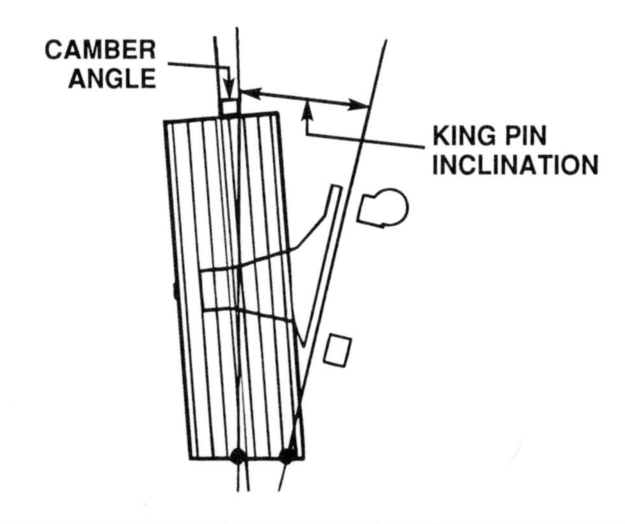

Figure 9-11 Kingpin inclination. (Courtesy of Meritor Automotive, Inc.)

line. With this type of front suspension severe tire scuffing would occur because the wheel and tire pivot around the KPI line during a turn.

With a 0 degree KPI line and a considerable distance between the KPI line and the tire vertical centerline at the road surface, greater steering effort is required during a turn and stress on the steering mechanism increases. This type of front suspension design causes excessive road shock and kickback on the steering wheel during a turn because the distance between the KPI line and the tire vertical line returns the wheels to the straight ahead position.

When the KPI line meets the road surface near the point where the tire vertical centerline meets the road surface, tire wear is reduced, steering effort is decreased, and stress on steering components is diminished.

See Chapter 9 in the Shop Manual for KPI measurement.

Kingpin Offset

Kingpin offset affects steering quality related to stability and returnability. However, kingpin offset is not an alignment angle, and it cannot be measured on conventional alignment equipment. Kingpin offset is the distance from the point where the tire vertical line contacts the road, and the location where the line through the center of the knuckle pin meets the road surface. If the line through the knuckle pin meets the road surface inside the tire vertical line, the kingpin offset is positive. Trucks with I-beam front suspension systems are designed with a positive kingpin offset. Refer to Figure 9-11. A negative kingpin offset is provided when the line through the kingpin center meets the road surface outside the tire and hub center line. (Refer to Chapter 7 for kingpin offset theory.)

Kingpin offset may be called scrub radius.

Kingpin offset is the distance between the KPI line and the true vertical centerline of the tire at the road surface.

WARNING: Installing larger tires or different rims than the ones specified by the vehicle manufacturer changes the kingpin offset, which may result in reduced directional control.

Positive kingpin offset occurs when the KPI line meets the road surface inside the true vertical centerline of the tire at the road surface.

Negative kingpin offset is present when the KPI line meets the road surface outside the true vertical centerline of the tire at the road surface.

Set Back

Set back is a condition in which one wheel is moved rearward in relation to the other front wheel.

Set back will not affect handling unless it is extreme. Collision damage may drive one front wheel rearward and bend the I-beam axle, causing excessive set back. Set back can also occur on rear wheels if the rear drive axle housing is bent, but it is more likely to occur on front wheels because of collision damage. Some computer-type wheel aligners have set back measuring capabilities.

Set back refers to a condition in which one wheel is moved rearward in relation to the opposite wheel.

Toe Definition

When the distance between the rear inside tire edges is greater than the distance between the front inside tire edges, the front wheels have a **toe-in** setting. **Toe-out** occurs when the distance between the inside front tire edges exceeds the distance between the inside rear tire edges (Figure 9-12). When the front wheel toe is not adjusted to manufacturer's specifications, front tire tread wear is excessive.

Toe Setting

On rear wheel drive trucks front tire friction on the road surface moves the wheels toward the toe-out position when the truck is driven. On this type of vehicle, manufacturers usually specify a slight toe-in on the front wheels. The front wheels are adjusted to a slight toe-in with the vehicle at rest so the wheels will be parallel to each other when the vehicle is driven on the road. A slight amount of lateral movement always exists in steering linkages. Forces acting on the front wheels will try to

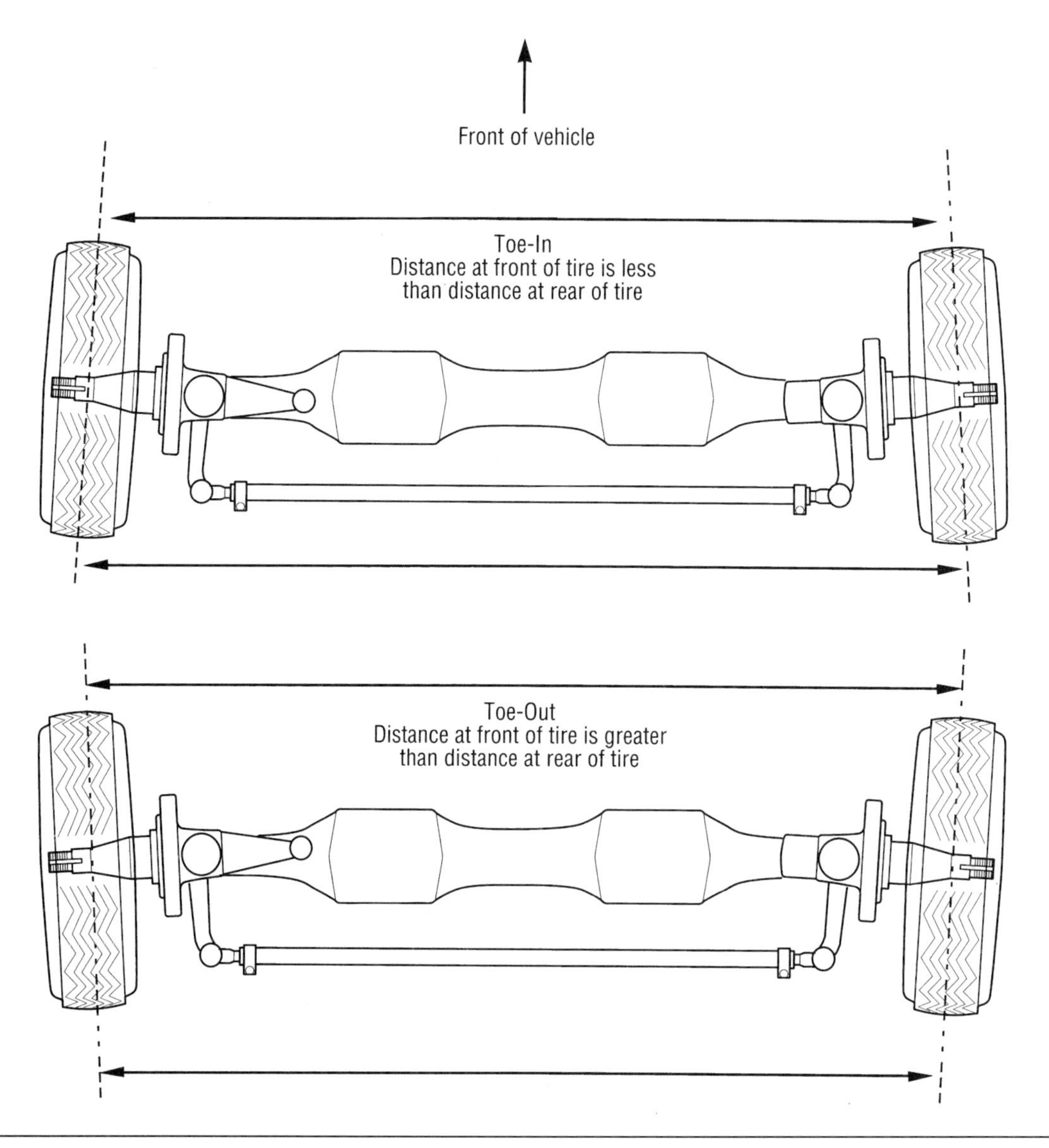

Figure 9-12 Front wheel toe-in and toe-out. (Courtesy of Eaton Corporation.)

Front wheel toe is the distance between the inside front edges of the tires compared with the distance between the inside rear edges of the tires measured at spindle height viewed from the top.

compress or stretch the steering linkages when the vehicle is driven. Because most medium- and heavy-duty trucks have the steering linkage behind the front wheels, the forces on the front wheels while driving the truck tend to compress the steering linkage. This action moves the front wheels slightly toward a toe-out position.

Toe Adjustment and Tire Wear

WARNING: Improper toe adjustment causes rapid tire wear, which may result in tire failure, collision damage, and personal injury.

Ford Motor Company has calculated that a toe-in error of $\frac{1}{8}$ inch (in) (3.17 millimeters [mm]) is equivalent to dragging the tires crosswise for 11 feet (3.3 meters) for each mile the vehicle is driven. This crosswise movement causes severe feathered tire tread wear (Figure 9-13). Improper toe adjustment is the most common cause of rapid tire tread wear.

Excessive toe-out causes wear on the inside of the tire tread ribs and a sharp feathered edge on the outside of the tread ribs (Figure 9-13). If excessive toe-in is present, the tire tread wear is reversed.

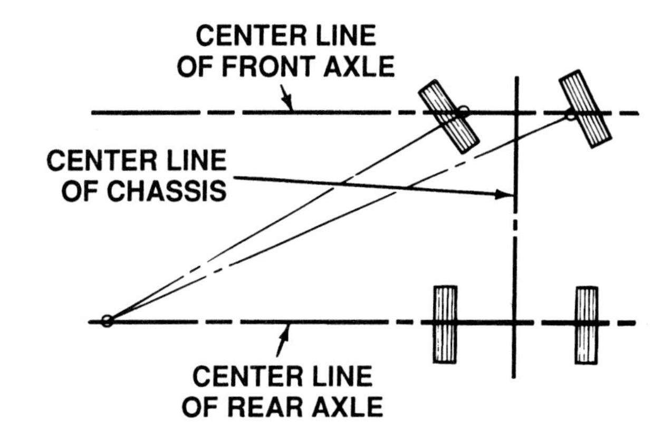

Figure 9-13 Tire tread wear on right front tire caused by excessive toe-in. (Courtesy of General Motors Corporation, Service Technology Group.)

Figure 9-14 Front and rear wheels turning around a common center. (Courtesy of Meritor Automotive, Inc.)

CAUTION: Worn steering components may become disconnected suddenly causing complete loss of steering control, collision damage, and personal injury.

Worn steering linkage components such as tie rod ends cause incorrect and erratic toe-in settings. If the front springs become weak, the front suspension height is lowered. When this occurs, the pitman arm and the idler arm move downward with the chassis. This action moves the tie rods to a more horizontal position, which tends to push outward on the steering arms and increase front wheel toe-in. The toe-in change we have just described occurs when the steering linkage is located at the rear of the front wheels.

See Chapter 9 in the Shop Manual for toe measurement.

Turning Radius

Front and Rear Wheel Turning Action

TRADE JARGON: **Turning radius** may be referred to as **cornering angle** or **Ackerman angle.**

When a vehicle turns a corner, the front and rear wheels must turn around a common center with respect to the turn radius (Figure 9-14). On a single rear axle vehicle this common center is located at the center of the rear wheels. This common center is positioned at the center between the tandem axles on a truck or tractor with a tandem rear axle.

Turning radius or cornering angle is the amount of toe-out on turns.

On most front suspension systems the front wheels pivot independently at different distances from the center of the turn, and therefore the front wheels must turn at different angles. The inside front wheel must turn at a sharper angle compared with the outside wheel. This action is necessary because the inside wheel is actually ahead of the outside wheel. When this turning action occurs, both front wheels remain perpendicular to their turning radius, which prevents tire scuffing.

Steering Arm Design

An understanding of a lever moving in a circle is necessary before an explanation of steering arm design and operation. If a lever moves from point *A* to *B,* it pivots around point *O* and moves through a horizontal distance *A* to *B* (Figure 9-15).

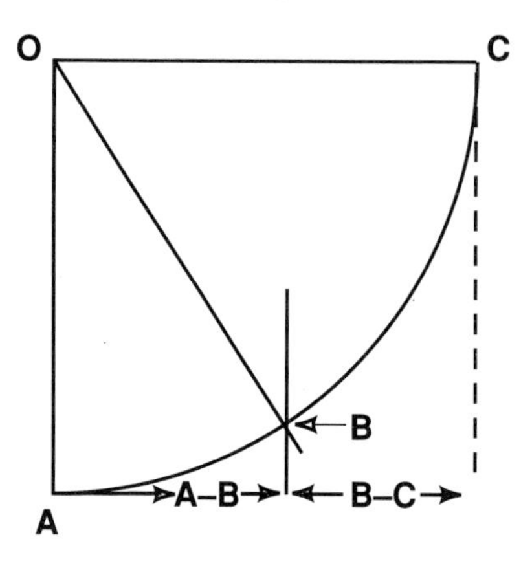

Figure 9-15 Lever movement in a circle.

Lever movement through arc *B* to *C* is much greater than the movement through arc *A* to *B*. However, during arc *B* to *C* the lever moves through horizontal distance *B* to *B* and this distance is the same as horizontal distance *A* to *B*.

The steering arms are connected from the tie rod ends to the front spindles. Steering arms and linkages maintain the front wheels parallel to each other when the vehicle is driven straight ahead. However, the steering arms are not parallel to each other. If the steering linkage is at the rear edge of the front wheels, the steering arms are closer together at the point where the tie rods connect than at their spindle pivot point (Figure 9-16). When the steering linkage is positioned at the front edge of the front wheels, the steering arms are closer together at their spindle pivot point than at the tie rod connecting point.

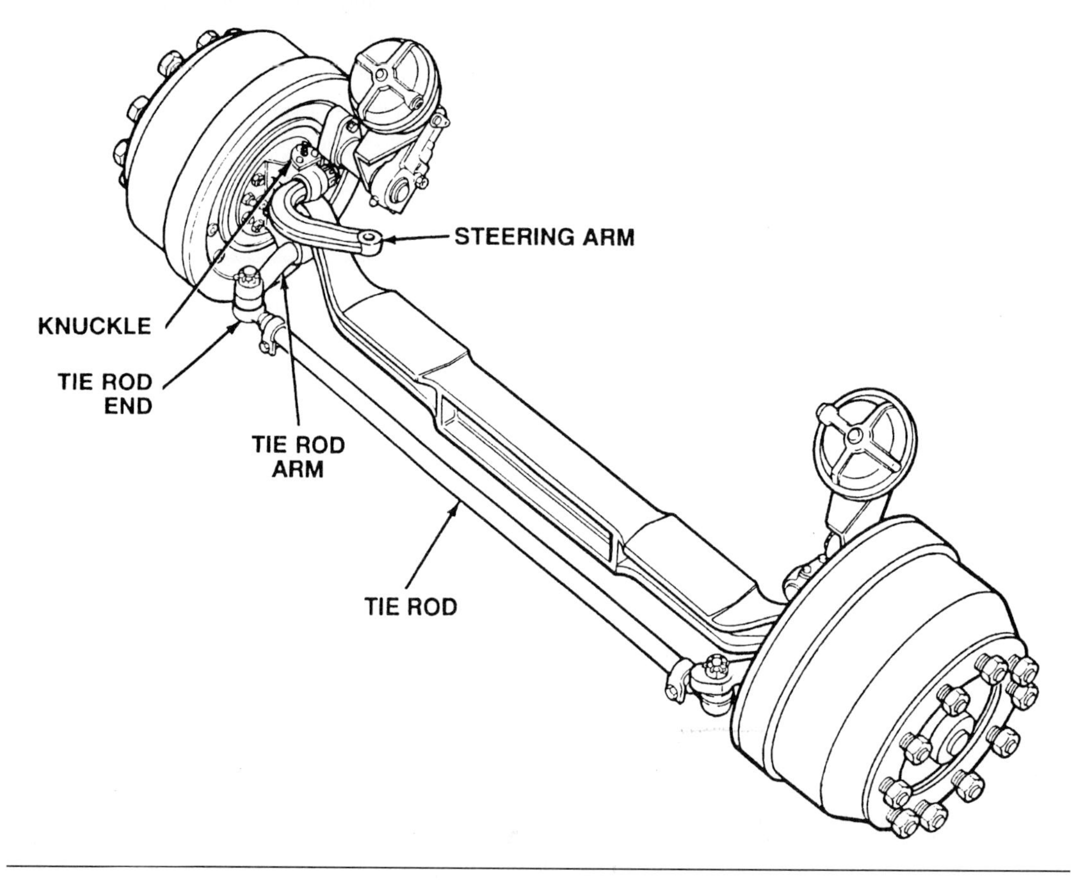

Figure 9-16 Steering arms are closer together at the point where the tie rods connect than at their spindle pivot points. (Courtesy of Meritor Automotive, Inc.)

When the front wheels are turned on a vehicle with the steering linkage at the rear edge of the front wheels, the angle formed by the inside steering arm and linkage increases, whereas the angle of the outside steering arm and linkage decreases. The inside steering arm moves through the longer arc *X,* and the outside steering arm moves through shorter arc *Z*.

Therefore the inside wheel turns at a sharper angle compared with the outside wheel. Because both steering arms are designed to have the same angle in the straight ahead position, the inside front wheel always has a sharper angle regardless of the turning direction. The sharper inside wheel angle during a turn causes the inside wheel to "toe-out." Therefore the term **toe-out on turns** is used for this steering action. During a turn if the front wheel turning angle is increased, the amount of toe-out on the inside wheel increases proportionally.

See Chapter 9 in the Shop Manual for turning radius diagnosis and measurement.

Slip Angle

During a turn, centrifugal force causes all the tires to slip a certain amount. The amount of tire slip increases with speed and the sharpness of the turn. This tire slip action causes the actual center of the turn to be considerably ahead of the theoretical turn center (Figure 9-17).

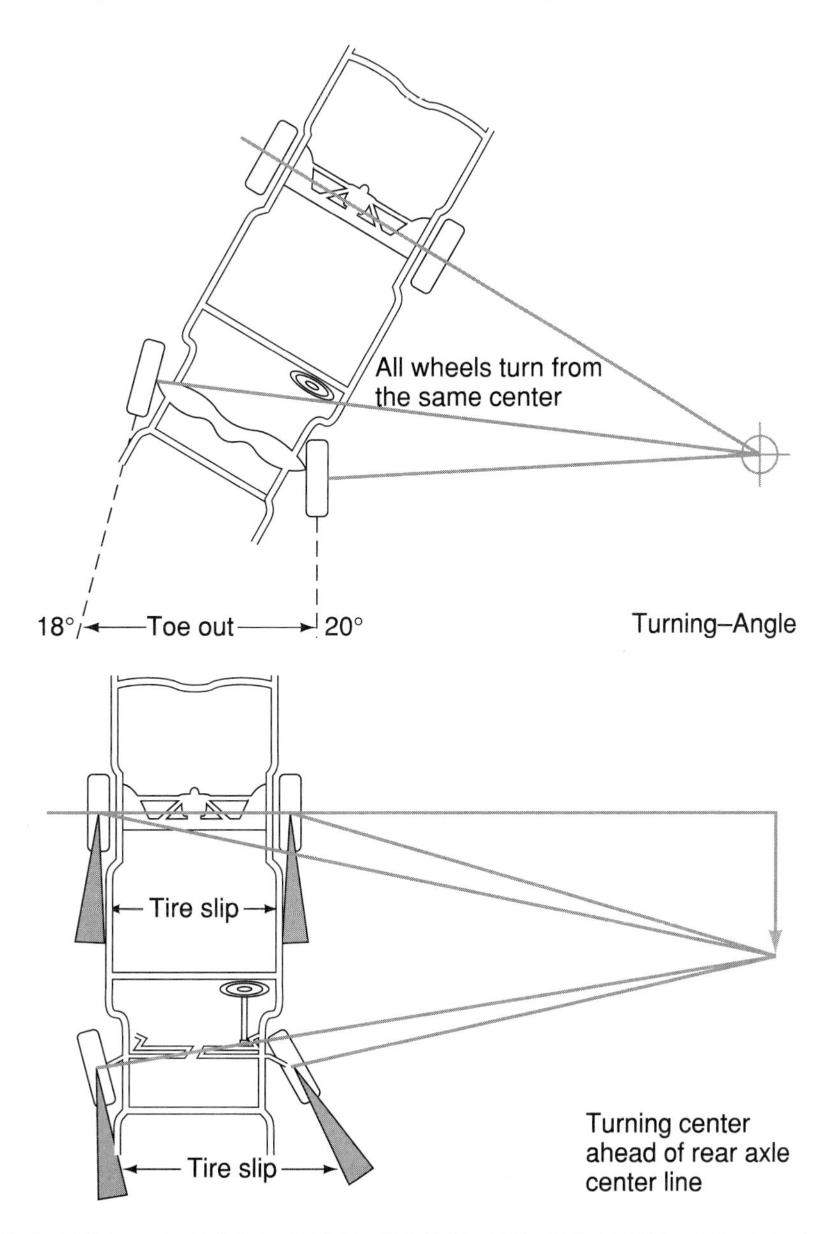

Figure 9-17 Slip angle during a turn. (Courtesy of Ammco Tools.)

Slip angle is the actual angle of the front wheels during a turn compared with the turning angle of the front wheels with the vehicle at rest.

The **slip angles** on different vehicles vary, depending on such factors as vehicle weight and type of suspension.

Rear Axle Alignment

Rear axle alignment is extremely important to maintain proper directional stability. The **thrust line** is a line positioned at a 90 degree angle to the center of the differential housing. This thrust line must be positioned at or very close to the geometric centerline of the vehicle. If the vehicle has tandem rear axles, the thrust line from each rear axle must be positioned at the geometric centerline (Figure 9-18).

If one of the forward rear axles has a thrust line that is positioned to the left of the geometric centerline, this axle position causes the steering to pull to the right. As a result, the driver turns the steering wheel to the left to try and correct this steering pull. In a tandem rear axle the angle formed by the thrust angles of the differentials may be called the tandem scrub angle.

See Chapter 9 in the Shop Manual for rear axle alignment with trammel gauge.

See Chapter 9 in the Shop Manual for rear axle alignment with laser aligner.

In Figure 9-19 the forward rear drive axle has a thrust line that is positioned to the left of the geometric centerline, and the rearmost drive axle has at thrust line that is moved to the right of the geometric centerline. This condition forms a scrub angle between the thrust lines of the two rear axles. The vehicle tends to steer in the direction of the scrub angle. Because the scrub angle is positioned closer to the right side of the frame, the steering pulls to the right and the driver has to turn the steering wheel to the left to keep the tractor going straight ahead. This condition results in reduced steering control and directional stability plus driver fatigue. Excessive tire wear also results from improperly positioned rear axle thrust lines because the tires are actually scrubbing sideways on the road surface.

Trailer axle thrust lines must also be positioned at the geometric centerline of the trailer. If trailer axle thrust lines are not at the geometric centerline of the trailer, they have basically the same effect as improper thrust line position on a truck or tractor. A combination of improperly positioned trailer and tractor thrust lines causes steering pull, wander, and erratic steering.

CAUTION: If a truck or tractor develops a steering pull and wander problem, it should be repaired as soon as possible. Driving a loaded vehicle with this problem may cause a collision, resulting in personal injury and vehicle damage.

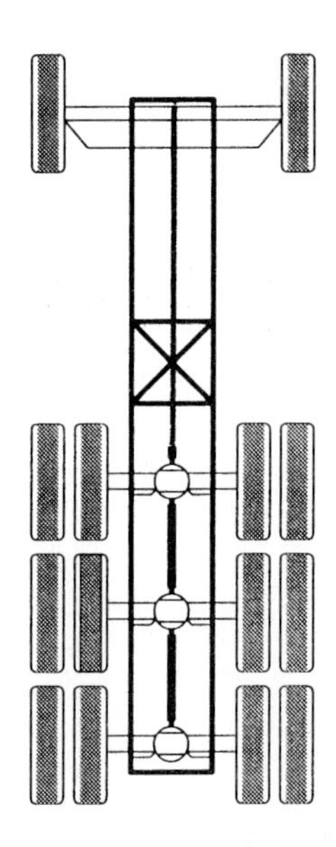

Figure 9-18 The thrust lines from each rear axle must be positioned at the geometric centerline. (Reprinted by permission of Hunter Engineering Company.)

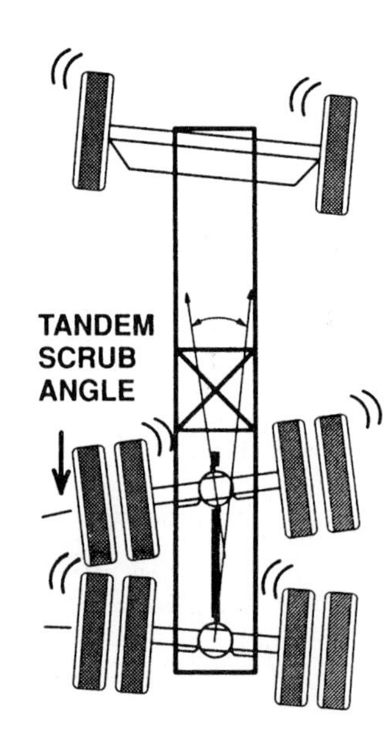

Figure 9-19 Improperly positioned rear drive axle thrust lines cause steering pull. (Reprinted by permission of Hunter Engineering Company.)

Truck and Tractor Wheel Alignment Equipment

Wheel Alignment Runways

Wheel alignment runways may be mounted above ground or recessed into a pit in the shop floor. The wheel aligner runways include turntables. The front wheel must be centered on the turntables. An above-ground alignment runway is illustrated in Figure 9-20, with turntables at the front of the runway. A drive-through wheel alignment pit runway may be used with the runway extending the full length of the pit and the turntables positioned near one end of the runway (Figure 9-21). This runway design allows the tractor to be driven ahead to exit the runway. Many wheel alignment runways have adjustable widths to accommodate various tread widths. A wide variety of above-ground and pit-type alignment runways are available. The entire runway and turntables must be in a level position.

Turntables

The turntables allow the front wheels to be rotated to measure some front wheel alignment angles such as caster and turning radius (Figure 9-22). The turntables have a tire stop that prevents the front tires from rolling forward off the turntables. The stops must be in the upward position when driving the tractor onto the runway. These stops may be lowered to drive off drive-through runways. A degree scale on each turntable informs the technician when the front wheels are in the 0-degree position or the number of degrees the front wheels are turned in either direction

Figure 9-20 Above-ground alignment runway. (Courtesy of Bee Line Company.)

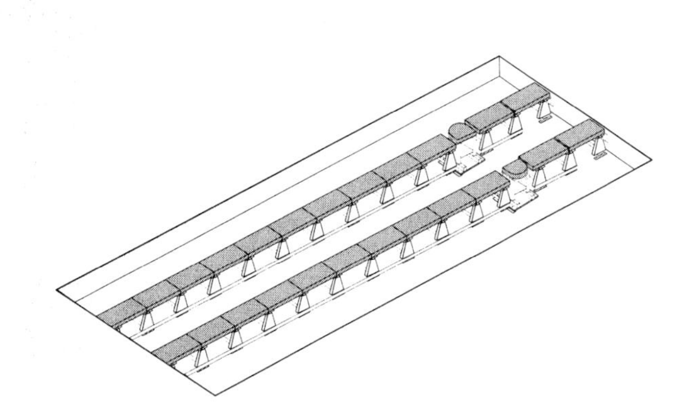

Figure 9-21 Drive-through, pit-type alignment runway. (Courtesy of Bee Line Company.)

Figure 9-22 Turntables for heavy duty alignment runways. (Courtesy of Bee Line Company.)

(Figure 9-23). The top turntable plates are supported on five permanently lubricated bearings to withstand heavy duty truck loads.

Wheel Units

Wheel units are designed to attach to the front tractor wheels. Some wheel units contain a mechanical camber and caster gauge, and this unit is clamped to the outside edge of the wheel rim (Figure 9-24). With this type of wheel unit, an adjustable bar is used to measure front wheel toe. Many wheel aligners now have electronic wheel units with optical toe gauges. These units are con-

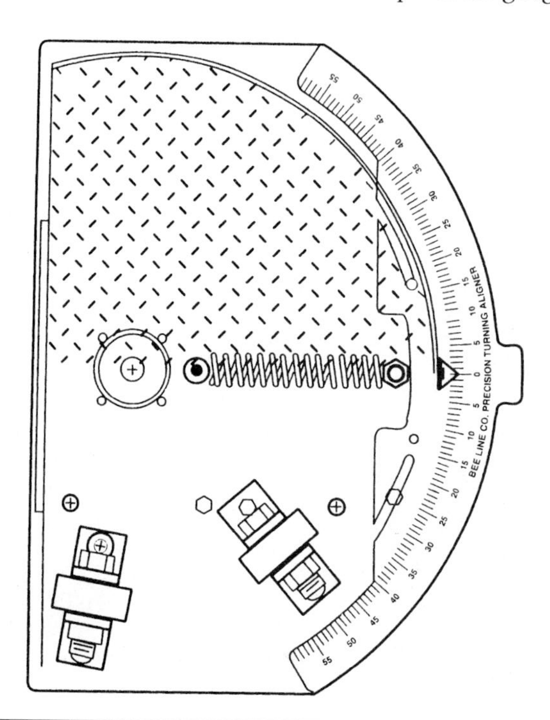

Figure 9-23 Turntable degree scales. (Courtesy of Bee Line Company.)

Figure 9-24 Wheel unit with mechanical camber and caster gauge. (Courtesy of Bee Line Company.)

nected by cables to the wheel aligner monitor, and wheel alignment angles are displayed on the monitor (Figure 9-25).

Laser Wheel Aligners

Laser wheel aligners provide very accurate tandem axle alignment on tractors or trailers. The laser tandem axle aligner may be used with the tractor on the alignment runway or parked on the shop floor (Figure 9-26). The laser unit is attached to a fixture that is mounted against the tires on the

Figure 9-25 Electronic wheel unit and computer wheel aligner. (Courtesy of Bee Line Company.)

Figure 9-26 Laser wheel aligner. (Courtesy of Bee Line Company.)

rearmost axle. This fixture is self-centering. Two laser beam targets are mounted so they are at the tractor centerline behind the rearmost axle and in front of the front (steer) axle (Figure 9-27). The laser aligner must be properly calibrated before measuring the rear axle alignment. If the rearmost axle alignment is correct, the laser beam is projected through the opening in the rear target to the vehicle centerline on the front target. When the rearmost axle is not properly aligned, the laser beam is not projected to the vehicle centerline. The laser unit may be connected to the computer in the computer wheel aligner, and the tandem axle data are displayed on the aligner monitor. Be-

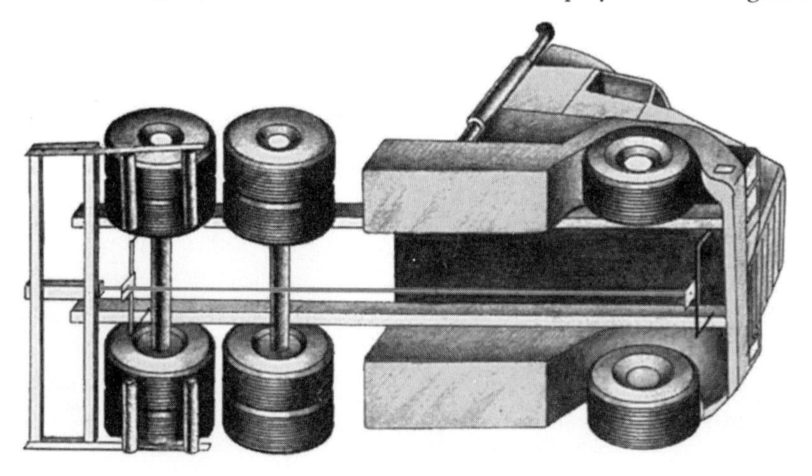

Figure 9-27 Laser wheel aligner and targets. (Courtesy of Bee Line Company.)

Figure 9-28 Portable wheel aligner. (Courtesy of Bee Line Company.)

Figure 9-29 Laser portable wheel aligner. (Courtesy of Bee Line Company.)

cause the laser aligner aligns the rearmost axle to the tractor centerline, it provides very accurate wheel alignment.

Portable Wheel Aligners

Portable wheel aligners are available so the alignment equipment may be transported to a truck terminal rather than driving the tractor a long distance to the alignment shop (Figure 9-28). This procedure allows the wheel alignment to be measured and the tractor returned quickly to the road. The portable aligner components are contained in a convenient carrying case. The data from the wheel units on the portable aligner may be displayed on a laptop computer or on a PC if AC power is available. The laser tandem axle aligner is also available in portable form (Figure 9-29). A power unit is available to operate the laser aligner.

See Chapter 9 in the Shop Manual for wheel alignment with computer wheel aligner.

A BIT OF HISTORY

The trucking industry has an excellent safety record. Commercial trucks are driven in excess of 165 billion miles each year on highways in the United States. In the last ten years, the fatal accident rate for medium and heavy duty trucks has decreased 37%, and the miles driven have increased 41%. Proper vehicle maintenance, including proper wheel alignment, is extremely important to maintain and improve this safety record.

Terms to Know

Ackerman angle
Bump steer
Cornering angle
Directional stability
Kingpin inclination
Kingpin offset
Memory steer
Negative camber
Negative caster
Positive camber
Positive caster
Road crown
Road variables
Set back
Slip angle
Steering pull, or drift
Steering wander
Thrust line
Toe-in
Toe-out
Toe-out on turns
Torque steer
Turning radius
Wheel alignment
Wheel jounce
Wheel rebound
Wheel shimmy

Summary

- ❑ Directional stability is the tendency of a vehicle to travel straight ahead without being steered.
- ❑ Suspension and steering systems are designed to provide satisfactory vehicle control, with acceptable driver effort and road feel, plus minimum tire tread wear.
- ❑ Proper wheel alignment and suspension component condition are extremely important to maintain vehicle driving safety.
- ❑ Many road variables such as bumps and holes, road crown, and heavy vehicle loads affect wheel alignment.
- ❑ Wheel alignment angles are designed to compensate for road variables.
- ❑ Camber is the tilt of a line through the center of the tire and wheel in relation to the vertical centerline of the tire and wheel.
- ❑ Positive camber is obtained when the centerline of the tire and wheel are tilted outward in relation to the vertical centerline of the tire and wheel.
- ❑ Negative camber is present when the centerline of the tire and wheel are tilted inward in relation to the vertical centerline of the tire and wheel.
- ❑ Excessive positive or negative camber concentrates the vehicle weight on one side of the front tire, and the tire edge on which the weight is concentrated has a smaller diameter compared with the other side of the tire. Because the side of the tire with the smaller diameter makes more revolutions to go the same distance, this side of the tire becomes worn and scuffed.
- ❑ A wheel turns in the direction it is tilted.
- ❑ Road crown causes vehicle steering to drift to the right.
- ❑ Caster is the tilt of a line through the center of the kingpins in relation to the vertical centerline of the wheel and spindle viewed from the side.
- ❑ Positive caster occurs when the centerline of the kingpins is tilted rearward in relation to the centerline of the wheel and spindle viewed from the side.
- ❑ Negative caster is obtained when the centerline of the kingpins is tilted forward in relation to the centerline of the wheel and spindle.
- ❑ Positive caster increases directional stability, steering effort, and steering wheel returning force.
- ❑ Excessive positive caster results in harsh ride quality.
- ❑ Excessive positive caster may cause front wheel shimmy.
- ❑ Negative caster reduces directional stability and steering effort, while improving ride quality.

- ❑ If the rear suspension height is lowered, the front wheel caster becomes more positive.
- ❑ Proper caster adjustment is very important to maintain vehicle directional control and safety.
- ❑ Kingpin inclination (KPI) is the angle of a line through the center of the kingpins in relation to the true vertical centerline of the tire and wheel viewed from the front of the vehicle.
- ❑ The KPI line is always tilted toward the center of the vehicle.
- ❑ The included angle is the sum of the KPI and positive camber angles.
- ❑ If the camber is negative, the camber must be subtracted from the KPI angle to obtain the included angle.
- ❑ KPI causes the front spindles to move through an arc when the front wheels are steered to the right or left.
- ❑ Because the front spindles move through an arc, the chassis lifts as the front wheels are turned, and this lifting action helps return the front wheels to the straight ahead position after a turn.
- ❑ KPI also helps to maintain the front wheels in the straight ahead position.
- ❑ KPI is not adjustable. If the KPI does not equal manufacturer's specifications, the front axle may be bent.
- ❑ A suspension system with a 0-degree KPI line would have increased tire wear, greater steering effort, stress on suspension and steering components, and excessive road shock and kickback on the steering wheel.
- ❑ When the KPI line intersects the true vertical tire centerline at or near the road surface, tire life is improved, stress on steering and suspension components is reduced, steering effort is decreased, and road shock and kickback on the steering wheel are minimized.
- ❑ Scrub radius is the distance between the KPI line and the true vertical tire centerline at the road surface.
- ❑ A front suspension has positive scrub radius when the KPI line contacts the road surface inside the true vertical tire centerline.
- ❑ A front suspension has negative scrub radius when the KPI line contacts the road surface outside the true vertical tire center line.
- ❑ If the front tires are larger than the original tires specified by the vehicle manufacturer, a change occurs in scrub radius that may affect steering control.
- ❑ The installation of larger than specified front tires may change positive scrub radius to a negative scrub radius.
- ❑ Set back is a condition in which one front wheel is moved rearward in relation to the opposite front wheel.
- ❑ Front wheel toe is the distance between the front edges of the tires compared with the distance between the rear edges of the tires.
- ❑ Toe-in is present when the distance between the front edges of the tires is less than the distance between the rear edges of the tires.
- ❑ Toe-out occurs when the distance between the front edges of the tires is greater compared with the distance between the rear edges of the tires.
- ❑ Trucks and tractors have a slight toe-in setting because driving forces tend to move the front wheels to a toe-out position.
- ❑ The front wheel toe is adjusted with the vehicle at rest so the front wheels are straight ahead when the vehicle is driven.

- ❑ Improper toe adjustment results in feathered tire wear.
- ❑ Toe-out on turns is the turning angle of the wheel on the inside of the turn compared with the turning angle of the wheel on the outside of the turn.
- ❑ When the front wheel on the inside of a turn has turned 20 degrees outward, the front wheel on the outside of the turn may have turned 18 degrees.
- ❑ Turning radius is the amount of toe-out on turns.
- ❑ Toe-out on turns prevents tire scuffing, and this angle is determined by the steering arm design.
- ❑ During a turn centrifugal force causes all the tires to slip a certain amount, depending on vehicle speed and the sharpness of the turn.
- ❑ Because the tires slip during a turn, the actual vehicle turning center is moved ahead of the theoretical turning center.
- ❑ Slip angle is the actual angle of the front wheels during a turn compared with the turning angle of the front wheels with the vehicle at rest.
- ❑ The thrust line of each rear axle must be positioned at or very close to the geometric centerline of the vehicle.

Review Questions

Short Answer Essays

1. Explain why the front wheels are not placed in the exact vertical position for proper wheel alignment.
2. Explain why excessive positive camber wears the outside edge of the tire tread.
3. Explain why positive caster provides increased directional stability.
4. Describe how positive caster provides increased steering wheel returning force.
5. Explain why positive caster causes harsh riding quality.
6. Explain how the included angle is calculated with positive and negative camber.
7. Describe how KPI affects the front spindle movement during a turn.
8. Define setback on a front suspension system.
9. Describe the type of tire tread wear caused by excessive toe-in on the front wheels.
10. Explain why toe-out on turns is necessary.

Fill-in-the-Blanks

1. Directional stability refers to the tendency of a vehicle to travel straight ahead without being ____________________.
2. When the front wheel camber is negative, the centerline of the wheel and tire is tilted ____________________ in relation to the true vertical centerline of the wheel and tire.
3. Excessive negative front wheel camber results in tread wear on the ____________________ edge of the front tire.
4. Road crown causes vehicle steering to pull toward the ____________________.
5. Negative front wheel caster decreases ____________________ and steering wheel ____________________.

6. Front wheel shimmy may be caused by excessive ____________________ caster.

7. Front wheel caster becomes more positive if the rear suspension height is ____________________.

8. When a front suspension has a positive scrub radius, the KPI line meets the road surface ____________________ the true vertical centerline of the tire.

9. During a turn the ____________________ front wheel turns at a sharper angle.

10. During a turn centrifugal force causes all the tires to slip a certain amount, and the actual center of the turn is shifted ____________________ of the theoretical center of the turn.

ASE Style Review Questions

1. While discussing road variables related to wheel alignment:
Technician A says road crown is a variable that affects wheel alignment.
Technician B says vehicle loads and unequal weight distribution are variables that affect wheel alignment.
Who is correct?
A. A only **C.** Both A and B
B. B only **D.** Neither A or B

2. While adjusting front wheel camber:
A. If a front wheel has a positive camber angle, the camber line is tilted inward from the true vertical centerline of the wheel and tire.
B. Excessive positive camber on a front wheel causes premature wear on the inside edge of the tire tread.
C. Excessive negative camber on a front wheel causes premature wear on the outside edge of the tire tread.
D. Improper camber angle on an I-beam front suspension may be caused by a bent axle or spindle.

3. While discussing front wheel caster and directional stability:
Technician A says a vehicle has better directional stability when the front suspension caster is positive.
Technician B says a vehicle has improved ride quality when the positive caster is increased on the front suspension.
Who is correct?
A. A only **C.** Both A and B
B. B only **D.** Neither A nor B

4. While adjusting front wheel caster:
A. Excessive positive caster decreases steering effort.
B. Excessive positive caster may cause front wheel shimmy.
C. The front suspension caster becomes more negative when the rear suspension height is lowered.
D. Excessive negative caster results in harsh ride quality.

5. While discussing kingpin inclination (KPI):
Technician A says the KPI line is tilted inward in relation to the true vertical tire centerline viewed from the front.
Technician B says the KPI is adjustable on an I-beam front suspension.
Who is correct?
A. A only **C.** Both A and B
B. B only **D.** Neither A nor B

6. While diagnosing KPI and front spindle movement:
A. When the steering wheel is turned, the front spindle movement is parallel to the road surface.
B. The KPI angle has no effect on steering wheel returning force.
C. The KPI angle tends to maintain the wheels in a straight ahead position.
D. An increase in the KPI angle decreases steering effort.

7. While discussing the included angle:
Technician A says the included angle is the sum of the KPI angle and the positive camber angle.
Technician B says if the positive camber is increased the included angle decreases.
Who is correct?
A. A only **C.** Both A and B
B. B only **D.** Neither A nor B

8. All these statements about front wheel toe are true *except:*
A. Driving forces tend to move the front wheels toward a toe-out position on an I-beam front suspension.
B. Improper front wheel toe causes feathered tread wear on the front tires.
C. Adjusting the front wheels on an I-beam front suspension to a toe-in position improves directional stability.
D. Front wheel toe setting on an I-beam front suspension does not affect steering effort.

9. While discussing turning radius:
Technician A says when a vehicle is making a left turn the left front wheel turns at a sharper angle than the right front wheel.
Technician B says during a right turn the left front wheel turns at a sharper angle than the right front wheel.
Who is correct?
A. A only
B. B only
C. Both A and B
D. Neither A nor B

10. While diagnosing turning radius:
A. The turning radius is affected by the length of the tie rod.
B. The turning radius is determined by the steering arm design.
C. Improper turning radius has no effect on tire tread wear.
D. During a turn the inside tire is actually behind the outside tire.

GLOSSARY

Note: Terms are highlighted in bold followed by Spanish translation in bold italic.

Ackerman angle The turning angle of one front wheel compared to the opposite front wheel. The Ackerman angle may be called turning angle or turning radius.
Angulo ackerman El ángulo de viraje de una rueda delantera comparado al de la rueda delantera opuesta. El ángulo ackerman tambien se puede llamar el ángulo de viraje o el radio de giro.

Ackerman arm The lower, left steering arm that connects the spindle to the tie rod.
Brazo ackerman El brazo de mando inferior del lado izquierdo que acopla el husillo con el tirante de tracción.

Air assisted steering system A steering system in which steering assist is provided by air pressure supplied to one side of a power cylinder piston.
Sistema de dirección de asistencia neumática Un sistema de dirección en el cual la asistencia a la dirección se provee por presión de aire suministrado en un lado del pistón del cilindro motor.

Air lift A lift axle with an air-operated lift mechanism.
Inyección de aire comprimido Un eje de apoyo con un mecanismo montacargas operado por aire comprimido.

Air springs Replace leaf springs in some front and rear suspension systems.
Muelles neumáticos Reemplazan los muelles de lámina en algunos sistemas de suspensión delantera y trasera.

Air suspended seat A driver's seat that is supported by an air cushion for increased driver comfort.
Asiento suspendido por aire Un asiento del conductor que se apoya por medio de un cojín de aire para mayor confort del conductor.

Air suspension system A suspension system in which the leaf springs are replaced with air springs to support the vehicle weight.
Sistema de suspensión neumática Un sistema de suspensión en el cual los muelles de láminas se reemplazan con los meulles neumáticos para sostener el peso del vehículo.

American Petroleum Institute (API) An organization responsible for oil classifications.
Instituto Petrolero American (API) Una organización responsable por la clasificación de los aceites.

Angular load A bearing load applied at an angle between 0° and 90°.
Carga angular Una carga en el cojinete puesta en un ángulo de entre el 0° y 90°.

Applied moment A measurement of a specific load placed in a certain location on the frame. The applied moment is based on a stationary vehicle.
Momento de aplicación Una medida de una carga específica puesta en un lugar cierto en el bastidor. El momento de aplicación está basado en un vehículo inmóvil.

Atmospheric pressure The pressure exerted on the earth by the atmosphere.
Presión atmosférica La presión puesta sobre la tierra por la atmósfera.

Auxiliary spring A leaf spring mounted above the lower rear leaf springs to help support the vehicle load.
Muelle auxiliar Un muelle de lámina montado sobre los muelles de láminas traseros inferiores para ayudar en sostener la carga del vehículo.

Barrier-type hub seal A wheel bearing seal with ribbed inner and outer mounting surfaces.
Sello de cubo tipo barrera Un obturador del cojinete de rueda cuyos superficies de montaje interiores e exteriores tienen costillas.

Bearing preload The amount of tension placed on a bearing.
Precarga de cojinete La cantidad de tensión puesto en un cojinete.

Bearing seals A seal that prevents contaminants from entering the wheel hub, and prevents lubricant leaking from the hub.
Sellos de cojinete Un obturador que prohibe que los contaminantes entran al cubo de la rueda, y previene la fuga de lubricantes del cubo.

Bearing shields Metal shields on one or both sides of a bearing that help to keep contaminants out of the bearing.
Escudos del cojinete Los escudos de metal en uno o ambos lados de un cojinete que ayudan excluir los contaminantes del cojinete.

Bending moment A load applied to the frame that is distributed across a given section of the frame.
Momento flector Una carga puesta en el bastidor que se distribuya por una sección específica del bastidor.

Bias ply tire A type of tire construction with the cord plies wound at an angle to the center of the tire.
Neumático de bandas diagonales Un tipo de construcción de neumático en el cual los pliegues las capas se montan en un ángulo al centro del neumático.

Box frames This type of frame is shaped like a rectangular metal box.
Bastidor tipo de caja Este tipo de bastidor toma la forma de una caja de metal rectangular.

Buckle A frame condition that refers to the frame or one side rail that is bent upward from the original position.
Alabeo Una condición del bastidor que refiere al bastidor o un larguero lateral que está torcido hacia arriba de la posición original.

Bump steer The tendency of the steering to veer suddenly in one direction when one or both of the front wheels strike a bump.
Desviación La tendencia en la dirección a desviarse repentinamente en una dirección cuando una o ambas ruedas pegan un tope.

Cab air suspension system A cab suspension system that has two air springs and two shock absorbers to support the cab.

Sistema de suspensión de la casilla por aire Un sistema de suspensión de la casilla del camion que tiene dos muelles neumáticos y dos amortiguadores de choque para sostener la casilla.

Cam ring The rotor and vanes or rollers in a power steering pump rotate inside the cam ring.

Anillo de leva El rotor y las aletas o los rodillos en una bomba de dirección hidraúlica giran dentro del anillo de levas.

C-channel A type of frame with a vertical web section between the top and bottom flanges.

Canal en C Un tipo de bastidor con una sección de alma en nervadura vertical entre las bridas superiores e inferiores.

Compensating fifth wheel Provides front-to-rear and side-to-side oscillation.

Rueda quinta compensadora Provee la oscilación de frente-hacia-atrás o de lado-a-lado.

Compressible The ability of a material to become smaller when pressure is applied.

Compresible La habilidad de una materia a devenirse más pequeño cuando se le aplica la presión.

Control valve A driver operated valve that works with relay valves to release air from the air suspension system.

Válvula de control Una válvula operado por el conductor que coordina con las válvulas de relée para soltar el aire del sistema de suspensión neumática.

Converter dolly A single axle with a fifth wheel mounted on it. The converter dolly is used to couple the lead trailer to the rear trailer.

Carro convertidor Un eje sencillo que tiene montada una quinta rueda. El carro convertidor se usa para acoplar el remolque de avance con el remolque trasero.

Cornering angle The turning angle of one front wheel in relation to the opposite front wheel.

Ángulo de ataque El ángulo de viraje de una rueda delantera en relación a la rueda delantera opuesta.

Corrosive A material that dissolves metals or burns the skin.

Corrosivo Una materia que disuelve los metales o quema al piel.

Cross channel The lower side of the air spring on each side of the rear suspension system is bolted to the cross channel.

Travesaño acanalado El lado inferior de un muelle neumático de cada lado de un sistema de suspensión trasera se emperna al travesaño acanalado.

Crossmembers Are steel members connected between the two sides of a frame.

Travesaños Son los miembros de acero conectados entre los dos lados de un bastidor.

Cylindrical ball bearing A bearing with round steel rolling elements.

Cojinete de bolas cilíndricas Un cojinete con elementos rodantes redondos de acero.

Cylindrical roller bearing A bearing with roller-type rolling elements.

Cojinete de rodillos cilíndricos Un cojinete con elementos rodantes tipo rodillos.

Diamond frame A frame that is diamond-shaped from collision damage.

Bastidor rómbico (diamante) Un bastidor que queda en forma rómbica debido a los daños de una colisión.

Directional stability The tendency of the vehicle steering to remain in the straight-ahead position when driven straight ahead on a smooth, level road surface.

Estabilidad direccional La tendencia de la dirección de un vehículo de quedarse en la posición de línea recta cuando está en marcha en una superficie lisa y plana.

Disc wheel A truck wheel with a one-piece casting in which the tire sealing lips are integral with the wheel.

Rueda maciza Una rueda de camion funidido en una pieza en la cual las pestañas para sellar los neumáticos son íntegras con la rueda.

Double-row ball bearing A cylindrical ball bearing with two rows of rolling elements.

Cojinete de bolas de dos hileras Un cojinete de bolas cilíndricas con dos hileras de elementos rodantes.

Drag link A steering linkage connected from the pitman arm on the steering gear to the left-front upper steering arm.

Biela de acoplamiento Un eslabón de dirección conectado del brazo pitman en el mecanismo de dirección al brazo de dirección delantero superior del lado izquierdo.

Draw bar May be used to connect a rear trailer to a lead trailer. The draw bar is attached to the rear trailer at two points, and has one pivot point connected to the lead trailer.

Barra de enganche Se puede usar para conectar un remolque trasero con un remolque delantero. La barra de enganche se junta con el remolque trasero en dos puntos, y tiene un punto de pivote conectado al remolque delantero.

Dump valve A valve that exhausts air from an air suspension system. This valve may be integral with the height control valve, or mounted separately.

Válvula de descarga Una válvula que vacía el aire del sistema de suspensión de aire. Esta válvula puede ser integral con la válvula de control de altura, o ser montado aparte.

Dynamic balance The balance of a wheel in motion.

Equilibrado dinámico La equilibración de una rueda en movimiento.

Electronically controlled air suspension system A computer-controlled air suspension system.

Sistema de suspensión neumática controlado electrónicamente Un sistema de suspensión controlado por computadora.

Electronic control unit (ECU) The computer that controls the operation of an air suspension system.

Unedad de control electrónico (ECU) La computadora que controla la operación de un sistema de suspensión neumática.

Elevating fifth wheel An elevating fifth wheel is used to change an over-the-road tractor to a tractor that is used to spot, switch, and haul trailers in the parking and warehouse area.

Rueda quinta elevador Una rueda quinta elevador se usa para convertir un tractor de carretera en un tractor que sirve para localizar, cambiar, y remolcar los remolques en el área de estacionamiento y del almacén.

Energy The ability to do work.
Energía La habilidad de efectuar el trabajo.

Environmental Protection Agency (EPA) The federal government agency in charge of air and water quality in the United States.
Agencia de Protección del Medio Ambiente (EPA) La agencia gubernamental federal de la calidad del aire y del agua en los Estados Unidos.

Equalizer Connects the front and rear leaf springs to the center spring hanger on some tandem rear axle suspension systems.
Compensador Conecta los muelles de lámina delanteros y traseros a la silleta central del muelle en algunos sistemas de suspensión de eje trasero tándem.

Equalizing beams Connected from the tandem rear axles to the lower side of the leaf springs on some tandem rear axle suspension systems.
Vigas compensadoras Conectadas de los ejes traseros tándem al lado inferior de los muelles de lámina en algunos sistemas de suspensión de ejes traseros tándem.

External dock lock system An air operated system that maintains the trailer at dock height, and prevents the trailer from walking forward.
Sistema externa de cerrojo del muelle Un sistema operado por aire que mantiene al remolque en la altura del muelle, y previene que el remolque se mueva hacia adelante.

Fifth wheel The mechanism attached to the tractor frame to which the trailer is coupled.
Rueda quinta El mecanismo conectado al bastidor del remolque al cual se acopla el remolque.

Fifth wheel jaws These jaws lock the trailer king pin to the fifth wheel on the tractor.
Mordaza de la rueda quinta Estas mordazas enclavan el pivote central a la quinta rueda en el tractor.

Fishplate A flat steel plate bolted to a truck frame for reinforcement.
Eclisa de desahogo Una placa plana de acero empernado al bastidor del camión como refuerzo.

Fixed steering column A steering column with no tilting mechanism.
Columna de dirección fija Una columna de dirección que no tiene mecanismo de desplazamiento.

Flow control valve A valve in a power steering pump that controls fluid movement in relation to system demands.
Válvula de control de flujo Una válvula en una bomba de dirección hidráulica que controla el movimiento de los fluidos en relación a las exigencias del sistema.

Fluted lip seal A seal with flutes positioned at an angle on the lip surface.
Sello de borde acanalado Un sello con acanaladuras posicionadas en un ángulo a la superficie del borde.

Frame reinforcements Steel plates with various configurations that are bolted to a truck frame to provide extra strength.
Refuerzos del bastidor Placas de acero de varias configuraciones que son empernadas al bastidor del camión para proveer mayor fuerza.

Frame sag Occurs when the frame or one side rail is bent downward from the original position.
Hundimiento del bastidor Ocurre cuando el bastidor o un larguero lateral está curvado hacia abajo de la posición original.

Frame twist Occurs when the end of one frame rail is bent upward or downward in relation to the opposite frame rail.
Bastidor torcido Ocurre cuando la extremidad de un larguero está torcido hacia arriba o abajo en relación al larguero opuesto.

Free air Air at atmospheric pressure.
Aire libre El aire de presión atmosférica.

Friction The resistance to movement between two objects in contact with each other.
Fricción La resistencia al movimiento entre dos objetos en contacto.

Front bracket A bracket that connects the main support beam to the frame in a rear axle air suspension system.
Soporte delantera Un soporte que conecta la apoyadera principal al bastidor en un sistema de suspensión neumática de eje trasero.

Fulcrum A pivot point between two objects.
Fulcro Un punto de pivote entre dos objetos.

Fully-oscillating fifth wheel Allows side-to-side and front-to-rear oscillation of the fifth wheel.
Rueda quinta de oscilación completa Permite la oscilación de lada-a-lado y de frente-a-atrás en la rueda quinta.

Garter spring A spring surrounding the outer side of a seal lip to provide more force on the seal lip and improve sealing quality.
Resorte de liga Un resorte alrededor del exterior de un borde sellante para proveer más fuerza en el borde sellante y mejorar la calidad sellante.

Gas-filled shock absorbers Contain a gas charge and hydraulic fluid.
Amortiguadores llenos de gas Contienen una carga de gas y un fluido hidráulico.

Gear ratio The ratio between the drive gear and driven gear in a gear set.
Relación de engranajes La relación entre el engranaje propulsor y el engranaje arrastrado en un juego de engranajes.

Gear tooth lash The amount of free-play between the gear teeth in two meshed gears.
Juego entre los dientes del engranaje La cantidad del juego entre los dientes de engranaje en dos engranajes endentados.

Grit guard A steel guard on some wheel bearing seals that helps to keep dirt out of the seal.
Guardabarros Una guarda de acero en algunos sellos de cojinetes de rueda que ayudan en excluir el barro del sello.

Halogen and halon fire extinguishers Fire extinguishers that are effective on class B fires, but give off toxic gases.
Extinctor de fuego halógeno y balon Los extinctores de fuego que son eficazes en apagar los fuegos de clase B, pero que crean los gases tóxicos.

Hazard Communication Standard The beginning of the right-to-know laws published by the Occupational Safety and Health Administration (OSHA).

Norma de Comunicación de Peligro Los primeros de los leyes derecho-en-saber publicados por la Administración de Seguridad y Salud en las Ocupaciones (OSHA).

Heavy duty shock absorbers Compared to conventional shock absorbers, heavy duty shock absorbers have improved seals, a single tube to reduce heat, and a rising rate valve for precise spring control.

Amortiguadores de uso pesado Comparado con los amortiguadores convencionales, los amortiguadores de uso pesado tienen los empaques mejorados, un tubo sencillo para disminuir el calor, y una válvula de proporción ascendente para el control preciso del muelle.

Height control valve A valve that controls the air pressure in the air springs to maintain the proper ride height.

Válvula de control de altura Una válvula que control la presión en los muelles neumáticos para mantener la altura de manejo adecuada.

Highly incompressible A material that exhibits very little decrease in size when pressure is applied.

Altamente incompresible Una materia que demuestra muy poca disminución en tamaño cuando se le aplica la presión.

Hub-piloted wheel A type of truck wheel in which the opening in the wheel is designed to center the wheel on the wheel hub.

Rueda piloteada en cubo Un tipo de rueda de camión en la cual la apertura de la rueda es diseñada para centrar la rueda en el cubo de la rueda.

Hydropneumatic spring A spring with a diaphragm in the center. Fluid is contained above this diaphragm, and a permanent gas charge is installed below the diaphragm. A hose connects the fluid chamber in each spring to an accumulator.

Muelle hidroneumático Un muelle con una diafragma en el centro. El fluido se contiene arriba de la diafragma, y una carga permanente de gas se instala abajo de la diafragma. Una manguera conecta la cámara de fluidos de cada muelle con una acumuladora.

I-beam axles Front axles with a I-beam shape.

Ejes de hierro en I Los ejes delanteros de hierro en forma de I.

I-beam frames Side members with an I-beam shape.

Bastideros de hierro en I Los largueros de hierro en forma de I.

Idler arm A short, pivoted steering arm bolted to the vehicle frame and connected to the steering center link.

Brazo loco Un brazo de dirección corto con pivote que se emperna al bastidor del vehículo y conectado al eslabón central de dirección.

Ignitable If a material burns when contacted by a spark, flame, or a certain degree of heat, it is ignitable.

Encendiente Si una materia quema al ponerse en contacto con una chispa, una llama o un grado determinado del calor, es encendiente.

Inertia The tendency of an object at rest to remain at rest.

Inercia La tendencia de un objeto en reposo a quedarse en reposo.

Inner race The inner component of a ball bearing or roller bearing.

Pista interior El componente interior de un cojinete de bolas o un cojinete de rodillos.

Integral power steering pump reservoir A fluid reservoir that is a part of a power steering pump.

Suministro íntegro de la bomba de dirección hidráulica Un depósito de fluido que es parte de la bomba de dirección hidráulica.

Interference fit A precision fit between two components that provides a specific amount of friction between the components.

Asiento fijo Un ajuste preciso entre dos componentes que provee una cantidad específica de fricción entre los componentes.

Inverted springs Rear leaf springs in some tandem rear axle suspension systems that are mounted upside down so outer ends of the spring are lower than the center of the spring.

Muelles invertidos Los muelles de láminas traseros en algunos sistemas de suspensión de eje trasero tándem que son montados de arriba abajo para que las extremidades exteriores del muelle quedan inferior al centro del muelle.

Jounce travel Refers to upward wheel and suspension movement.

Sacudo Refiere al movimiento hacia arriba de la rueda y suspensión.

Kick-back A force transmitted to the steering wheel that opposes the rotational force supplied to the steering wheel by the driver.

Reculada de la dirección Una fuerza transmitida al volante de dirección que opone la fuerza giratoria que proporciona el conductor en el volante de dirección.

Kingpins Heavy steel pins that retain the front spindles to the I-beam axle.

Pivotes centrales Las clavijas gruesas de acero que retienen los husillos delanteras al eje de hierro en I.

Kingpin inclination The angle of a line through the center of the king pins in relation to the true vertical centerline of the tire and wheel.

Inclinación del pivote central El ángulo de una línea que atraviesa el centro de los pivotes centrales en relación al eje mediano vertical verdadero de la rueda y el neumático.

King pin offset The distance between the centerline of the tire and the king pin inclination (KPI) line at the road surface.

Descentrado del pivote central La distancia entre la línea central de la rueda y la línea de la inclinación del pivote central (KPI) en la superficie del camino.

Lift axle An axle on a tractor or trailer that may be lifted so the tires are off the road surface when the axle is not required to support the vehicle load.

Eje montacarga Un eje en un tractor o en un remolque que puede ser levantado para que los neumáticos se levantan de la superficie del camino cuando no se requiere que el eje sostenga la carga del vehículo.

Lithium based grease A type of wheel bearing grease recommended in some wheel bearings.

Grasa a base de litio Un tipo de grasa para el conjunto de los cojinetes de la rueda que se recomienda en algunos conjuntos de cojinetes de la rueda.

Low profile tire A tire in which the section width is considerably more than the section height.

Neumático de bajo perfil Un neumático cuyo perfil de anchura es considerable más grande que el perfil de su altura.

Main support beams The main support beams support the axles and air springs in an air suspension system.

Apoyaderas centrales Las apoyaderas centrales sostienen los ejes y los muelles neumáticos en un sistema de suspensión neumática.

Mass, weight, and volume Mass is the measurement of an object's inertia, weight is the measurement of the earth's gravitational pull on an object, and volume is the length, width, and height of the space occupied by an object.

Masa, peso, y volumen La masa es la medida de la inercia de un objeto, el peso es la medida de la atracción de gravedad en el objeto, y el volumen es la longitud, la anchura, y la altura del espacio que ocupa un objeto.

Material Safety Data Sheets (MSDS) Sheets that provide information regarding hazardous materials.

Hojas de Datos de Seguridad de la Material (MSDS) Las hojas que proveen la información en cuanto a las materiales peligrosas.

Memory steer Occurs when the steering does not return to the straight-ahead position after a turn, and the steering attempts to continue turning in the original turn direction.

Desviación persistente Ocurre cuando la dirección no regresa a la posición de línea recta después de efectuar una vuelta, y la dirección persiste en girar en la dirección del giro original.

Momentum An object gains momentum when a force overcomes static inertia and moves the object.

Impulso (momento) Un objeto gana impulso cuando una fuerza vence la inercia estática y mueva al objeto.

Multileaf shackle-type spring A leaf spring with several leaves that is attached to the chassis at each end via shackles.

Muelle de múltiples láminas tipo grillete Un muelle de láminas con varias hojas que se conecta al chasis en cada extremidad con grilletes.

Multipurpose dry chemical fire extinguisher A common type of fire extinguisher that may be used on various types of fires.

Extinctor de múltiple uso de productos químicos secos Un tipo de extinctor de fuego común que se puede usar para apagar varios tipos de fuegos.

Needle roller bearing A bearing with small diameter steel rolling elements.

Cojinete de rodamientos de aguja Un cojinete cuyos elementos rodantes son de acero de un diámetro pequeño.

Negative camber Occurs when the camber line through the center of the tire is tilted inward in relation to the true vertical centerline of the tire viewed from the front of the vehicle.

Comba negativa Ocurre cuando la línea de comba que atraviesa el centro del neumático se inclina hacia adentro en relación al eje mediano verdadero del neumático vista desde el frente del vehículo.

Negative caster Occurs when the caster line through the center of the kingpins is tilted forward in relation to the vertical centerline of the wheel and tire viewed from the side.

Ángulo de inclinación negativo Ocurre cuando la línea del ángulo de inclinación que atravieza por el centro del pivote central se inclina hacia afrente en relación al eje mediano vertical de la rueda y del neumático vista de un lado.

Negative scrub radius Present when the king pin inclination line meets the road surface outside the true vertical centerline of the tire at the road surface.

Radio estregadero negativo Presente cuando la línea de inclinación del pivote central toca la superficie del camino afuera de la línea central vertical verdadera del neumático en la superficie del camino.

Non-tilt convertible fifth wheel Can be changed from a rigid fifth wheel to a semi-oscillating fifth wheel.

Rueda quinta convertible rígido Se puede cambiar de una rueda quinta rígida a una rueda quinta semi-oscilatorio.

Occupational Safety and Health Administration (OSHA) The federal government agency in the United States in charge of safety and healthful working conditions.

Administración de Seguridad y Salud en la Ocupación (OSHA) La agencia federal gubermental de los Estados Unidos que se encarga de la seguridad y las condiciones saludables en el trabajo.

Outer race The outer component in a ball bearing or roller bearing.

Pista exterior El componente exterior en un cojinete de bolas o en un cojinete de rodillos.

Outlet fitting venturi A narrow passage in the outlet fitting that supplies fluid to one side of the flow control valve.

Montaje venturi de salida Un pasaje estrecho en el montaje de salida que suministra el fluido a un lado de la válvula de control del flujo.

Oversteer Occurs when the vehicle tends to steer toward the inside of a curve when accelerating around the curve. The driver must turn the steering wheel toward the outside of the curve to maintain the original vehicle direction.

Sobrevirador Ocurre cuando el vehículo tiene la tendencia de girar hacia la parte interior de una curva al acelerar durante una vuelta. El conductor tiene que accionar al volante hacia la parte exterior de la curva para mantener la dirección original del vehículo.

Parabolic spring A leaf spring with leaves that are narrower at the outer ends compared to the center.

Muelle parabólico Un muelle de láminas en el cual las láminas más estrechas quedan en las extremidades exteriores comparadas con las en el centro.

Parabolic taperleaf spring A leaf spring with leaves that are narrower and thinner at the outer ends compared to the center.

Muelle parabólico de láminas cónicas Un muelle de lámina en el cual las láminas más estrechas y delgadas quedan en las extremidades exteriores comparadas con las del centro.

Parallelogram steering linkage A steering linkage in which the steering arms are parallel to the lower control arms.

Eslabón de dirección en paralelogramo Un eslabón de dirección en la cual los brazos de dirección son paralelos a los brazos de control inferiores.

Pintle hook A hook attached to a lead trailer to which the draw bar on the rear trailer is connected.
Gancho timón Un gancho conectado al remolque delantero al cual se conecta la barra de enganche.

Pitman arm A short steel arm connected from the steering gear sector shaft to the drag link.
Brazo pitman Un brazo corto de acero conectado del eje del sector del mecanismo de la dirección a la biela de acoplamiento.

Poppet valves Valves mounted in some recirculating ball steering gears that relieve power assist pressure when the steering wheel is turned near the full-right or full-left position.
Válvulas levadizas Las válvulas montadas en algunos engranajes de dirección de bolas recirculatorias que alivian la presión de asistencia hidráulica cuando se acciona al volante de dirección en la posición totalmente a la derecha o totalmente a la izquierda.

Positive camber Occurs when the camber line through the centerline of the tire is tilted outward in relation to the true vertical tire centerline viewed from the front.
Comba positiva Ocurre cuando la línea de comba que atraviesa el eje mediano del neumático está inclinada hacia afuera en relación al eje mediano del neumático verdadero vista de frente.

Positive caster Occurs when the caster line through the center of the king pins is tilted rearward in relation to the vertical centerline of the tire and wheel viewed from the side.
Angulo de caster positivo Ocurre cuando la línea del ángulo del caster que atraviesa el centro del pivote central se inclina hacia atrás en relación al eje mediano vertical del neumático y la rueda vista del lado.

Positive scrub radius Occurs when the KPI line through the center of the king pins meets the road surface inside the true vertical centerline of the tire and wheel when viewed from the front.
Radio estregadero positivo Ocurre cuando la línea KPI que atravieza el centro del pivote central se reune con la superficie del camino dentro del eje mediano vertical de la rueda y del neumático vista de frente.

Power A measurement of the rate at which work is done.
Potencia Una medida del régimen en el cual se efectúa el trabajo.

Power cylinder A cylinder and piston assembly in which hydraulic pressure is applied to help move a component.
Cilindro motor Una asamblea de cilindro y pistón en la cual la presión hidráulica se emplea para asistir en mover un componente.

Pressure protection valve Protects the air supply in the brake system if an air leak occurs in the air suspension system.
Válvula de protección de presión Proteja el suministro de aire en el sistema de frenos si ocurre una fuga de aire en el sistema de suspensión neumática.

Pressure reducing valve Reduces the air pressure to some air operated components such as an air seat.
Válvula reductor de presión Disminuya la presión del aire en algunos componentes operados por aire tal como el asiento de aire.

Pressure relief ball A spring-loaded ball in the center of the flow control valve that opens to limit the maximum power steering pressure and protect system components.
Bola reductor de presión Una bola cargada por resorte en el centro de la válvula de control de flujo que se abre para limitar la potencia máxima de la presión de dirección y protejer los componentes del sistema.

Radial load A bearing load applied in a vertical direction.
Carga radial Una carga del cojinete puesta en una dirección vertical.

Radial ply tire A tire in which the carcass plies are positioned at 90° in relation to the center of the tire.
Neumático de bandas radiales Un neumático en el cual las capas del armazón se posicionan a 90° en relación al centro del neumático.

Reactive The capability of a material to react violently when it comes in contact with another material.
Reactivo La capacidad de una materia de reaccionar con violencia al ponerse en contacto con otra materia.

Rebound travel Downward wheel and suspension movement.
Caída Un movimiento hacia abajo de la rueda y la suspensión.

Recirculating ball steering gear A steering gear with recirculating balls positioned in a cage between the wormshaft grooves and the sector shaft teeth.
Mecanismo de dirección de bola recirculatoria Un mecanismo de dirección con las bolas recirculatorias posicionadas en una jaula entre las muescas del árbol del sin fin y los dientes del árbol del sector.

Relay valves Are used with a driver operated control valve to release air pressure from an air suspension system.
Válvulas de relée Se usan con una válvula controlada por el conductor para dejar escapar la presión de aire de un sistema de suspensión de aire.

Remote power steering pump reservoir A reservoir that is mounted separately from the power steering pump.
Suministro remota de la bomba de dirección hidraúlica Un suministro montado aparte de la bomba de dirección hidraúlica.

Resisting bending moment (RBM) The most accurate indication of frame strength. RBM is calculated by multiplying the yield strength and the section modulus.
Resistencia al momento flector (RBM) La indicacíon más precisa de la fuerza del bastidor. La RBM se calcula multiplicando el límite elástico y el módulo del sección.

Resource Conservation and Recovery Act (RCRA) An act that controls hazardous waste disposal in the United States.
Acta de Conservación y Recobro de Recursos (RCRA) Una acta que controla la disposición de los deshechos tóxicos en los Estados Unidos.

Right-to-know laws Laws that state workers have a right to know when the materials they use at work are hazardous.
Leyes de derecho en saber Los leyes que declaran que los obreros tienen el derecho de ser informados de cuando los materiales con los cuales trabajan son peligrosos.

Rigid beams Support the axles and air springs in some air suspension systems.
Vigas rígidas Sostienen los ejes y los muelles neumáticos en algunos sistemas de suspensión neumáticas.

Rigid fifth wheel Is permanently fixed and does not oscillate in any direction.
Rueda quinta rígida Se fija permanentamente y no oscila en ninguna dirección.

Road crown The higher center of the road surface in relation to the edges of the road.
Bombeo del camino La parte central de la superficie del camino que es más alta en relación a las orillas del camino.

Road variables Variables such as vehicle weight and road surface that affect wheel alignment when driving.
Variables del camino Las inconstancias tal como el peso del vehículo y la superficie del camino que afecta la alineación de las ruedas durante el manejo.

Rolling elements The elements between the inner and outer races in a bearing.
Elementos rodantes Los elementos entre las pistas interiores e exteriores en un cojinete.

Rotary valve A valve mounted with the spool valve in a power steering gear. The position of these two valves directs power steering fluid to the appropriate side of the steering gear piston.
Válvula rotativa Una válvula montado con la válvula rodete en un mecanismo de dirección hidráulica. La posición de estas dos válvulas dirige el fluido de dirección hidráulica hacia el lado indicado del pistón del mecanismo de dirección.

Rubber cushions Are used in some rear tandem axle suspension systems in place of leaf springs.
Cojines de caucho Se usan en algunos sistemas de suspensión del eje trasero tándem en vez de los muelles de lámina.

Safety valve A valve that limits the maximum pressure to protect system components.
Válvula de seguridad Una válvula que limita la presión máxima para proteger los componentes del sistema.

Scrub radius The distance between the KPI line and the true vertical centerline of the tire at the road surface viewed from the front.
Radio estregadero La distancia entre la línea KPI y el eje mediano vertical verdadero del neumático en la superficie del camino vista de frente.

Section modulus Is an indication of frame strength based on the height, width, thickness, and shape of the frame side rails.
Módulo de la sección Es una indicación de la fuerza del bastidor basado en la altura, lo ancho, el espesor y el perfíl de los largueros laterales del bastidor.

Sector shaft A shaft in a recirculating ball steering gear that is connected to the pitman arm. The gear teeth on the sector shaft are meshed with the worm gear.
Eje de la sección Un eje en un mecanismo de dirección de bolas recirculatorias que se conecta al brazo pitman. Los dientes del engranaje en el eje del sector se endentan con el engranaje sin fin.

Semi-oscillating fifth wheel Oscillates around an axis perpendicular to the tractor centerline.
Rueda quinta semi-oscilatoria Oscila alrededor de un eje que es perpendicular al eje mediano del tractor.

Separator A separator keeps the rolling elements equally spaced apart in a bearing.
Separador Un separador mantiene un espacio parejo entre los elementos rodantes de un cojinete.

Set back Occurs when one front wheel is moved rearward in relation to the opposite front wheel.
Desplazamiento trasero Ocurre cuando una rueda delantera se mueve hacia atrás en relación con la rueda delantera opuesta.

Shock absorbers Tubular hydraulic devices connected from the suspension to the chassis to help control spring action.
Amortiguadores Los dispositivos tubulares hidráulicos conectados de la suspensión al chasis para ayudar en el control de la acción de los muelles.

Shock absorber compression stroke Occurs when the shock absorber becomes shorter during wheel jounce.
Golpe de compresión del amortiguador Ocurre cuando el amortiguador se acorta durante un sacudo.

Shock absorber extension stroke Occurs when the shock absorber becomes longer during wheel rebound.
Golpe de extensión del amortiguador Ocurre cuando el amortiguador se alarga durante una caída de la rueda.

Shock absorber ratios The amount of extension control compared to the amount of compression control.
Relación de los amortiguadores La cantidad de control en extensión comparado a la cantidad de control en compresión.

Short-and-long arm (SLA) front suspension system A front suspension system with lower control arms that are longer than the upper control arms.
Sistema de suspensión delantera de brazos cortos y largos Un sistema de suspensión delantera cuyos brazos de control inferiores son más largos que los brazos de control superiores.

Sidesway Occurs when one or both frame rails are bent inward or outward.
Desviaje lateral Ocurre cuando uno o ambos travesaños del bastidor son torcidos hacia adentro o hacia afuera.

Single-row ball bearing A cylindrical ball bearing with only one row of rolling elements.
Cojinete de bola simple Un cojinete de bolas cilíndricas que sólo tiene una hilera de elementos.

Slider-type trailer air suspension system A trailer air suspension system that may be slid forward and rearward on the trailer frame.
Sistema de suspensión neumática de remolque tipo deslizador Un sistema de suspensión que se puede deslizar hacia afrente y atrás sobre el bastidor del remolque.

Sliding fifth wheel Designed to move forward or rearward on its mounting plate.
Rueda quinta deslizante Diseñada a mudar hacia afrente o hacia atrás en su placa de montaje.

Slip angle The actual angle of the front wheels during a turn compared to the turning angle of the front wheels with the vehicle at rest.
Ángulo de deslizamiento El ángulo actual de las ruedas delanteras durante un viraje comparada al ángulo de viraje de las ruedas delanteras cuando el vehículo ésta parado.

Snap rings A circular steel ring with some tension designed to snap into a groove and retain a component.
Anillo elástico Un anillo circular de acero con un poco de tensión diseñado a quedarse en una muesca para retener un componente.

Society of Automotive Engineers (SAE) A society of professional engineers dedicated to improving and advancing mobility technology.
Asociación de Ingenieros Automotrices (SAE) Una sociedad de ingenieros profesionales dedicados a mejorar y promover la tecnología móvil.

Sodium based grease A special lubricant with a sodium base that may be required in some steering and suspension components.
Grasa a base de sodio Un lubricante especial de un base de sodio que puede requerirse en algunos componentes de dirección y suspensión.

Spoke wheel A truck wheel with spokes between the wheel hub and the rim.
Rueda con rayos Una rueda de camión con los rayos entre el cubo de la rueda y la llanta.

Spool valve A valve in a power steering gear that works with the rotary valve to direct fluid to the proper side of the steering gear piston to provide steering assist in the proper direction.
Válvula rodete Una válvula en un mecanismo de dirección hidráulica que trabaja con la válvula rotativa para dirigir el fluido al lado adecuado del pistón de engranaje de dirección para proveer asistencia de dirección en la dirección indicada.

Spring cradle Connects the center of the rear leaf springs to the frame on rear tandem axle suspension systems with inverted springs.
Muelle cuna Conecta el centro de los muelles de lámina traseros al bastidor en los sistemas traseros de suspensión del eje tándem con muelles invertidos.

Springless seal A seal lip with no spring behind the lip.
Cierre sin resorte Un reborde de un cierre que no trae un resorte atrás del reborde.

Spring loaded seal A seal lip with a garter spring behind the lip to increase lip tension.
Sello cargado por muelle Un reborde para sellar con un resorte de liga atrás del reborde para aumentar la tensión en el reborde.

Spring windup The tendency of the axle to twist the spring in a circular motion during severe braking and acceleration.
Torsión del resorte La tendencia del eje de torcer el muelle en un movimiento circular durante casos severos del frenado o acceleración.

Sprung weight Is the vehicle weight supported by the springs.
Peso suspendido Es el peso del vehículo sostenido por los muelles.

Stabilized fifth wheel The top of this fifth wheel oscillates with the trailer.
Rueda quinta estabilizado La parte superior de esta rueda quinta oscila con el remolque.

Static balance Refers to the balance of a wheel and tire at rest.
Equilibrio estático Refiere a la equilibración de una rueda y un neumático que no están en movimiento.

Stationary fifth wheel Permanently attached to the tractor frame and cannot be moved forward or rearward.
Rueda quinta estacionario Conectada permanentemente al bastidor del remolque y no se puede mover hacia afrente o atrás.

Steering arm A heavy steel arm that is connected from the drag link to the left front spindle. Steering arms are also connected from each front spindle to the tie rod.
Brazo de dirección Un brazo grueso de acero conectado de la biela de dirección al husillo delantero del lado izquierdo. Los brazos de dirección también se conectan de cada husillo delantero al tirante de tracción.

Steering knuckles Connected by knuckle or king pins to the front I-beam axle, and the wheel hubs, bearings, and wheels are mounted on these knuckles.
Muñones de dirección Conectados por los muñones o los pivotes centrales al eje delantero de viga en I, y los cubos de las ruedas, los cojinetes y las ruedas son montandos en estos muñones.

Steering pull, or drift The tendency of the steering to pull or drift to one side when the vehicle is driven straight ahead on a smooth, straight road surface.
Desviación o arrastre de la dirección La tendencia de la dirección de arrastrarse o desviarse a un lado cuando el vehículo es conducido en línea recta en un camino cuyo superficie es liso y plano.

Steering shaft The shaft that is positioned in the center of the steering column and connects the steering wheel to the steering gear.
Eje de dirección El eje quc se posiciona en el centro de la columna de dirección y conecta el volante de dirección al engranaje de dirección.

Steering wander The tendency of the steering to pull to the right or left when the vehicle is driven straight ahead on a smooth road surface.
Divagación de la dirección La tendencia de la dirección de desviar hacia la derecha o a la izquierda cuando el vehículo se maneja en línea recta en un camino de superficie lisa.

Steering wheel free-play The amount of steering wheel movement before the front wheels begin to turn.
Juego de volante de dirección La cantidad de movimiento en el volante de dirección antes de que comienzan a girar las ruedas delanteras.

Steering wheel kickback A force transmitted to the steering wheel that tends to rotate this wheel in the opposite direction to the rotational force supplied to the steering wheel by the driver.
Reculada del volante de dirección Una fuerza transmitida al volante de dirección que suele girar al volante en el sentido opuesto a la fuerza rotativo que el conductor ha implementado en el volante de dirección.

Stud-piloted wheel A truck wheel with tapers on the outside of the stud openings. Tapers on the inside of the wheel nuts fit in the tapers around the stud openings to center the wheel.
Rueda piloteada por espigas Una rueda de camión cuyos orificios de las espigas son cónicos. La forma cónica en la parte interior de las tuercas de la rueda queda en los ahusamientos alrededor de las aperturas de las espigas así centrando la rueda.

Swinging shackle A spring shackle that is designed to pivot forward or rearward.

Grillete oscilante Un grillete de muelle diseñado para pivotear hacia afrente o atrás.

Tapered caster shim Tapered metal shims installed between the axle I-beam and front springs to adjust front wheel caster.

Laminilla correctora achaflanada de inclinación del eje Las laminillas metálicas instaladas entre el eje de viga en I y los muelles delanteros para ajustar el ángulo de inclinación de las ruedas delanteras.

Tapered roller bearing A bearing with tapered rolling elements between the inner and outer races.

Cojinete de rodillos cónicos Un cojinete que tiene elementos rodantes cónicos entre las pistas interiores e exteriores.

Thrust bearing load A bearing load applied in a horizontal direction.

Carga del cojinete de empuje Una carga del cojinete puesta en una dirección horizontal.

Thrust line A line positioned at a 90° angle to the rear axle and projected towards the front of the vehicle.

Línea de empuje Una línea posicionada en un ángulo de 90° al eje trasero y proyectada hacia la parte delantera del vehículo.

Tie rod A long rod connected between the two front steering arms.

Tirante de tracción Una barra larga conectada entre los dos brazos de dirección delanteros.

Tilt and telescoping steering column A steering column in which the steering wheel may be tilted upward and downward and also moved closer to or further from the driver.

Columna de dirección desplazable y telescópica Una columna de dirección en la cual el volante de dirección puede inclinarse hacia arriba o abajo y también se puede acercar o alejar del conductor.

Tire matching The process of matching tire diameters on the tires on both ends of an axle.

Emparejar los neumáticos El proceso de igualar los diámetros de los neumáticos de ambas extremidades del eje.

Toe-in A condition that is present when the distance between the front edges of the front or rear wheels is less than the distance between the rear edges of the wheels.

Convergencia Una condición presente cuando la distancia entre las orillas delanteras de las ruedas delanteras o traseras es menos que la distancia de las orillas traseras de las ruedas.

Toe-out A condition that is present when the distance between the front edges of the front or rear wheels is more than the distance between the rear edges of the wheels.

Divergencia Una condición presente cuando la distancia entre las orillas delanteras de las ruedas delanteras o traseras es más que la distancia de las orillas traseras de las ruedas.

Toe-out on turns The steering angle of the wheel on the inside of a turn compared to the steering angle of the wheel on the outside of the turn. Toe-out on turns may be called Ackerman angle, turning angle, or turning radius.

Divergencia de viraje El ángulo de viraje de la rueda en la parte interior de un viraje comparado al ángulo de dirección de la rueda en la parte exterior de un viraje. La divergencia en viraje puede llamarse el ángulo Ackerman, el ángulo de viraje o el radio de giro.

Torque A twisting force.

Torsión Una fuerza torcedora.

Torque rod Heavy steel rods connected from the rear axle to the chassis to control axle movement.

Varilla de torsión Las varillas gruesas de acero conectadas del eje trasero al chasis para controlar el movimiento del eje.

Torque steer The tendency of the steering on a front wheel drive vehicle with unequal length drive axles to pull to one side during hard acceleration.

Dirección torcida La tendencia que tiene la dirección de un vehículo de propulsión delantera con los ejes propulsores de longitudes disparejos de jalar hacia a un lado durante una aceleración fuerte.

Torque valve A valve in an air-assisted steering system that senses the direction of steering wheel rotation and supplies air pressure to the appropriate side of the power cylinder piston.

Válvula de torsión Una válvula en un sistema de dirección asistida por aire que detecta la dirección de rotación del volante de dirección y provee la presión de aire al lado indicado del pistón del cilindro motor.

Torsion bar A bar that replaces the leaf spring or coil spring in a suspension system. The twisting action of this bar supports the vehicle weight.

Barra de torsión Una barra que reemplaza el muelle de láminas o el muelle espiral en un sistema de suspensión. La acción torcedora de esta barra sostiene el peso del vehículo.

Toxic Poisonous to the human body.

Tóxico Venenoso al cuerpo humano.

Tracking Refers to the position of the rear wheels in relation to the front wheels.

Seguimiento Refiere a la posición de las ruedas traseras en relación a las ruedas delanteras.

Trailer air suspension systems A trailer suspension system in which air springs support the vehicle weight in place of leaf springs.

Sistemas de suspensión neumática de remolque Un sistema de suspensión neumática en el cual los muelles neumáticos sostienen el peso del vehículo en vez de los muelles de lámina.

Trailer bolster plate The plate on the trailer that contacts the fifth wheel on the tractor.

Placa reforzadora de remolque La placa en el remolque que se pone en contacto con la rueda quinta en el tractor.

Transverse beam Connected between the lower sides of the air springs on each side of an air suspension system.

Viga transversal Se conecta entre las partes inferiores de los muelles neumáticos de cada lado del sistema de suspensión neumática.

Transverse rod Connected from the rear axle housing to the frame to prevent lateral rear axle movement in an air suspension system.

Varilla transversal Se conecta del envoltura del eje trasero al bastidor para prevenir el movimiento lateral del eje trasero en un sistema de suspensión neumática.

Tubular steel front axles Front axles manufactured from tubular steel rather than an I-beam.
Ejes delanteras de acero tubular Los ejes delanteros fabricados del acero tubular en vez de una viga en I.

Turning radius The turning angle of one front wheel in relation to the opposite front wheel during a turn.
Radio de giro El ángulo de viraje de una rueda delantera en relación a la rueda delantera opuesta durante un viraje.

Understeer Occurs when the vehicle tends to steer towards the outside of a curve when accelerating around the curve. The driver must turn the steering wheel toward the inside of the curve to maintain the original vehicle direction.
Subvirador Ocurre cuando el vehículo tiene la tendencia de dirigirse hacia el exterior de una curva al accelerarse efectuando una vuelta. El conductor debe dar la vuelta al volante de dirección hacia el interior de la curva para mantener la dirección original del vehículo.

Unitized hub assemblies A hub assembly that is permanently lubricated, sealed and non-serviceable.
Ensamblado unetizado del cubo Un ensamblado que es lubricado y sellado permanentemente al cual no se puede hacer mantenimiento.

Unitized one-piece seal A one-piece wheel bearing seal.
Sello unetizado de una pieza Un sello del cojinete de la rueda en una sóla pieza.

Unsprung weight The weight of the axle, suspension system, tires and wheels that is not supported by the suspension system springs.
Peso no suspendido El peso del eje, el sistema de suspensión, los neumáticos y la ruedas que no es sostenido por los muelles del sistema de suspensión.

Vacuum A pressure less than atmospheric pressure.
Vacío Una presión que es menor de la presión atmosférica.

Vari-rate spring Rather than having a standard spring deflection rate, these springs have an average spring rate based on load at a predetermined deflection.
Muelle de tasa variable En vez de tener una tasa de deflección del muelle normativo, estos muelles tienen un régimen promedio de muelle basado en la carga con una deflección predeterminada.

V-belt A V-shaped belt that acts as a drive belt between the crankshaft pulley and the power steering pump.
Correa en V Una correa con perfíl de V que funciona como una correa motríz entre la polea del cigüeñal y la bomba hidráulica de la dirección.

Vehicle wander The tendency of the vehicle steering to pull in either direction when driving straight ahead on a smooth, straight road surface.
Divagación del vehículo La tendencia de la dirección del vehículo a jalar hacia una dirección o la otra cuando se maneja en línea recta en una superfice de camino lisa y recta.

Wheel alignment May be defined as the adjustment and refitting of suspension parts to original specifications that ensures design performance.
Alineación de las ruedas Puede definirse como el reglaje y reparación de las partes de suspensión para recuperar las especificaciones originales que asegura la ejecución del diseño.

Wheel jounce Upward wheel and suspension movement.
Sacudo de la rueda Un movimiento de la rueda y la suspensión hacia arriba.

Wheel rebound Downward wheel and suspension movement.
Caída de la rueda Un movimiento de la rueda y la suspensión hacia abajo.

Wheel shimmy Rapid inward and outward oscillations of the front wheels.
Oscilación anormal de las ruedas Las oscilaciones rápidas hacia adentro y afuera de las ruedas delanteras.

Wiper ring A metal ring pressed onto some truck spindles that provides a sealing surface for the wheel bearing seal.
Anillo estregador Un anillo de metal prensado en algunos husillos de camiones que provee una superficie obturador para el sello del cojinete de ruedas.

Workplace hazardous materials information systems (WHMIS) Sheets that provide information regarding hazardous materials.
Sistemas de información de materiales peligrosas del taller (WHMIS) Las hojas que proveen la información tocante a las materiales peligrosas.

Worm and roller steering gear A steering gear developed in the early 1900's that required high steering effort because of internal friction.
Mecanismo de dirección sin fin y de rodillos Un mecanismo de dirección desarrollado en los primeros años de 1900 que requería un esfuerzo de maniobra enorme en el volante a causa de la fricción interna.

Wormshaft The gear meshed with the pitman shaft sector in a recirculating ball steering gear.
Árbol sin fin El engranaje endentado con el sector del árbol pitman en un mecanismo de dirección de bola recirculatoria.

Yield strength A measure of the steel strength used in the frame.
Límite elástico Una medida de la fuerza del acero incorporado en el bastidor.

INDEX

Note: Numbers in italics refer to illustrations; numbers in boldface refer to tables.